HEALTH AND SAFETY AT HAZARDOUS WASTE SITES

HEALTH AND SAFETY AT HAZARDOUS WASTE SITES

AN INVESTIGATOR'S AND REMEDIATOR'S GUIDE TO HAZWOPER

Steven P. Maslansky
Carol J. Maslansky

JOHN WILEY & SONS, INC.
New York • Chichester • Weinheim • Brisbane • Singapore • Toronto

Library of Congress Cataloging-in-Publication Data

    Health and safety at hazardous waste sites: the investigator's and remediator's guide to HAZWOPER/Stephen P. Maslansky, Carol J. Maslansky
       p.   cm.
    Includes bibliographical references and index.
    ISBN 0-471-28805-5
    1.   Hazardous waste sites—Safety measures. 2. Hazardous waste sites—Health aspects. I. Maslansky, Carol J.   II. Title
    TD1050.S24M37   1997
    628.5—dc21   97-5909
                                       CIP

10 9 8 7 6 5 4

This book is dedicated to Kathy Butcher and the staff of the National Ground Water Association, for their kind support and commitment to a safe industry. We also dedicate this text to all who participated in our 40-hour HAZWOPER classes, for their insightful questions and comments. We learned from you as you learned from us— and you always kept us on our toes.

CONTENTS

PREFACE

This book provides basic information that one should have before performing investigative or remedial activities at hazardous waste sites. Since 1983, we have been presenting basic 40-hour Hazardous Waste Operations [20 CFR 1910.120(e)] Safety Courses sponsored by the National Ground Water Association and the National Drilling Federation; the New York State Department of Environmental Conservation became a sponsor in 1987. Because we could find no manual that suited our needs, we developed this text and have been using various draft versions of it for several years. We are thankful for the criticisms, comments, and corrections received from our students.

This text is not intended to make the reader an instant "expert"; rather it provides information pertaining to health, safety, and operations at projects involving hazardous materials. Although many examples are derived from the ground-water industry, the text should be useful to anyone working at hazardous waste sites.

Subjects covered in this manual include hazard recognition, properties of hazardous materials, toxicology and chemical exposure, sources of information, respiratory protection, air monitoring instrumentation, personal protective equipment, levels of protection, site control, decontamination, organizational health and safety programs, contingency planning, and special operations. A generic site-specific health and safety plan and an emergency response plan are included in the appendix, along with applicable OSHA regulations and a select bibliography.

The information and recommendations contained herein have been compiled from several sources, including the U.S. Environmental Protection Agency, the Occupational Health and Safety Administration, the National Institute for Occupational Safety and Health, the U.S. Department of Transportation, the U.S. Army Corps of Engineers, and the U.S. Coast Guard. Some recommendations also are drawn from our combined 35 years of field experience. We believe that the information and recommendations are reliable and in accord with regulatory requirements, and that they represent the best consensus at the time of publication.

References to products and manufacturers are for illustrative purposes only and do not imply an endorsement by the authors, course-sponsoring organizations, or any government agency mentioned in the text.

1

Hazard Recognition

One of the most important aspects of hazardous waste investigation is hazard recognition. A dictionary definition for hazard is danger or something that causes danger, peril, risk, or difficulty. Hazard also is defined as the absence of predictability, or uncertainty, or an unexpected or unpredictable event. The opposite of hazard is safety.

Hazardous waste site operations can pose a number of real or potential hazards to workers. The hazards present are a function of the nature of the site, contaminants found on-site, and the work being performed. The use of personal protective equipment (PPE) subjects workers to additional stress.

Although no two sites are alike in terms of the contaminants present and the site conditions, hazardous waste sites share several distinguishing features. One feature is the uncontrolled condition of the site; the number, identity, and type of hazards present are usually unknown. Another factor is the variety of hazards present; most sites present multiple hazards to site workers. There is also a relatively large number of chemical contaminants present on-site. Determining the identity and the concentration of these materials is extremely difficult, especially during the initial phases of investigation. Finally, workers are potentially exposed not only to chemical hazards but to other dangers presented by site conditions, work activities, and the use of PPE.

RISK VERSUS HAZARD

Most sites pose multiple hazards to workers. Many activities associated with site investigation or remediation, such as subsurface investigations and monitoring/recovery well installation, are in themselves hazardous. In discussing a hazardous substance, the degree of hazard refers to the inherent characteristics that define it as being hazardous (e.g., corrosive, flammable, reactive, radioactive, or toxic). The degree of hazard is a function of the specific hazardous properties of the material and, most important, the concentration of the material. For instance, concentrated acetic acid is corrosive and flammable; but the vinegar in salad dressing, which is dilute acetic acid, is not considered a hazardous material.

An important part of hazard recognition, then, is determining the amount of material present. In many cases, the lower the concentration present, the lower the degree of hazard.

The potential for a particular hazard to cause harm is known as the degree of risk. It is a function not only of the hazards present but also of the likelihood that the worker will encounter those hazards. A good definition of risk is that it is the probability that an accident will occur, which can be expressed by a simple equation:

$$\text{Degree of risk} = \text{Likelihood of an accident} \times \text{Degree of hazard}$$

Engaging in activities that are "accident-prone" increases the worker's degree of risk. For example, a 55-gallon drum of benzene represents a potential toxic and flammability hazard, but there is little risk of fire or toxic overexposure if the drum is intact and not leaking (a good analogy is the gas tank of a car). The risk increases if the drum is damaged, or if a worker opens the drum to take a sample. Risk is a function not only of what the worker is doing, but also of what else is being done on-site. (See Figure 1–1.)

HAZARD CLASSES

Hazardous waste sites present a variety of health and safety hazards, many of which can cause serious illness, injury, or even death. These hazards are a function of the nature of the site and the activities performed. Major hazard classes include:

Figure 1–1. Workers equipped with personal protective equipment (PPE) have increased risk of injury while working around heavy equipment.

- Electrical hazards
- Physical hazards
- Mechanical hazards
- Structural hazards
- Noise
- Weather and temperature stress
- Radiation hazards
- Oxygen deficiency
- Confined space hazards
- Biological hazards
- Flammable hazards
- Chemical hazards

ELECTRICAL HAZARDS

Electrical hazards may be encountered at hazardous waste sites whenever electrical equipment or power lines are present. Shock to personnel is the primary hazard from equipment such as power tools or electric pumps; severe electric shock, or electrocution, can be fatal. Although electric shock does not always cause death, it can cause burns or injury-producing falls.

Electrocution due to inadvertent contact with buried cable or overhead wires is the most frequent cause of job-related deaths for drillers and their helpers (NWWA, 1980, 1997). Electrocutions caused 15% of all construction-related fatalities during the period 1980–1989 (Kisner, 1994). Line arcing has also resulted in electrocution.

Electrical equipment can overheat and cause thermal burns to the user; electrical fires or explosions also may occur. Lightning is a frequently overlooked hazard, particularly for workers using metal equipment. The presence of water on-site compounds these hazards.

Electrical Safety

To minimize electrical hazards, low voltage equipment with ground-fault interrupters and watertight corrosion-resistant connecting cables should always be used. Another way to protect against shock is to use double-insulated tools and overcurrent devices, which prevent overload or short-circuiting. A minimum of 20 feet should be maintained between drilling equipment and overhead wires. The minimum distance may be increased according to local utility requirements or state and local regulations. Distance requirements often are based on line voltage. Most drill rig manufacturers recommend a minimum distance of 100 feet.

Local utilities should be contacted for information regarding the location of buried cables. In many states, an underground utility protective organization, such as One Call or Dig Safe, will notify local utilities about proposed digging activities. Remote sensing or controlled pit excavations may be required at sites where information is limited. It should be remembered that more than one underground line has been found "where it shouldn't have been."

PHYSICAL HAZARDS

Physical hazards are often referred to as general safety hazards. A hazardous waste site can be physically hazardous because it contains unstable slopes, uneven terrain, holes and ditches, steep grades, and slippery surfaces. Sharp debris, such as broken glass and jagged metal, may litter the site. (See Figure 1–2.) Drilling and excavation usually increases the slip/rip hazard by creating wet working surfaces. Drilling and excavation operations may not allow clearance on all sides, and avenues of egress may be limited; extra care must be taken to avoid being struck or caught between equipment.

Compressed gas cylinders can present a physical hazard to site workers. Cylinders under pressure can be overpressurized and explode if heated. A cylinder can become a missile if its valve is sheared off, and the pressure is suddenly released. To prevent damage to the valve stem and the cylinder, cylinders should be placed in a secure position, and screw caps or valve covers should always be used when the cylinders are not in use. Cylinders of compressed breathing air used by site workers may not have valve covers and should always be handled carefully. (See Figures 1–3 and 1–4.)

MECHANICAL HAZARDS

Mechanical hazards exist whenever there is machinery operating. Hazards with machines are created when there are moving parts that can catch, cut, shear, or crush all or part of the human body. The hazard may exist at the point where work is performed, or at other moving parts of the machine, or both.

Machine guarding is the most common means of preventing mechanical injury. Guarding prevents access to a dangerous part by enclosing it within a barrier; the barrier may be a cage, a bar, a rod, a wire, or an electric-eye beam. Eliminating loose clothing or long hair that can be caught in machinery is another way of preventing injury. Interlocking devices automatically stop a machine if the barrier is removed; similar devices are employed to stop a machine if the operator's body contacts or crosses the barrier.

STRUCTURAL HAZARDS

Structural hazards may exist whenever there are damaged or unstable building elements, such as walls, flooring, stairways, or ceilings. Ladders with missing rungs, gangways without railings, missing manhole covers, or other structural deficiencies, such as rotted wood or corroded metal, also represent structural hazards. Overhead hazards, such as falling masonry or debris loosened by other workers are also structural hazards.

Always check and make sure that ladders are in good repair; be sure that extension ladders are tied off to prevent slippage, and that they extend at least three feet above the level to be reached. Wear head protection when working around overhead hazards.

NOISE

A commonly overlooked hazard at hazardous materials sites, noise can produce potential hazards because it interferes with normal communication between workers and equipment operators. It also may startle or distract the worker, thus leading to injury. Noise can cause

Figure 1-2. Working around heavy equipment is hazardous. Allowing a work site to be littered with tools and equipment increases the risk of injury.

Figure 1–3. Compressed gas cylinders should be stored upright with the valve covers in place to protect the valve stems. Note that these cylinders are not secured in place, an OSHA violation.

physical damage to the ear, which may result in temporary or permanent hearing loss. The effect of noise on hearing depends on the amount and the type of noise as well as the duration of exposure (Plog, 1988).

There are three general classes of noise, all of which are found around drilling and ground-water monitoring operations: (1) continuous noise is noise heard when the drill rig or heavy equipment is operating (e.g., engine noise); (2) intermittent noise can be heard over continuous noise, such as when a compressor or pumping equipment is in use; (3) impact noise is produced by hammers or driving tools. Loudness and frequency or pitch affect the perception of sound. Most sounds contain a mixture of frequencies; high frequency sounds, such as fingernails scratching across a blackboard, are generally more annoying than low frequency sounds. High frequency noise also has a greater potential for causing hearing loss.

Sound, a form of acoustic energy, varies in loudness or intensity and is measured in decibels (dB). Low intensity sounds such as quiet conversation measures about 40 to 50 dB, heavy equipment measures about 85 to 90 dB, and a jackhammer produces 100 to 120 dB. The 120 to 130 dB sound from a jet engine or a heavy metal rock group can produce discomfort and temporary hearing loss. A single exposure to a rifle blast can cause pain and permanent hearing loss.

Prolonged exposure to loud noises from heavy equipment (85–90 dB) can produce permanent hearing loss characterized by the inability to hear sounds as well as difficulty in

Figure 1-4. Compressed gas cylinders encountered at hazardous waste sites usually represent an increased risk because they are often in poor condition.

understanding and distinguishing sounds. Excessive noise can increase the heart rate and the blood pressure, which may place an added burden on the heart. Noise also may stress other body organs, causing increased hormone output and muscle tension, which may lead to nervousness, insomnia, and fatigue.

Continuous or Intermittent Noise

OSHA (Occupation Health and Safety Administration) and ACGIH (American Conference of Governmental Industrial Hygienists) have established guidelines to prevent occupational hearing loss. Whenever continuous or intermittent noise exposures equal or exceed 85 dB, recorded on a sound level meter on the A-weighted scale (dBA), per 8-hour day, employers must implement a hearing conservation program as described in the OSHA regulation 29 CFR Part 1910.95. If engineering controls cannot reduce sound levels to acceptable limits, workers should use hearing protectors, such as ear plugs which fit into the ear and/or ear muffs which fit over the ear.

Manufacturers supply Noise Reduction Ratings (NRRs), based on a system that indicates how much noise reduction is attained with each type of protector. Hearing protectors reduce high intensity sound pressure levels reaching the ear; they also make it easier to hear human speech over loud background noise.

Workers may be exposed to higher noise levels if the exposure interval is shortened. ACGIH time limits for shorter intervals are listed in Table 1-1. OSHA time limits are less

Table 1–1. Selected Threshold Limit Values (TLVs) for noise

	Exposure limit (per day)	Sound level (dBA)
Hours	16	82
	8	85
	4	88
	2	91
	1	94
Minutes	30.00	97
	15.00	100
	1.88	109
Seconds	14.06	118
	1.76	127
	0.11	139

Adapted from ACGIH (1996).

protective than those of ACGIH; most industries follow the ACGIH-recommended threshold limit values. Sound meters use an A-weighted scale (dBA) designed to respond to different sound frequencies in the same manner as the human ear.

Impulsive or Impact Noise

Impact noise is a sudden burst of sound that lasts only a fraction of a second and repeats no oftener than once per second. Such noise is produced by hammer blows, punch press strokes, or gunshots. Impact noise should be automatically included in the noise measurement obtained by using an integrating-averaging sound level meter.

Noise Exposure Guidelines to Prevent Hearing Loss

No unprotected ear should be exposed to any impact, continuous, or intermittent noise above 140 dB. When daily noise exposure is composed of more than one exposure interval of different sound levels, the combined effect must be considered. This can be achieved by taking the duration of the exposure at each specific noise level (C), and dividing it by the total duration of exposure permitted at that level (T), and adding up all the exposure fractions:

$$\frac{C_1}{T_1} + \frac{C_2}{T_2} + \frac{C_3}{T_3} \cdots \frac{C_n}{T_n} = X$$

X should equal 1.0 or less. If X is greater than one, the daily allowable noise level limit has been exceeded. ACGIH threshold limit values (T values) for selected noise levels are listed in Table 1–1.

For example, suppose a worker is exposed to 88 dBA for one hour, 80 dBA for seven hours, and 85 dBA for two more hours. Then $C_1 = 1$ hour, $C_2 = 7$ hours, and $C_3 = 2$ hours.

Using Table 1–1, T_1 for 88 dBA = 4 hours, T_2 for 80 dBA = 16, and T_3 for 85 dBA = 8. The final equation would look like this:

$$\frac{1}{4} + \frac{7}{16} + \frac{2}{8} = \frac{4}{16} + \frac{7}{16} + \frac{4}{16} = \frac{15}{16}$$

As 15/16 is less than 16/16 or 1.0, the worker's sound level exposure does not exceed the daily limit.

WEATHER AND TEMPERATURE STRESS

Severe weather is an often overlooked factor in planning for hazardous waste site work. It is unreasonable to expect workers to function under severe weather conditions. Severe weather conditions include: torrential rain, which limits visibility and makes footing hazardous; lightning; high wind conditions, which can jeopardize equipment stability; sleet, snow, freezing rain, and hail, which increase the risk of cold stress and slip hazards; and dust storms, which limit visibility as well as communication.

Temperature stress includes heat stress as well as cold injury. Both types of temperature stress can occur on the same site and at the same time; the ambient air temperature, wind velocity, humidity, and activity level of the worker, as well as the type of protective equipment employed, all play a role in producing temperature stress.

The human body has a thermoregulator mechanism that maintains the body temperature at approximately 98.6°F. A significant portion of the body's energy is devoted to keeping the body temperature within a relatively narrow range. When the body overheats, heat loss is facilitated by evaporative heat loss via sweating and by radiant heat loss by dilation of blood vessels in the skin. When the body temperature drops, involuntary muscle contractions or shivering produces heat while constriction of blood vessels in the skin reduces radiant heat loss.

Personal protective clothing designed to protect against chemical vapors and splash interferes with the thermoregulator mechanism. Heat stress develops rapidly because perspiration cannot evaporate inside the clothing's microenvironment, which is saturated with water vapor. Cold stress can also occur when the metabolic heat produced by intense activity is reduced when work stops. Chemical protective clothing offers no insulation and little protection against the cold; it does not retain body heat. The water-saturated atmosphere within protective clothing loses heat more rapidly than the ambient air. Radiant heat loss can occur very rapidly. *During episodes of heavy work, it is possible for a worker to suffer symptoms of heat stress, only to experience cold stress when activity levels decrease or the work stops.*

Heat Stress

Symptoms of heat stress can vary, depending on its severity. Heat-related problems usually occur in workers who are unaccustomed to heavy workloads and heat, or who are in poor physical condition. Obesity, alcohol or drug use, age, and the presence of other complicating factors, such as acute and chronic diseases, also affect the way an individual responds to hot working conditions.

Heat rash or prickly heat is caused by prolonged exposure to heat and humid air, aggravated by close-fitting or chafing clothes. Heat cramps are painful spasms that occur in the skeletal muscles of workers who perspire profusely and drink large quantities of water but fail to replace salts or electrolytes lost through perspiration. The treatment for heat cramps is rest in a cool area and replacement of electrolytes.

Heat exhaustion is characterized by extreme weakness, fatigue, dizziness, nausea, and headache. In serious cases, a worker may vomit and lose consciousness. The skin is clammy and moist; perspiration is profuse; the body temperature is normal or may be slightly above normal. The treatment is rest in a cool environment and replacement of lost body water and electrolytes. Severe cases of heat exhaustion may require several days of rest before recovery is complete.

Heat stroke is a serious, potentially fatal condition caused by the breakdown of the body's thermoregulator mechanisms. The skin is very dry and hot; the pulse is strong and rapid; there is no perspiration. Nausea, dizziness, and confusion are followed rapidly by unconsciousness and convulsions; death or brain damage can occur unless the body is cooled. Body heat should be reduced by moving the person to a cool environment and artificially cooling the body with cool-packs or cool water. Heat stroke should be considered a medical emergency; medical attention should be obtained *immediately* for heat stroke victims.

Cold Stress

Hypothermia, or a lowering of the body's internal or core temperature, may occur in workers wearing protective clothing, especially after episodes of heavy work. Cold injuries are increased under damp or wet conditions, or when there is a significant wind. Tanks containing breathable air should not be stored outside during cold weather. Breathing cold air can rapidly lead to hypothermia and can also cause lung damage. Symptoms of hypothermia include shivering followed by numbness, drowsiness, weakness, and progressive loss of coordination. The body may feel warm, even when the core temperature is low. Severe hypothermia results in loss of consciousness and eventual death.

Frostbite, or localized cooling of the body, is the most common injury resulting from exposure to cold. Affected areas are usually the feet, hands, nose, and ears. Numbness is the first sign of frostbite (also called frostnip). The affected area remains soft to the touch but becomes reddened and then waxy-white; pain is felt initially but then subsides. Deep frostbite occurs when the skin and underlying tissues freeze, and ice crystals form; the surface of the skin becomes hard to the touch and turns mottled white or gray.

Victims of mild hypothermia should be removed from the cold and slowly warmed. In cases of severe hypothermia or frostbite, medical advice and assistance should be obtained *immediately.*

Wind chill is a significant factor in frostbite and hypothermia. The wind chill index is used to determine the risk of cold injury by estimating the cooling effect of wind as an equivalent temperature (Table 1–2). Unprotected flesh may freeze within one minute when exposed to wind chill temperatures at or below –25°F. At wind chill temperatures above –25°F, there is less hazard of rapid freezing but a greater hazard of eventual hypothermia over time, due to worker complacency and failure to recognize the significance of wind chill.

Table 1–2. Cooling effect of wind expressed as equivalent temperatures

Estimated wind speed in mph	*Actual Temperatures in Degrees Fahrenheit*							
	40	30	20	10	0	–10	–20	–30
	Equivalent Wind-Chill Temperature							
calm	40	30	20	10	0	–10	–20	–30
5	37	27	16	6	–5	–15	–26	–36
10	28	16	4	–9	–24	–33	–46	–58
15	22	9	–5	–18	–32	–45	–58	–72
20	18	4	–10	–25	–39	–53	–67	–82
25	16	0	–15	–29	–44	–59	–74	–88
30	13	–2	–18	–33	–48	–63	–79	–94
35	11	–4	–20	–35	–51	–67	–82	–98
40	10	–6	–21	–37	–53	–69	–85	–100

Adapted from ACGIH (1996).

Evaporative heat loss can exacerbate the effects of wind chill. At temperatures below 36°F, workers who are immersed in water or whose clothing becomes wet should immediately change to dry clothing and be treated for hypothermia.

RADIATION HAZARDS

Radiation is a form of energy, specifically electromagnetic energy, that can be described as a series of waves that differ in length, frequency, and energy. The spectrum of electromagnetic energy wavelengths extends over a broad range, from less than 10^{-12} cm to greater than 10^{10} cm.

Some types of radiation can be seen (visible light) or felt (heat); the effect of too much ultraviolet radiation in sunlight can be seen as a sunburn. Most forms of radiation cannot be seen or felt, including cosmic rays, radar, radio and TV waves, and X rays.

Radiation can be broadly classified into two categories, ionizing and nonionizing. Ionizing radiation has sufficient energy to change the normal electrical balance in an atom of matter, producing charged particles or ions that can be highly reactive with other atoms in their vicinity. The process of producing ions is called ionization. Nonionizing radiation has less energy and a longer wavelength than ionizing radiation and does not produce ions.

Ionizing radiation hazards can include radioactive materials from industry, laboratories, and hospital wastes. These radioactive materials may be found as solids or liquids in drums or lab packs. Landfills and disposal sites that have been used by hospitals, universities, and research centers are prime candidate locations for radioactive materials. Hazardous waste sites should be carefully monitored to rule out the presence of detectable amounts of radioactivity.

Radioactive materials or radionuclides are unstable because they contain excess energy; they release this energy as particles or high energy waves in order to return to their normal or stable state. Ionizing radiation can cause significant injury to biological tissues.

Table 1-3. Types of ionizing radiation

	Symbol	Character	Pathlength
Alpha	α	particle	up to 10 cm in air
Beta	β	particle	1 to 9 meters in air
Gamma	γ	wave	meters to kilometers

Three types of ionizing radiation may be found at hazardous materials sites: alpha particles, beta particles, and gamma waves (Table 1–3).

Types of Ionizing Radiation

Alpha radiation consists of positively charged heavy particles that release a significant amount of energy as they pass through matter, which in turn produces more ions. Alpha particles are heavy and travel only a short distance. The range of alpha particles is at most about 4 inches (10 cm) in air; alpha particles can be stopped by a piece of paper, a film of water, or intact skin. Radioactive materials that emit alpha particles are called alpha emitters and are said to undergo radioactive decay. Alpha emitters are not considered especially hazardous as long as they are kept outside the body, but are dangerous if ingested or inhaled, especially when they are concentrated in bone, liver, or kidney tissues.

Beta particles are very small, fast, negatively charged particles that can penetrate wood, clothing, and skin. Beta radiation has a velocity near the speed of light and may have a range up to 60 feet (18 m) in air. Skin burns can result from overexposure to beta radiation. As a beta particle loses energy and slows down, it emits X radiation, which can also cause damage to biological tissue. Beta particles and beta emitters are hazardous by ingestion, inhalation, and skin contact.

Gamma radiation or gamma rays are high energy electromagnetic energy with a very short wavelength that originates in the nucleus of an atom. Gamma radiation travels at the speed of light, has an extended range, and can penetrate several inches of steel. Gamma rays readily penetrate the body, where they are scattered or absorbed; this process produces secondary radiation in the form of charged particles and X rays. X rays are similar to gamma rays except that they do not originate in the nucleus. Because of its energy and penetration capabilities, gamma radiation is considered the most hazardous form of radiation.

Radiation Terminology

The rate at which a radioactive substance or a radionuclide loses its radioactivity and stabilizes is its half-life. After one half-life interval, half the original radioactivity remains; after two half-life intervals, only 25% of the radioactivity remains; after ten half-lives, less than 0.1% remains. Radioisotopes vary in their half-lives. For example, uranium-238 has a half-life of 4.5 billion years, whereas iodine-131 has a half-life of only eight days.

All forms of ionizing radiation transmit energy to matter; in living tissue, the amount of energy transmitted per kilogram of irradiated material is defined by using a unit called a gray (Gy). The conventional unit of absorbed dose is the radiation absorbed dose or rad. One gray (1 Gy) equals 100 rads; 1 centigray (cGy) equals 1 rad.

The biological effect of equally absorbed radiation doses, however, varies according to the type and the energy of the radiation. Quality factors (Q) are used to account for these differences and to determine dose equivalency; X rays, gamma radiation, and beta particles have a quality factor of 1.0; alpha particles have a quality factor of 20. The conventional unit for a radiation dose equivalent for humans is the radiation equivalent for man or the rem. The roentgen (R) is a unit of exposure for X rays and gamma waves; because their quality factor is 1.0, one R equals one rem. A rem can therefore also be defined as a roentgen equivalent in man. Radiation doses in R, rem, or millirem (mR) are expressed on a per hour basis (e.g., mR/hr).

Reducing Radiation Exposure

Protection from sources of radioactivity can be achieved by time, distance, and shielding:

- Keeping the time of exposure as short as possible.
- Maintaining a safe distance from the source.
- Using protective barriers or shielding specifically designed for the type of radioactive materials present.

Limiting the time that workers are exposed to a radiation hazard can minimize the biological effect of absorbed radiation. The longer the exposure, the greater the absorbed dose, and the greater the potential for injury. There is a direct relationship between time and exposure; doubling the time doubles the exposure, whereas halving the time decreases exposure by half. A worker exposed to a 1 mR/hr radiation source for 8 hours receives a total exposure of 8 mR; a worker exposed to the same source for 4 hours receives an exposure of 4 mR. A worker who suddenly encounters a 100 mR/hr source, and leaves the area within 5 minutes, receives an exposure of only approximately 8 mR.

The total absorbed radiation dose can be calculated by using a simple formula, in which exposure time, if in minutes, must be converted to hours—one minute being equivalent to 0.0167 hour (1 minute divided by 60 = 0.0167):

$$\text{Total exposure} = \text{Radiation source} \times \text{Time in hours}$$

In the above example:

$$
\begin{aligned}
\text{Total exposure} \quad &= \quad 100 \text{ mR/hr} \times (5 \text{ minutes} \times 0.0167) \\
&= \quad 100 \text{ mR/hr} \times 0.0835 \text{ hour} \\
&= \quad 8.35 \text{ mR}
\end{aligned}
$$

Increasing the distance between the worker and the radiation source can also decrease the hazard. The relationship between distance and radiation dose follows the inverse square law: the radiation dose decreases by the square of the distance from the source (Figure 1–5). Doubling the distance decreases the dose to one-fourth; if the distance is tripled, the dose is decreased to one-ninth the value measured at the source.

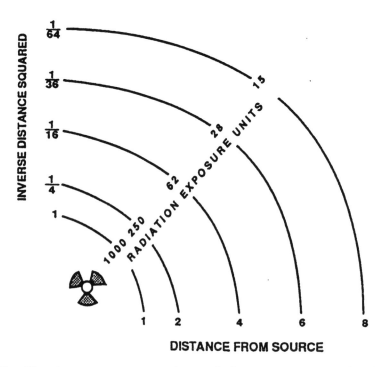

Figure 1–5. The inverse square rule: radiation exposure is the inverse of the square of the distance from the source. (*Source:* Maslansky and Maslansky, 1993. Courtesy of Van Nostrand Reinhold.)

For example, a worker standing 5 feet from a radiation source rated at 50 mR/hr is exposed to a radiation dose of 2 mR/hr:

$$\text{Radiation dose} = \frac{\text{Radiation source}}{\text{Distance}^2}$$

$$= \frac{50 \text{ mR/hr}}{5^2} = \frac{50 \text{ mR/hr}}{25}$$

$$= 2 \text{ mR/hr}$$

Shielding or protective barriers placed between workers and the source also can be used to reduce radiation exposure (Figure 1–6). Shielding is commonly used to protect against X rays, gamma radiation, and high energy beta radiation. The density and the volume of the shielding affect its effectiveness; lead, steel, concrete blocks, earthen berms, or containers filled with water are commonly used shielding materials.

Radiation Exposure Guidelines

Occupational and nonoccupational radiation exposure guidelines (Table 1–4) have been developed by the Nuclear Regulatory Commission (NRC) and the National Council on Radiation Protection and Measurement (NCRP). The recommended maximum whole-body occupational exposure dose is currently 5 rem per year.

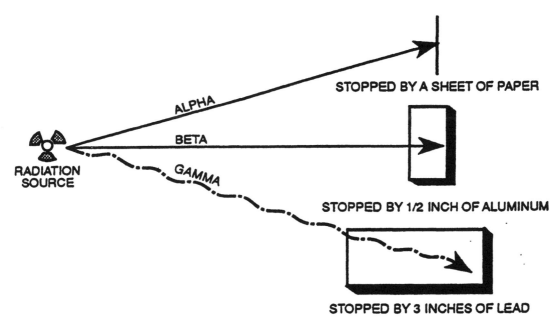

Figure 1–6. Shielding materials can be used to prevent penetration of radioactivity. (*Source:* Maslansky and Maslansky, 1993. Courtesy of Van Nostrand Reinhold.)

Action Levels for Radiation Exposure

Naturally occurring or background radiation has always been present on earth in the form of cosmic rays and radioactive elements such as uranium. Background radiation levels vary, depending on the amount of radioactive materials present in soil, rock, and building materials, with typical background radiation levels being between 0.01 and 0.05 mR/hr. In the United States, the annual dose equivalent from naturally occurring radiation (excluding radon) is estimated to be approximately 125 mR.

Table 1–4. Annual maximum permissible radiation dose

Occupational Therapy	
Whole body	5 rem
Lens of eye	15 rem
Hands (per quarter)	25 rem
Forearms (per quarter)	10 rem
Other organs	15 rem
Public Exposures	
Frequent exposure	0.1 rem
Infrequent exposure	0.5 rem
Embryo–Fetal Exposures	
Total dose limit	0.5 rem
Monthly limit	0.05 rem

Radiation survey meters used at hazardous materials sites usually display radiation levels on a scale marked R/hr, rems/hr, or mR/hr. These meters indicate the amount of radiation present at the time of measurement, whereas dosimeters are used to determine the accumulated dose over a period of time, usually an eight-hour workday. Personal dosimeters are the instrument of choice to record individual exposures. Thermal luminescent dosimeters (TLDs) are useful for detecting beta and gamma radiation. Alpha particles cannot be counted directly with a dosimeter; radiation survey meters equipped with flattened or "pancake" detectors can be used for surface alpha contamination.

An action level is a specific amount or level measured that triggers a particular response or action. The EPA recommended action level for radiation is 1.0 mR/hr (USEPA, 1992). If the source of the emitted radiation is greater than or equal to 1.0 mR/hr, action should be taken to reduce the radiation dose by increasing the distance between the source and the worker, or by using shielding. If the source cannot be located, or the measured radiation dose is particularly high, it may be necessary to leave the area. If radiation levels are above background but less than 1.0 mR/hr, workers may remain in the area as long as continuous radiation monitoring is conducted.

A radiation dose of 1 mR/hr is an extremely safe dose; a worker would have to be continuously exposed to 1 mR/hr per workday for an entire year before reaching the maximum recommended annual occupational exposure limit of 5 rems. At radiation levels of 1 mR/hr or less, workers can remain on-site as long as radiation monitoring continues. A potential radiation hazard is considered present when radiation survey instruments indicate levels greater than 1 mR/hr above background.

Whenever possible, however, every effort should be made to comply with the current ACGIH recommendation of ALARA (as low as reasonably achievable) for ionizing radiation exposure.

Although workers are most concerned with ionizing radiation, it is important to recognized that nonionizing radiation also can be hazardous. Ultraviolet, infrared, microwave, and laser radiation can cause injury, especially to the skin and eyes.

OXYGEN DEFICIENCY

Oxygen-deficient atmospheres can present a very real hazard to waste site workers. At sea level, the concentration of oxygen in normal air is approximately 21%. Although the physiological effects of oxygen deprivation are readily apparent when oxygen concentrations drop below 16%, the presence of an oxygen-deficient atmosphere is not detectable to workers until it is too late for them to escape from it. The physiological effects of oxygen deficiency include impaired judgment, lack of coordination, increased breathing rate, and increased heart rate. Severe oxygen deficiency can cause nausea, vomiting, unconsciousness, brain damage, and death.

Causes and Locations of Oxygen Deficiency

Oxygen deficiency may occur when oxygen in air is displaced by another gas. Heavier-than-air gases may seep into low-lying areas and replace normal air. Locations that are particularly susceptible to oxygen displacement are natural valleys, caves, ditches, trenches, pits, basements of buildings, sewers, electrical conduits or vaults, silos, and tanks.

Consumption of oxygen by a chemical reaction is another cause of oxygen deficiency. Fire is the most common cause of oxygen consumption; and chemical oxidation, such as rusting, is another potential cause of oxygen deficiency. Biological consumption of oxygen by bacteria and other microbes also can cause oxygen deficiency. Suspect locations for oxygen deficiency caused by oxygen consumption include any enclosed area during or after a fire, the insides of rusting metal tanks, and compost pits.

Exposure Limits for Oxygen Concentrations

An oxygen meter should be used whenever there is a potential for oxygen deficiency. These meters are portable and give almost immediate readings, in percent oxygen by volume in air. OSHA defines oxygen-deficient atmospheres as those containing less than 19.5% oxygen. The action level for oxygen deficiency, then, is 19.5%. In atmospheres containing less than 19.5% oxygen, workers must use supplied air respirators if the area cannot be ventilated. In very unusual circumstances, oxygen concentrations may be above normal, a situation that increases the risk of fire. The action level for oxygen excess is 23.5%.

It is acceptable for workers to remain in atmospheres containing less than 20.9% and more than 19.5% oxygen *if the cause of oxygen deficiency is known and does not represent a health or safety hazard* to workers. For example, if the oxygen concentration is decreased because of the presence of an inert, nontoxic gas such as nitrogen or helium, workers may remain in the area as long as oxygen levels are monitored and remain at or above 19.5%. Workers should not remain, however, if the displacing gas is toxic or flammable, or if the cause of the oxygen deficiency is unknown.

CONFINED SPACE HAZARDS

OSHA (29 CFR 1910.146) defines a permit-required confined space as any location that is large enough to enter, is not designed for human occupancy, and has limited or restricted means for entry and exit. A permit-required confined space also may contain either a hazardous atmosphere, a material that could engulf a worker, internal parts that could trap the worker, or other unrecognized hazards such as energized electrical wires or moving, unguarded equipment. Commonly encountered confined spaces include sewers, culverts, wells, large-diameter pipes, sumps, pipe galleries, above- and below-ground storage tanks, tank trucks, rail tank cars and barges, silos, hoppers, bins, stacks, shafts, ducts, electrical conduits, and vaults.

Many confined spaces have very limited or no ventilation, and there is the potential for oxygen deficiency as well as accumulation of toxic, corrosive, or flammable gases. Confined spaces also offer a variety of other potential hazards, including physical, mechanical, electrical, biological, and radiation hazards, temperature extremes, and poor visibility. At ground-water monitoring sites, confined spaces include pits and excavations; even the toe of a steep slope could be considered a confined space if there is a close fence or other obstacle that limits one's mobility.

Before entering a confined space, workers should consider an important question: is the entry really necessary? If the entry is required, then precautions must be taken to safeguard the health and safety of the workers. Prior to entry, the confined-space atmosphere must be tested with air-monitoring instruments for oxygen deficiency, as well as for excess combustible gases and toxic or corrosive materials. A hazard evaluation also should be conducted to determine if other conditions exist with the potential to cause injury.

An emergency plan should always be in place in case of power or equipment failure, injury, or other circumstances that might require immediate evacuation or rescue of personnel. The equipment required to rescue confined-space workers, such as pulleys or tripods, should be readily available before the initial entry.

BIOLOGICAL HAZARDS

Biological hazards or biohazards are living organisms, or their products or by-products, that can cause illness or death of an exposed individual. For example, biohazards include hospital, medical office, and laboratory material that may contain infectious wastes. Such materials may contain microorganisms that cause hepatitis, acquired immune deficiency syndrome, influenza, tuberculosis, and other viral or bacterial diseases. Special care should be exercised around landfills in which biological wastes from hospitals or laboratories may be present. Municipal landfills operating prior to RCRA (the Resource, Conservation, and Recovery Act) (i.e., before 1980) should also be suspect. Fungal spores are often found in and around landfills; one variety causes histoplasmosis, a respiratory disease that can produce severe symptoms, even death, in susceptible individuals.

Not all biohazards are deposited by humans. Many disease-producing microorganisms occur naturally and often require a host or a carrier to transmit the disease to humans. Such carriers may be insects or mammals. Commonly encountered diseases transmitted by carriers include: rabies, which is carried by dogs, cats, foxes, raccoons, skunks, and other mammals; plague, transmitted by rodent-borne fleas; Rocky Mountain Spotted Fever (typhus fever), Lyme Disease, Q Fever, and Colorado Tick Fever, carried by ticks; tularemia, carried by rabbits and usually transmitted by ticks; and a variety of encephalopathies (inflammation of the brain), such as St. Louis encephalitis and equine encephalitis, which are carried by birds as the intermediate host but transmitted by mosquitoes.

Biological organisms that produce allergic skin reactions in sensitive individuals also should be considered biohazards. Allergic reactions to plants, such as poison ivy, poison oak or sumac, or stinging nettles, can be painful. Even when not causing disease or inducing an allergic reaction, organisms such as bees, wasps, fire ants, biting flies, spiders, chiggers, ticks, and scorpions should be considered biohazards.

Many wild animals are attracted to hazardous waste sites, or they may be already present on-site, especially if the site has been undisturbed for some time, or if it is located in a remote or unpopulated area. Bears, wolves, foxes, and wild dogs will investigate equipment left overnight and may still be present when workers arrive in the morning. Vibrations and water discharge associated with drilling and excavation activities can disturb snakes and other reptiles. Snakes are often attracted to buckets, mud tubs, and other containers left empty. In hot desert environments, the shade and the protection offered by equipment are attractive to a variety of the desert inhabitants, including snakes, scorpions, and Gila monsters.

FLAMMABLE HAZARDS

Flammable hazards can be encountered at nearly every hazardous material site. They are often classified as strictly chemical hazards, but this assumption is not always correct. Fire, whether caused by natural forces or by human activity, can be a very real hazard for persons working in drought-stricken areas or in grasslands or forests during the dry season. A carelessly tossed cigarette or match, or heat from mufflers, stoves, space heaters, or electrical equipment, can ignite dry tinder. Distant lightning strikes can also start fires; workers should watch the horizon for telltale smoke, which might indicate an approaching wildfire. Similarly, many materials used at hazardous materials sites are flammable, including compressed gases such as acetylene and propane, flammable fuels such as gasoline and kerosene, and solvents such as acetone and hexane.

Flammability is the capacity of a material to ignite and burn rapidly. The more readily a material will ignite, the greater its flammability hazard. Flammable gases are gases that at ambient temperature and pressure can form a flammable mixture in air that can be ignited in the presence of an ignition source. Flammable liquids readily give off ignitable vapors at ambient temperature. Flammable solids ignite readily and continue to burn vigorously. Flammable solids can be ignited by friction, by chemical reaction, or upon contact with air or water.

Combustible liquids must be heated slightly before they release sufficient vapors to form a flammable mixture in air. A typical combustible liquid often encountered at hazardous waste sites is diesel fuel.

Fire Classifications

The most common fire classification system (NFPA, 1991) is based upon the type or the class of material that is burning. This system is important for determining what type of fire suppressant or extinguisher should be used. Some extinguishers are designed solely for one category and are ineffective on other types of fires.

Class A fires involve ordinary combustible materials, such as wood, paper, and cotton, which burn readily under normal conditions. Noncombustible materials cannot be readily ignited and do not burn.

When a fire involves a flammable or combustible liquid, it is considered a Class B fire. Gasoline, diesel fuel, kerosene, turpentine, acetone, and alcohol are all flammable or combustible liquids that can act as fuel in a Class B fire.

A Class C fire involves electrical equipment or originates from an electrical source. Common sources of electrical fires include faulty wiring, burned-out motors, or short circuits. Although most electrical fires include other classes of combustibles, they should still be classified as Class C.

Class D refers to fires that involve reactive metals, such as sodium, potassium, or magnesium. These materials burn very hot and are incompatible with water.

The Fire Tetrahedron

Combustion is a chemical reaction that requires three basic items: fuel, oxygen, and heat. When these ingredients are present, they combine and enter into the second stage of the fire process:

$$\text{Fuel} + \text{Oxygen} \xrightarrow{\text{Heat}} \text{Free radical formation}$$

During the second stage of combustion, the fuel breaks down into very reactive particles called free radicals. The free radicals react with each other and the rest of fuel, to produce more and more heat. An ongoing chain reaction results, which persists as long as adequate fuel and oxygen are present:

$$\text{Fuel} + \text{Oxygen} \xrightarrow{\text{Heat}} \text{Chain reaction} = \text{Products of combustion}$$

Figure 1–7 shows the fire tetrahedron.

CHEMICAL HAZARDS

Chemical hazards are routinely encountered at hazardous waste site investigations and remediations. Hazardous chemicals may be toxic, flammable, explosive, reactive, or corrosive; and sites with chemical hazards often have more than one hazard associated with them. A good, informal definition of a hazardous substance, which certainly applies to

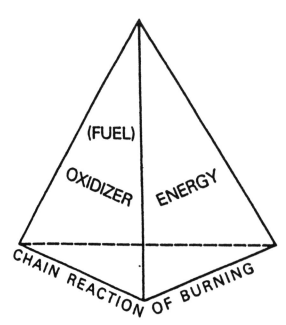

Figure 1–7. Sustained rapid burning involves four parameters, fuel, oxygen, heat, and the chain reaction; this is the basis of the fire tetrahedron. Elimination of one side of the tetrahedron results in fire extinguishment.

chemical hazards, was offered by Ludwig Benner, a hazardous materials specialist formerly with the National Transportation Safety Board in Washington, DC:

> *Any substance that jumps out of its container at you when something goes wrong and hurts or harms the thing it touches.*

Preventing exposure to hazardous chemicals is a primary concern at waste sites. It is important to remember, however, that although the primary focus may be on chemical hazards, many other hazards also must be recognized.

Not all chemical hazards are found buried in the ground or in abandoned drums; many chemical hazards are actually brought onto the site by workers. These hazards include gasoline, diesel and kerosene fuels, hypochlorites or concentrated bleaching solutions to kill pathogens in water wells, muriatic acid used in well maintenance work, solvents or concentrated solutions for equipment decontamination, explosives used for downhole fracturing and some remote sensing techniques, compressed gases such as acetylene used for cutting and welding, and concentrated acids used as preservatives for water samples. Material Safety Data Sheets (MSDS) for chemical hazards known to exist on-site should be available at the site.

Explosives

An explosion is a sudden, violent release of energy from a restricted area. The energy is usually released in the form of a pressure or shock wave, heat, projection of missiles or debris, and gas. Energy released as the result of an explosion can be extremely destructive to life and property. Most explosives contain nitrogen compounds, whose chemical names frequently include "nitrate" or "nitro."

High explosives detonate as a result of shock or heat. Mass detonation occurs at speeds greater than the speed of sound (greater than 1087 feet/second), and is preceded by a shock wave (that is, you feel it before you hear it). The most infamous high explosive is 2,4,6-trinitrotoluene (TNT). Other high explosives include picric acid (trinitrophenol), ammonium picrate, and nitroglycerin; dynamite is nitroglycerin mixed with an inert material such as diatomaceous earth. The "plastic" explosives, such as C-3 and C-4, are also high explosives.

High explosives are usually flammable and toxic by one or more routes of exposure. For example, nitroglycerin and TNT are toxic by skin absorption, ingestion, and inhalation.

The majority of high explosives used in the main charge are relatively insensitive to heat and shock; detonation is induced by small amounts of detonators or initiators, also known as primary explosives or blasting caps. In some cases, another type of shock-sensitive high explosive or secondary explosive is used to add additional punch to the main charge. An explosive train consists of an initiator (detonator), a secondary explosive (primer and booster), and a main charge. The initiator detonates the secondary explosive, which in turn detonates the main charge.

Blasting caps contain small amounts of primary explosives that are extremely sensitive to heat or shock, such as fulminate of mercury or silver, or lead azide. Commonly used secondary explosives include RDX (cyclotrimethylenenitramine), pentaerythritol tetranitrate (PETN), water gel explosives (Dynagel), and lead styphnate. Detonating cord is round, flexible cord containing a core of high explosive, usually PETN, which detonates at a velocity of approximately 22,000 feet/second.

Low-order explosives do not detonate—they deflagrate; when an explosive deflagrates, it burns extremely rapidly. Deflagration proceeds at speeds less than the speed of sound, and a shock wave does not occur unless the deflagration is confined. A confined or restricted deflagration can result in detonation. It is wrong to consider low explosives as low hazards. The most common low explosive, black powder, is extremely sensitive to heat, shock, and friction. Because black powder is so sensitive to shock, it is placed in the highest explosion hazard category.

A blasting agent is a mixture of materials designed to explode with sufficient force for a particular application but so insensitive to shock that it takes a powerful high explosive to set it off. Blasting agents do not contain any material that alone would be classified as an explosive. Do not confuse blasting caps, which contain small amounts of high explosives, with blasting agents. Ammonium nitrate and fuel oil (ANFO) constitute a familiar blasting agent.

Reactive Materials

Reactive substances undergo transformation or chemical change spontaneously or as a result of contact with air, water, or other materials. Energy is released during the reactive chemical change, which can result in a buildup of pressure, heat, or formation of chemical vapors. Many reactive materials release hydrogen gas, which is ignited by the heat of the chemical reaction; for this reason, they are also considered flammable solids or substances.

Water-Reactive Materials

Water-reactive materials react violently with water to release energy. The alkali metals, which include lithium, sodium, potassium, and cesium, all react violently when they come into contact with water. These metals are so reactive that they extract oxygen from water, releasing hydrogen and heat; sufficient heat is liberated actually to ignite the hydrogen. Because of their extreme reactivity with water, the alkali metals do not exist as pure metals in nature but rather as metallic salts.

Alkaline-earth metals, including magnesium, calcium, and strontium, are also water-reactive, but are less reactive than the alkali metals. Magnesium is used in industry and was once used in automobile engine blocks because of its strength and lightness. To prevent water-reactivity, magnesium is often coated with a metallic oxide or a waterproof material.

Metallic hydrides are toxic, irritant, flammable, and water-reactive substances. When metallic hydrides such as lithium hydride, sodium hydride, potassium hydride, and boron hydride contact water, oxygen is extracted and hydrogen liberated. The corresponding metallic hydroxide (an alkaline corrosive) is also formed.

Metallic carbides are chemical compounds that contain carbon and a metal; however, not all carbides are water-reactive. The best-known carbide is calcium carbide (used in miners' lamps), which liberates highly flammable acetylene gas when it is wetted. Magnesium carbide liberates flammable propyne gas.

Many other materials are water-reactive, including the inorganic peroxides, nitrides,

Table 1-5. Water-reactive materials

acetic anhydride	lithium
acetyl chloride	lithium nitride
aluminum carbide	lithium peroxide
aluminum hydride	magnesium
aluminum nitride	magnesium borohydride
aluminum phosphide	magnesium carbide
aluminum selenide	nitryl chloride
arsenic hydride	oleum
barium arsenide	pentaborane
benzoyl chloride	phosphorus pentasulfide
bismuth nitride	phosphorus oxychloride
bismuth oxalate	potassium
borane (boron hydride)	potassium amide
calcium	rubidium hydride
calcium carbide	silane (silicon hydride)
calcium cyanide	sodium
calcium hypochlorite	sodium hydrosulfide
cerium hydride	sodium selenide
chromium trioxide	stannic chloride
diethyl zinc	tellurium hydride
ethyl sodium	trichlorosilane
iodine pentafluoride	vanadium tetrachloride

phosphides, inorganic chlorides, calcium oxide, and acetic anhydride. Table 1–5 lists some water-reactive materials.

Air-Reactive or Pyrophoric Materials

Materials that ignite spontaneously in air below 130°F are pyrophoric (Table 1–6). Some pyrophoric materials ignite so rapidly that they may detonate upon exposure to air. Elemental white or yellow phosphorus is a commonly encountered pyrophoric; it is usually stored under water to prevent contact with air. When dry phosphorus contacts air (at a temperature of 86°F or greater) it begins to burn. Red phosphorus is not pyrophoric.

Finely ground or powdered alkali metals, such as cesium, lithium, sodium, and potassium, react with moisture in air, liberate hydrogen, and burn. Although the ignition of these materials is due to hydrogen liberation, they will react with air and burn; so they should be considered a pyrophoric hazard.

There are only a few organo-metallic compounds that are pyrophoric; they were developed for a specific chemical process and are not very commonly encountered.

Table 1-6. Pyrophoric materials

cesium (powdered)	phosphorus (white/yellow)
diethyl cadmium	potassium (powdered)
dimethyl arsine	sodium (powdered)
dimethyl zinc	titanium dichloride
diethyl aluminum	triethyl aluminum
lithium (powdered)	trimethyl aluminum

Oxidizers

An oxidizer is a material that readily yields oxygen to stimulate the oxidation (combustion) of organic matter. The classic definition of an inorganic oxidizer is that it is a material that does not burn but supports or facilitates combustion. Organic oxidizers, however, not only support combustion but can also be flammable.

The halogens (chlorine, fluorine, iodine, bromine) do not contain oxygen, but they are powerful oxidizers because they support combustion. Halogens alone, in the absence of oxygen, can support fire; when oxygen is also present, the fire could become explosive. Peroxides contain additional oxygen, are powerful oxidizers, and are often unstable. Many inorganic peroxides react with water and liberate oxygen and heat. Several strong inorganic acids are also potent oxidizers, including chromic acid, nitric acid, and perchloric acid (Table 1–7).

The chemical name of the material will often indicate if oxygen is present. Inorganic substances that contain oxygen often have names ending in "-ate" or "-ite"; these materials are often oxidizers. A chemical name that ends in "-oxide" or "peroxide" also suggests that the material is an oxidizer. The prefix "per-" or "peroxy-" indicates additional oxygen is present, and that the material probably is a potent oxidizer.

Organic oxidizers are usually peroxides that have been synthesized for use as catalysts or initiators in chemical reactions such as polymerization. Organic peroxides are unstable, flammable, and highly reactive, and may also be explosive, corrosive, and/or toxic. Organic peroxides often contain "per-," "peroxy-," or "peroxide" in their chemical name; the name often ends in "-ate" or "-ite" (Table 1–8). It is not so important to determine the exact composition of a potential organic or inorganic oxidizer as it is to recognize the potential hazard it presents. If the material is a suspected oxidizer, keep it away from heat and combustible materials. Obtain additional hazard information from the supplier or the manufacturer.

Incompatible Materials

Hazardous materials that can remain in contact with each other indefinitely without undergoing a chemical reaction are considered compatible. Incompatible materials (Table 1–9) undergo a chemical reaction when they come into contact with each other. The chemical

Table 1–7. Inorganic oxidizing agents

ammonium nitrate	nitric acid
ammonium perchlorate	nitrogen peroxide
ammonium permanganate	perchloric acid
bromine trifluoride	potassium nitrate
calcium chlorate	potassium nitrite
calcium hypochlorite	potassium permanganate
chromic acid	sodium bromate
copper nitrate	sodium chlorite
lead nitrate	sodium nitrite
lead peroxide	strontium nitrate
magnesium perchlorate	zinc chlorate
mercuric nitrate	zinc peroxide

Table 1-8. Organic oxidizing agents

acetyl benzoyl peroxide	methane hydroperoxide
benzoyl peroxide	methylethylketone peroxide
butyl hydroperoxide	p-nitrobenzoic acid
cumene hydroperoxide	paramethane hydroperoxide
dicumyl peroxide	peracetic acid
guanidine nitrate	peroxyacetic acid
isopropyl percarbonate	propyl nitrate
lauryl peroxide	tetranitromethane

reaction generates heat and gases, thus increasing the pressure and accelerating and perpetuating the chemical reaction. Incompatible materials, when mixed, have the potential to produce one or more hazardous aftereffects, such as generation of toxic gases or vapors, fire, explosion, formation of shock- or friction-sensitive compounds, or sudden, violent polymerization.

Polymerization is a chemical reaction that combines one or more small units (monomers) to form a long chain of units linked together (polymer). A commonly encountered polymer is polyurethane, which is formed by linking many urethane molecules together. Other common polymers include polyethylene, polypropylene, polystyrene, and nylon. Wool and cellulose are naturally occurring polymers. Synthetic polymers are usually made under controlled conditions that limit the speed and the extent of polymerization. Hazardous polymerization takes place when the chemical reaction occurs at an accelerated rate that generates enough heat to cause a fire or an explosion, or when the reaction occurs within the confines of a drum or a tank, which may rupture or burst. Hazardous polymerization also may occur when monomers are contaminated or subjected to excessive heat or pressure.

Chemical Toxicants

Chemical toxicants are substances capable of producing a toxic or adverse health effect in living organisms. Acute toxicity refers to the ability of a chemical to produce adverse effects as a result of a single exposure of short duration. A poisonous material is a liquid or a solid substance considered to be a significant health hazard because of its acute toxicity. Irritants produce pain, swelling, redness, and inflammation of exposed tissues. Extreme irritants induce irreversible tissue damage and are corrosive.

Chronic toxicity results when a chemical produces damage after repeated exposure. Some chronic effects, such as cancer, pulmonary fibrosis, emphysema, and organ damage, require years of repeated exposure or take a long time to develop. For example, inhalation exposure to asbestos fibers over a prolonged interval has caused chronic lung disease (asbestosis) and cancer (mesothelioma). Systemic toxicants induce damage to specific organs or organ systems. It is not unusual for a material to produce multiple organ effects;

Table 1-9. Some incompatible hazardous materials

acids and water	oxidizers and fuels
acids and caustics	ammonia and acrylonitrile
chlorine and ammonia	acids and metals

for example, chlorinated hydrocarbons may affect the liver, kidney, nervous system, and heart.

Sensitizers are substances that induce allergic reactions in sensitive individuals. Typical allergic reactions can include: skin rashes with itching and blister formation; watery, itching eyes; nasal discharge and sneezing; or asthma-like symptoms that make breathing difficult. The most commonly encountered sensitizers, such as poison ivy and ragweed pollen, are naturally occurring.

Carcinogens are substances that cause cancer. Asbestos, benzene, and vinyl chloride are recognized human carcinogens. Mutagens cause mutations, that is, changes in the genetic material or DNA. Teratogens produce birth defects by damaging the fetus while it develops in the womb. Some chemicals cause lethal damage and cause the fetus to be aborted. The most commonly encountered human teratogen is ethyl alcohol, which causes fetal alcohol syndrome. Reproductive toxicants affect male or female reproductive systems and can cause impotence, loss of sexual drive, infertility, or growth problems in the developing fetus.

Pesticides are substances designed to inhibit growth or to kill specific types of pests, such as weeds, fungi, rodents, insects, ticks, snails, or other life forms. These substances include insecticides, herbicides, fungicides, arachnidicides, and rodenticides. A material whose name ends in "-icide" is usually a pesticide. Pesticides are regulated by the U.S. Environmental Protection Agency (USEPA) under FIFRA (the Federal Insecticide, Fungicide and Rodenticide Act).

Pesticides are classified into broad categories based on their mechanism of action. The chlorinated hydrocarbon insecticides Chordane, DDT, Aldrin, Dieldrin, Mirex, and Kepone are no longer used because of their persistence in the environment and their effects on wildlife. Organophosphate pesticides such as Malathion and Parathion are more toxic to humans than the chlorinated hydrocarbons, but they are short-lived and do not persist in the environment. Botanical (plant-derived) insecticides include nicotine, rotenone, and pyrethrums.

Corrosives

A chemical that destroys living tissue by chemical action at the site of contact is corrosive. Chemical corrosive injuries are also called chemical burns and can only be repaired by the body through the formation of scar tissue. A liquid or a solid material that causes severe degradation in steel is also considered corrosive.

Inorganic acids and bases are commonly encountered corrosives. Strong inorganic acids include hydrochloric acid, hydrofluoric acid, sulfuric acid, and nitric acid; the so-called weak acids include carbonic acid and boric acid. Acetic acid is an organic acid found in dilute solution in vinegar; when very concentrated, it is known as glacial acetic acid, a flammable, corrosive material. Concentrated solutions of both weak and strong acids will produce corrosive damage. Inorganic bases or caustics are also highly corrosive; commonly encountered inorganic caustics include sodium hydroxide, potassium hydroxide, calcium hydroxide or slaked lime, and calcium oxide or quick lime.

The pH scale (an indicator of hydrogen ion concentration) is often used to determine if a material is an acid or a base. A solution with a pH of 7.0 is neutral. Solutions with a pH of less than 7.0 are acidic; the lower the pH is, the more acidic the material. Solutions with a pH greater than 7.0 are alkaline or basic; the higher the pH is, the more alkaline the material.

The pH scale is logarithmic; a solution with a pH of 2 is ten times more acidic than a pH 3 solution, and 100 times more acidic than a solution with a pH of 4. It is not always possible to determine the corrosion hazard by using pH alone. For example, concentrated lemon juice has a pH of 2; household vinegar and lemonade have a pH of 3; tomato juice and flavored carbonated beverages have a pH around 4; and strong black coffee and tea have a pH of about 5 (Table 1–10). Although hydrofluoric acid has a pH close to 6, this acid—even when diluted—is extremely corrosive and can cause severe tissue and bone injury.

A common misconception is that all corrosive materials are measurable on the pH scale. The pH scale measures the acidity or the alkalinity of aqueous solutions; so an organic acid or alkali that contains no water (i.e., a nonaqueous solution) is not measurable on the pH scale.

DEFINING AND CLASSIFYING HAZARDOUS MATERIALS

Many organizations and agencies have developed definitions for the term hazardous material. There is no single definition that covers all circumstances.

U.S. Environmental Protection Agency (USEPA)

The USEPA uses the term hazardous substances for specific chemicals regulated under CERCLA (the Comprehensive Environmental Response, Compensation and Liability Act, or Superfund), chemicals that, when released, may represent a significant threat to the environment. When a reportable quantity (RQ) is released, the USEPA must be notified. A list of USEPA CERCLA hazardous substances can be found in 40 CFR Part 302, in Table 302.4. These substances are listed alphabetically, with synonyms; the RQ for each material is also provided. An alternate listing by CASRN or CAS # (Chemical Abstract Service Registry

Table 1–10. pH of aqueous solutions

pH scale	Solution
1	stomach acid
2	lemon juice
3	vinegar
4	tomato juice
5	seltzer water
6	milk
7	Ivory liquid in water
8	egg whites
9	baking soda
10	window washing solution
11	household cleaners
12	household ammonia
13	household bleach
14	drain cleaners

Table 1-11. CASRN and RQ values for selected chemicals

Hazardous substance	CASRN	RQ (lbs)
Acetone	67641	5000
Benzene	71432	10
Chlorine	7782505	10
Creosote	8001589	1
DDT	50293	1
Mitomycin C	50077	10
Naphthalene	91203	100
PCBs	1336363	1
Warfarin	81812	100
Xylenes	1330207	1000

Number) is also available. CASRN and RQ values for selected chemicals are presented in Table 1–11.

When CERCLA was reauthorized in 1986, the USEPA published a list of extremely hazardous substances. The list can be found in the Superfund Amendments and Reauthorization Act (SARA) in 40 CFR Part 355, Appendix A. The list contains CASRN, chemical names, and the threshold planning quantity (TPQ). Facilities that store or use chemicals in excess of the TPQ must comply with Title III of SARA (Community Right-to-Know).

SARA Title III requires industry and government to collaborate and plan for chemical emergencies. SARA also requires industry to identify chemicals of concern and their quantity, the consequences of a release, and preventive measures taken to avoid a chemical release.

USEPA defines hazardous wastes as substances that are regulated under RCRA (the Resource, Conservation, and Recovery Act) as defined in 40 CFR Part 261. Although the USEPA regulates their storage and use, the transportation of hazardous wastes is regulated by the USDOT (U.S. Department of Transportation).

U.S. Department of Transportation (USDOT)

The USDOT is empowered under 49 CFR Parts 170–179 to regulate the transportation of hazardous materials. The term hazardous material defines any commodity (gas, liquid, or solid) capable of causing harm to people, the environment, or property. Hazardous materials are classified according to their primary hazard and/or use (Table 1–12).

Table 1-12. Classification of hazardous materials

UN Number	Classification	Commodity
1005	Nonflammable gas	Compressed air
1075	Flammable gas	LPG
1203	Flammable liquid	Gasoline
1428	Flammable solid	Sodium
1689	Poisonous substance	Sodium cyanide
1830	Corrosive substance	Sulfuric acid
1978	Flammable gas	Propane
2188	Poison gas	Arsine

Occupational Safety and Health Administration (OSHA)

OSHA defines hazardous substance as any chemical or biological material that is designated or listed by the USEPA under CERCLA, SARA, or RCRA; this includes all materials that are considered hazardous substances, extremely hazardous substances, and hazardous wastes. Substances regulated by DOT as hazardous materials or hazardous wastes are also considered hazardous substances under OSHA. This very broad definition is found in the definition section of the final OSHA rule, Hazardous Waste Operations and Emergency Response (HAZWOPER, 29 CFR 1910.120) which was promulgated on March 6, 1989. The HAZWOPER rule regulates the safety, health, and training of workers involved in hazardous waste site operations.

OSHA defines hazardous chemical as any substance or mixture that is capable of producing adverse health effects in exposed workers. OSHA has established occupational exposure limits for approximately 600 hazardous chemicals or processes. The OSHA Hazard Communication rule (29 CFR 1910.1200) requires chemical manufacturers and importers to evaluate the health and safety hazards associated with their products. Hazard evaluation information is made available through the use of material safety data sheets (MSDS), labels, and product stewardship. Employers are required to ensure that such hazard information is made available to employees.

HAZARD CLASSIFICATION SYSTEMS

A variety of agencies and organizations have developed classification systems to aid in the identification of various types of hazardous materials. Each system was established for a particular application or to meet the needs of a particular group.

U.S. Department of Transportation (DOT)

The USDOT uses the term hazardous materials, which covers nine hazard classes; some hazard classes have subclasses called divisions (Table 1–13). The USDOT system, as outlined in 49 CFR, is used for materials in transportation, and consists of placards for vehicles and labels for containers.

The placard is a $10\frac{3}{4}$-inch-square diamond that is applied to each side and end of a motor vehicle, railcar, freight container, or portable tank container containing hazardous materials. Most placards are not required in highway transportation unless 1000 pounds or more of hazardous material is being transported.

A label is a 4-inch-square diamond that is affixed to the container being shipped. Any material classified as hazardous by DOT should carry the appropriate label or labels.

Labels and placards provide general hazard information by using specific background colors and symbols for specific hazard classes. For example, flammable liquids have a red-colored background and a flame symbol. The hazard class number and division should also be present (Figure 1–8). A four-digit UN (United Nations) identification number is usually present in the center of the placard.

Identification numbers are listed in the North American Emergency Response Guide-

Table 1–13. USDOT Hazard classes and divisions (placard description)

Divisions	Example
Class 1: Explosives (orange background, bursting ball symbol)	
1.1 Mass detonating	Dynamite
1.2 Mass detonating w/fragments	Incendiary ammunition
1.3 Fire hazard, minor blast	Propellant explosives
1.4 No significant blast hazard	Common fireworks
1.5 Very insensitive explosives	Blasting agents
1.6 Extremely insensitive explosive	
Class 2: Compressed Gases (green or red background, cylinder symbol)	
2.1 Flammable gases	Propane, acetylene
2.2 Nonflammable gases	Chlorine, ammonia
2.3 Poisonous gases*	Arsine, phosgene
(*white background, skull and crossbones symbol)	
Class 3: Flammable and Combustible Liquids (red background, flame symbol)	
3.1 Flash point <0°F	Gasoline
3.2 Flash point >0°F to <73°F	Acetone, toluene
3.3 Flash point >73°F to <141°F	Turpentine
Combustible liquids FP >141°F	Fuel oils
Class 4: Flammable Solids or Substances (all with flame symbol, backgrounds that vary; see below)	
4.1 Flammable solids (red and white vertical stripes)	Powdered magnesium
4.2 Spontaneously combustible, pyrophoric (white over red)	White phosphorus
4.3 Dangerous when wet (blue)	Calcium carbide
Class 5: Oxidizing Substances (yellow background, circle with flame symbol)	
5.1 Oxidizing substances	Ammonium nitrate
5.2 Organic peroxides	Benzoyl peroxide
Class 6: Poisonous and Infectious Substances (white background, skull and crossbones symbol)	
6.1 Poisons	Arsenic, aniline
6.2 Infectious substances	Botulism, tetanus
Class 7: Radioactive Substances (yellow over white background, propeller symbol)	
Radioactive material	Plutonium, uranium
Class 8: Corrosives (white over black background, test tube over hand and metal symbol)	
Corrosive material	Acids, alkalies
Class 9: Miscellaneous Hazardous Material (black and white vertical stripes over white)	
	Bleaching powders
	Hazardous substances

book (USDOT, 1996). The *Guidebook* is a reference book containing information that can help emergency responders identify hazardous materials by using visible placards or labels and the identification number. Approximately 2500 chemicals are listed alphabetically in one section; another section lists the same materials by their four-digit code number. The dual listing system allows the user to look up either the shipping or chemical name of the material or its identification number.

The NAERG96 contains action guides, which describe proper initial response proce-

HAZARD CLASS SYMBOL

HAZARD CLASS DESIGNATION OR FOUR-DIGIT IDENTIFICATION NUMBER

1090

3

COLORED BACKGROUND

UNITED NATIONS HAZARD CLASS NUMBER

Figure 1-8. A USDOT placard with the four-digit identification number.

dures during early stages of a hazardous materials emergency. After the hazard class or the actual identity of the commodity has been determined, the *Guidebook* can be used to determine appropriate safe evacuation distances, fire-fighting procedures, and medical emergency treatment for exposed personnel.

USDOT hazard classes reflect the major hazard that a material presents during transportation. Many substances have multiple hazards; for example, liquid morpholine is toxic, corrosive, and flammable. Hazard Class 9 materials are those substances that present a hazard during transport but are not included in another hazard class. This class includes materials that, when released, may be irritating, have anesthetic effects, or have a noxious odor. Hazardous wastes also fall into Class 9 unless they have specific characteristics that place them in another category.

National Fire Protection Association (NFPA)

The National Fire Protection Association (NFPA) has developed a rapid hazard identification system for chemicals stored or used at fixed facilities. This system, known as NFPA 704, is widely recognized and used. The NFPA 704 system uses a diamond symbol that is divided into four smaller diamonds with colored backgrounds and numerical ratings to indicate the general hazard and degree of severity for toxicity, flammability, and reactivity.

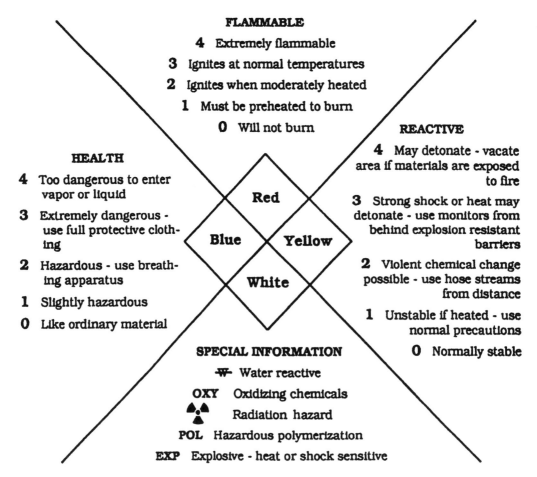

Figure 1–9. The NFPA 704 system is used to indicate the presence of hazardous materials at fixed facilities. The 704 system does not identify the specific hazardous material present, but indicates the general degree of severity for toxicity, flammability, and reactivity.

A special hazard section gives information regarding water reactivity, radiation, oxidation, polymerization, or other unusual hazards (Figure 1–9).

The color code for individual small diamonds is:

- Blue for health hazard.
- Red for flammability hazard.
- Yellow for reactivity hazard.
- White for special hazards.

Within the blue, red, or yellow diamond, a number is assigned from 0 to 4, which indicates the degree of hazard; 0 indicates no hazard, whereas 4 indicates severe hazard. The white area will contain a hazard symbol or letters to indicate any special hazards:

Table 1-14. NFPA numerical hazard classification system

Number	HEALTH Hazard Description (blue background)
4	Material is too dangerous to health for fire fighters to be exposed. Very brief exposure may cause death, or vapor or liquid may be fatal if it penetrates the fire fighter's normal protective clothing. Normal turnouts and self-contained breathing apparatus (SCBA) insufficient to protect against inhalation and skin contact.
3	Material is extremely hazardous to health, but area may be entered with extreme care. No skin should be exposed. Full protective ensemble, including SCBA, full protective clothing, rubber gloves, and boots; bands or tape around arms, legs, and waist should be provided.
2	Material is hazardous to health, but area may be entered with SCBA.
1	Material is only slightly hazardous to health.
0	Exposure under fire conditions offers no health hazard beyond that of ordinary combustibles.

Number	FLAMMABILITY Hazard Description (red background)
4	Materials are very flammable gases or very volatile flammable liquids. Shut off flow of gas or liquid; keep cooling water streams on exposed containers or tanks. Withdrawal may be necessary.
3	Material can be ignited under almost all normal conditions. Water may be ineffective because of low flash point.
2	Material must be subject to moderate heat before ignition will occur. Water spray may be used to extinguish fire because material can be cooled below its flash point.
1	Material must be preheated before ignition can occur. Water may cause frothing if it mixes with material and turns to steam. Apply water fog to surface to extinguish fire.
0	Material will not burn.

Number	REACTIVITY Hazard Description (yellow background)
4	Material is capable of detonation, or it can undergo explosive decomposition at normal temperature and pressure. If involved in massive fire, evacuate the area.
3	Material is capable of detonation or explosive decomposition when heated and may react violently with water. Conduct fire-fighting activities from behind explosion-proof locations.
2	Material can undergo violent chemical change at elevated temperatures, but does not detonate. Use caution during fire-fighting activities.
1	Material is normally stable but many become unstable when mixed with other materials, or when subjected to elevated temperature and pressure. Use normal precautions when approaching fire.
0	Material is normally stable and does not present any reactivity hazard to fire fighters.

- Propeller for radiation hazard.
- OXY or OX for oxidation hazard.
- W with a bar through it to caution against using water.

The NFPA 704 system does not, however, identify the materials present. It was developed to assist fire fighters in determining general fire, health, and safety hazards when they first arrive at a facility (Table 1-14). The degree of hazard rating assigned to the material may be affected by storage conditions. For example, a drum containing a flammable liquid stored indoors presents a different fire-fighting hazard from that of one stored outside on a protected storage pad.

Health hazard ratings are determined for a single exposure; the length of exposure may be seconds or minutes, up to one hour. NFPA 704 health ratings thus are for acute toxicity only and should not be used to determine the hazard to workers repeatedly exposed for prolonged intervals.

Reactivity hazard ratings are based upon the likelihood of the material to release ener-

Figure 1–10. U.S. military symbols for detonation and chemical hazards.

gy by itself or in combination with other materials. The reaction of the material when exposed to fire, along with shock and pressure, is also considered.

U.S. and Canadian Military

The U.S. military has established its own system for identifying hazardous materials. The system uses seven symbols to mark structures that store hazardous materials. Figure 1–10 shows symbols used to indicate explosion and chemical hazards in this system.

The Canadian military also has a unique system of symbols for identifying hazardous materials (Figure 1–11). Under this system, explosive hazards are divided into nine classes (Table 1–15).

OTHER SOURCES OF HAZARD INFORMATION

Chemical Transportation Emergency Center (CHEMTREC)

CHEMTREC is a public service established by the Chemical Manufacturers' Association (CMA) in 1971 in response to the increasing need for information and assistance during hazardous materials transportation emergencies. CHEMTREC operates continuously and can be reached through its toll-free emergency telephone number (800-424-9300). A separate number for international emergencies is also available (703-527-3887; call collect). CHEMTREC has a library containing information on more than one million chemicals and chemical products. CHEMTREC has enlarged its focus to provide nonemergency information about hazardous materials. A toll-free telephone number (800-262-8200) is available Monday through Friday, from 9:00 A.M. to 6:00 P.M., eastern time. The service is free to the public, chemical users, and providers of emergency services who require information about the health, safety, and environmental fate of chemicals and chemical products. An E-mail address (chemtrec@mail.cmahq.com) is also available for computer access.

Product Labels

Product labels found on packages and containers contain information provided by the manufacturer for hazard identification and assessment. The USEPA regulates labeling of pesticides and agricultural chemicals. The Consumer Product Safety Commission oversees labeling of consumer products. The labeling of hazardous chemicals in the workplace is regulated by OSHA.

In most cases, the label should contain the following information:

- Product name.
- Active ingredients.
- Physical and chemical hazards statements, such as flammability, reactivity, and corrosivity.
- First aid statement.
- One of three hazard signal words, usually prominently displayed in the center of the label, which indicates one of three hazard levels:

  Danger = highest hazard.

  Warning = moderate hazard.

  Caution = low hazard.

Signal words are used to indicate a toxic or flammable hazard. The nature of the haz-

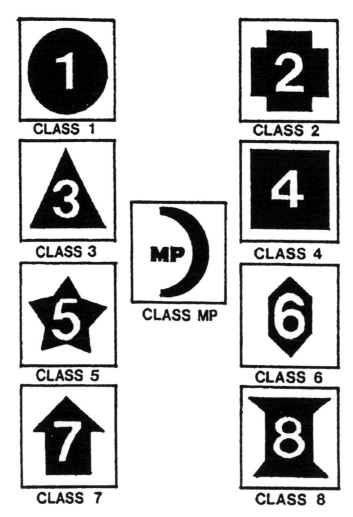

Figure 1–11. Canadian military markings for explosive hazards.

Table 1–15. Canadian explosive hazard class identification system

Class	Description
1	Expected to detonate upon contact with fire.
2	Readily deflagrates but does not detonate.
3	Can explode with fragments after exposure to fire.
4	Burns and gives off dense smoke; will not detonate.
5	Contains toxic substances.
6	May be exposed to fire for a period of time before sporadic explosions occur; fragment hazard but not mass detonation hazard.
7	Flammable, toxic, and corrosive hazards; may be exposed to fire before explosions occur; fragment hazard but no mass detonation hazard.
8	Explosion and radiation hazard.
MP	Substances containing pyrophoric metallic powders in ammunition or bulk.

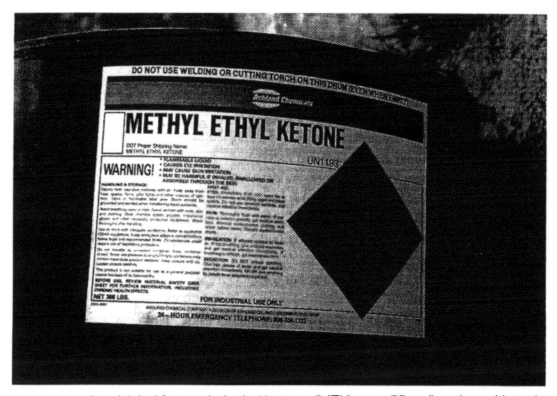

Figure 1–12. A label for methyl ethyl ketone (MEK) on a 55-gallon drum. Note the UN number 1193, which is specific for MEK. The 4-inch USDOT diamond indicates that the drum contents are flammable. The signal word WARNING! indicates an intermediate flammability hazard.

ard should also be specified. Figure 1–12 shows a label on a 55-gallon drum; it carries the signal word WARNING, indicating an intermediate flammability hazard.

Labels for USEPA-regulated pesticides must also carry a product-specific EPA Registration Number and a USEPA Establishment Number, which indicates where the pesticide was manufactured. The EPA Registration Number has two or three sections (e.g., 1111-222-33 or 1111-2222) and is a useful way to positively identify a particular pesticide product. Pesticides labels also carry antidote information and/or a Note to Physician statement.

Material Safety Data Sheets (MSDS)

MSDS are information sheets prepared by chemical manufacturers for specific products, which describe the products and the hazards associated with their use, storage, transportation, and disposal. Although MSDS may look different, they all should contain the following information:

- Identity of the product, including chemical and common names; many MSDS also contain the Chemical Abstract Service Registry Number.
- All hazardous ingredients, in some cases (carcinogens) even those present at concentrations less than 0.1%.

- A physical description of the material, including physical state, color, viscosity, odor, or other physical characteristics that may aid the user in verifying that the material is indeed the product described.
- Physical and chemical data such as boiling point, water solubility, viscosity, melting point, evaporation rate, and flash point.
- All physical and chemical hazards associated with the material, such as flammability, corrosivity, and reactivity.
- How the material enters the body (inhalation, skin or eye absorption, ingestion), and if it produces acute or chronic health effects.
- Emergency and first aid procedures in case of worker overexposure, or in the event of a spill or an unexpected release.
- Worker exposure limits that are required by OSHA or recommended by other organizations.
- Precautions to be taken during storage and use, and safety equipment that should be available.
- Identification of the organization responsible for creating the MSDS and the date of issue.
- Telephone number to call in case of a health emergency.

The OSHA Hazard Communication Standard (29 CFR 1910.1200) requires employers to have MSDS available for all chemicals and products used by workers. MSDS vary considerably in their quality; there should be no spaces or missing information. Sheets should be legible and understandable. A standardized MSDS format developed by CMA has gained acceptance and has been adopted by the American National Standards Institute (ANSI 1993); hopefully this will help ensure better-quality MSDS in the very near future. A more detailed description of information found in an MSDS can be found in Chapter 4. An MSDS for isopropyl alcohol, prepared according to ANSI guidelines, can also be found in Chapter 4, courtesy of the Genium Publishing Corporation.

2

Properties of Hazardous Materials

The physical state of a material, as well as its physical and chemical properties, determines its behavior in the environment, the manner in which it moves or migrates, and the hazard it may present. An understanding of the basic properties of any chemical will give the waste site worker an idea of where to look for the material, what instruments to use to detect it, and what type of protective equipment may be warranted.

Most hazardous materials have multiple hazards associated with their manufacture, transport, use, storage, and disposal. Hazards associated with a material are dependent on how it is used, where it is, and how much is present in terms of volume and concentration. For example, the chlorinated solvent trichloroethylene (TCE) is usually considered a toxic material. Ground water contaminated by TCE is considered undrinkable; exposure to TCE in the workplace is regulated by OSHA. It is less well recognized, however, that TCE is also a combustible liquid.

Physical and chemical properties of a material are usually defined at normal temperature (68°F or 20°C) and pressure (760 mm Hg or 1 atm).

PHYSICAL STATE

A chemical may exist as a solid, a liquid, a gas, a vapor, or an aerosol. Temperature, pressure, and concentration may affect the physical state of a material. Most chemicals can exist in more than one state. For example, water is most commonly encountered as a liquid, but it can also exist as a solid (ice), a mist (rain), and a vapor (steam).

A solid is a substance that has definite size and shape. A liquid has a definite volume, but no shape; the shape of a liquid will conform to the shape of its container. Gases are materials that are neither solids nor liquids. Gases have no shape or volume, and are completely airborne at room temperature. Gases can be changed to liquids or solids, but only when subjected to increased pressure or decreased temperature. The liquefied gases propane and butane are kept liquid because they are kept under pressure. Vapors are the evaporation products of materials that are liquids or solids under normal temperature

(68°F or 20°C). Water, acetone, isopropyl alcohol, and methylene chloride are all liquids that release vapors; paradichlorobenzene and naphthalene are solids used in mothballs because they slowly release vapors that repel moths.

An aerosol is a mixture of solid or liquid particles dispersed in a gas. Aerosols commonly found in households are usually generated by spray cans or pump spray bottles and include spray paints, bug killers, deodorants, cleaners, and disinfectants. Dusts, smoke, mists, and fumes are all different types of aerosols.

Dusts are produced from solids by mechanical means such as grinding, crushing, or blasting; they may be inorganic (lead, silica) or organic (cotton, grain, flour). Dusts are usually spherical in shape and differ from fibers, which are not spherical.

Mists are small liquid droplets suspended in air or another gas. Mists may be formed mechanically by spraying or splashing; they also can be formed when a vapor is cooled and condenses into the liquid state. Highly visible mists are often called fog.

Fumes occur when hot gases from heated solids cool and condense to form minute solid particles in air. Fumes can be formed by heating metals, waxes, polymers, and plastics. The most commonly encountered fumes are those produced by the evaporation of metal during welding. Breathing metal fumes and polymer fumes can cause a flu-like illness (metal fume fever or polymer fume fever).

Smoke is a complex mixture of particles, vapors, and gases produced as a result of incomplete combustion of organic materials. Smog is a combination of smoke and fog.

PHYSICAL/CHEMICAL PROPERTIES

Physical/chemical properties are determined under standardized laboratory conditions and published in reference texts or manuals.

Vapor Pressure

Vapor pressure (VP) determines the ability of a liquid or a solid to evaporate into the air. It is dependent on temperature; as the temperature increases, the vapor pressure increases dramatically. Vapor pressures for liquids and solids are usually expressed in millimeters of mercury or mm Hg (Table 2–1). Substances that exist as gases at room temperature have vapor pressures defined in pounds per square inch (psi) or atmospheres (atm). Normal atmospheric pressure is 760 mm Hg, 14.7 psi, or 1 atm. Atmospheric pressure compresses liquids and solids and inhibits their evaporation. Materials that exist as gases at room temperature have pressures greater than 1 atm and are therefore able to overcome the compressive force of normal atmospheric pressure.

Vapor pressure is really a measure of how much the material "wants" to become airborne (i.e., its tendency to become a vapor or a gas). Substances with high vapor pressures are highly volatile and quickly evaporate into the air. Many liquids, such as isopropyl alcohol, acetone, ethyl acetate (nail polish remover), paint thinner or turpentine, and gasoline, evaporate very quickly; these liquids feel cool when spilled on the skin because of their rapid evaporation.

Temperature dramatically affects vapor pressure. As temperature increases, vapor pressure increases. Water, for example, has a vapor pressure of 18 mm Hg at 68°F. When heat-

Table 2-1. Vapor pressures (mm Hg) of some chemicals*

Acetone	180
Benzene	75
Chloroform	160
Ethyl acetate	74
Ethyl ether	440
Ethyl alcohol	40
Fuel oil	2
Isopropyl alcohol	33
Methyl alcohol	92
Methylene chloride	350
PCBs	0.001
Toluene	20
Trichloroethylene	60
Water	18
Xylene	9

*At 68°F.

ed on a stove, it quickly evaporates. A puddle of water also evaporates quickly on a hot sunny day, but it may linger several days during cold weather.

Most mobile liquids (i.e., liquids that readily pour at room temperature) have vapor pressures of at least 1 mm Hg. Compounds with vapor pressures greater than 100 mm Hg

Figure 2-1. The vapor pressure and the evaporation rate of chemicals are affected by temperature. Warm gases and vapors from a landfill quickly condense when they contact cold air.

are highly volatile; materials with vapor pressures between 1 and 100 mm Hg will release significant vapors into the atmosphere. Viscous materials or solids also have vapor pressures! Organic materials that are solid at room temperature may have vapor pressures of 0.1 mm Hg or more. For example, paradichlorobenzene and naphthalene have vapor pressures of 0.44 and 0.08 mm Hg; both are solids at room temperature and are often found in the home as mothballs or mothflakes.

Liquids and solids are usually kept in a closed or sealed container to minimize evaporation. Imagine a closed bottle half-full with a liquid; the air trapped in the container above the liquid is called the headspace. A certain amount of the liquid will evaporate into the headspace; eventually the vapor in the headspace will exert so much pressure on the liquid and the sides of the container that some vapor will be forced back into the liquid. When this happens, the container and its contents are considered to be in equilibrium. The maximum pressure exerted by the vapor against the walls of the container and the liquid is its vapor pressure.

It is possible to calculate roughly how much vapor will be given off by a liquid or a solid on the basis of its vapor pressure. There is a simple equation used to estimate the part per million (ppm) concentration of vapors that collect in the headspace of a container at equilibrium at 68°F (20°C), the so-called 1300 rule:

$$\text{Vapor pressure (mm Hg)} \times 1300 = \text{Headspace concentration (ppm)}$$

For instance, the vapor pressure of 1,1,1-trichloroethane (methyl chloroform) is 100 mm Hg. The headspace concentration of a container above the liquid can be calculated as:

$$100 \text{ mm Hg} \times 1300 = 130,000 \text{ ppm}$$

This value is an estimation (±1–2% of the calculated value) of the actual concentration within the container. Remember that temperature and barometric pressure (altitude) affect the vapor pressure and thus the calculated value. However, even though the calculated value is not identical to the ppm concentration at 68°F, the 1300 rule can still give an indication of the potential concentration that may be present in a closed container or a confined space.

The 1300 rule cannot be used for materials that exist as gases or vapors at room temperature. These materials have vapor pressures greater than 760 mm Hg or 1 atm and are already completely airborne. *The 1300 rule can only be used for materials that are normally liquids or solids at room temperature.*

Boiling Point

Boiling point (BP), is a property that goes hand-in-hand with vapor pressure. Boiling point is the temperature at which the vapor pressure of a liquid or a solid equals atmospheric pressure; that is, it is the temperature at which the substance's vapor pressure reaches 760 mm Hg. At 760 mm Hg, the atmosphere no longer exerts any pressure on the material to remain a liquid or a solid; therefore, it is free to "boil away" and become completely airborne. The boiling point of water is 100°C or 212°F; if kept at its boiling point, water will boil away.

Most materials that are liquid or solid at room temperature have boiling points greater than room temperature; otherwise, they would quickly evaporate into the air. Liquids with relatively low boiling points, such as ethyl ether (BP 94°F), are often kept refrigerated to reduce evaporation. Chemicals that exist as gases at normal temperature and pressure have, by definition, low boiling points. Liquefied gases with boiling points less than normal temperature, such as propane (BP −44°F), butane (BP +32°F), and ethylene oxide (BP +55°F), are maintained under pressure as compressed liquids.

Evaporation Rate

The rate at which a solid or a liquid evaporates into the air, compared to a reference material, is the evaporation rate. The reference material used is butyl acetate, which is assigned an evaporation rate of 1.0. Materials with evaporation rates greater than 3.0 are classified as rapid evaporators, whereas those with rates between 0.8 and 3.0 are considered to have medium evaporation rates. Slow evaporators are those with rates less than 0.8. The rates in Table 2–2 are based on how fast a thin film of liquid evaporates off a nonporous surface. Actual evaporation rates for liquids under field conditions may differ; the rates given in the table should be used to make comparisons.

Melting Point

The temperature at which a solid changes phase into a liquid is its melting point. This temperature is also the freezing point (i.e., the temperature at which the liquid changes phase into a solid). Whether the temperature is regarded as the freezing point or the melting point depends on the direction of the phase change.

Vapor Density

Vapor density (VD) is the weight of a vapor or a gas relative to an equal volume of air at comparable temperature and pressure. The vapor density of air is 1. Materials with vapor densities less than 1 are lighter than air, and, if released, will tend to rise and dissipate. Materials with vapor densities greater than 1 are heavier than air and will tend to sink or settle when released (Table 2–3).

Table 2-2. Evaporation rates of some liquids

Ethyl ether	37.5	Methyl ethyl	3.8
Methylene chloride	27.5	Isopropyl alcohol	2.8
Carbon tetrachloride	12.8	Toluene	2.2
Hexane	8.3	VMP naphtha	1.4
Trichloroethylene	6.2	Butyl acetate	1.0
Acetone	6.0	Isobutyl alcohol	0.6
Methyl alcohol	5.9	Water	0.3
Mineral spirits	0.1		

Table 2-3. Relative weight of some gases

Lighter than air	Heavier than air
Ammonia	Butane
Helium	Chlorine
Hydrogen	Nitrogen dioxide
Methane	Propane
Neon	Sulfur dioxide

Some chemical reference manuals give vapor density values. Other chemical guides furnish the specific gravity of the vapor or the gas relative to dry air, which is usually designated as specific gravity (gas); this value and vapor density are one and the same and can be used interchangeably. Other chemical references, usually oriented toward chemists, do not offer vapor density but rather give the weight of the gas in grams per liter (g/l). This value is of little use unless one knows how much air weighs. At 32°F (0°C) and normal atmospheric pressure, air weighs 1.29 g/l. The vapor density of a material is determined by dividing the weight of the vapor in g/l by 1.29. For example, one liter of propane at 0°C weighs 2.02 g/l; the vapor density of propane is 2.02 divided by 1.29, which equals 1.57.

Vapors from materials that are liquid or solid at normal temperature and pressure will have vapor densities greater than 1.0. The vapor densities of gases depend on their chemical composition. Carbon monoxide and ethane have vapor densities approximately equal to that of air.

If the reference source does not offer information regarding vapor density or weight of the vapor in grams per liter, it is still possible to estimate the vapor density. This can be done by finding the molecular weight (MW) of the compound of interest, and dividing it by the molecular weight of air, which is 29. For example, the MW of propane is 44; therefore, the approximate vapor density is 1.52 (44/29 = 1.52). Note how similar this value is to the vapor density calculated earlier using the weight of the vapor in grams per liter. Most liquids release vapors that are heavier than air because of their high molecular weight; a notable exception is water vapor, which is lighter than air.

Vapor density is somewhat dependent on the temperature of the vapor or the gas. Vapor density and vapor pressure decrease as temperature decreases. Lighter-than-air gases are cooler than ambient air temperatures when released from pressure; these gases often initially behave like heavier-than-air gases until they warm up. Likewise, heavier-than-air gases or vapors that are heated may rise on hot air thermals.

Vapor density is most important in situations where high concentrations of vapors or gases are known or anticipated to occur, as when a compressed gas cylinder ruptures or a large volume of a very volatile chemical is released, or in confined spaces where vapors may concentrate or stratify. Vapor density is of less concern if vapors or gases are present in ppm concentrations, and the predominant gas present is air with a vapor density of 1.0.

Specific Gravity

Specific gravity is the ratio of the density or the weight of a substance to that of a reference material. With liquids and solids, the density of the material in grams per cubic centimeter (g/cc) or grams per milliliter (g/ml) is compared to that of water. The specific gravity of water is 1.0 g/cc. Materials with specific gravities less than 1 are lighter than water and will float on it; materials with specific gravities greater than 1 are heavier than water and will tend to sink in it.

Specific gravity varies with temperature. The temperature of the water reference and the material are usually indicated, along with the density or the specific gravity value. The density of ethyl alcohol, for instance, is listed in a chemical reference as d20/4 = 0.789. The "d" stands for density and 20/4 indicates the temperature in °C, with the numerator or top number of the fraction referring to the temperature of ethyl alcohol and the denominator or lower number to the temperature of the reference material, water.

It is important not to confuse specific gravity or density of liquids and solids with vapor density. Remember that vapor density may also be listed as specific gravity (gas) or specific gravity (vapor).

Solubility

Water solubility is the amount of material that will mix or dissolve completely in water. Materials that are miscible in water will mix freely, regardless of the quantities present. Some chemical references consider a material to have negligible solubility or to be insoluble in water if its solubility is less than 1 part in 1000 parts water, or 0.1%; in other references, the term practically insoluble may be used to indicate the same thing. Although a solubility of 0.1% may be negligible by some standards, it may be significant in terms of potential environmental impact, such as ground-water contamination. Always determine the criterion used when a chemical reference indicates that a material is insoluble in water.

Water solubility is typically measured by the weight or the volume of material that will dissolve in a volume of water. Solubility can be stated as one part material in so many parts water, or grams of material per gram of water, or grams of material per milliliter water, or grams of material per cubic centimeter of water. Relax! All of these indicate the same thing, as 1 cc (and 1 ml) of water weighs 1 g.

For instance, the solubility of table salt may be expressed correctly as:

1.0 g salt in 2.8 g water; or

1.0 part in 2.8 parts water; or

1.0 g in 2.8 cc water; or

1.0 g in 2.8 ml water.

Water solubility is often presented as a percentage. A material with a solubility of 1 part in 100 has a solubility of 1% (Table 2–4). In another example, a commonly used decontam-

Table 2–4. Water solubilities by percentage

10%	10 parts in 100 parts water or 1 part in 10
1%	1 part in 100
0.1%	1 part in 1,000
0.01%	1 part in 10,000
0.001%	1 part in 100,000
0.0001%	1 part in 1,000,000

ination solvent, *n*-hexane, has a solubility of 0.002%. This means that 0.002 part hexane will dissolve in 100 parts or water, or 2 parts hexane will dissolve in 100,000 parts water, which is the same as 1 part hexane dissolving in 50,000 parts water.

In some cases, it may be necessary to convert solubility data given in parts to a percentage. This is easily accomplished by dividing the parts of material by the parts of water and then multiplying by 100. Recall that 1 part table salt is soluble in 2.8 parts water; thus it has a solubility of approximately 36% (1/2.8 = 0.357; 0.357 × 100 = 35.7%). One part benzene is soluble in 1430 parts of water, it therefore has a solubility of approximately 0.07% (1/1430 = 0.000699; 0.000699 × 100 = 0.0699 or 0.07%).

A material that is miscible will mix with water, whereas other materials will sink, float, and/or mix with water according to the limits of their solubility. Water solubility is obviously important in predicting the behavior of materials in water. Solubility under field situations will differ from that observed in the laboratory. In many cases, "field solubility" is higher than anticipated, often because of adsorption or dispersal of the material in the presence of other dissolved and solid organic compounds in water. For example, PCB solubility in water is enhanced by sorption onto silt or clay particles, sawdust, or carbon soot.

Viscosity

Viscosity is a measure of how well a liquid pours. Liquids that are thick and do not flow are considered very viscous. Thin liquids that are easy to pour are considered nonviscous or mobile liquids. Gasoline, isopropyl alcohol, and acetone are all mobile liquids. Molasses and lubricating oils are viscous liquids.

Viscosity is typically measured in SUS or Saybolt Universal Seconds. The SUS rating of a liquid is determined by preheating the liquid to a standard temperature and then pouring it through an opening. The number of seconds required to fill a 60 ml flask is its SUS viscosity rating.

FLAMMABILITY

Flammability can be defined as the capacity of a material to ignite and burn rapidly. The more susceptible a material is to burning, the greater the flammability hazard.

Flash Point

Flash point (FP), is a parameter that is defined only for liquids and solids. Flash point is the temperature to which a liquid or a solid must be raised in order to emit sufficient vapors to form an ignitable mixture with air. At the flash point, the concentration of vapors present will ignite or flash over in the presence of an ignition source, such as a lighted match, a cigarette, a road flare, an electrical spark, or static electricity. The flammability hazard of a material is directly related to its flash point. The lower the flash point is, the greater the hazard.

Current USDOT regulations (49 CFR 173.117-119) define a flammable material as a liquid or a solid with a flash point of less than 141°F, whereas a combustible material has a flash point at or above 141°F and less than 200°F. To accommodate heating fuels and other similar materials, for domestic shipments (i.e., not transported by vessel or aircraft) a material with a flash point at or greater than 100°F that does not meet the definition of another transportation hazard can be reclassified as combustible. The domestic exception allows the continued use of previous USDOT definitions of combustible materials that are not reactive, toxic, or corrosive.

The USEPA and OSHA definition of an ignitable material is similar to that of USDOT: an ignitable material has a flash point of less than 140°F; solids that spontaneously combust are also considered ignitable by USEPA.

A liquid or a solid must be raised to its flash point temperature before ignition or flashover can occur. Remember that liquids and solids do not burn; the vapors or the gases that are emitted actually burn. In order to release sufficient vapors, the substance must be heated. Think about how the flame on a candle flickers when it is first lit. Initially, the flame is quite small; then the wax becomes soft and begins to melt. As the wax melts, it releases paraffin vapor, which burns; the flame then becomes brighter and higher.

Substances that are already gases at normal temperature and pressure do not have flash points, the flash point being the temperature to which a liquid or a solid must be heated in order to release sufficient vapors to burn. No heat is necessary to release gases; they are already airborne.

Unless a material is subjected to an external heat source, such as a fire, ambient temperature will be an important factor to consider when assessing the potential flammability hazard. For example, on a cool (60°F) cloudy day, there is little likelihood that a material with a flash point of 90°F will release sufficient vapors to flash over unless the vapors are confined. If it is a hot (80°F) sunny day, however, and the material is spilled on a hot asphalt surface, there is a good chance for a flashover if a source of ignition is available. Substances with flash points less than ambient temperature are always fire hazards. A good example of such a material is gasoline, which has a flash point of –45°F.

Flammable Range

The flammable range is the concentration range of a vapor or a gas that will support combustion. Within the flammable range, the vapor or the gas mixes with oxygen in the air and burns. Outside the flammable range, the mixture is too lean or too rich to burn. The flammable range is defined in terms of the percent by volume of gas or vapor in air. Every chemical capable of burning has a flammable range.

The lower explosive limit (LEL) or lower flammable limit (LFL) is the minimum concentration of gas or vapor in air that will ignite in the presence of a source of ignition; concentrations that are less than the LEL are too lean to burn and will not ignite. The upper explosive limit (UEL) or upper flammable limit (UFL) is the maximum concentration of gas or vapor in air that will ignite; higher concentrations are too rich to burn. The concentration range between the LEL and the UEL is the flammable range.

Chemicals vary significantly in their flammable range. Some materials have a narrow flammable range, whereas others have a very wide range (Table 2–5). In general, the wider the flammable range is, the greater the flammability hazard. It is foolhardy to assume that there are no other potential hazards when workers are operating outside the flammable range; vapors or gases may still pose a toxic or asphyxiation hazard to unprotected indi-

Table 2-5. Flammable range of liquids and gases*

	LEL	UEL
Acetone	2.5	13
Acetylene	2.5	100
Butane	1.6	8.4
Cylohexane	1.3	8
Ethyl acetate	2.0	11.5
Ethyl alcohol	3.3	19
Ethyl ether	1.9	36
Gasoline (92 octane)	1.5	7.6
Hydrogen	4.0	75
Jet fuel JP-4	1.3	8
Methyl alcohol	6.0	36
Methane	5.0	15
Propane	2.1	9.5
Toluene	1.2	7.1

*LEL and UEL values in percent by volume.

viduals. In addition, vapors or gases that are initially too rich or too lean may be diluted or concentrated and eventually reach the flammable range.

Other Flammability Definitions

The minimum temperature at which a substance ignites without the presence of a source of ignition is its autoignition temperature, also called the ignition temperature or self-ignition temperature. Once the autoignition temperature is reached, self-sustaining combustion occurs.

The fire point is the minimum temperature at which a liquid or a solid produces sufficient vapors to flash over and then continue to burn. The fire point of a substance is always higher than its flash point.

Flash back occurs when vapors distant from their source are ignited; the flash then travels back along the vapor trail until it reaches and ignites the source.

PARTS PER MILLION

In discussing chemical contamination in the soil, water, or air, the most commonly used term to describe concentration is parts per million, or ppm. A ppm is an incredibly small amount; as a volume, it is equivalent to one ounce of vermouth in 7800 gallons of gin (a really dry martini!). On a weight basis, it is equivalent to one grain of salt in 150 pounds of potato chips. A part per billion (ppb) is an even smaller amount, 1 ppm divided by 1000.

In some cases it may be necessary to convert percent by volume concentrations to ppm. This is easy to do because 100% percent by volume is equal to a million parts per million, or 1,000,000 ppm. It follows, then that:

$$
\begin{aligned}
50\% \text{ by volume} &= 500,000 \text{ ppm} \\
10\% &= 100,000 \text{ ppm} \\
5\% &= 50,000 \text{ ppm}
\end{aligned}
$$

1%	=	10,000 ppm
0.1%	=	1,000 ppm
0.05%	=	500 ppm
0.01%	=	100 ppm
0.005%	=	50 ppm
0.001%	=	10 ppm
0.0005%	=	5 ppm

This information allows us to reexamine percentages given for water solubility and convert them directly to ppm. Hexane, for example, had a water solubility of 0.002%, which is the same as 2 parts per 100,000 or 20 ppm.

3

Toxicology and Chemical Exposure

All substances are poisons; there is none which is not a poison. The right dose differentiates a poison and a remedy.

This astute observation about the toxicity of chemicals was made by Paracelsus (1493–1541). He recognized that an inherent characteristic of all chemicals is their ability to produce injury or death under some exposure conditions. There is, then, no such thing as a "safe" chemical. Conversely, there is no chemical that cannot be used safely by limiting the dose (i.e., the amount and the duration of exposure).

Toxicology is the study of adverse effects of chemicals on biologic systems. The biologic systems may be humans or other animals, plants, or the environment. The study of mammalian toxicology is quite complex; however, some of its basic concepts are easy to understand and necessary for making decisions about personnel protection at hazardous waste sites.

Whether or not a particular chemical or combination of chemicals will produce toxicity, or cause a harmful effect, is determined by a variety of factors, including the chemical agent involved, the route of exposure, the concentration or the dose of the chemical present, the duration and the frequency of exposure, and the susceptibility of the exposed subject. Simply recognizing a toxic hazard is not enough; it is also necessary to know the type of toxic effect elicited and the dose required to produce such an effect.

In addition to basic concepts and terminology of toxicology, this chapter addresses health hazards and exposure limits.

ROUTES OF EXPOSURE

There are four major exposure pathways through which a chemical agent can enter the body: inhalation or breathing, skin and/or eye contact and absorption, ingestion, and injection.

Inhalation

The primary route of exposure at hazardous waste sites is inhalation; when present in the air, chemical agents and particulates are carried into the lungs with every breath in an unprotected worker. The lungs have an enormous surface area for absorption. Chemicals carried into the lungs are absorbed into the bloodstream and transported to other organs in the body.

The water solubility of a vapor or a gas is an important determinant of how much of the inhaled material actually reaches the lung tissue (alveolor sacs). Highly water-soluble gases and vapors, such as ammonia, alcohols, and chlorine, dissolve readily in the upper respiratory tract; these materials affect the trachea and bronchi, and can cause irritation. Symptoms of irritation, such as severe coughing, sore throat, and difficulty breathing, alert the worker to the exposure. Gases such as nitrogen dioxide and phosgene are not water-soluble and easily pass into the lungs, where they produce severe irritation that can result in life-threatening pulmonary edema.

Many chemicals may be present in the air at toxic concentrations but may not be detected by human senses because they are colorless and odorless. Many chemicals may enter the lungs without producing any immediate symptoms, such as irritation or difficulty breathing. In many cases, the lungs serve as the route of entry while the toxicity elicited is in another organ, such as the liver, kidney, or brain.

Skin and Eye Contact and Absorption

Direct contact of the skin and eyes is another potential route of exposure to chemical agents. Skin and eye contact may be made with a solid, a liquid, a vapor, an aerosol, a mist, or a fume. Skin and eye exposure with vapors and gases are frequently overlooked as a route of exposure. Skin contact is also called dermal exposure.

Although the skin is an effective barrier against particulates and large molecular weight chemicals, lipid-soluble chemicals can be readily absorbed through the skin and into the bloodstream. Skin that is abraded, lacerated, sunburned, under- or overhydrated, or otherwise compromised is particularly susceptible to chemical absorption and damage. Chemicals in the gas or vapor stage can be absorbed by the skin; this is an often overlooked route of chemical exposure.

The eye can absorb lipid- and water-soluble organic materials; the secretions that bathe the eyes contain water, proteins, and lipids that solubilize airborne chemicals. Small blood vessels close to the surface of the eye, as well as vessels in the mucous membranes surrounding the eye, carry absorbed chemicals to the rest of the body.

The skin and eyes also can be directly damaged upon contact with chemical agents, many of which can cause itching, redness, or irritation. Corrosives can destroy skin and eye tissues, producing chemical burns and scarring; severe eye damage can produce permanent blindness.

Ingestion

Ingestion or swallowing of chemicals at a hazardous waste site should be the least likely route of exposure. Although deliberate ingestion may be unlikely, personal habits—such as chewing gum or tobacco, smoking cigarettes, biting the fingernails, applying cosmetics or lip balm, and eating or drinking on-site—can facilitate ingestion or oral exposure to chemicals. This is especially true when protective clothing is not removed before such activities, or when workers do not wash their hands and faces before eating or drinking.

Injection

Injection, or introduction of chemicals directly into the bloodstream, may occur as a result of a puncture wound, or if a fresh laceration or an open cut is contaminated with a chemical agent. The amount injected is usually of little significance.

AMOUNT OR CONCENTRATION

Almost all chemicals will produce a toxic effect if present in sufficient amounts. Accidental human exposures due to spills or workplace accidents, childhood poisoning, suicide attempts, and laboratory experiments provide toxicity information, as they represent exposure to pure products or high concentrations of chemicals.

Laboratory experiments that utilize animals frequently use high dose levels deliberately to cause visibly toxic effects. The results of these tests are used to estimate the impact of similar exposure on humans.

Epidemiology studies follow a group of workers exposed to low concentrations of chemicals. Such studies last for many years and compare the exposed group to another, nonexposed group. If the incidence of a disease or a physical complaint is increased in the exposed group, the role of workplace exposure is carefully examined.

Chemical reference texts can provide information about the chemical and physical properties of chemicals as pure products, that is, at 100% concentration. For example, acetic acid is listed as a flammable and corrosive liquid when present as a pure, undiluted product; but the acetic acid present in vinegar is part of a well-known, beneficial food product and is not hazardous.

Dose–Response Relationship

The dose–response relationship is a basic tenet of toxicology, that is, that there is a direct relationship between concentration (dose) and response. Put another way, as the concentration or the dose increases, the response (good or bad) increases. As illustrated in Figure 3–1, at very low doses, there is a negligible or unrecognizable response; as the concentration increases, the response increases, until a maximum response is achieved. For example, the elicited response may be a reduction in pain with an increasing dose of aspirin, or an increased nasal irritation with increasing concentrations of chlorine gas.

Another accepted tenet, which is not so easily demonstrated, is that at some concentration, however small, a chemical will have no adverse effect. The term threshold is often used to describe the dividing line between no-effect and effect levels of exposure.

DURATION AND FREQUENCY

The potential for a chemical agent to produce a toxic effect is increased as the duration and the frequency of exposure increase. Duration can be defined as the length of time of each exposure in minutes, hours, days, or weeks. Frequency is the number of such exposures.

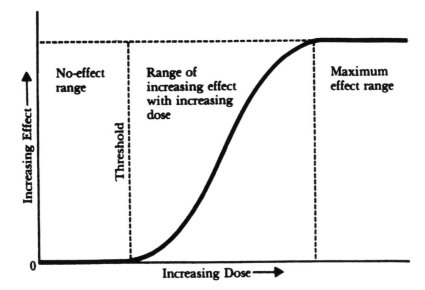

Figure 3-1. A dose–response curve. (*Source:* Ottoboni, 1991. Courtesy of Van Nostrand Reinhold.)

The duration and the frequency of exposure are classified into two categories, acute and chronic.

Acute Exposure

Acute exposure is a single, usually brief, contact with a chemical agent. Acute toxicity refers to the ability of a chemical to elicit adverse effects as a result of a one-time exposure. Acute effects vary, depending on the chemical and the dose. In many cases, acute effects occur during or soon after exposure, and are often reversible. Such effects may include sneezing, coughing, watery eyes, headache, red itchy skin, fatigue, weakness, nausea, vomiting, diarrhea, loss of appetite, and dizziness.

Chronic Exposure

Chronic exposure is repeated exposure over a prolonged period of time. Chronic exposures occur in the workplace and at home and consist of exposure to low concentrations over a period of months or years. Chronic toxicity is the ability of a chemical to produce damage as a result of repeated exposures. Usually chronic toxicity is not detected until exposure has continued for some time. Many chronic effects develop gradually over a long period of time, making it difficult to establish a cause-and-effect relationship.

Many chronic effects are irreversible; they do not disappear when exposure ceases. Examples of irreversible chronic effects include emphysema, asbestosis, and cancer.

Combined Effects of Exposure

Repeated overexposure to chemical agents can produce acute and chronic effects at the same time. For example, exposure to high concentrations of *n*-hexane may result in scratchy throat, headache, and dizziness. Prolonged exposure can cause permanent peripheral nerve damage (numbness and tingling in the fingers and toes).

It is also possible for a material to elicit short-term and long-term effects after a single exposure. For example, exposure to a severe respiratory irritant may result in severe coughing, difficulty breathing, and chest pain. If sufficient concentrations reach the lung tissue, scarring of the tissues (pulmonary fibrosis) may occur, resulting in decreased oxygen delivery and exchange and reduced pulmonary function.

OTHER FACTORS AFFECTING SUSCEPTIBILITY

Not all members of an exposed population respond to the same dose in the same way. Variability in response is caused by one or many factors and can make some segments of the population more susceptible than others.

Individual Sensitivity

Individual sensitivity to chemicals may be affected by a variety of factors, including age, sex and hormonal status, pregnancy, overall physical and emotional health, diet and nutritional status, obesity, inherited traits or genetic makeup, and use of medications, drugs, alcohol, and tobacco. Individual sensitivity may enhance or reduce the toxic effect elicited by a particular chemical.

Multiple Chemical Exposure

Exposure to more than one chemical at a time can affect an individual's susceptibility to chemical agents. Multiple chemical exposures can result from exposures in the workplace. Also chemical exposure at home is an often overlooked source of such exposure. Medications, solvents and glues used in hobbies, pesticides used in the garden, paints, gasoline, and cleaning supplies are all sources of multiple chemical exposure.

Some combinations of chemicals produce different effects from those attributed to the individual chemicals. Additive effects occur when the combined effect of two chemicals is equal to the sum of the effects of each agent acting alone (1 + 1 = 2). When two organophosphate insecticides (Malathion and Parathion) are used together, the effects are usually additive. A synergistic effect occurs when two chemicals have an effect that is greater than additive (1 + 1 = 4). The liver damage produced as a result of exposure to ethyl alcohol and carbon tetrachloride is an example of synergism.

Potentiation is a type of synergism where one chemical, the potentiator, is usually not toxic by itself, but has the ability to markedly increase the toxicity of another chemical (0 + 1 = 5). Isopropyl alcohol, for example, does not normally produce liver damage; but when combined with chloroform, it enhances the liver injury produced.

Antagonism occurs when exposure to two chemicals produces less toxicity than that found when only one of the agents is present. Antagonism is the basis for many antidotes used to treat poisoning cases. Functional antagonists produce the opposite effect to that of the toxic chemical on the same physiological system; for example, phosphates decrease lead absorption by forming insoluble lead phosphates in the gastrointestinal (GI) tract. Chemical antagonists react with a toxic chemical to form a less toxic material; for example, chelating agents used to treat lead poisoning are chemical antagonists.

Another type of antagonism is dispositional antagonism, which alters the absorption, metabolism, or excretion of a chemical. A good example of a dispositional antagonist is ethyl alcohol when used to treat methanol and ethylene glycol poisoning. Receptor antagonists are used to block the action of a chemical agent at a crucial receptor in the body; atropine is used to block neurochemical receptors responsible for the nervous system effects elicited during organophosphate poisoning.

TERMINOLOGY

LD$_{50}$

One way to measure the acute toxicity of a chemical is to expose a group of test organisms and then observe them for signs of toxicity. One test performed is the LD$_{50}$, which is used to determine what dose is required actually to cause death. Although it is unpleasant to conduct such tests, and such tests are sometimes controversial, they are necessary to determine the potential hazard a material may pose to humans. Lethal dose test data are especially important in cases involving accidental childhood poisoning or intentional adult overdoses.

LD means lethal dose, and 50 refers to 50% of the animals for whom the dose was lethal. The LD$_{50}$ is the quantity of material required to kill 50% of the exposed organisms after a one-time exposure by ingestion or skin absorption. The LD$_{50}$ thus is a measure of the acute toxicity of a chemical substance. The concentration required to kill half the exposed organisms by inhalation is usually designated as the LC$_{50}$.

Chemical references may contain other similar types of data, including TD (toxic dose) or TC (toxic concentration) for harmful chemicals and ED (effective dose) for beneficial medicines or pharmaceuticals. The subscript number following LD, LC, TD, TC, or ED always refers to the percentage of the test group affected by the exposure. TD$_{10}$, for example, indicates that 10% of the test organisms suffered toxic effects.

Lethal or toxic dose information is usually given in milligrams of chemical per kilogram (mg/kg) body weight of the test animal. Lethal or toxic concentration refers to a concentration of chemical dissolved in air, and may be presented in ppm or milligrams per cubic meter (mg/m^3). It is important to remember that the smaller the LD$_{50}$, the more toxic the material because it takes less to produce a lethal effect (Table 3–1). Chemical references usually indicate the species used as well as the route of exposure used. LD$_{50}$ (rat, oral) 450

Table 3–1. Examples of LD$_{50}$ descriptions

How toxic?	LD$_{50}$ (mg/kg)	Example	Lethal dose
Super	<5	TCDD*	10 drops
Extremely	5–50	Strychnine	1 teaspoon
Very toxic	50–500	DDT	1 tsp–1 oz
Moderately	500–5,000	Table salt	1 oz–1 pt
Slightly	5,000–15,000	Ethanol	1 pt–1 qt
Harmless	100,000+	Water	3 gallons

*2,3,7,8-Tetrachlorodibenzo-*p*-dioxin.

mg/kg denotes that the rat was the test species, and the chemical was administered by the oral or ingestion route.

LD_{50} and TD_{50} information provides only a crude estimation of the toxicity of a particular substance and should not be extrapolated to determine the risk to humans at lower concentrations. This is so because the shape of the dose–response curve varies between species and between individual chemical agents. Two materials with the same LD_{50} may have very different LD_{10} and LD_0 values.

NOAEL

The concentration or the dose that produces no evidence of toxicity is the NOAEL, or the No Observable Adverse Effect Level. The lowest concentration or dose that produces some evidence of toxicity is the LOAEL, or the Lowest Observable Adverse Effect Level.

Regulatory Definitions of Toxicity

OSHA (29 CFR 1910.120, Appendix A) defines a toxic substance as a chemical that has an oral rat LD_{50} between 51 and 500 mg/kg, or has an dermal rabbit LD_{50} between 201 and 1000 mg/kg when administered to the skin over a period of 24 hours, or has a rat LC_{50} between 201 and 2000 ppm when administered by continuous inhalation for one hour. OSHA highly toxic substances have similar definitions except that their doses are lower: an oral LD_{50} of 50 mg/kg or less, a dermal LD_{50} of 200 mg/kg or less, or an inhalation LC_{50} of 200 ppm or less.

USDOT defines a poisonous material as a material, other than a gas, that is known to be toxic to humans so as to present a health risk during transportation. USDOT poisonous materials are also defined on the basis of LD_{50} data. Any liquid or any solid with an oral LD_{50} not more than 500 mg/kg or 200 mg/kg, respectively, is considered a poisonous material. Any material with a dermal LD_{50} not more than 1000 mg/kg, or any mist or dust with an LC_{50} not more than 10 mg/l is also classified as a poisonous substance.

Toxicity Ratings

Toxicity ratings are assigned to substances on the basis of their LD_{50} or LC_{50}. These ratings are usually given in numeric or descriptive terms. Ratings can differ between references; for example, the same material may be assigned a rating of slightly toxic, moderately toxic, and toxic (Table 3–2).

HEALTH HAZARDS

OSHA defines a variety of health hazards posed by chemical agents. These substances include irritants, corrosives, and toxic and highly toxic agents, as well as systemic or target organ toxins. Systemic toxins are substances that produce toxic effects in specific organs, such as the liver or the kidney.

It is important to recognize that the toxicity of a chemical agent is dependent on the concentration or the dose that *reaches* susceptible tissue or the target organ. Even when absorbed into the bloodstream, chemical agents do not automatically cause damage. Body defenses can biotransform chemicals into nontoxic forms that can be excreted through the

Table 3-2. A comparison of toxicity scales

Oral LD_{50} in mg/kg	Relative toxicity scales			Examples
100,000	relatively harmless	practically nontoxic	nontoxic	water
				sugar
1 quart*				ethyl alcohol
10,000	practically nontoxic	slightly toxic		
1 pint*				methoxychlor table salt liquid bleach aspirin
	slightly toxic	moderately toxic	toxic	
1,000 / 1 ounce*				2,4-D
	moderately toxic	very toxic	moderately toxic	DDT
100 / 1 teaspoon*				nicotine strychnine
	highly toxic	extremely toxic	highly toxic	
10 / 7 drops*				Parathion
1.0 / 1 drop*		super toxic		
	extremely toxic			sodium fluoroacetate
0.1				

*Probable lethal dose for 150 lb human male.

lungs, kidneys, and feces. It is only when the rate of absorption exceeds the body's capacity for excretion that toxic effects may be observed.

Systemic Toxins

Systemic toxins are substances capable of producing toxicity in specific organs. Many chemicals produce multiple target organ effects.

Central nervous system (CNS) damage may occur, due to compounds that cause damage to neurons or inhibit their function, including lead, fluoroacetate (a rodenticide), triethyltin (an ingredient in insecticides and fungicides), hexachlorophene, thallium (used in pesticides), tellurium, and organic mercury compounds.

Peripheral nervous system (PNS) damage may be caused by chemicals that produce weakness of the extremities and abnormal sensations (numbness and tingling), including acrylamide, carbon disulfide, n-hexane, methyl butyl ketone, and organophosporus compounds such as triorthocresyl phosphate, Leptofor, and Mipafox.

Liver injury can be categorized according to the type of response elicited by the hepatoxin. Acute liver damage occurs when liver cells die as the result of exposure. Continuous exposure may lead to chronic liver damage, which may include fatty liver or cirrhosis of the liver. Some recognized hepatotoxins are carbon tetrachloride, chloroform, trichloroethylene, tetrachloroethylene, bromobenzene, tannic acid, aflatoxins, and ethyl alcohol.

The kidney is especially susceptible to chemical agents; because it receives 20 to 25% of the blood flow, renal toxicants reach the kidney quickly. Chemical agents capable of causing renal or kidney damage include heavy metals (mercury, lead, arsenic, chromium), chloroform, carbon tetrachloride, carbon disulfide, ethylene glycol, and ethylene dibromide.

Chemical agents can cause damage to blood components (red blood cells, white blood cells, platelets), the blood-forming system (bone marrow), or the oxygen-carrying capacity of red blood cells. Chemicals that affect bone marrow include arsenic, bromine, methyl chloride, and benzene. Platelets help prevent blood loss by facilitating the formation of blood clots. Chemicals that affect platelets are aspirin, which inhibits clotting; benzene, which decreases the number of platelets; and tetrachloroethylene, which increases the number of platelets.

White blood cells or leukocytes are responsible for defending against foreign organisms, and usually the number of white blood cells present in the blood is carefully regulated. Chemicals that can decrease the number of leukocytes include benzene and phosphorus. Red blood cells or erythrocytes transport oxygen in blood. Chemicals that destroy red blood cells include arsine, naphthalene, and warfarin.

Oxygen transport can be affected by substances that affect hemoglobin, found in red blood cells. Carbon monoxide combines with hemoglobin to form carboxyhemoglobin; hemoglobin has an affinity for carbon monoxide 200 times greater than that for oxygen. Methylene chloride is biotransformed to carbon monoxide in the body, and can also cause carboxyhemoglobinemia. Some chemicals can change hemoglobin to methemoglobin, a form that does not transport oxygen very well. Methemoglobinemia reduces oxygen delivery to cells and can be serious if not recognized. Some amine compounds can cause methemoglobinemia; excessive concentrations of nitrites and nitrates in well water used for drinking have caused blue-baby syndrome in infants. Other compounds that can cause methemoglobinemia include aniline, nitrobenzene, dinitrobenzene, trinitrobenzene, and 2-nitropropane.

The respiratory tract can be divided into three sections: the trachea, bronchi, and pulmonary or lung tissue. Irritants and asphyxiants affect the entire respiratory tract; but some materials exert their toxicity primarily on lung tissue, including paraquat, ozone, asbestos, silica, nitrogen dioxide, and beryllium.

The male and the female reproductive systems can be affected by exposure to certain chemicals; sterility, infertility, an abnormal or low sperm count, or changes in hormone levels may occur. Some agents that have been implicated as male reproductive toxins include the anesthetic halothane, vinyl chloride, cadmium, mercury, lead, boron, methyl mercury, PCBs, dioxin, a variety of pesticides, ethyl alcohol, dibromochloropropane, and heat. Female reproductive toxins include diethylstilbestrol (DES), PCBs, cadmium, methyl mercury, hexafluoroacetone, and halothane.

Asphyxiants

Simple asphyxiants are gases that displace oxygen in air. Inert simple asphyxiants are nonflammable and nontoxic at low concentrations: carbon dioxide, nitrogen, argon, neon, and helium. Other simple asphyxiants are flammable but do not display any toxicity at nonasphyxiating concentrations: acetylene, ethane, methane, and LPG.

Chemical asphyxiants are substances which interfere with the transport or the utilization of oxygen (see above). These include nitrogen compounds, which produce methemoglobinemia, and carbon monoxide (Figure 3–2), which produces carboxyhemoglobinemia. Cyanides poison cellular enzyme systems and prevent cells from using oxygen.

Irritants

Skin or dermal irritants act directly on normal skin at the site of contact. Acute skin irritation or acute dermatitis is a local reversible inflammation caused by a single application; it can produce a rash, reddening, swelling, and discomfort. Chronic skin irritation or chronic dermatitis is a reversible irritation caused by repeated or continued exposure; it results in dry, cracked skin that bleeds easily. In many cases, the substance producing chronic dermatitis does not produce significant acute skin irritation.

Absolute irritants are strong irritants that can produce chemical burns upon contact. Strong acids and bases can rapidly produce second and third degree chemical burns upon brief contact with skin. These materials are considered corrosive; their direct chemical action on the skin results in irreversible alteration, manifested by ulceration, necrosis, and subsequent scar formation.

Eye irritants can produce inflammation of the conjunctiva (the "white of the eye"), iris, and cornea. Corneal damage, if severe, can result in permanent vision changes and even blindness. Acids, alkalis, and organic corrosives can cause severe chemical burns to the eye.

Figure 3–2. Exhaust from heavy equipment contains carbon monoxide, a chemical asphyxiant.

Alkali burns initially may appear mild but later may lead to ulceration, perforation, and destruction of the cornea. Some noncorrosive organic solvents, such as acetone, toluene, and isopropyl alcohol, dissolve fats, cause pain, and dull the cornea. Some materials, such as naphthalene and methanol, can cause retinal damage when they reach the eye through the bloodstream.

Upper respiratory tract irritants irritate the trachea and bronchi, causing constriction of the airways and excess secretion of mucus. Pulmonary irritants can produce pulmonary edema as a result of lung tissue irritation and subsequent escape of fluids into the lung tissue. The onset of pulmonary edema may be immediate or delayed for up to 72 hours. Severe pulmonary edema may be life-threatening. Severe pulmonary irritants include ammonia, bromine, chlorine, fluorine, hydrogen chloride, hydrogen fluoride, nitric acid, nitrogen dioxide, phosgene, ozone, and sulfur dioxide.

Sensitizers

Sensitizers are substances that produce an allergic reaction in sensitive individuals. Allergic individuals demonstrate a reaction when exposed to very small doses that do not affect other people. Not all chemicals cause allergies; a person must have been previously exposed to a chemical before an allergic reaction can occur.

Some sensitizers produce symptoms of skin inflammation. Skin rashes may develop, and there may be blister formation. Itching frequently accompanies allergic skin reactions. Hives is a form of allergic skin reaction manifested by hot, itchy, raised red wheals. Eye irritation also is commonly caused by allergic reactions, and is usually seen as watery, itchy, reddened eyes.

Respiratory sensitizers produce asthma-like reactions consisting of constriction of the small airways, increased mucus production, and difficulty breathing. Severe allergic reactions caused by exposure to bee or wasp venom or sulfites in food can be life-threatening.

Carcinogens

Carcinogens are substances that cause cancer. Fewer than 50 chemicals or chemical processes have been recognized as carcinogens in humans; for example, benzene, vinyl chloride, and asbestos are known human carcinogens. Fewer than two hundred other chemicals are carcinogenic in animals and should be considered potential human carcinogens. NIOSH Occupational Carcinogens are substances that have been demonstrated to produce tumors in humans or animals. The International Agency for Research on Cancer (IARC) carefully examines the relevant animal and human data to determine if a substance is an animal or a human carcinogen.

Mutagens

Mutagens are substances that cause genetic change by damaging genes or chromosomes. This type of change is called a mutation, and a chemical that produces such change is called a mutagen. Mutations affect the way cells function and reproduce; they can be passed to new cells and eventually to groups of cells that do not function or reproduce properly.

Some kinds of mutations can result in cancer; many carcinogens are also mutagens. If a mutation occurs in the egg or the sperm, it can be transmitted to the offspring; such mutations are called germ cell mutations. High doses of gamma radiation can cause germ cell mutations.

Teratogens

Chemicals that cause birth defects by damaging the fetus while it develops in the mother's womb are called teratogens. Some chemicals may produce lethal damage and cause the fetus to be aborted. Ethyl alcohol and lead are recognized human teratogens.

EXPOSURE LIMITS

Standards or guidelines have been developed by various organizations and government agencies to prevent chemical overexposure of workers. Exposure limits are based on the concept that there is a threshold dose or concentration below which no adverse health effects will be elicited in the majority of exposed workers. Each exposure limit defines or recommends the maximum concentration in air that is acceptable for a specified period of time. The duration of exposure is an important factor in defining overexposure.

Susceptibility and response to chemical exposure vary between individuals. Although safety factors are incorporated into workplace exposure limits, they cannot be expected to protect everyone. Sensitive individuals exist in every population, who may find odors objectionable or experience slight effects such as headache or mild irritation. Workplace exposure limits have traditionally been based on the anticipated response of an average young male worker who is in good health and good physical condition. Many factors are not considered, including allergies, preexisting medical conditions, and the use of drugs, alcohol, and tobacco.

Exposure limits should not be used as a relative index of toxicity or toxic hazard, as each limit is determined on the basis of the characteristics of each particular substance. One exposure limit may be based upon irritancy and another on target organ toxicity or carcinogenicity.

Most exposure limits are formulated on the basis of: historical information from workplace experiences; experimental animal studies; experimental human studies; analogy to similar chemicals; or a combination of one or more of the above.

Permissible Exposure Limits (PELs)

PELs were established by the Occupational Safety and Health Administration (OSHA) in 1971. PELs are defined in 29 CFR 1910.1000 for nearly 600 materials; there are individual standards for another 26 substances, including benzene, ethylene oxide, asbestos, lead, and formaldehyde. PELs for 164 substances were revised in 1989; although the revised PELs were overturned by the federal courts (on a procedural technicality), they are still observed by the majority of affected industries.

The PEL is the maximum average concentration that a worker may be exposed to for 8

Figure 3-3. Chemical decontamination agents and sorbents brought on-site can also be hazardous.

hours a day, 40 hours a week. Note that the PEL is an *average* concentration for an 8-hour workday; a worker may be exposed to concentrations above and below the PEL as long as the average does not exceed the PEL. This time weighted average or TWA for 8 hours or another time interval is the basis for most occupational exposure limits.

Ceiling Limits

An OSHA ceiling limit is used when the concentration in air cannot exceed a specific maximum concentration at any time during exposure. Ceiling limits are often used for materials that produce respiratory or eye irritation or a rapid toxic response. The ceiling limit is the *maximum* concentration that an individual should be exposed to at any time during the workday.

Short Term Exposure Limits (STELs)

A STEL is the peak concentration that a worker may be exposed to for a short period of time during the workday. Exposure to concentrations above the STEL may result in irritation, drowsiness, dizziness, or other symptoms that may affect worker efficiency or safety. STEL exposures should not exceed 15 minutes each and should be followed by a 60-minute interval without exposure. There should be no more than four STEL exposures per workday.

Threshold Limit Values (TLVs)

The American Conference of Governmental Industrial Hygienists (ACGIH), founded in 1938, provides TLVs for approximately 700 substances. The ACGIH TLVs are recommended exposure limits that are similar to OSHA limits. (Note that although ACGIH TLVs are recommended guidelines, the OSHA exposure limits are required by law. TLVs are reviewed, updated, and published on a yearly basis by ACGIH. New chemicals are added, and previous TLVs are adjusted in response to new information. OSHA exposure limits can only be changed by amending the original law passed by Congress. Not surprisingly, ACGIH TLVs are usually more current and are often lower than OSHA limits. When one is evaluating exposure limits, it is prudent to examine both OSHA and ACGIH and to select the lower value.)

The TLV-TWA is analogous to the OSHA PEL and refers to the maximum average concentration that is acceptable for an 8-hour-day, 40-hour work week. ACGIH also recommends STELs (TLV-STEL) and ceiling limits (TLV-C). ACGIH is currently phasing out STELs and replacing them with excursion limits, which are multiples of the TWA. The excursion limit is the peak concentration that workers should be exposed to for no more than a total of 30 minutes during any workday. Three times the TWA is the recommended excursion limit; the maximum excursion limit should never exceed five times the 8-hour TWA.

Workplace Environmental Exposure Levels (WEELs)

WEEL Guides have been developed by the American Industrial Hygiene Association (AIHA) for those substances that do not have an OHSA exposure limit or an ACGIH TLV. WEELs have been established for 61 substances; an additional 40 are being prepared. Most WEELs are 8-hour TWAs; AIHA has also established STELs and ceiling values for some substances.

Skin Notation

OSHA, ACGIH, and AIHA use a skin notation to designate chemicals that can be absorbed through the skin, mucous membrane, and eye. Overall exposure to substances that carry a skin notation can be significantly increased by direct contact with the chemicals, or by contact with gases, vapors, or airborne aerosols. Care should be taken to minimize skin and eye exposure to substances that carry a skin notation.

National Institute for Occupational Safety and Health (NIOSH)

NIOSH is an agency within the Department of Health and Human Services given the responsibility to identify workplace hazards, conduct research and field investigations within the workplace, develop and publish air sampling and analysis methods for chemicals, and recommend workplace exposure limit standards to OSHA. Each NIOSH recom-

Figure 3–4. Perimeter monitoring is one method used to ensure that off-site receptors are not inadvertently exposed to on-site chemical hazards.

mended exposure limit or REL is initially published as part of a criteria document on a specific chemical; NIOSH RELs may be found in the *NIOSH Pocket Guide to Chemical Hazards* (NIOSH, 1994). In most cases, the NIOSH REL is based on a 10-hour workday and a 40-hour work week. NIOSH RELs are recommendations to OSHA; it is up to OSHA to determine the feasibility of each recommendation and propose realistic exposure limits to Congress.

NIOSH has also established Immediately Dangerous to Life and Health or IDLH values for some hazardous materials. The NIOSH IDLH is the concentration in air of any toxic, corrosive, or asphyxiant substance that poses an immediate threat to life or would cause irreversible or delayed adverse health effects. The IDLH also represents the maximum concentration in air from which one could escape within 30 minutes without suffering any escape-impairing symptoms, such as dizziness, drowsiness, or eye or respiratory tract irritation. NIOSH also defines a concentration equal to 10% of the LEL as an IDLH. The NIOSH IDLH is an important value to consider in selecting the type of respiratory protection to use. This subject is covered in detail in chapter 5.

4

Sources of Information

Many reference texts and online computer systems are available to assist hazardous waste site workers, safety officers and others in assessing potential site hazards and determining proper control of associated risks. Most references have target audiences, such as chemists, industrial hygienists, toxicologists, or emergency responders; however, these texts contain information that can be utilized by persons in other disciplines as well.

All reference texts and computer data bases contain errors—some typographical, others derived from incorrect "original" sources. For this reason, at least two references should be consulted, and the latest editions should be utilized, especially for research on industrial hygiene standards or toxicity information. The use of several sources of information often results in discrepancies between texts regarding chemical or physical parameters and acute toxicity information. In these cases, it is prudent to accept the most conservative value.

BASIC REFERENCES

Although a large number of reference texts are available, the following enjoy particularly wide use:

• *North American Emergency Response Guidebook*, Department of Transportation, Washington, DC. The *NAERG* is used primarily as a guide for first responders to identify the specific or generic hazards of a material involved in a transportation incident. The 1996 *NAERG* is organized into four sections. The first section (yellow) lists materials by a four-digit UN or NA identification number found on a vehicle placard, package label, or shipping papers. The second section (blue) contains an alphabetic listing of materials; this section has been expanded to include 100 waste materials listed under Canadian regulations. The third section (orange) describes spill and leak procedures and contains advice on initial procedures to stabilize a spill. The fourth section (green) provides initial isolation and downwind evacuation distances by UN number.

• *Chemical Hazard Response Information Systems (CHRIS Manual)*, Government Printing Office, Washington, DC. The U.S. Coast Guard's *CHRIS Manual* consists of four volumes. Volume 1, the *Condensed Guide to Chemical Hazards*, is intended for use by first responders

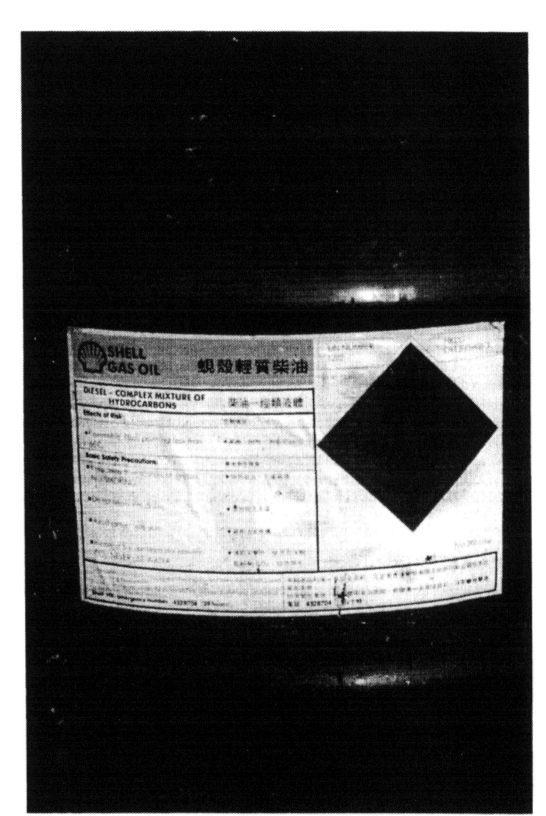

Figure 4–1. A multilingual drum label for a petroleum product sold in Hong Kong. Note the diamond-shaped UN label for flammable liquids.

on the scene of an incident. The chemicals involved must be known, however, before the appropriate information can be obtained. This volume also contains a list of questions one should be able to answer to properly use Volume 3. All information in Volume 1 is repeated in Volume 2.

Volume 2, the *Hazardous Substance Data Manual*, is the most useful of these volumes for responding to spills or hazardous waste sites. It contains information on hazardous chemicals shipped in large volume by water and is intended to be used by port security and other responders who may be first to arrive at the scene of a hazardous materials incident. The information presented includes chemical, physical, and toxicologic properties.

Volume 3, the *Hazard Assessment Handbook*, describes methods for estimating the quantity of chemicals that may be released during an incident, their rate of dispersion, and the methods for predicting any potential toxicity, fire, and explosion hazard. Volumes 2 and 3 are designed to be used together. The chemical-specific hazard assessment codes in Volume 2 are used in Volume 3 to select the appropriate procedures for estimating the degree of hazard and risk.

Volume 4, the *Response Methods Handbook*, contains information on existing methods for handling spills of hazardous materials. The appendix lists manufacturers of spill response and cleanup equipment. It also describes methods of containment for oil spills. This volume was intended for use by Coast Guard on-site coordinators (OSCs) who have had some training or experience in hazardous materials response.

• *Documentation of Threshold Limit Values and Biological Exposure Indices*, 6th Edition, American Conference of Governmental Industrial Hygienists (ACGIH), Cincinnati, OH. The *Documentation of TLVs and BEIs* presents the basic rationale for establishing workplace TLVs for chemical substances and physical agents. Information is also presented on OSHA PELs, NIOSH RELs, carcinogen designations from various governmental and international organizations, and exposure limits adopted by other countries.

• *Fire Protection Guide to Hazardous Materials*, 11th Edition, National Fire Protection Association, Quincy, MA. The *Guide* provides hazardous materials emergency response personnel with information on chemical emergencies, such as fires, accidental spills, and highway accidents. Information presented includes: flash points and fire hazard properties of flammable liquids, gases, and solids; chemical data sheets for specific substances; quantitative health hazard ratings; protective equipment requirements; and hazardous reactivity data.

• *Hawley's Condensed Chemical Dictionary*, 12th Edition, edited by Richard J. Lewis, Sr., Van Nostrand Reinhold, New York. This dictionary contains nearly 20,000 entries and is a compendium of technical data and descriptive information covering thousands of chemicals and chemical reaction products. Although designed for use in the chemical industry, it can be helpful in several ways, as it provides: technical descriptions of compounds, raw materials, and chemical process; expanded definitions of chemical substances and terminology; and descriptions or identification of a wide range of trade-name products used in the chemical industry.

• *Handbook of Environmental Data on Organic Chemicals*, 3rd Edition, by Karl Verschueren, Van Nostrand Reinhold, New York. This *Handbook* offers quick access to environmental fate data as well as information of the effects of organic chemicals on the environment for over 1300 substances.

• *Hazardous Chemicals Desk Reference*, 3rd Edition, edited by Richard J. Lewis, Sr., Van Nostrand Reinhold, New York. The *Desk Reference* is a shortened, condensed version of *Sax's Dangerous Properties of Industrial Materials*, also published by Van Nostrand Reinhold. The *Desk Reference* contains approximately 6000 entries on substances encountered in the chemical, pharmaceutical, and manufacturing industries, as well as the home and the environment. If *Sax's DPIM* has already been examined, the *Desk Reference* should not be used as a backup reference because it obtains its data from Sax.

• *Hazardous Materials Emergency Action Guides*, Association of American Railroads, Washington, DC. The *EAG* provides information about hazardous materials most commonly transported by rail. Information on 142 commodities is presented, including chemical/physical data, synonyms and trade names, color, odor, use, and potential hazards. Hazard potential is assessed in terms of flammability, explosion, and health; protective clothing and equipment are recommended.

• *Merck Index*, 12th Edition, Merck and Co., Inc., Rahway, NJ. The *Index* is a comprehensive, interdisciplinary encyclopedia of chemicals, drugs, and biological substances. Information is presented in alphabetical order by chemical name. A formula index, two CAS number indexes (alphabetical and numerical), and a chemical name cross-index make the manual easy to use, even for nonchemists. Although intended for chemists, pharmacists, physicians, and members of allied medical professions, the *Merck Index* provides readily available information on physical and chemical properties, and limited toxicity data, for 10,100 substances.

• *NIOSH/OSHA Occupational Safety and Health Guidelines for Chemical Hazards*, U.S. Government Printing Office, Washington, DC. This three-volume work provides technical data for many substances listed in the *NIOSH Pocket Guide to Chemical Hazards*. The information is much more detailed than that in the *Pocket Guide*, and is designed primarily for use by industrial hygienists and medical surveillance personnel. The *Guidelines* provide information on chemical names and synonyms, permissible exposure limits, chemical and physical properties, signs and symptoms of overexposure, environmental and medical monitoring procedures, and recommended respiratory and personal protective equipment.

The *Guidelines* were originally published in 1981; Supplements I and II were published in 1988 and 1989, respectively. The permissible exposure limits cited are frequently outdated, and an up-to-date source should be used for this and other potentially time-sensitive information.

• *NIOSH Pocket Guide to Chemical Hazards*, Government Printing Office, Washington, DC. The *Pocket Guide* presents information in a tabular format and was originally designed as a ready reference for industrial hygienists and medical surveillance personnel. Information is presented on chemical names and synonyms, CAS Registry Numbers, OSHA permissible limits and ACGIH TLVs, NIOSH RELs, signs and symptoms of overexposure, target organs, treatment procedures, environmental monitoring techniques, and recommended personal protective equipment. The *Pocket Guide* is updated every few years; the latest *Guide* was published in 1994.

• *Registry of Toxic Effects of Chemical Substances (RTECS)*, Government Printing Office, Washington, DC. This publication is sponsored by NIOSH and contains toxicity data, providing toxic doses and routes of exposure for multiple species, including humans. It is an invaluable source of information on chemical synonyms and trade names, including trade names no longer in use. *RTECS* is available as a soft-cover multiple-volume set and on microfiche; it is also available through on-line computer access.

• *Pestline: Material Safety Data Sheets for Pesticides and Related Chemicals*, Occupational Health Services, New York. This two-volume set offers MSDS for approximately 1200 pesticides. Each MSDS offers information on chemical/physical properties, toxicity data, health effects and first aid, reactivity, storage and safe handling, disposal, spill procedures, exposure limits, and recommended personal protective equipment.

• *Quick Selection Guide to Chemical Protective Clothing*, 3rd Edition, by K. Forsberg and S. Z. Mansdorf, Van Nostrand Reinhold, New York. This inexpensive, pocket-sized guide covers both chemical protective gloves and suit materials and lists permeation rates and breakthrough times for 600 substances.

• *Sax's Dangerous Properties of Industrial Materials*, 9th Edition, edited by Richard J. Lewis, Sr., Van Nostrand Reinhold, New York. This massive, three-volume text contains 4300 pages and 20,000 chemical entries. Information on specific chemicals includes synonyms, trade names, description, chemical formula, and physical/chemical properties. Also available are data to assist readers in hazard analysis in terms of toxicity, flammability, and explosion. Countermeasures to reduce risk or prevent future spills include information on shipping, handling, storage, fire fighting, first aid, spill control, and cleanup procedures. Van Nostrand Reinhold also offers annual updates, as well as annually updated lists of chemicals considered hazardous by one or more regulatory agencies.

• *Threshold Limit Values and Biological Exposure Indices*, ACGIH, Cincinnati, OH. All current ACGIH TLVs and BEIs are listed in a handy, pocket-sized booklet that is updated and published yearly. TLVs are recommended for approximately 700 chemical substances and physical agents. Substances are listed in alphabetical order; CAS Registry Numbers are also presented.

COMPUTER-ACCESSED RESOURCES

There are a variety of online interactive computer data bases, as well as menu-driven data bases available on CD-ROM or multiple disks. Most data bases require a minimum of 640K RAM and take up a significant amount of hard-disk space. Many agencies also have web sites or can be accessed via the Internet.

• *Toxicology, Occupational Medicine and Environmental Series (TOMES)*, Micromedex Inc., Denver, Co. *TOMES* and *TOMES Plus* are available on CD-ROM and include a variety of data bases (see below) oriented to first responders, emergency room personnel, hazardous waste site managers, industrial hygienists, toxicologists, and occupational physicians. Quarterly updates are available; a CD-ROM disc drive is required. The data bases are as follows:

- *MEDITEXT* presents information on evaluation and treatment of persons exposed to hazardous chemicals.
- *SARATEXT* lists acute and chronic health effects of hazardous substances to assist with SARA Title III reporting.
- *INFOTEXT* presents lists of regulated chemicals by category.

- *REPRORISK* is a reproductive risk data base for chemicals and environmental agents.
- *RTECS* offers the NIOSH toxicology data base for over 100,000 chemicals.
- *HSDB* (*Hazardous Substance Data Base*) offers relatively short, concise reviews of the use, hazardous properties, and toxicity of over 4000 chemicals. This is a valuable resource that offers a fast, reliable means of obtaining much information in a short time.
- *IRIS* (*Integrated Risk Information System*) from USEPA offers risk assessments for over 450 chemicals.
- *CHRIS* from the U.S. Coast Guard provides information on hazardous properties for over 1200 substances.
- *OHM/TADS* from USEPA lists environmental effects of over 1200 petroleum products and hazardous materials.
- *First Medical Response Protocols* offer emergency first aid procedures for workplace injuries and overexposure.
- *New Jersey Hazardous Substance Fact Sheets* from the New Jersey Department of Health present employee-oriented chemical fact sheets and exposure information for over 700 substances.
- *USDOT North American Emergency Response Guides* from the U.S. Department of Transportation, are the on-line version of the NAERG (see p. 67).

- *National Library of Medicine (NLM)*, Bethesda, MD. NLM provides a variety of interactive data bases, most offering excerpts or literature references to help individuals conduct scholarly library research. The NLM data bases *MEDLINE* and *TOXLINE* provide literature references, key words, excerpts, and abstracts from books and scientific articles. *CHEMLINE, TOXNET,* and *TOXLIT* contain excerpts from published sources, as well as data from other sources of information such as *CHRIS* and *RTECS*. Selecting and using the appropriate NLM data base take training and a good deal of practice.

- *Computer-Aided Management of Emergency Operations (CAMEO)*, National Oceanographic and Atmospheric Agency, Seattle, WA. The *CAMEO* data base includes information on first aid and emergency response at hazardous materials releases and was originally designed to assist communities in pre-planning for a large-scale chemical emergency. It helps the user to map community hazards, calculate dispersal rates of release of airborne contaminants, and manage information on chemical hazards.

- *Hazardline*, Occupational Health Services, New York. *Hazardline* is an interactive data base containing information on chemical/physical properties, exposure limits, toxicity, symptoms of exposure, and first aid for thousands of substances. The user can select the information needed or simply view the entire document about the selected chemical. OHS also offers other on-line services, including preparation of complete Material Safety Data Sheets (MSDS).

- *Comprehensive Chemical Contaminant (CCC) Series on CD-ROM*, Van Nostrand Reinhold, New York. The *CCC Series on CD-ROM* contains five data base texts that contain information on over 20,000 regulated chemical substances, including *Sax's Dangerous Properties of Industrial Materials, Hawley's Condensed Chemical Dictionary, Environmental Contaminant Reference Databook, Handbook of Environmental Data on Organic Chemicals, and Hazardous Materials Handbook.*

• *Internet Access—U.S. Government.* Fedworld, a government bulletin board of on-line services, has the following URL: http://www.fedworld.gov/

Internet addresses of federal agencies are as follows:

- Directory of Federal Agencies: http://www.lib.lsu.edu/gov/fedgov.html
- Centers for Disease Control (CDC): http://www.cdc.gov/
- Department of Defense Environmental Restoration Small Business Link: http://www.dtic.dla.mil/environdod.html
- Government Printing Office (GPO): http://www.access.gpo.gov/su_docs/
- Library of Congress: http://lcweb/loc.gov/
- Occupational Health and Safety Administration (OSHA): http://www.osha.gov
- U.S. Environmental Protection Agencies (USEPA): http://www.epa.gov
- U.S. Geological Survey (USGS): http://www.usgs.gov

• *Internet Access—Organizations.* The following organizations may be contacted at their Internet addresses:

- American National Standards Institute (ANSI): http://www.ansi.org/docs/home.html
- Chemical Manufacturing Association (CMA): http://www.cmahq.com
- International Organization for Standardization (ISO): http://hike1.hike.te.chiba-u.ac.jp/ikeda/ISO/home.html
- National Fire Protection Association (NFPA): http://www.wpi.edu/~fpe/nfpa.html
- American Conference of Governmental Industrial Hygienists (ACGIH): http://www.ACGIH.org

TELEPHONE-ACCESSED RESOURCES

A variety of governmental agencies and private organizations provide 24-hour telephone access to hazardous materials information resources in cases of emergency. Most emergency numbers are toll-free. Nonemergency telephone numbers often are not toll-free and have limited hours. The telephone numbers provided here were correct at the time of this book's publication.

- CHEMTREC (Chemical Transportation Emergency Center): 800-424-9300. CHEMTREC is a 24-hour *emergency* service provided by the Chemical Manufacturers Association (CMA) to assist responders during chemical transportation emergencies. CHEMTREC assists in the identification of unknown substances, provides proper initial response procedures for specific chemicals, and can make contact with shippers, carriers, manufacturers, and special product response teams, such as CHLOREP for chlorine emergencies or the Pesticide Safety Team Network.

- CMA Chemical Referral Center: 800-262-8200 (9:00 A.M. to 6:00 P.M. eastern time). CMA expanded the public service function of CHEMTREC by adding a toll-free *nonemergency* number to give the public, chemical users, and providers of emergency services health, safety, and environmental information on chemicals and chemical products. The center can also refer individuals to manufacturers, industry interest groups, research institutions, and regulatory agencies for more information.

- National Response Center (NRC): 800-424-8802. Operated by the USDOT and the U.S. Coast Guard, the NRC must be notified whenever a serious transportation incident has occurred, involving serious injury, death, property damage, impact on navigable waters, or continuing danger to life and property. The NRC offers 24-hour assistance in chemical identification and response procedures during hazardous materials emergencies.

- Association of American Railroads (AAR): 202-693-2222. The AAR provides 24-hour assistance to emergency responders at chemical accidents involving railroad transportation.

- Centers for Disease Control (CDC): 404-633-5313. The CDC provides 24-hour assistance to responders at transportation incidents involving biohazards, such as bacteria, viruses, laboratory specimens, and potentially contaminated blood.

- National Pesticide Telecommunications Network (NPTN): 800-858-7378. The NPTN is operated out of Texas Tech University and is also known as the Texas Tech University Pesticide Hotline. It provides information on pesticide-related exposures, human toxicity, and minor cleanup and decontamination procedures to emergency departments, physicians, veterinarians, and other interested parties.

- Hazardous Technical Information Services (HTIS): 800-848-4847. HTIS is operated by the Defense Logistics Agency (DLA) to provide consultation services to Department of Defense (DOD) personnel worldwide. HTIS provides information on safety, health, transportation, storage, handling, disposal, and environmental considerations of hazardous materials and hazardous wastes. Other agencies may utilize HTIS services if the hazardous material or waste involved was formerly DOD-owned or DOD-managed.

- U.S. Government Agency Hotlines. Agency hotlines are also available to provide information and guidance pertaining to specific federal regulations. These include the USDOT Hotline (202-335-4488) and the USEPA RCRA Hotline (800-424-9346).

- Regional Poison Control Centers. Numerous regional poison control centers throughout the United States provide assistance to emergency departments, physicians, veterinarians, and emergency responders dealing with poisonings or hazardous materials exposure. Telephone numbers can be obtained through directory assistance or by referring to the local telephone directory. Poison control centers offer information on signs and symptoms of overexposure, initial first aid, emergency room treatment, and subsequent follow-up procedures.

MATERIAL SAFETY DATA SHEETS (MSDS)

The OSHA Hazard Communication Standard (29 CFR 1910.1200), also known as right-to-know regulation, requires that manufacturers and distributors of hazardous chemicals provide a Material Safety Data Sheet or MSDS, as well as a warning label, for each chemical or chemical product.

MSDS are detailed information sheets that summarize what is known about the material or the mixture. Each section of the MSDS contains a specific type of information about the chemical or the product. Although there is no single mandatory MSDS format, all MSDS must contain a minimum amount of information, as described below.

Chemical Information

This section identifies the material by product name, trade name, and the manufacturer's name and address. Emergency phone numbers may also appear here.

Hazardous Ingredients

This section lists the chemical ingredients in the product that are hazardous, as well as their concentration. It also lists the recommended safe workplace exposure levels, as either OSHA exposure limits, ACGIH TLVs, or both.

Physical/Chemical Data

This information includes the chemical product's appearance, odor, and other characteristics, such as boiling point, vapor pressure, vapor density, evaporation rate, solubility, and specific gravity. The percent of the product that is volatile (how much will evaporate at room temperature) may also be listed.

Fire and Explosion Data

Information on the flammability of the material, in terms of flash point and flammable range, is presented in this section. It also should clearly indicate if the material is flammable or combustible. Extinguishing methods, such as water spray, foam, or fire extinguisher, are recommended.

Health Hazards

Signs and symptoms of overexposure by skin and eye contact, inhalation, and ingestion are detailed in this section. First aid procedures to follow in case of overexposure, medical conditions that may be worsened by overexposure, and health effects of chronic exposure also are discussed.

Reactivity/Instability Data

Materials that cause the product to burn, explode, release dangerous gases, polymerize, or release excessive amounts of heat are given. An instability section lists environmental conditions, such as heat or sunlight, that may cause a dangerous reaction.

Spill or Leak Procedures

The user is told what methods to employ when cleaning up a small spill or leak, including the use of personal protective equipment.

Personal Protection

Specific types of personal protective equipment, such as respiratory protection and chemical-resistant clothing, are recommended.

Special Precautions

Details are given of any special precautions to follow in handling, using, or storing the chemical product. Other safety information not covered elsewhere may also appear here.

Regulatory Compliance

This section tells if the product or one or more ingredients are subject to specific regulations. It should list the applicable USDOT hazard class and UN number, any SARA Title III ingredients and their concentrations, and the RCRA hazardous waste classification for the product used.

A representative MSDS can be found at the end of this chapter, reproduced with permission from the *Material Safety Data Sheet Collection*, published by Genium Publishing Corporation, One Genium Plaza, Schenectady, NY 12304-4609, phone 800-2-GENIUM.

PRODUCT LABELS

Affixed to the product container, the label identifies the product and contains basic information on health effects, first aid, fire and spill procedures, and handling, storage, and disposal. Labels can be considered a concise summary of the warnings and information contained in the MSDS, but should never be used in place of an MSDS.

Labels for hazardous materials usually carry one of three hazard words, indicating the level of hazard: "Danger!" indicates the highest hazard; "Warning!" and "Caution" indicate moderate and slight hazards, respectively.

SITE INFORMATION SOURCES

Aerial Photography

Aerial photography can be useful in researching a hazardous waste site because a plane can fly over an area and record buildings, site layout design, and natural features, such as vegetation, topography, and drainage. Historical photos can trace a facility from its inception

Figure 4-2. Although somewhat informal, the label on this drum is informative. In this case, "slop" refers to bilge waste.

to the present, pinpointing past activities that may indicate the locations of potential chemical contamination. Infrared photography can help characterize waste sites by highlighting moisture and temperature gradients and stressed vegetation.

EROS Data Center, Sioux Falls, SD

The EROS system, run by the U.S. Geological Survey, uses remote-sensing techniques to inventory, monitor, and manage natural resources. EROS includes research and training in the interpretation and application of remotely sensed data and provides these data at nominal cost.

A central computer complex controls the EROS data base of over 6 million images and photographs of surface features, searches for geographic data on areas of interest, and serves as a management tool for the entire data reproduction process. EROS data can provide a chronological overview of a specific area, establishing the extent of damage over time.

U.S. Geological Survey Maps

Topographic quadrangle maps are useful in that they show land contours, the network of water features, and elevations. They also show cities and urban areas and can be used to determine the proximity of a waste site to sensitive ecosystems, bodies of water, or population centers.

Hydrologic maps show water in or beneath the land surface. They are very useful in evaluating the water supply and water-related hazards such as flooding. They also show drainage areas, depth to ground water, and the thickness of water-bearing formations. In the case of a spill or waste site, a hydrologic map can indicate potential contamination of the ground water and/or drainage area.

Land use and cover maps have been prepared by using the standard topographic quadrangle maps or larger-scale, low altitude photographs as a base. These maps provide detailed information about the use of land or about vegetative cover. This information can be useful at hazardous waste sites.

USGS maps are available through the Branch of Distribution office, in Denver, Colorado, for western states, Alaska, Hawaii, Guam, and American Samoa. The Arlington, Virginia, office serves areas east of the Mississippi, Puerto Rico, and the Virgin Islands.

Genium Publishing Corp.
One Genium Plaza
Schenectady, NY 12304-4690
(518) 377-8854

Material Safety Data Sheets Collection

Isopropyl Alcohol   **MSDS No. 324**

Date of Preparation: 9/85   Revision: A, 10/93

Section 1 - Chemical Product and Company Identification   42

Product/Chemical Name: Isopropyl Alcohol
Chemical Formula: $(CH_3)_2CHOH$
CAS No.: 67-63-0
Synonyms: Dimethyl carbinol, 2-hydroxypropane, IPA, Isohol, Lutosol, isopropanol, Petrohol, 2-propanol, *sec*-propyl alcohol, rubbing alcohol, Spectrar.
Derivation: Treating propylene with sulfuric acid and then hydrolyzing or direct hydration of propylene using superheated steam. Most commonly available as rubbing alcohol (70% IPA).
General Use: As a solvent for gums, shellac, and essential oils, chemical intermediate, dehydrating agent, vehicle for germicidal compounds, de-icing agent for liquid fuels; for denaturing ethyl alcohol, preserving pathological specimens; in extraction of alkaloids, quick-drying inks and oils, and an ingredient of skin lotions, cosmetics, window cleaner, liquid soaps, and pharmaceuticals.
Vendors: Consult the latest *Chemical Week Buyers' Guide.* [73]

Section 2 - Composition / Information on Ingredients

Isopropyl alcohol, 100% vol. Most commonly sold as 70% isopropyl alcohol (rubbing alcohol).

OSHA PELs
8-hr TWA: 400 ppm (980 mg/m³)
STEL: 500 ppm (1225 mg/m³) *

ACGIH TLVs
TWA: 400 ppm (983 mg/m³)
STEL: 500 ppm (1230 mg/m³)

* Vacated 1989 Final Rule Limits

NIOSH REL
10-hr TWA: 400 ppm (980 mg/m³)
STEL: 500 ppm (1225 mg/m³)

IDLH Level
12,000 ppm

DFG (Germany) MAK
TWA: 400 ppm (980 mg/m³)
Category II: Substances with systemic effects
Half-life: < 2 hr

Peak Exposure Limit: 800 ppm, 30 min. average value, 4/shift

Section 3 - Hazards Identification

☆☆☆☆☆ Emergency Overview ☆☆☆☆☆

Isopropyl alcohol is a highly flammable, volatile liquid. It is considered more toxic than ethyl alcohol, but less toxic than methyl alcohol. Inhalation can cause irritation of the eyes and respiratory tract and central nervous system depression at high concentrations. Repeated skin contact may cause dermatitis. Systemic toxicity appears to occur mostly in cases of heavy ingestion or inhalation. There is recent evidence that skin absorption may be more likely to cause systemic effects than previously thought.

Wilson Risk Scale	
R	1
I	2
S	2*
K	3

*Skin absorption

HMIS	
H	1
F	3
R	0

PPE†
†Sec. 8

Potential Health Effects

Primary Entry Routes: Inhalation, ingestion, skin contact/absorption.
Target Organs: Eyes, skin, respiratory system.
Acute Effects
 Inhalation: Vapor inhalation is irritating to the respiratory tract and can cause central nervous system depression at high concentrations. Volunteers exposed to 400 ppm for 3 to 5 min experienced mild eye and respiratory irritation. At 800 ppm, irritation was not severe, but most people found the air uncomfortable to breathe.
 Eye: Exposure to the vapor or direct contact with the liquid causes irritation and possible corneal burns.
 Skin: Some irritation may occur after prolonged exposure.
 Ingestion: Accidental ingestions have provided the most information on isopropyl alcohol toxicity. Symptoms include nausea and vomiting, headache, facial flushing, dizziness, lowered blood pressure, mental depression, hallucinations and distorted perceptions, difficulty breathing, respiratory depression, stupor, unconsciousness, and coma. Kidney insufficiency including oliguria (reduced urine excretion), anuria (absent urine excretion), nitrogen retention, and edema (fluid build-up in tissues) may occur. One post-mortem examination in a case of heavy ingestion showed extensive hemorrhagic tracheobronchitis, broncho-pneumonia, and hemorrhagic pulmonary edema. Death can occur in 24 to 36 h post-ingestion due to respiratory paralysis.
Carcinogenicity: NTP and OSHA do not list isopropyl alcohol as a carcinogen. The IARC has studied IPA and has classified it as Class-3 (unclassifiable, inadequate human and animal evidence). There appears to be an association between the *manufacture* (strong acid process, rather than the alcohol itself) of isopropanol and parasinus cancer, but this may be due to the diisopropyl sulfate or isopropyl oil by-products.
Medical Conditions Aggravated by Long-Term Exposure: Dermatitis or respiratory or kidney disorders.
Chronic Effects: Repeated skin contact can cause drying of skin and delayed hypersensitivity reactions in some individuals.

MSDS No. 324	Isopropyl Alcohol	10/93

Other: Isopropyl alcohol is oxidized in the body to acetone where it is excreted by the lungs or kidneys. Some acetone may be further metabolized to acetate, formate, and finally carbon dioxide. Probable oral lethal dose is 240 mL.

Section 4 - First Aid Measures

Inhalation: Remove exposed person to fresh air and support breathing as needed.

Eye Contact: *Do not* allow victim to rub or keep eyes tightly shut. Gently lift eyelids and flush immediately and continuously with flooding amounts of water until transported to an emergency medical facility. Consult a physician immediately.

Skin Contact: *Quickly* remove contaminated clothing. Rinse with flooding amounts of water for at least 15 min. Wash exposed area with soap and water. For reddened or blistered skin, consult a physician.

Ingestion: Never give anything by mouth to an unconscious or convulsing person. Contact a poison control center. Unless the poison control center advises otherwise, have the *conscious and alert* person drink 1 to 2 glasses of water to dilute. Vomiting may be contraindicated because of the rapid onset of central nervous system depression. Gastric lavage is preferred.

After first aid, get appropriate in-plant, paramedic, or community medical support.

Note to Physicians: Diagnostic test: acetone in urine.

Section 5 - Fire Fighting Measures

Flash Point: 53 °F (12 °C)
Flash Point Method: CC
Burning Rate: 2.3 mm/min.
Autoignition Temperature: 750°F (399°C)
LEL: 2 % v/v
UEL: 12.7 % v/v at 200 °F

NFPA
3
1 0

Flammability Classification: Class 1B Flammable Liquid

Extinguishing Media: Carbon dioxide, dry chemical, water *spray* (solid streams can spread fire), alcohol-resistant foam, or fog.

Unusual Fire or Explosion Hazards: Container may explode in heat of fire. Vapors may travel to an ignition source and flash back. Isopropyl alcohol poses an explosion hazard indoors, outdoors, and in sewers.

Hazardous Combustion Products: Carbon oxides and acrid smoke.

Fire-Fighting Instructions: If possible without risk, move container from fire area. Apply cooling water to container side until well after fire is out. Stay away from ends of tanks. For massive fire in cargo area, use monitor nozzles or unmanned hose holders; if impossible, withdraw and let fire burn. Withdraw immediately if you hear a rising sound from venting safety device or notice any tank discoloration due to fire. *Do not* release runoff from fire control methods to sewers or waterways.

Fire-Fighting Equipment: Because fire may produce toxic thermal decomposition products, wear a self-contained breathing apparatus (SCBA) with a full facepiece operated in pressure-demand or positive-pressure mode. Structural firefighters' protective clothing provides only limited protection.

Section 6 - Accidental Release Measures

Spill /Leak Procedures: Notify safety personnel, isolate and ventilate area, deny entry, and stay upwind. Shut off ignition sources. Cleanup personnel should protect against vapor inhalation and skin/eye contact. Water spray may reduce vapor, but may not prevent ignition in closed spaces.

Small Spills: Take up with earth, sand, vermiculite, or other absorbent, noncombustible material and place in suitable containers.

Large Spills
  Containment: For large spills, dike far ahead of liquid spill for later disposal. Do not release into sewers or waterways.
  Regulatory Requirements: Follow applicable OSHA regulations (29 CFR 1910.120).

Section 7 - Handling and Storage

Handling Precautions: Use non-sparking tools to open containers.

Storage Requirements: Store in a cool, dry, well-ventilated area away from heat, ignition sources, and incompatibles (Sec 10). Install electrical equipment of Class 1, Group D.

Section 8 - Exposure Controls / Personal Protection

Engineering Controls: To prevent static sparks, electrically ground and bond all equipment used with and around IPA.

Ventilation: Provide general or local exhaust ventilation systems to maintain airborne levels below OSHA PELs (Sec. 2). Local exhaust ventilation is preferred since it prevents contaminant dispersion into the work area by controlling it at its source. [103]

Administrative Controls: Consider preplacement and periodic medical exams of exposed workers with emphasis on the skin, kidneys, and respiratory system. Be extra cautious when using IPA concurrently with carbon tetrachloride because animal studies have shown it enhances carbon tetrachloride's toxicity.

Protective Clothing/Equipment: Wear chemically protective gloves, boots, aprons, and gauntlets to prevent prolonged or repeated skin contact. Nitrile rubber (breakthrough time > 8 hr), Neoprene and Teflon (breakthrough time > 4 hr) are suitable materials for PPE. Do not use PVA, PVC or natural rubber (breakthrough time < 1 hr). Wear protective eyeglasses or chemical safety goggles, per OSHA eye- and face-protection regulations (29 CFR 1910.133). Because contact lens use in industry is controversial, establish your own policy.

Respiratory Protection: Seek professional advice prior to respirator selection and use. Follow OSHA respirator regulations (29 CFR 1910.134) and, if necessary, wear a MSHA/NIOSH-approved respirator. For < 1000 ppm, use any powered, air purifying respirator with organic vapor cartridges or any chemical cartridge respirator with a full facepiece and organic vapor cartridge(s). For < 10,000 ppm, use any supplied-air respirator (SAR) operated in continuous-flow mode. For < 12,000 ppm, use any air-purifying, full facepiece respirator (gas mask) with a chin-style, front-or back-mounted organic vapor canister or any SCBA or SAR with a full facepiece. For emergency or entrance into unknown concentrations, use any SCBA or SAR (with auxiliary SCBA) with a full facepiece and operated in pressure-demand or other positive-pressure mode. For emergency or nonroutine operations (cleaning spills, reactor vessels, or storage tanks), wear an SCBA. *Warning! Air-purifying respirators do not protect workers in oxygen-deficient atmospheres.* If respirators are used, OSHA requires a written respiratory protection program that includes at least: medical certification, training, fit-testing, periodic environmental monitoring, maintenance, inspection, cleaning, and convenient, sanitary storage areas.
Safety Stations: Make available in the work area emergency eyewash stations, safety/quick-drench showers, and washing facilities.
Contaminated Equipment: Separate contaminated work clothes from street clothes. Launder before reuse. Remove isopropyl alcohol from your shoes and clean personal protective equipment.
Comments: Never eat, drink, or smoke in work areas. Practice good personal hygiene after using isopropyl alcohol, especially before eating, drinking, smoking, using the toilet, or applying cosmetics.

Section 9 - Physical and Chemical Properties

Physical State: Liquid
Appearance and Odor: Colorless with a slight odor and bitter taste.
Odor Threshold: 22 ppm*
Vapor Pressure: 44 mm Hg at 25 °F (77 °C)
Saturated Vapor Density (Air = 1.2 kg/m^3, 0.075 lb/ft^3): 1.274 kg/m^3 or 0.080 lb/ft^3
Formula Weight: 60.09
Density (H$_2$O=1, at 4 °C): 0.78505 at 68°F (20 °C)
Water Solubility : > 10 %
Ionization Potential: 10.10 eV

Other Solubilities: Soluble in alcohol, ether, chloroform, and benzene. Insoluble in salt solutions.
Boiling Point: 180.5 °F (82.5 °C)
Freezing Point: -129.1 °F (-89.5 °C)
Viscosity: 2.1 cP at 77 °F (25 °C)
Refraction Index: 1.375 at 68 °F (20 °C)
Surface Tension: 20.8 dyne/cm at 77 °F (25 °C)
Critical Temperature: 455 °F (235 °C)
Critical Pressure: 47 atm
Octanol/Water Partition Coefficient: log Kow = 0.05

* References range from 1 to as high as 610 ppm.

Section 10 - Stability and Reactivity

Stability: Isopropyl alcohol is stable at room temperature in closed containers under normal storage and handling conditions.
Polymerization: Hazardous polymerization does not occur.
Chemical Incompatibilities: Include acetaldehyde, chlorine, ethylene oxide, acids and isocyanates, hydrogen + palladium, nitroform, oleum, phosgene, potassium *t*-butoxide, oxygen (forms unstable peroxides), trinitromethane, barium perchlorate, tetrafluoroborate, chromium trioxide, sodium dichromate + sulfuric acid, aluminum, aluminum triisopropoxide, and oxidizers. Will attack some forms of plastic, rubber, and coatings.
Conditions to Avoid: Exposure to heat, ignition sources, and incompatibles.
Hazardous Decomposition Products: Thermal oxidative decomposition of isopropyl alcohol can produce carbon oxides and acrid smoke.

Section 11- Toxicological Information

Toxicity Data:*

Eye Effects:
Rabbit, eye: 100 mg caused severe irritation.

Skin Effects:
Rabbit, skin: 500 mg caused mild irritation.

Reproductive:
Rat, inhalation: 3500 ppm/7 hr given from 1 to 19 days of pregnancy caused fetotoxicity.

Acute Oral Effects:
Human, oral, TD$_{LO}$: 223 mg/kg caused hallucinations, distorted perceptions, lowered blood pressure, and a change in pulse rate.
Human, oral, LD$_{LO}$: 3570 mg/kg caused coma, respiratory depression, nausea, and vomiting.
Rat, oral, LD$_{50}$: 5045 mg/kg caused a change in righting reflex, and somnolence (general depressed activity).

* See NIOSH, *RTECS* (NT8050000), for additional toxicity data.

Section 12 - Ecological Information

Ecotoxicity: Guppies (*Poecilia reticulata*) LC$_{50}$ = 7,060 ppm/7 days; fathead minnow (*Pimephales promelas*) LC$_{50}$ = 11,830 mg/L/1 hr. BOD = 133 %/5 days.

Environmental Degradation: On soil, IPA will volatilize or leach into groundwater. Biodegradation is possible but rates are not found in available literature. It will volatilize (est. half-life = 5.4 days) or biodegrade in water. It is not expected to bioconcentrate in fish. In the air, it reacts with photochemically produced hydroxyl radicals with a half-life of one to several days. Because it is soluble, removal by rain, snow or other precipitation is possible.

Section 13 - Disposal Considerations

Disposal: Microbial degradation is possible by oxidizing isopropyl alcohol to acetone by members of the genus *Desulfovibrio*. Spray waste into incinerator (permit-approved facilities only) equipped with an afterburner and scrubber. Isopropyl alcohol can be settled out of water spills by salting with sodium chloride. Note: Salt may harm aquatic life, so weigh the benefits against possible harm before application. Contact your supplier or a licensed contractor for detailed recommendations. Follow applicable Federal, state, and local regulations.

Container Cleaning and Disposal: Triple rinse containers.

Section 14 - Transport Information

DOT Transportation Data (49 CFR 172.101):

Shipping Name: Isopropanol or isopropyl alcohol
Shipping Symbols: –
Hazard Class: 3
ID No.: UN1219
Packing Group: II
Label: Flammable Liquid
Special Provisions (172.102): T1

Packaging Authorizations
a) **Exceptions**: 173.150
b) **Non-bulk Packaging**: 173.202
c) **Bulk Packaging**: 173.242

Quantity Limitations
a) **Passenger, Aircraft, or Railcar**: 5 L
b) **Cargo Aircraft Only**: 60 L

Vessel Stowage Requirements
a) **Vessel Stowage**: B
b) **Other**: –

Section 15 - Regulatory Information

EPA Regulations:
Listed as a RCRA Hazardous Waste Number (40 CFR 261.21)
RCRA Hazardous Waste Classification (40 CFR 261.21): Characteristic of Ignitability
Listed (Unlisted Hazardous Waste, Characteristic of Ignitability) as a CERCLA Hazardous Substance (40 CFR 302.4) per RCRA, Sec. 3001
CERCLA Reportable Quantity (RQ), 100 lb (45.4 kg)
SARA 311/312 Codes: 1, 2, 3
Listed as a SARA Toxic Chemical (40 CFR 372.65); *only persons who manufacture by the strong acid process are subject; no supplier notification.*
SARA EHS (Extremely Hazardous Substance) (40 CFR 355): Not listed
OSHA Regulations:
Listed as an Air Contaminant (29 CFR 1910.1000, Table Z-1, Z-1-A)

Section 16 - Other Information

References: 73, 103, 124, 126, 127, 132, 136, 139, 148, 153, 159, 164, 167, 168, 176, 187

Prepared By .... M Gannon, BA
Industrial Hygiene Review .... PA Roy, MPH, CIH
Medical Review .... T Thoburn, MD, MPH

Disclaimer: Judgments as to the suitability of information herein for the purchaser's purposes are necessarily the purchaser's responsibility. Although reasonable care has been taken in the preparation of such information, Genium Publishing Corporation extends no warranties, makes no representations, and assumes no responsibility as to the accuracy or suitability of such information for application to the purchaser's intended purpose or for consequences of its use.

5

Respiratory Protection

The respiratory system is able to tolerate exposures to toxic gases, vapors, and particulates, but only to a limited extent. Air that looks dirty or has an obnoxious odor may pose no threat, whereas other gases without odor can cause serious injury. Some substances can damage or destroy portions of the respiratory tract, or they may be absorbed directly into the bloodstream and affect the function of other organs or tissues.

The respiratory system possesses a variety of anatomical and physiological mechanisms to protect itself from surprisingly high concentrations of particulates and noxious gases. These mechanisms can be overwhelmed, however, with resultant serious injury or permanent disability. The respiratory system should be protected by avoiding or minimizing exposure to harmful substances. Engineering controls and changes in workplace practices may provide sufficient protection. When these methods are insufficient or not feasible, respiratory protection must be used. A respirator may filter out gases, vapors, and particulates in the air; this type of device is called an air-purifying respirator or APR. Another type of respirator supplies clean breathing air to the user and is known as an atmosphere-supplying respirator. If the cylinder containing breathable air is worn by the user, the respirator is called a self-contained breathing apparatus or SCBA. When air is supplied from a source located some distance away and connected to the user by an airline hose, the device is called a supplied-air respirator.

The use of respirators is regulated by OSHA (29 CFR 1910.134). The OSHA regulation requires the development of a written respiratory protection program. An acceptable program should stipulate, as a minimum, the use, selection, cleaning, inspection, and fit-testing of approved respirators, as well as a medical surveillance program for personnel who utilize respiratory protection equipment.

STRUCTURE AND FUNCTION OF THE RESPIRATORY SYSTEM

Cells in the body need a continuous supply of oxygen to carry out their functions; many of the cell functions release significant amounts of carbon dioxide, which must be eliminated quickly and efficiently. The exchange of gases, oxygen and carbon dioxide, between the atmosphere and the cells is called respiration. Two body systems participate in the exchange of oxygen and carbon dioxide, the circulatory system and the respiratory system. Failure of either system results in rapid death of cells from oxygen starvation.

When air is inhaled, the chest muscles and diaphragm contract, lifting the rib cage and dropping the diaphragm. This action enlarges the chest cavity, which allows the lungs to expand and fill with air.

The respiratory system initially conducts air into the nose, mouth, and pharynx. The pharynx is a common port for both air and food. The mouth and the oral-pharynx are a globular, open chamber. The nasal passages in the nose and the nasopharynx contain ridges or projections that are bathed in mucous secretions; turbulent airflow across the moist surfaces facilitates deposition of particulates and absorption of gases.

The inhaled air passes through the pharynx and enters the trachea at the larynx. The trachea is a single tube or airway, which is also called the windpipe. The trachea divides into two bronchi, one leading to each lung. The bronchi then divide into a branching system of bronchioles, each branch becoming smaller and smaller. The trachea, bronchi, and bronchioles are all conducting tubes that carry air; they are all lined with a mucous membrane bathed by lubricating mucous secretions. The majority of the cells making up the mucous membrane have cilia, or small projections, that beat rhythmically and propel the mucous secretions upward toward the mouth. The mucous blanket and the mucociliary escalator allow the respiratory system to clean itself and remove inhaled particulates. The immune system also protects and cleanses the respiratory system. Particles and cellular debris are engulfed by specialized white blood cells or macrophages, which then pass into the lymphatic system.

At the end of the bronchioles are the alveoli, which are small sacs with very thin walls filled with bundles of minute blood vessels or capillaries. Oxygen from the inhaled air and carbon dioxide from the blood are exchanged in the alveoli. The alveoli and the capillaries are extremely fragile and susceptible to injury.

When air is exhaled, the chest muscles and the diaphragm relax, decreasing the size of the chest cavity; this action forces air out of the lungs along the same route by which it entered. Exhaled air contains approximately 16% oxygen and 5% carbon dioxide; normal air contains 21% oxygen, 78% nitrogen, 0.9% inert gases, and 0.04% carbon dioxide. A relaxed person breathes about 10 liters of air per minute. During exercise, the volume can increase to up to 80 liters per minute.

RESPIRATORY HAZARDS

Inhalation is a major route of exposure for toxic chemicals, irritating particulates, biological hazards, and alpha and beta radiation. An atmosphere containing toxic contaminants, even at very low concentrations, can present a hazard to the lungs and the body. A large concentration of gas or vapor can decrease the percentage of oxygen in the air, and can lead to death by asphyxiation, even if the displacing agent is an inert gas.

Oxygen Deficiency

The body's cells require oxygen to live; normal air contains 20.9% oxygen. Oxygen requirements increase with exercise or exertion. If the oxygen concentration decreases, the body reacts (Table 5–1). Death occurs rapidly when the oxygen concentration in inhaled air decreases to 6%.

In resting adults, the effects of oxygen deficiency usually are not evident until concentrations reach approximately 16%; but there is some variability in response, even between healthy young individuals. Oxygen requirements increase, however, during exercise, exertion, or exposure to a hot environment. Individuals with heart disease or chronic obstructive pulmonary diseases (bronchitis, asthma, emphysema, fibrosis) are particularly susceptible to oxygen deficiency. For this reason, OSHA has mandated that 19.5% oxygen in air is the lowest allowable "safe" concentration.

Oxygen deficiency may occur when oxygen is intentionally replaced by another gas; when a vapor or a gas builds up and eventually displaces oxygen; when oxygen is consumed in a chemical reaction, such as occurs in a fire; or when oxygen is consumed by a biological reaction, such as occurs during aerobic oxidation by microorganisms.

Particulates

Particulates are encountered everywhere in the form of fogs, aerosols, mists, sprays, fumes, dusts, smoke, and smog. In many cases, a liquid or a solid is suspended in a propellant under pressure to spray or apply paint, pesticides, cleaners, or other materials. Well-known responses to inhaled particulates are the sneeze caused by nasal irritation and the cough caused by irritation of the trachea, bronchi, or bronchioles. The sneeze and the cough are protective reflexes that help to expel mucus and foreign particles.

Particulates ranging in diameter between 5 and 30 microns are deposited in the nasal and pharyngeal passages. Although these particulates are extremely small (a micron is 1/1000 of a millimeter or 1/1,000,000 of a meter), they still may enter the respiratory system and cause injury. During nasal breathing nearly all particles greater than 10 microns in diameter are removed by nose hairs and mucous secretions. Particles less than 10 microns in diameter pass through the nose and are then deposited, depending on their size, in the trachea, bronchi, or bronchioles. The majority of particles larger than 3 microns in diameter are removed via the mucociliary escalator and never reach the sensitive alveoli. Smaller particles can reach the alveoli but may never be deposited because they remain sus-

Table 5–1. Oxygen deficiency in resting adults

% Oxygen	Effect
21–16	Nothing abnormal.
16–12	Loss of peripheral vision, increased rate and depth of breathing, increased heart rate, impaired attention and judgment, difficulty in concentrating, impaired coordination.
12–10	Very faulty judgment, poor muscle coordination, exertion causes extreme fatigue and may cause heart damage, intermittent respiration.
10–6	Nausea, vomiting, inability to perform vigorous movement or loss of ability to move, unconsciousness—followed by death.
<6	Breathing spasms, convulsions, death.

pended in air and diffuse back to the bronchioles and are exhaled. Particles that are deposited in the alveoli cannot be easily removed.

Particulates can pose a variety of hazards, and they are often classified by the respiratory system response they elicit (Table 5–2).

Gases and Vapors

Gases and vapors are filtered to some degree on their trip through the respiratory tract. Water-soluble gases and vapors readily dissolve in the mucous membrane, secretions in the conducting airways and often reach the alveoli at a much lower concentration than when they were originally inhaled. Once in the alveoli, gases and vapors are directly absorbed into the bloodstream.

Inhaled gases and vapors can pose a variety of hazards, and are frequently classified by their hazard or their effect on the respiratory system (Table 5–3).

Respiratory Response to Inhaled Hazards

All parts of the respiratory system can be damaged by inhaled particulates or toxic or irritating gases and vapors. A common response is a reflexive decrease in the diameter of the trachea and other conducting airways. This increases the resistance to airflow and makes breathing more difficult. The concentration of particulates or an irritant required to elicit an airway narrowing reaction usually is large except in persons who are allergic to the substance. The dose required to produce an asthmatic reaction in allergic individuals is usually very small.

Many gases and vapors and most particulates are irritating to the mucous membrane of the nose and the airways. If sufficient irritation occurs, there is a reversible increase in mucus secretion, which in turns causes coughing and increased production of sputum. Persistent stimulation by irritant gases such as cigarette smoke induces chronic hypersecretion

Table 5–2. Physiological classification of particulates

Nuisance	Do not cause direct injury but can impair function at high concentrations.
Fibrosis-causing	Can cause fibrotic nodules or diffuse fibrosis (asbestos, silica).
Irritants	Cause irritation, inflammation, or actual ulceration (destruction) of respiratory tissues, depending on severity (chlorine, fluorine).
Pulmonary sensitizers	Cause hypersensitivity reactions in lung tissue, producing chronic inflammation of the lung, which can lead to fibrosis (beryllium, moldy hay).
Upper respiratory sensitizers	Cause allergic hypersensitivity reactions including excess mucus production, sneezing, and difficulty breathing (pollen, molds).
Distant target organ effects	Cause tissue injury or disease in organs other than (or distant from) the lung (lead, metal oxide fumes).

Table 5–3. Physiological classification of gases and vapors

Irritants	Corrosive or irritating materials cause inflammation or actual destruction of tissue (acids, alkali).
Asphyxiants	These substances displace oxygen (simple asphyxiants: nitrogen, helium, methane) or prevent oxygen from being used (chemical asphyxiants: carbon monoxide, cyanide).
Anesthetics	These materials depress the central nervous system, causing loss of sensation or intoxication (chloroform, nitrous oxide).
Distant target organ effects	These substances produce effects in target organs distant from the lung (hexane, carbon tetrachloride, mercury).
Temperature effects	If inhaled, extremely hot or cold gases and vapors can cause damage to the nose, pharynx, airways, and lung tissue.

of mucus, which leads to chronic cough, accumulation of secretions in the airways, and increased resistance to airflow.

Gases such as nitrogen dioxide and sulfur dioxide can also paralyze the mucociliary escalator, thereby interfering with the removal of mucous secretions and causing excess accumulation of mucus in the conducting tubes and increased resistance to airflow. Excess mucus accumulation and inadequate cleansing of the respiratory system encourage bacterial growth, resulting in acute bronchitis or pneumonia.

Irritating gases and vapors that reach the alveoli may injure or destroy the alveolar surface cells, as well as capillary cells. When this occurs, blood plasma or whole blood may pour into the damaged area, leading to pulmonary edema. Severe pulmonary irritants cause an excessive outpouring of liquid, a condition sometimes called chemical pneumonia; death due to drowning in accumulated fluids can occur. In some cases, the death of lung tissue is slowed; delayed pulmonary edema can occur up to 72 hours after exposure.

Pulmonary fibrosis is a progressive reaction in response to prolonged exposure to materials that induce a chronic inflammatory reaction in lung tissue. Fibrosis occurs when dense connective tissue (scar tissue) replaces healthy lung tissue. Silica, coal dust, and asbestos can cause fibrosis.

Carcinogenesis, or the production of malignant tumors, is another response of the respiratory system to inhaled substances. Cancer may occur in the pharynx, the larynx, bronchi, bronchioles, or the lung itself. Lung cancer is increased in workers exposed to chromium, nickel, asbestos, and uranium. Cigarette smoke, a complex of irritating gases and particles, is responsible for more than 95% of the cases of lung cancer in the United States.

RESPIRATORY PROTECTION DEVICES

The basic function of a respirator is to reduce the risk of respiratory or other target organ injury due to inhalation of airborne contaminants. A respirator provides protection by removing the contaminants from the air, or by supplying an alternate source of clean breathable air.

All respirators are composed of two main parts: (1) the device that supplies or purifies the air; and (2) the facepiece, which covers the nose and the mouth and seals out contaminants. The first of these components defines the class of respirator (supplied air or air-purifying); the second determines the relative extent of protection that it provides.

Air-Purifying Respirators (APRs)

APRs remove air contaminants by passing the air to be breathed through a filter or a purifying device prior to its inhalation. APRs do not have a separate air source and *must not be used unless there is at least 19.5% oxygen* present in the air.

APRs can be further differentiated, as positive pressure or negative pressure devices, by the type of airflow supplied to the facepiece. Positive pressure respirators maintain a positive pressure in the facepiece during inhalation and exhalation. Powered air-purifying respirators (PAPRs) are positive pressure, continuous flow respirators that send a continuous stream of air into the facepiece at all times. At maximal breathing rates, however, it is possible that a negative pressure will be created in the facepiece. Some PAPRs are equipped with a hood or a helmet that fits over the head, rather than a facepiece.

Negative pressure respirators draw air into the facepiece by the negative pressure created when the wearer takes a breath. The primary disadvantage of negative pressure respirators is that if a leak develops because of a poor facepiece fit, or if another source of leakage exists anywhere in the system, contaminated air can be drawn into the facepiece.

Air-purifying respirators are discussed further in a later section of this chapter.

Atmosphere-Supplying Respirators

Respirators with an air source consist of two types: the self-contained breathing apparatus or SCBA, which supplies breathing air from tanks carried by the user; and supplied-air respirators or SARs, which supply air either through a hose from a stationary source located at a distance from the user or from a portable source.

Like APRs, atmosphere-supplying respirators may be further classified as either positive pressure or negative pressure respirators. Negative pressure or demand respirators use the negative pressure produced within the facepiece during inhalation to open an admission valve, which admits breathing air into the facepiece. Whenever negative pressure exists, air flows to the facepiece. There is no airflow into the facepiece except when the user takes a breath.

Positive pressure atmosphere-supplying respirators are either pressure-demand or continuous flow devices. In pressure-demand respirators, the admission valve is always partially open, so that positive pressure is maintained within the facepiece at all times. During inhalation, the valve opens wide, and airflow to the facepiece increases. The pressure regulator and an exhalation valve maintain positive pressure within the facepiece during exhalation. If a leak in the system develops, contaminated air cannot enter because the pressure in the facepiece is always greater than ambient pressure.

Positive pressure, continuous flow atmosphere-supplying respirators deliver a constant flow of breathing air to the facepiece at all times. Continuous flow systems use up the air supply much more rapidly than demand systems, and may allow negative pressure to exist in the facepiece if the breathing rate exceeds the airflow.

Atmosphere-supplying respirators are discussed further in a later section of this chapter.

Respirator Protection Factors

The amount of protection provided by a respirator is a function of how well the facepiece fits. No matter how efficient the purifying element of an APR, or how clean the supplied air of an atmosphere-supplying respirator, there will be little protection afforded if the respirator mask does not provide a leak-free face-to-mask seal. Facepieces are available in two basic configurations, which relate to their protective capacity.

1. Half-mask: This facepiece fits over the bridge of the nose, along the cheek, and under the chin. The headbands usually have a four-point suspension. Because they maintain a better seal and are less likely to be dislodged, half-masks afford more protection and have a greater protection factor than quarter-masks.

2. Full facepiece: This facepiece fits across the forehead, down over the temples and cheeks, and under the chin. The head harness usually has a five- or six-point suspension. These masks offer the greatest protection because they are held in place more securely than other types and because it is easier to maintain a good seal across the flat forehead than across the bridge of the nose. As an added benefit, the clear lens in the full facepiece protects the eyes from airborne contaminants.

Hoods or helmets that fit over the head and attach at the neck or the shoulders are used primarily with powered air-purifying respirators (PAPRs).

Limitations of Respirator Facepiece Fit

Not all respirators fit everyone; each individual must determine which facepiece fits properly and can be worn comfortably. Any given respirator will fit about 60% of the worker population. With many different respirator models and sizes available, however, at least one type should be found to fit each individual.

The use of respirators is prohibited when circumstances prevent a good facepiece fit. A number of conditions (described below) have been demonstrated to adversely affect the face-to-mask seal and should be avoided.

Facial hair and long hair that passes between the face and the sealing surface of the facepiece are prohibited because they interfere with respirator fit and wearer vision. Long hair must be effectively contained within protective hair coverings. Facial hair that affects facepiece fit includes beards, sideburns, mustaches, and beard stubble. Individuals wearing facepieces should be clean-shaven.

Eyeglasses with conventional temple bars are prohibited because they interfere with the facepiece-to-face seal of a full facepiece. A manufacturer-supplied spectacle-insert kit (Figure 5–1) should be installed in full facepieces of workers who require vision correction. Contact lenses may trap contaminants and/or particulates between the lens and the eye, causing irritation. Wearing contact lenses with a respirator in a contaminated atmosphere is prohibited by OSHA [29 CFR 1910.134 (e)(5)(ii)].

Gum and tobacco chewing should be prohibited during respirator use because they may cause ingestion of contaminants and may compromise respirator facepiece fit.

Other things that can affect facepiece fit include wearing skullcaps, hoods, or head coverings, which affect the suspension harness, and the use of cosmetics or lotions that make the face slippery and may cause excessive sliding of the mask on the face.

Determining the Protection Factor

A fit-test is used to determine if a respirator facepiece fits the wearer well enough to provide adequate protection. A quantitative fit-test is an analytical determination of the concentration of a test agent inside the facepiece compared to that outside the mask. The ratio is called the protection factor (PF). For example, if the ambient concentration outside the

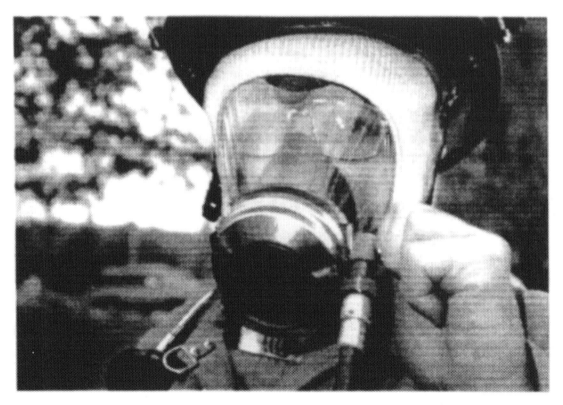

Figure 5–1. Full-facepiece respirators should be equipped with spectacle insert kits, provided by the manufacturer, for workers who must wear prescription glasses.

mask is 500 ppm, and the concentration inside the mask is 5 ppm, the protection factor for that particular respirator model and size is 100:

$$PF = \frac{\text{Concentration outside mask}}{\text{Concentration inside mask}}$$

$$= \frac{500 \text{ ppm}}{5 \text{ ppm}} = 100$$

Quantitative fit-tests are expensive and time-consuming, and they require special equipment. A special enclosure is required to hold the test atmosphere. The respirator tested must have a special adaptor to allow sampling of the air within the facepiece. The quantitative fit-test, however, is the best means of determining the maximum protection factor afforded by a particular respirator.

Qualitative fit-testing is often used as an alternative to quantitative fit-testing. A qualitative fit-test is a subjective, nonanalytical test used to determine if there is a good face-to-mask seal. Several harmless test atmospheres may be used, including saccharin (aerosol—see below), isoamyl acetate (banana oil), and irritant smoke. If the wearer does not respond by smelling or tasting the substance or coughing, the respirator passes the test and a standardized protection factor (PF) is assigned.

Current ANSI guidelines (Z88.2-1992), as well as draft changes to the OSHA standard (proposed in 1994 as an amendment to the OSHA Respiratory Protection standard, 29 CFR 1910.134), contain new PFs based upon the type of facepiece, hood, or helmet used (Table 5–4). According to the 1994 proposed rule, OSHA and NIOSH have agreed that the assignment of PFs should be made by NIOSH. Respirators should be selected in accordance with interim PFs assigned by NIOSH in the current NIOSH Respirator Decision Logic (NIOSH,

Table 5–4. Qualitative protection factors (PF) for protection against gas/vapor exposures

	ANSI	NIOSH
Negative pressure APR		
Half-mask	10	10
Full-facepiece	100	50
Powered APR		
Half-mask	50	50
Hood or helmet	1,000	25
Full-facepiece	1,000	50
Demand SAR		
Half-mask	10	10
Full-facepiece	100	50
Continuous flow SAR		
Half-mask	50	50
Hood or helmet	1,000	25
Full-facepiece	1,000	50
Positive pressure SAR		
Half-mask	50	1,000
Full-facepiece	1,000	2,000
Positive pressure SCBA		
Full-facepiece	≤10,000	10,000

Source: *ANSI Z88.2-1992*; NIOSH (1987).

1987). The NIOSH values, however, are not intended to replace more stringent PFs that have been assigned in individual substance-specific OSHA standards.

A separate section on fit-testing of APRs appears later in the chapter.

Using Protection Factors

A respirator protection factor is used to determine the Maximum Use Concentration (MUC) of a successfully fit-tested respirator. The MUC is the maximum concentration of contaminants in air that the worker can be exposed to without exceeding the protection factor. Put another way, the MUC is the protection factor multiplied by the workplace exposure limit:

$$MUC = PF \times TWA$$

For example, suppose that a full-facepiece APR with a qualitative protection factor of 50 is used to protect a worker in an atmosphere containing acetic acid. The OSHA TWA for acetic acid is 10 ppm; so the MUC for the worker is 50×10 ppm = 500 ppm. At concentrations above 500 ppm, the contaminant concentration *within* the facepiece can exceed the TWA. At this point the worker can be quantitatively fit-tested to determine if a higher PF is warranted for the APR used. If a higher PF cannot be assigned, at concentrations above 500 ppm a different type of respiratory protection with a higher protection factor must be used.

AIR-PURIFYING RESPIRATORS (APRS)

Air-purifying respirators or APRs selectively remove contaminants from air by filtration, absorption, adsorption, or chemical reaction. The air-purifying device is typically a particulate filter, or a cartridge or a canister containing sorbents for specific gases or vapors, or a filter and cartridge/canister combination (Figure 5–2).

Types of Air-Purifying Respirators

The two major components of an APR are the facepiece and the air-purifying element. In some cases, both components are combined into a single unit; more often, however, they are separate pieces. Several types of air-purifying respirators are available:

• Disposable dust respirators consist of cloth, paper, or synthetic fiber. Many disposable dust respirators are not NIOSH-approved. Disposable dust masks that are approved provide protection against nuisance dusts. This type of respirator is quite difficult to fit-test because it is very difficult to obtain and maintain a good facepiece-to-face seal.

• Mouthbit respirators are approved for *escape only*. The mouthpiece is held by the teeth, and a clamp is used to close the nostrils. A cartridge-type filter removes the contaminants from the air. This type of respirator must be approved for the specific hazard.

• Quarter-type respirators (type B half-mask) are used with cartridges or filters for toxic and nontoxic dusts with exposure limits greater than 0.05 mg/m^3. Dusts with lower exposure limits require a more efficient respirator than this.

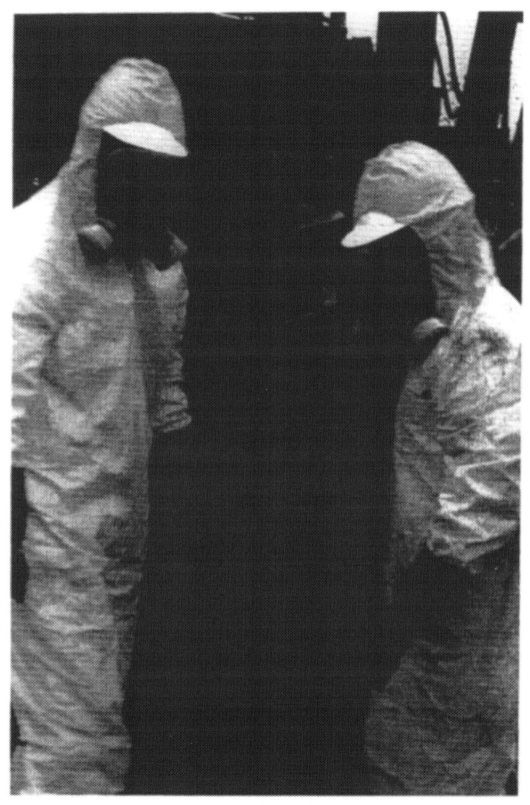

Figure 5–2. These workers are wearing full-facepiece air-purifying respirators equipped with cartridges and nosecups.

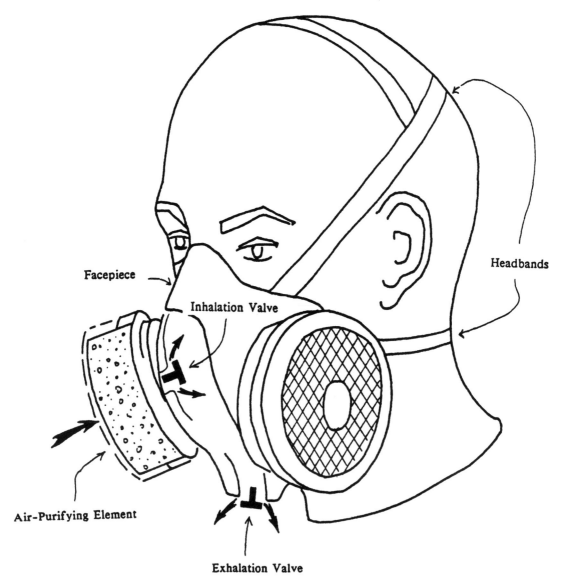

Facepiece

Inhalation Valve

Headbands

Air-Purifying Element

Exhalation Valve

Figure 5–3. Components of a half-mask air-purifying respirator. (*Source:* NIOSH, 1987.)

• Half-mask respirators (type A half-mask) fit better and have less breathing resistance than the quarter-type mask, which is approved only for dusts. The half-mask respirator (Figure 5–3) has approved cartridges for pesticides, organic vapors, dusts, mists, fumes, acid gases, ammonia, and several combinations.

• Full-face respirators (Figure 5–4) cover the whole face, including the eyes, and give a protection factor five times that of the half-mask (50 vs. 10). The full-facepiece respirator may be used with twin or single cartridges, chin-mounted cartridges, or chest-, belt-, or back-mounted canisters. Filters and sorbents are available for the same materials as the half-mask. Some additional sorbents are available for use only in canisters, such as those approved for protection against vinyl chloride and formaldehyde.

• Powered air-purifying respirators (PAPRs) give no breathing resistance and are available for use with half-mask, full-face, and special-fitting hoods and helmets.

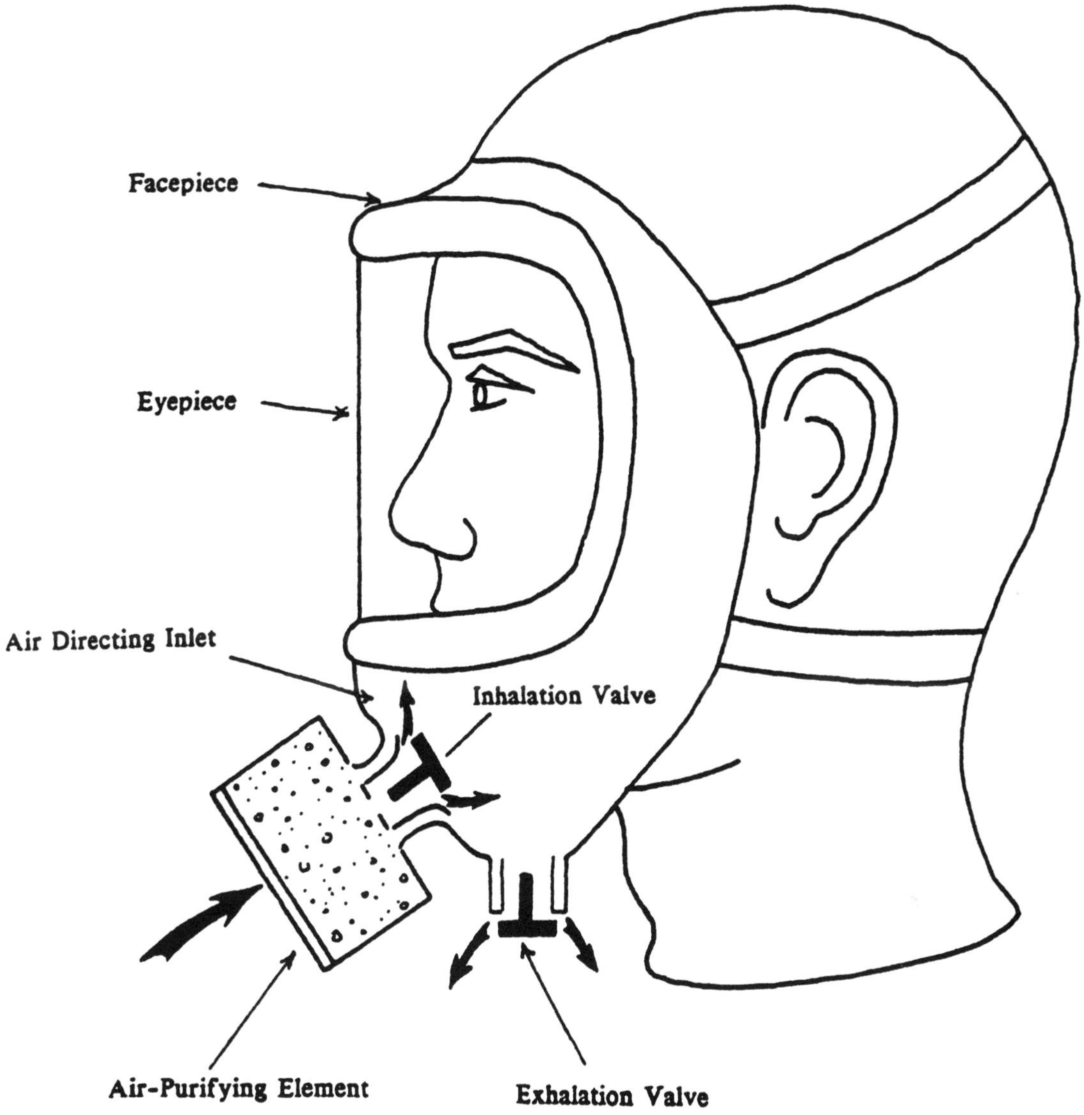

Facepiece

Eyepiece

Air Directing Inlet

Inhalation Valve

Air-Purifying Element

Exhalation Valve

Figure 5–4. Components of a full-facepiece air-purifying respirator. (*Source:* NIOSH, 1987.)

Facepiece Design

The facepiece is the means of sealing the respirator to the wearer; most facepieces are constructed of neoprene, PVC, or silicone rubber. Attached to the facepiece is a transparent lens (in the case of the full facepiece) and a head harness or suspension for holding the facepiece to the face. An adaptor provides an airtight connection between the facepiece and the cartridge or the canister. An inhalation check valve prevents exhaled breath from entering the cartridge or the canister. An exhalation valve permits exhaled breath to exit and prevents air from entering during inhalation. Many respirators have an integral speaking diaphragm.

Individual respirator manufacturers have their own methods of assembling and attach-

Full-facepiece,
dual cartridge

Half-mask, facepiece-
mounted cartridge.

Powered air-purifying respirator,
half-mask

Full-facepiece,
chin-mounted canister

Full-facepiece,
harness-mounted canister

Figure 5–5. Types of air-purifying respirators. (*Source:* NIOSH, 1985.)

ing parts. Never combine parts from different manufacturers; such practices void all warranties and approvals, and there is no guarantee that the resulting hybrid will provide adequate protection.

Although many configurations exist (Figure 5–5), only four types of facepiece-purifying element assemblies are satisfactory for use at hazardous waste sites:

- Half-mask with cartridge(s) or filter (or combination of both).
- Full-face with cartridge(s) or filter (or combination of both).
- Full-face with chin-mounted canister.
- Full-face with chest-, back-, or belt-mounted canister.

At hazardous materials sites, the recommended facepiece is the full facepiece. The full-face mask provides eye protection, is easier to fit than other types, and has a protection factor at least five times greater than that of the half-mask.

AIR-PURIFYING ELEMENTS

Respiratory hazards can be placed in one of two broad categories: particulates and vapors/gases. Particulates are filtered by mechanical means, whereas vapors and gases are removed by sorbents. Respirator elements using a combination of a mechanical filter and a chemical sorbent will effectively remove both types of hazards.

Particulate-Removing Filters

Particulates can occur as dusts, fibers, fumes, or mists. The particle size can range from macroscopic to microscopic, and the health effect from their inhalation can be severe or minimal. The relative health hazard posed by a particulate can be assessed by examining its exposure limit. A nuisance dust or particulate will have an OSHA or an ACGIH exposure limit of 10 mg/m^3, whereas a toxic dust may have a limit well below 0.05 mg/m^3.

Mechanical filters are classified according to the protection for which they are approved by NIOSH. Most particulate filters are approved only for dusts and/or mists with exposure limits equal to or greater than 0.05 mg/m^3. These dusts are usually considered capable of producing fibrosis and other lung disorders. These particulate filters have an efficiency of 80 to 90% for 0.6 micron particles.

Respirators approved for fumes are more efficient than those used for dusts and mists, removing 90 to 99% of the 0.6 micron particles. This type of respirator is approved for dusts, fumes, and mists with exposure limits equal to or greater than 0.05 mg/m^3.

Finally, there is a high efficiency particulate (HEPA) filter, which is 99.97% effective against particles 0.3 micron in diameter. The HEPA filter is approved for dusts, mists, and fumes with an exposure limit less than 0.05 mg/m^3.

Mechanical filters become loaded with particulates with use. As they are used, they become more efficient, but the resistance to airflow increases, and it becomes increasingly hard to breathe through them. When a mechanical filter becomes difficult to breathe through, it should be replaced.

Color Coding of Cartridges and Canisters

When a particulate-, gas-, or vapor-removing purifying element is chosen, it must be selected for protection against a specific type of contaminant. OSHA requires color coding for each type of chemical cartridge and canister (Table 5–5).

Table 5–5. Color coding of cartridges and canisters

Chemical class	Colors assigned
Acid gases	White
Hydrocyanic acid gas	White with green stripe
Chlorine gas	White with yellow stripe
Organic vapors	Black
Ammonia gas	Green
Acid gases + ammonia gas	Green with white stripe
Carbon monoxide	Blue
Acid gases + organic vapors	Yellow
Acid gases + organic vapors + ammonia gas	Brown
Radioactive particles	Purple (magenta)
Asbestos	Purple
Particulates* + other class	Class color with gray stripe

*Nonradioactive dust, mist, fog, smoke with an exposure limit equal to or greater than 0.05 mg/m^3.

Size and Style of Gas/Vapor-Removing Elements

Gas- and vapor-removing elements are available in various styles, which differ in size and the way that they attach to the facepiece. The smallest elements are cartridges that contain 50 to 200 cm^3 of chemical sorbent and are attached directly to the facepiece, usually in pairs. Chin canisters contain between 250 and 500 cm^3 of sorbent and also are attached directly to the facepiece. Gas masks, or industrial-sized canisters, are attached to the full facepiece by a breathing hose. The canister contains 1000 to 2000 cm^3 of sorbent and is worn mounted on the belt, or on a harness attached to the wearer's chest or back.

A maximum use concentration for cartridges and canisters is assigned by NIOSH/MSHA at the time of approval. For example, organic vapor purifying elements are approved for use in contaminated atmospheres as follows:

- Chemical cartridge: up to 1,000 ppm (0.1% by volume).
- Chin canister: up to 5,000 ppm (0.5% by volume).
- Gas mask canister: up to 20,000 ppm (2% by volume).

It is important to remember, however, that *no air-purifying respirator, regardless of its capacity, is permitted in IDLH, LEL, or oxygen-deficient atmospheres.*

Service Life

Each chemical sorbent has a limited capacity for removing chemical contamination from the air. When the limit is reached, the cartridge or the canister is said to be saturated. At this point, the purifying element *will* allow the contaminants to enter the facepiece and be inhaled by the wearer; this is called breakthrough. The length of time that a cartridge or a canister will effectively remove air contaminants is known as its service life, and depends on the breathing rate of the wearer, the concentration of the contamination present, and the efficiency of the chemical sorption.

If the breathing rate of the wearer is rapid, the flow rate of contaminated air through the purifying element is greater than that at slower rates. A higher airflow rate brings a larger amount of contamination into contact with the chemical sorbent, thus increasing the rate of sorbent saturation and shortening the service life.

As ambient contamination concentration increases, the service life decreases. As contamination increases, more contamination comes in contact with the chemical sorbent in a given period of time. For example, at a constant breathing rate, ten times as much contamination contacts the purifying element when the concentration is 500 ppm as compared to 50 ppm.

One way to compensate for increased rates or high concentrations of contamination is to use a larger canister. The expected service life of a canister is longer than that of a cartridge as the canister contains more chemical sorbent, which gives it more sorptive capacity than the cartridge has.

Chemical sorbents vary in their ability to remove contaminants from air. The organic vapor cartridge efficiency has been compared for a number of solvents by determining how much time elapsed before a 1% breakthrough concentration was measured in cartridge-filtered air. When the initial test concentration was 1000 ppm of solvent vapor, the breakthrough concentration was 10 ppm (1% of 1000 ppm = 10 ppm).

Cartridge efficiency, as measured by breakthrough time, varies considerably when tested against different organic materials (Table 5–6). Breakthrough can occur after 107 minutes for chlorobenzene, but after only 3.8 minutes for vinyl chloride. The sorbent in the organic vapor cartridge (activated charcoal) is much more effective for removing chlorobenzene than vinyl chloride. One must consider cartridge efficiencies and breakthrough times when selecting and using APRs.

Warning Properties

A warning property—which is a sign that the cartridge or the canister is beginning to lose its effectiveness, and breakthrough is occurring—can be detected as an odor, a taste, or an irritation. At the first sign of breakthrough, the old cartridge or canister must be exchanged for a fresh one. Without an adequate warning property, the respirator efficiency may decrease without the knowledge of the wearer, a situation that ultimately can lead to over-exposure and adverse health effects.

Most substances have warning properties at some concentration. A warning property detected at a dangerous level (i.e., a concentration greater than the exposure limit) is not adequate. An odor, taste, or irritation detected at extremely low concentrations (i.e., well below the exposure limit) also is not satisfactory; the warning will be given all the time or long before the purifying element is saturated. In such cases, the wearer would detect the substance all the time and never know when the purifying element actually became ineffective.

Warning properties are considered adequate when odor, taste, or irritant effects are noted and persist at concentrations *below* the OSHA PEL, NIOSH REL, or ACGIH TWA. An optimal warning property is first detected around the exposure limit. For example, tertiary butyl alcohol has an odor threshold range of 47 to 73 ppm and an irritation threshold of 100 ppm. Because the OSHA PEL is 100 ppm, tertiary butyl alcohol is considered to have good warning properties. Dimethylformamide, on the other hand, has an OSHA PEL of 10 ppm and an odor threshold range of 2.2 to 100 ppm. A substance with an odor threshold that is ten times the exposure limit does not have an adequate warning property. Table 5–7 lists odor thresholds for a variety of chemical compounds.

Some substances cause rapid olfactory fatigue; when this occurs, an odor that was once quite noticeable can no longer be smelled. Odor cannot be used as a warning property for materials that produce rapid olfactory fatigue.

Many chemical references indicate what a material smells like by listing its chemical odor class. The American Society for Testing and Materials (ASTM, 1970) has divided chemicals into eight major odor classes (Table 5–8).

Table 5–6. Organic vapor cartridge efficiency[1]

Chemical contaminant[2]	Breakthrough time (minutes)
Aromatics[3]	
Benzene	73
Toluene	94
Ethylbenzene	84
m-Xylene	99
Cumene	81
Mestiylene	86
Alcohols	
Methanol	00.2
Ethanol	28
Isopropanol	54
Allyl alcohol	66
n-Propanol	70
sec-Butanol	96
Butanol	115
2-Methoxyethanol	116
Isoamyl alcohol	97
4-Methyl-2-pentanol	75
2-Ethoxyethanol	77
Amyl alcohol	102
2-Ethyl-1-butanol	76
Monochlorides[3]	
Methyl chloride	00.05
Vinyl chloride	3.8
Ethyl chloride	5.6
Allyl chloride	31
1-Chloropropane	25
1-Chlorobutane	72
Chlorocyclopentane	78
Chlorobenzene	107
1-Chlorohexane	77
o-Chlorotoluene	102
1-Chloroheptane	82
3-Chloromethyl heptane	63
Dichlorides[3]	
Dichloromethane	10
trans-1,2-Dichloroethylene	33
1,1-Dichloroethane	23
cis-1,2-Dichloroethane	30
1,2-Dichloroethane	54
1,2-Dichloropropane	65
1,4-Dichlorobutane	108
o-Dichlorobenzene	109
Trichlorides[3]	
Chloroform	33
Methyl chloroform	40
Trichloroethylene	55
1,1,2-Trichloroethane	72
1,2,3-Trichloropropane	111
Tetra- and Pentachlorides[3]	
Carbon tetrachloride	77
Perchloroethylene	107
1,1,2,2-Tetrachloroethane	104
Pentachloroethane	93
Acetates[3]	
Methyl acetate	33
Vinyl acetate	55
Ethyl acetate	67
Isopropyl acetate	65

Table 5–6. (continued)

Chemical contaminant[2]	Breakthrough time (minutes)
Propyl acetate	79
Allyl acetate	76
sec-Butyl acetate	83
Butyl acetate	77
Isopentyl acetate	71
2-Methoxyethyl acetate	93
1,3-Dimethyl butyl acetate	61
Amyl acetate	73
2-Ethoxyethyl acetate	80
Hexyl acetate	67
Ketones[4]	
Acetone	37
2-Butanone (MEK)	82
2-Pentanone	104
3-Pentanone	94
4-Methyl-2-pentanone (MIBK)	96
Mesityl oxide	122
Cyclopentanone	141
3-Heptanone	91
2-Heptanone	101
Cyclohexanone	126
5-Methyl-3-heptanone	86
3-Methylcyclohexanone	101
Diisobutyl ketone	71
4-Methylcyclohexanone	111
Alkanes[4]	
Pentane	61
Hexane	52
Methylcyclopentane	62
Cyclohexane	69
2,2,4-Trimethylpentane	68
Heptane	78
Methylcyclohexane	69
Nonane	76
Decane	71
Amines	
Methyl amine	12
Ethyl amine	40
Isopropyl amine	66
Propyl amine	90
Diethyl amine	88
Butyl amine	110
Triethyl amine	81
Dipropyl amine	93
Diisopropyl amine	77
Cyclohexyl amine	112
Dibutyl amine	76
Miscellaneous materials[4]	
Acrylonitrile	49
Pyridine	119
1-Nitropropane	143
Methyl iodide	12
Dibromomethane	82
1,2-Dibromomethane	141
Acetic anhydride	124
Bromobenzene	142

[1] *Amer. Indus. Hyg. Assoc. J.* 33:745, 1972.

[2] Test atmosphere 1000 ppm, 50% RH, 22°C, flow rate 53 l/min.

[3] Mine Safety Appliances cartridges.

[4] American Optical cartridges.

Table 5-7. Odor thresholds

Chemical name	Odor threshold (ppm)
Acetaldehyde	0.031–2.3
Acetic acid	0.2–24
Acetic anhydride	0.14–81.2
Acetone	46.8–100
Acetonitrile	21.4–170
Acetophenone	0.002–0.60
Acetyl bromide	0.0005
Acetyl chloride	1.0
Acetylene	620
Acrolein	0.1–16.6
Acrylic acid	0.094–1.04
Acrylonitrile	17–100
Akrol	10
Allyl alcohol	0.75–7.2
Allylamine	6.3–28.7
Allyl chloride	0.21–1.2
Allyl chloroformate	1.4
Allyl disulfide	0.0012
Allyl isocyanide	0.018
Allyl isothiocyanide	0.15–0.42
Allyl mercaptan	0.00005–0.21
Allyl sulfide	0.000014–0.01
Ammonia	0.32–55
Ammonium hydroxide	50
iso-Amyl acetate	0.0028–0.11
n-Amyl acetate	0.0009–0.08
sec-Amyl acetate	0.0017–0.082
tert-Amyl acetate	0.0017
Amyl alcohol	0.0065–35
Amylene	0.0022–2.3
Amyl isovalerate	0.11
iso-Amyl mercaptan	0.0043–0.7
n-Amyl mercaptan	0.07
n-Amyl methyl ketone	0.0009
Anethole	0.0033
Aniline	0.05–70
Apiol	0.0063
Arsenic anhydride	1
Arsine	0.21–0.50
Benzaldehyde	0.003–0.69
Benzene	2.14–12
Benzyl alcohol	5.5
Benzyl chloride	0.01–0.31
Benzyl mercaptan	0.00019–0.037
Benzyl sulfide	0.0021–0.07
Biphenyl	0.00083
Bornyl acetate	0.0078
Boron trifluoride	1.0–1.5
Bromine	0.047–3.5
Bromoacetone	0.090
Bromoacetophenone	0.079
Bromoform	1.3–530
1,3-Butadiene	0.16–1.8
iso-Butane	1.2
n-Butane	5.5–2700
2-Butoxyethanol	0.10
n-Butyl acetate	0.037–20
iso-Butyl acetate	0.002–7
sec-Butyl acetate	4.0–7

Table 5–7. (continued)

Chemical name	Odor threshold (ppm)
tert-Butyl acetate	0.004
n-Butyl acrylate	0.035
n-Butyl alcohol	0.83–15
iso-Butyl alcohol	40
sec-Butyl alcohol	2.6–43
tert-Butyl alcohol	47–73
n-Butylamine	0.24–5
Butyl cellosolve	0.48
Butyl cellosolve acetate	0.20
n-Butyl chloride	13
1-Butene (butylene)	0.07–26
2-Butene	0.57–22
Butyl ether	0.24–0.47
Butylene oxide	0.71
n-Butyl formate	17
n-Butyl acetate	0.00054–0.00097
iso-Butyl mercaptan	0.00082–0.38
n-Butyl mercaptan	0.00009–0.06
tert-Butyl mercaptan	0.015–0.18
Butyl sulfide	0.015–0.18
p-tert-Butyltoluene	5
n-Butyraldehyde	0.0046–0.039
Butyric acid	0.00047–0.001
iso-Butyric acid	0.001
Calcium hypochlorite	3.5
Calcium phosphide	0.13–13.4
Camphor	0.018–200
Caprolactam	0.065
Carbitol acetate	0.157–0.263
Carbon dioxide	74,000
Carbon disulfide	0.0011–7.7
Carbon monoxide	100,000
Carbon tetrachloride	10–200
Carvacrol	0.0023
Chloral	0.047
Chlorine	0.01–5
Chlorine dioxide	0.1–9.4
Chloroacetaldehyde	<1.0
Chloroacetic acid	0.045
Chloroacetophenone	0.01–1.35
Chlorobenzene	0.21–60
Chlorobromomethane	400
Chloroform	50–307
Chloromethane	10–100
Chlorophenol	0.034
o-Chlorophenol	0.0036
p-Chlorophenol	1.2–30
Chloropicrin	0.78–1.1
Chloroprene	15
Chlorosulfonic acid	1.0–5
Chlorotoluene	0.32
Chlorovinyl arsine	1.6
Cinnamaldehyde	0.0026
Coumarin	0.0033–0.2
m-Cresol	0.00028–0.68
o-Cresol	0.28–0.68
p-Cresol	0.00047–0.001
Crotonaldehyde	0.035–7.35
Crotyl mercaptan	0.00016–0.0099

Table 5–7. *(continued)*

Chemical name	Odor threshold (ppm)
Crude oil	0.1–0.5
Cumene	0.047–1.2
Cyanogen chloride	1.0
Cyclohexane	0.41–300
Cyclohexanol	0.15–160
Cyclohexanone	0.12–0.88
Cyclohexene	0.18
Cyclopentadiene	1.9–250
Decaborane	0.05–0.35
Decanal	0.0064–0.168
Decanoic acid	0.0020
1-Decylene	0.12
Diacetone alcohol	0.28–1.7
Diacetyl	0.025
Diallyl ketone	9.0
Diborane	2.5–4.0
Di-N-butyl amine	0.27–0.48
Dichlorobenzene	0.005
o-Dichlorobenzene	0.30–50
p-Dichlorobenzene	0.18–30
Dichlorodiethyl sulfide	0.19
Dichloroethane	120
Dichloroethylene	0.085–500
Dichloroethyl ether	0.049–35
Dichloroisopropyl ether	0.32
Dichloromethane	25–320
2,4-Dichlorophenol	0.21–0.008
1,2-Dichloropropane	50
2,2-Dichloropropionic acid	428
Dicyclopentadiene	0.003–0.020
Dieldrin	0.041
Diesel fuel (#1)	0.25
Diesel fuel (#2)	0.08
Diesel fuel (#4)	0.01
Diethanolamine	0.011–0.27
Diethylamine	0.06–0.498
Diethylaminoethanol	0.011–0.04
Diethylene triamine	10
Diethyl ketone	2.0
Diethyl selenide	0.00014
Diethyl succinate	0.021
Diglycidyl ether	5.0
Diisobutyl carbinol	0.048–0.16
Diisobutyl ketone	0.11–0.31
Diisopropylamine	0.38–1.8
Dimethyl acetamide	21.4–46.8
Dimethylamine	0.021–6.0
Dimethylaminoethanol	0.045
Dimethylaniline	0.013
Dimethylformamide	2.2–100
1,1-Dimethylhydrazine	1.7–14
Dimethyl sulfide	0.001–0.020
2,6-Dinitrophenol	0.21
Dioxane	1.8–170
Dioxolane	64–128
Diphenyl	0.01–0.06
Diphenylcyanoarsine	0.3
Diphenyl ether	0.001–1.0
Diphenyl sulfide	0.0021–0.0047

Table 5–7. (*continued*)

Chemical name	Odor threshold (ppm)
Diphosgene	1.2
Dipropylamine	1.0
Dithioethylene glycol	0.031
Dodecanol	0.0064
Epichlorohydrin	0.93–16
Ethane	150–120,000
1,2-Ethanedithiol	0.0042
Ethanol	4.68–5100
Ethanolamine	2–3
2-Ethoxyethanol	0.55–2.7
2-Ethoxyethanol acetate	0.056–50
Ethyl acetate	0.056–50
Ethyl acrylate	0.00024–1.0
Ethylamine	0.021–0.95
Ethylamyl ketone	6.0
Ethyl benzene	0.25–200
Ethyl bromide	3.1
2-Ethylbutanol	0.77
Ethyl butyrate	0.0082–0.015
Ethyl chloride	4.2
Ethyl decanoate	0.00017
Ethyldichloroarsine	0.14–1.4
Ethyl disulfide	0.0028
Ethylene	290–700
Ethylenediamine	1.0–11.2
Ethylene dibromide	10–25
Ethylene dichloride	6.2–100
Ethylene glycol	0.08
Ethylene imine	1.5
Ethylene oxide	0.84–700
Ethyl ether	0.33–8.9
Ethyl formate	31–330
Ethyl glycol	25
Ethyl hexanol	0.138
Ethyl hexanoate	0.0056
Ethyl hexyl acetate	0.21
Ethyl hexyl acrylate	0.18
Ethylidene norbornene	0.007–0.073
Ethyl isothiocyanate	1.6–10.7
Ethyl mercaptan	0.00051–0.075
Ethyl methacrylate	0.0067
n-Ethylmorpholine	0.25–25
Ethyl selenide	0.0012–0.014
Ethyl selenomercaptan	0.0003
Ethyl silicate	17–85
Ethyl sulfide	0.0006–0.068
Ethyl undecanoate	0.00054
Ethyl *iso*-valerate	0.12
Ethyl *n*-valerate	0.06
Eugenol	0.0046
Fluorine	0.035–3.0
Formaldehyde	0.1–1.0
Formic acid	21–49
Fuel oil (#1, kerosene)	0.082–1.0
Fuel oil (#2, diesel)	0.082
Fuel oil (#4)	0.5
Fuel oil (#6, bunker)	0.13
Furfural	0.078–5.0
Furfuryl alcohol	8.0

Table 5-7. (continued)

Chemical name	Odor threshold (ppm)
Gasoline	0.005–10
Glycol diacetate	0.312
Halothane	33.0
n-Heptal chloride	0.06
Heptachlor	0.02
Heptaldehyde	0.05
n-Heptane	50–220
Heptanol	0.057–20
Hexachlorocyclopentadiene	0.03–0.33
Hexachloroethylene	0.15
Hexamethylenediamine	0.0009
n-Hexane	130
Hexanoic acid	0.0061
Hexanol	0.005–0.09
sec-Hexyl acetate	100
Hexylene glycol	50
Hydrazine	3–4
Hydrochloric acid gas	10
Hydrocinnamyl alcohol	0.00027
Hydrogen bromide	2
Hydrogen chloride	0.77–10
Hydrogen cyanide	0.00027–5
Hydrogen fluoride	0.042
Hydrogen selenide	0.3–3
Hydrogen sulfide	0.00001–0.8
Iodoform	0.005–0.5
Ionone	0.000012
Isoamyl acetate	0.002–7
Isoamyl alcohol	0.42–35
Isobutyl acetate	0.002–7
Isobutyl alcohol	1.8–40
Isobutyl acrylate	0.009–0.012
Isobutyl cellosolve	0.114–0.191
Isobutyl mercaptan	0.00054–0.00097
Isobutyraldehyde	0.047–0.336
Isobutyric acid	0.001
Isocyanochloride	0.98
Isodecanol	0.031–0.042
Isopentanoic acid	0.015–0.026
Isophorone	0.20–0.54
Isoprene	0.005
Isopropyl acetate	0.9
Isopropyl alcohol	7.5–200
Isopropylamine	0.71–10
Isopropyl ether	0.017–300
Isopropyl glycidyl ether	300
Isopropyl mercaptan	0.00025
Kerosene	0.082–1.0
Lactic acid	0.0000004
Lauric acid	0.0034
Light gasoline	800
Linoleyl acetate	0.0016
LPG	20,000 (propane)
Maleic anhydride	0.25–0.5
Menthol	1.5
2-Mercaptoethanol	0.64
Mesitylene	0.27–0.55
Mesityl oxide	0.051–12
2-Methoxyethanol	0.22–60

Table 5-7. (*continued*)

Chemical name	Odor threshold (ppm)
Methoxynaphthalene	0.00012
Methyl acetate	4.6–200
Methyl acetylene–propadiene mix	100
Methyl acrylate	0.0048–20
Methylacrylonitrile	2–14
Methyl alcohol	53.3–5900
Methylamine	0.021–10
Methyl amyl acetate	0.23–0.40
Methyl amyl alcohol	0.52–50
Methyl amyl ketone	0.35
N-Methylaniline	1.7
Methyl anthranilate	0.00066–0.06
Methyl bromide	<20
2-Methyl-2-butanol	0.23–2.3
2-Methyl *n*-butyl ketone	0.076
Methyl *n*-butyrate	0.0026
Methyl cellosolve	0.22–60
Methyl cellosolve acetate	0.64–50
Methyl chloride	10–100
Methyl chloroform	20–500
Methyl 2-cyanoacrylate	1–3
Methylcyclohexane	500–630
Methylcyclohexanol	500
Methyl dichloroarsine	0.11
Methylene chloride	25–320
Methyl ethanolamine	3.4
Methyl ethyl ketone (MEK)	4.68–25
Methyl ethyl pyridine	0.008–0.050
Methyl formate	1500–2000
Methyl glycol	60
5-Methyl-3-heptanone	6
Methyl hydrazine	1–3
Methyl isoamyl alcohol	0.20
Methyl isoamyl ketone	0.28
Methyl isobutyl ketone (MIBK)	0.28–8
Methyl isocyanate	2.0
Methyl isopropyl ketone	1.9
Methyl mercaptan	0.0021–1.1
Methyl methacrylate	0.05–0.34
2-Methylpentaldehyde	0.136
2-Methyl-1-pentanol	0.024–0.082
2-Methylpropene	0.57
Methyl *n*-propyl ketone	11.0
Methyl salicylate	0.096
a-Methyl styrene	0.156–200
Methyl sulfide	0.001–0.020
Methyl thiocyanate	0.25–3.2
Methyltrichlorosilane	1.0
Methyl vinyl ketone	0.2
Methyl vinyl pyridine	0.040
Mineral spirits	30
Morpholine	0.01–0.14
Naphtha (coal tar)	4.68–100
Naphthalene	0.003–0.3
2-Naphthol	1.3
Nickel carbonyl	0.30–1.3
p-Nitroaniline	0.3–1.0
Nitrobenzene	0.0047–6.0

Table 5-7. (continued)

Chemical name	Odor threshold (ppm)
o-Nitrochlorobenzene	0.002
Nitroethane	2.1–200
Nitrogen dioxide	0.39–5.0
Nitrogen tetroxide	5
Nitromethane	3.5–100
1-Nitropropane	11–300
2-Nitropropane	70–300
Nitrotoluene	0.045–1.74
Nonane	47
n-Octane	4–150
Ocatanoic acid	0.0014
1-Octanol	0.0021
2-Octanol	0.0026
Osmium tetroxide	0.0019
Oxygen difluoride	0.1–0.5
Ozone	0.005–0.5
Pelargonic acid	0.00086
Pentaborane	0.8–0.96
n-Pentane	2.2–1000
2,4-Pentanedione	0.020
Pentanol	0.0065–35
Pentanone	8.0
Pentene	2.2
iso-Pentyl acetate	0.0028–0.11
n-Pentyl acetate	0.00090–0.08
1-Pentyl mercaptan	0.00021
Perchloroethylene	4.68–50
Perchlormethyl mercaptan	<0.1
Perchloryl fluoride	10
Petroleum distillate	<500
Phenol	0.021–5
Phenyl ether	0.001–0.10
Phenyl ether–biphenyl mix	0.1–1.0
Phenyl isothiocyanate	0.43
Phenyl mercaptan	0.0094
Phosgene	0.125–1.0
Phosphine	0.02–3.0
Phosphorus pentasulfide	0.0047
Phosphorus trichloride	0.7
Phthalic anhydride	0.05–0.12
2-Picoline	0.023–0.046
Propane	1000–20,000
Propionaldehyde	0.04–1.0
Propionic acid	0.034–0.16
n-Propyl acetate	0.15–200
Propyl alcohol	0.08–200
Propylene	67.6–76
Propylene diamine	0.048–0.067
Propylene dichloride	0.25–130
Propylene oxide	35–200
Propyl mercaptan	0.00075–0.02
Propyl nitrate	44–90
Propyl sulfide	0.011–0.17
Pyridine	0.012–5
Quinoline	0.16–71
Quinone	0.084–0.5
Resorcinol	40
Safrole	0.0032
Stoddard solvent	1.0–30

Table 5-7. (continued)

Chemical name	Odor threshold (ppm)
Styrene	0.047–200
Styrene oxide	0.40
Sulfoxide	91
Sulfur dichloride	0.001
Sulfur dioxide	0.47–5
Sulfur chloride	0.001
Tannic acid	2.0–4.0
1,1,2,2-Tetrachloroethane	1.5–5
Tetrachloroethylene	4.68–50
Tetraethyl-o-silicate	5.0–7.2
Tetrahydrofuran	2.0–50
Tetramethylbenzene	0.0029
Thiocresol	0.0027–0.02
Thiophenol	0.014
Thymol	0.00086
Toluene	0.17–40
Toluene diisocyanate (TDI)	0.17–2.14
o-Toluidine	0.0048–20
Toxaphene	0.0052
1,2,4-Trichlorobenzene	1.4
1,1,1-Trichloroethane	20–400
Trichloroethylene	21.4–400
Trichlorofluoromethane	5.0–209
Trichlorophenol	0.1–0.667
1,2,3-Trichloropropane	100
1,1,2-Trichloro-1,2,2- trifluoroethane	68–135
Triethylamine	0.28–0.48
Trimethylamine	0.00021–1.7
Trimethylbenzene	0.027–0.55
Trinitrobutylxylene	0.0000065–0.0008
Turpentine	200
n-Undecane	0.12
n-Valeraldehyde	0.028
Valeric acid	0.0006
iso-Valeric acid	0.0018
Vanillin	0.000000032
Vinyl acetate	0.12–0.55
Vinyl chloride	260–3000
Vinylidene chloride	190
Vinyl toluene	10–50
Xylene	0.05–200
m-Xylene	1.1–3.7
o-Xylene	1.8
p-Xylene	0.47–0.53
2,4-Xylidine	0.0048–0.056

Table compiled by using the following sources: Amoore and Hautala (1983); Billings and Jonas (1981); Arthur D. Little, Inc. (1965); Ruth (1986); and USEPA *Course Manual* (1987a).

REQUIREMENTS FOR AIR-PURIFYING RESPIRATOR USE

An air-purifying respirator or APR is used because the concentration of one or more contaminates in the atmosphere is high enough to cause some type of health effect. Effects may range from respiratory or eye irritation to organ damage. Exposure limits delineated by

Table 5–8. Odors classified by chemical types

Odor class	Chemical types	Examples
Estery	Esters Ethers Lower ketones	Lacquer, solvents, most fruits, many flowers
Alcoholic	Phenols, cresols Alcohols Hydrocarbons	Creosote, tars, smoke, alcohol, liquor, rose and spicy flowers, spices and herbs
Carbonyl	Aldehydes Higher ketones	Rancid fats, butter, stone fruits and nuts, violets, grasses and vegetables
Acidic	Acid anhydrides Organic acids Sulfur dioxide	Vinegar, perspiration, rancid oils, resins, body odor, garbage
Halide	Quinones Oxides and ozone Halides Nitrogen compounds	Insecticides, weed killer, musty and moldy odors, husks, medicinal odors, earth and peat
Sulfury	Selenium compounds Arsenicals Mercaptans Sulfides	Skunks, bears, foxes, rotting fish and meat, cabbage, onion, sewage
Unsaturated	Acetylene derivatives Butadiene Isoprene	Paint thinners, varnish, kerosene, turpentine, essential oils, cucumber
Basic	Vinyl monomers Amines Alkaloids Ammonia	Fecal odors, manure, fish and shellfish, stale flowers such as lilac, lily, jasmine, and honeysuckle

Source: ASTM (1970).

OSHA, ACGIH, NIOSH, AIHA, or the chemical manufacturer are often used to determine the need for respiratory protection. Concentrations greater than a recognized exposure limit require respiratory protection.

The use of an air-purifying respirator is contingent upon a number of criteria; if the criteria listed below cannot be met, then the use of an APR is prohibited. A flow diagram can also be used to select appropriate respiratory protection (Figure 5–6).

Air purifying respirators can be used only if *all* of the following conditions are met:

1. The identity and the concentration of the contaminants in air are known.
2. The oxygen content is at least 19.5%.
3. The contaminants have sufficient warning properties when sorbent or filter capacity has been exceeded.
4. There is periodic monitoring for contaminants in air.
5. There are no IDLH or LEL concentrations present.
6. The respirator assembly is approved for protection against the specific contaminants and their concentration levels.

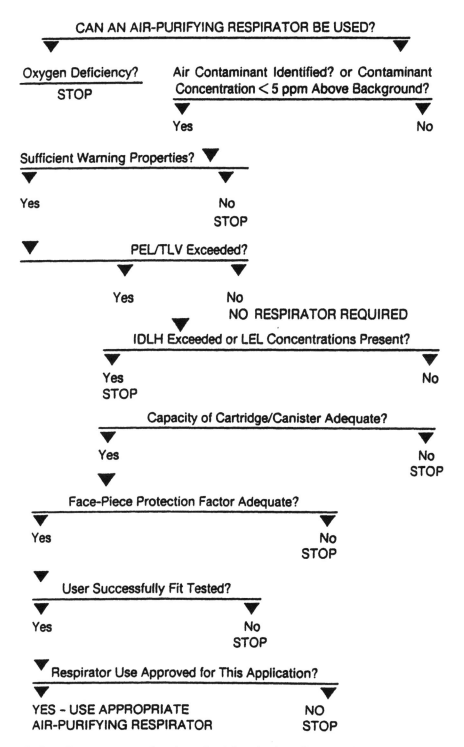

Figure 5–6. Respirator selection decision logic. (*Source:* Maslansky and Maslansky, *Water Well Journal,* December 1988.)

7. The respirator used has been successfully fit-tested by the user.
8. The user has been certified as medically fit and capable of wearing the respirator and performing assigned duties while using respiratory protection.
9. The work area is not an unmonitored confined space.

Oxygen Concentration in Air

Normal air contains 20.9% oxygen. The physiological effect of reduced oxygen begins to be evident at 16% oxygen in a resting individual. Regardless of the type of air contamination present, APRs may not be used in atmospheres containing less than 19.5% oxygen or in unmonitored confined spaces. In confined spaces or in oxygen-deficient atmospheres containing less than 19.5% oxygen, atmosphere-supplying respirators must be used.

Identification of Contaminants

Air-purifying elements are specific for certain types of air contaminants; therefore, the identity and the concentration of the hazardous materials present must be known, as must the efficiency of the respirator against the contaminants. The mere fact that a cartridge or a canister says it is for use against organic vapors does *not* mean that it is good for *all* organic vapors. Information on cartridge/canister efficiency can be obtained from the manufacturer.

It is imperative that the contaminants be identified so that:

1. Adverse health effects due to overexposure can be determined.
2. Appropriate particulate filters or cartridges/canisters can be selected.
3. It can be determined if adequate warning properties exist.
4. The appropriate facepiece (half-mask or full-face) can be selected.
5. The maximum use concentration ($MUC = PF \times$ exposure limit) can be determined.
6. The maximum use concentration for the selected APR can be determined (for organic vapors: 1,000 ppm for cartridges, 5,000 ppm for chin canister, 20,000 ppm for gas mask canister).
7. The expected service life of the purifying element can be determined.
8. Appropriate air monitoring devices can be selected and used.

Periodic Monitoring of Hazards

The identification of contaminants and their approximate concentration is an important consideration in determining the type and the frequency of air monitoring at the work area. Work-site monitoring must occur periodically or continuously during the workday. Monitoring is required to ensure that no significant changes have occurred, and that the respirators being used are adequate for the work conditions.

FIT-TESTING OF AIR-PURIFYING RESPIRATORS

As noted above, the use of an APR is prohibited when conditions prevent a good facepiece fit. These conditions include beards, large mustaches, long sideburns, scars, and eyeglass temple bars. Currently, the wearing of contact lenses is prohibited. Because maintaining a leak-free seal is important to the health and safety of the user, all personnel who wear negative pressure or demand-type respiratory protection should be fit-tested to verify the integrity of the seal of the particular make, model, and size of the device used.

It is prudent to fit-test on several respirators, including both half-mask and full facepiece types; this is especially important to do when a respirator becomes increasingly uncomfortable with continued use.

Procedures Prior to Fit-Testing

Prior to fit-testing, the respirator wearer should ensure that the facepiece and the head harness are in good condition, and that the cartridge or the canister is appropriate for the anticipated test atmosphere. Tests should be performed in an area with no noticeable air movement. Prior to testing, the wearer should be exposed to a very low concentration of the test substance to ensure that it can be detected by taste or odor; a very weak concentration of irritant smoke should be used so that the wearer can become familiar with its characteristic odor.

After the wearer dons the respirator, the tester should visually inspect the facepiece-to-face seal and the head straps. If the facepiece fit appears unsatisfactory, or if the wearer is uncomfortable, the mask should be removed and the test ended.

The wearer should perform a negative pressure test prior to the fit-test. The negative pressure test is performed by closing off the inlet of the canister or the cartridge by covering it with the palm of the hand, covering it with a latex or vinyl glove, or squeezing the breathing tube so that no air can pass into the facepiece. The wearer inhales gently so that the facepiece collapses slightly, and holds the breath for about 10 seconds. If the facepiece remains slightly collapsed, and no inward leakage is detected, this indicates that the fit is probably sufficient for the wearer to proceed.

The wearer then performs a positive pressure test, which is conducted by closing off the exhalation valve and exhaling gently into the facepiece. The fit is satisfactory if a slight positive pressure can exist inside the facepiece without evidence of outward leakage.

Qualitative Fit-Test Procedures

Fit-testing should be conducted in a room separate from the area used for odor and taste discrimination and respirator selection. The fit-test room should be well ventilated to prevent excessive contamination of room air with the fit-test substances.

After donning and adjusting the respirator, the wearer enters a transparent test chamber, such as an 85-gallon drum liner inverted over an inflexible frame, which is suspended over the wearer's head. The wearer stands in the bag and breathes normally for 30 to 60 seconds. If no leakage is detected, the wearer performs various exercises simulating, as nearly as possible, work conditions. These exercises include: turning the head from side to side; moving the head up and down; talking (reciting the Rainbow Passage); making facial movements, such as smiling and frowning; bending over; and jogging in place. The sequence of exercises and a copy of the Rainbow Passage (see below) should be affixed to the inside of the test chamber.

If the respirator wearer detects the test atmosphere, the respirator fit is unsatisfactory, and the wearer should exit the test chamber. If any doubt exists about the respirator or the cartridge, one should test a duplicate to assure that leakage was due to an inadequate face-piece-to-face seal and not a saturated cartridge or a damaged sealing surface.

Test Atmospheres

The most commonly used test atmosphere is isoamyl acetate (isopentyl acetate) or banana oil, which is used with organic vapor cartridges or canisters. Isoamyl acetate is available in small sealed ampules (similar to an ammonia ampule or smelling salts), which must be crushed prior to use. Isoamyl acetate also can be obtained as a pure liquid from a chemical distributor. Pure, undiluted isoamyl acetate should be used for fit-testing; approximately 0.75 cc should be introduced into the chamber by wetting an absorbent towel with the liquid or crushing an ampule.

Irritant smoke from a ventilation smoke tube containing stannic oxychloride is recommended for testing HEPA filters. Irritant smoke can be irritating to the eyes; so in testing half-mask respirators, airtight chemical goggles that do not interfere with the fit of the respirator should be used (we do not recommend keeping the eyes closed during the test to avoid eye irritation). To avoid unnecessary respiratory tract irritation, the respirator fit should not be tested with irritant smoke unless the wearer has already passed the banana oil test.

A solution of saccharin currently is the only recommended test material used for testing the fit of disposable dust respirators not equipped with HEPA filters. The saccharin test solution consists of 83 g saccharin dissolved in 100 cc warm water. A medication nebulizer is used to spray the solution into the test chamber through a small hole positioned directly in front of the wearer. To ensure an adequate concentration of saccharin aerosol, the test atmosphere should be replenished every 30 seconds.

The Rainbow Passage

This passage was developed to stimulate the range of facial expressions elicited by normal conversation during work activities. A copy of the Rainbow Passage accompanies the example of fit-test procedures given below.

Recording Results

The employer must keep a record of the fit-test and its results. The record for each employee tested should contain, as a minimum, the name of the employee, date of testing, name of tester, type of respirator tested (i.e., full-facepiece, half-mask, disposable dust mask), manufacturer and model number, size, test atmospheres used, type of test (i.e., qualitative or quantitative), and test results for each test atmosphere. A written protocol for testing also should be included, as well as pass/fail criteria. Special conditions or difficulties encountered should be noted.

Example of Fit-Test Procedure

APR FIT-TEST PROCEDURE**

1. Select an APR and ensure that it has organic vapor/particulate cartridges/canister.
2. Don the respirator, and adjust it until it is comfortable yet snug; use a mirror to make sure the mask is on properly and the straps are lying flat against your head.
3. Perform positive and negative pressure checks; if the mask leaks, readjust it and try again. If the mask is uncomfortable, select another and start over.
4. Enter the fit-test chamber containing isoamyl acetate (banana oil). *Do not* enter the irritant smoke chamber *unless* you have passed the banana oil test! Inside the test chamber:

   - Breathe normally for about 10 seconds.
   - Breathe deeply for about 10 seconds.
   - Move your head from side to side for about 10 seconds.
   - Move your head up and down for about 10 seconds (do not bump the mask on your chest).
   - Count or recite for at least 10 to 15 seconds, *or* recite the Rainbow Passage (see below).
   - Bend over and move your head around, and smile and frown, for at least 10 to 15 seconds.
   - Jog in place for 10 to 15 seconds.
   - Breathe normally for about 10 to 15 seconds.

5. If no odor is detected during the test, you pass. (If an odor detection test was not performed prior to the test, the tester must crack the mask slightly to ensure that the wearer actually can smell the banana oil.)
6. If you pass the banana oil test, enter the irritant smoke chamber and repeat Step 4. If you are testing a half-mask, use vaportight goggles to prevent eye irritation.

**Modified OSHA/USEPA method.

RAINBOW PASSAGE

When the sunlight strikes raindrops in the air, they act like a prism and form a rainbow. The rainbow is a division of white light into many beautiful colors. These take the shape of a long round arch, with its path high above, and its two ends apparently beyond the horizon. There is, according to legend, a boiling pot of gold at one end. People look, but no one ever finds it. When a man looks for something beyond reach, his friends say he is looking for the pot of gold at the end of the rainbow.

ATMOSPHERE-SUPPLYING RESPIRATORS

Atmosphere-supplying respirators or ASRs provide breathable-grade air, not oxygen, to the facepiece via a supply line from a stationary or portable source, or from a source carried by the wearer (Figure 5–7). When air is supplied from a source through a long airline or hose, it is called a supplied-air respirator or an SAR; when the air source is carried by the wearer, it is called *a self-contained breather* apparatus or an SCBA (Figure 5–8).

Types of Atmosphere-Supplying Respirators

Several types of ASRs are available. Although not all types are used at hazardous waste sites, it is worthwhile to understand the differences between them.

• Closed circuit rebreathers use a chemical reaction or an oxygen supply to replenish oxygen in the user's exhaled breath. Excess carbon dioxide and water vapor are removed, and reoxygenated air then is returned to the user. Closed-circuit rebreathers are often used in situations where extended use is required.

• A hose mask respirator consists of a facepiece attached to a large-diameter hose no more than 75 feet long. The hose transports clean air from a distant source. The wearer breathes the air, or it is forced in by a blower.

• An airline respirator is similar to a hose mask except that air is delivered to the wearer under pressure through a high pressure hose (Figure 5–9); the hose length usually is lim-

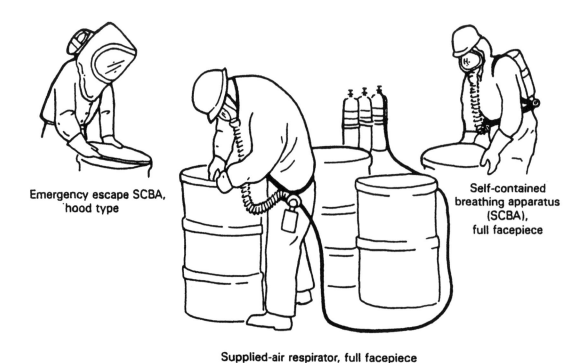

Emergency escape SCBA, hood type

Self-contained breathing apparatus (SCBA), full facepiece

Supplied-air respirator, full facepiece

Figure 5–7. Types of atmosphere-supplying respirators. (*Source:* NIOSH, 1985.)

Figure 5-8. An open-circuit SCBA (self-contained breathing apparatus).

ited to 300 feet. An air compressor or a compressed air cylinder is the source of breathable air (Figure 5–10). The airflow may be continuous, demand, or pressure-demand.

• A self-contained breathing apparatus or an SCBA consists of a facepiece and a regulator connected to a cylinder of compressed air that is carried by the user. The airflow in SCBAs may be demand or pressure–demand. There are two basic types of SCBA: closed-circuit and open-circuit.

Closed-Circuit SCBA

The closed-circuit SCBA or rebreather was developed for oxygen-deficient situations where the use of heavy, bulky cylinders of compressed air is not feasible, as, for example, in mines and on board ships. Because the rebreather recycles exhaled breath and carries only a small oxygen supply, the service time can be considered greater than that of an open-circuit device, which must supply all of the breathing air.

Air for breathing is mixed in a flexible breathing bag; when air is inhaled, the bag deflates. The deflation depresses the admission valve, allowing oxygen to enter the bag. Exhaled air then enters the bag; while excess carbon dioxide is removed, the oxygen mixes with exhaled breath.

Although most closed-circuit SCBAs operate in the demand mode, several rebreathers are designed to provide positive pressure in the facepiece. Regardless of the type of airflow delivered to the facepiece, all closed-circuit SCBAs are approved as demand-type respirators.

Rebreathers use either compressed oxygen or liquid oxygen. To assure the quality of air to be breathed, the oxygen must be at least medical-grade breathing oxygen, which meets all requirements as defined in the *U.S. Pharmacopeia*.

Closed-circuit SCBAs tend to generate heat when carbon dioxide is scrubbed from exhaled air. Unlike the case of open-circuit SCBAs, air entering the facepiece is not cool; heat stress is often a greater problem than oxygen deficiency in using rebreathers. Some closed circuit systems have cooling systems to minimize heat generation.

Oxygen-generating respirators (chemical rebreathers) use a canister of potassium superoxide to replenish the oxygen in the user's exhaled breath; potassium superoxide reacts with exhaled carbon dioxide and water vapor to produce oxygen. The reoxygenated air is then returned to the user. Oxygen-generating respirators have been used by the military and for escape purposes in mines. They generally are not used in hazardous materials applications because of the chemical reaction taking place within the respirator itself.

Open-Circuit SCBA

The open-circuit SCBA requires a supply of compressed breathing air. The user simply inhales and exhales; air is inhaled once, and exhaled air is exhausted from the system. Because the air is not recycled, the wearer must carry a full air supply; the size and the weight of the air cylinder limit the amount of air that can be easily carried. Open-circuit SCBAs can last from 5 to 60 minutes; units with 5- to 15-minute supplies are used only for escape purposes.

Figure 5-9. This worker is dragging the hose of his airline respirator. His facepiece also is equipped with cartridges, to allow him to work in other areas where supplied air is not required. The small cylinder at his side contains an emergency supply of breathable air.

Figure 5-10. A manned airline supply station. A low pressure alarm system signals the attendant to switch over to the next tank.

Components of an Open-Circuit SCBA

A cylinder containing 45 cubic feet of Grade D air, at a pressure of 2216 psi, is needed for a 30-minute air supply; 90 cubic feet at a pressure of 4500 psi is required for a 60-minute supply. Cylinders are filled by using a compressor or a cascade system of several large cylinders of breathing air. The cylinder pressure is indicated on a cylinder gauge at the cylinder valve, and should be accurate within ±5%. If the cylinder is overfilled, a relief diaphragm releases the pressure.

A low pressure warning alarm is located near the cylinder valve. The alarm sounds when 75 to 80% of the air supply has been consumed, to alert the user that only 20 to 25% of the original supply is available for retreat. Entry into a hazardous waste site should consume no more than 20% of the initial supply, to allow enough for exit when the alarm sounds.

A high pressure hose connects the cylinder and the regulator. The hose should be connected to the cylinder only by hand, never with a wrench. An O-ring inside the connector assures a good seal. (It is always a good idea to carry spare O-rings.) Air travels from the cylinder through the high pressure hose to the regulator. At the regulator it can take one of two paths: if the bypass valve is open, the air "bypasses" the regulator and travels under high pressure directly to the facepiece. If the mainline valve is open, the high pressure air passes through the regulator, where the air pressure is reduced to approximately 50 to 100 psi. The mainline valve may be external and controlled by the user, or it may be internal.

In a pressure-demand SCBA, a spring keeps the admission valve partly open so there is always positive pressure into the facepiece; during inhalation, the valve opens completely to allow more air to enter the facepiece.

Some SCBA models have the regulator in the facepiece; other have a low pressure breathing hose that connects the facepiece to the regulator. The hose is usually made of neoprene and is corrugated to allow stretching. A check valve prevents exhaled air from entering the breathing hose.

The facepiece is normally constructed of neoprene or silicone rubber. The head harness usually has a five- or six-point suspension; some manufactures use a net-type head harness instead of straps. The visor lens is made of polycarbonate or other clear material. An exhalation valve is located at the bottom of the facepiece; an airtight speaking diaphragm may also be present to facilitate communication.

A back-pack and a harness support the cylinder and the regulator, allowing the user to move freely. The weight of the cylinder should be supported on the hips, not on the shoulders.

The compressed air cylinder of an SCBA must meet DOT specifications as defined in *General Requirements for Shipments and Packaging* (49 CFR 173), and *Shipping Container Specifications* (49 CFR 178). A hydrostatic test must be performed on a cylinder at regular intervals: on all steel or all aluminum, every 5 years; on composite glass fiber and aluminum, every 3 years. Composite tanks have a short service life and must be taken out of service after 15 years.

Breathing Air Quality

Compressed air used in atmosphere-supplying respirators must be of high purity. Compressed air should at least meet requirements for Type I, Grade D breathing air, as described by the Compressed Gas Association (CGA). Grade D air must contain between 19.5 and 23.5% oxygen, with the balance being predominantly nitrogen. Contaminant concentrations are limited to 5 mg/m^3 hydrocarbons, 20 ppm carbon monoxide, and 1000 ppm carbon dioxide. An undesirable odor is prohibited. Air quality can be checked by using an oxygen meter, a carbon dioxide monitor, and detector tubes. Grade E through Grade H air is also acceptable, with Grade H (hospital grade) having the best quality.

RESPIRATOR USE AND SELECTION

User Requirements

Users of respirators should be aware that wearing a respirator has limitations and is not a substitute for following good work practices. When respirators are necessary, specific procedures are required to overcome potential deficiencies and ensure equipment effectiveness.

A variety of agencies and organizations have prepared standards or guidelines on respiratory protection practices. The American National Standards Institute (ANSI) developed an *American National Standard for Respiratory Protection-Z88*. Z88.2-1980 was issued in 1980 as a voluntary standard; proposed revisions to ANSI Z88.2-1980 were released in 1988. The final version, ANSI Z88.2-1992, was published in 1992. The ANSI standard addresses all phases of respirator use and is recommended as a guide to respiratory protection.

OSHA used the ANSI standard as a source of information for its respiratory protection regulation (29 CFR 1910.134) issued in 1975. The Mine Safety and Health Administration (MSHA) also cites ANSI Z88.2-1980 in its regulations (30 CFR 11.2-1), stating that "in order to insure the maximum amount of respiratory protection, approved respirators will be selected, fitted, used, and maintained in accordance with the provisions of the American National Standards Practices for Respiratory Protection, Z88.2."

OSHA, MSHA, and ANSI all require a minimum respiratory protection program to ensure sound respiratory protection practices. The regulations and the standards also discuss specific requirements for respirator use.

Respiratory Protection Program

Employers are responsible for establishing an effective respiratory protection program, and employees are responsible for wearing the respirators and complying with the program. In large companies, the industrial hygiene, safety, or medical department often administers the program. In small companies or plants, the program may be administered by a supervisor, safety manager, or foreperson. Responsibility for the program should rest with one person; that person should have sufficient knowledge of the subject to properly administer and supervise the program.

Any respiratory protection program should stress training. Workers must understand that using a respirator does not eliminate the hazard. If the respirator fails, overexposure can occur. To reduce the possibility of failure, respirators should fit properly and be kept clean and serviceable.

An effective respiratory protection program should include the following:

- Written standard operating procedures (SOPs) governing selection and use of respirators.
- Selection of respirators based on the hazards to which the workers will be exposed.
- Instruction and training in the proper use of respirators and their limitations.
- Procedures for regular inspection, cleaning, maintenance, and storage.
- Medical examinations to determine if personnel are physically able to perform work and use respirators. This includes a periodic review of the medical status of each worker.
- Observation and supervision of work area conditions and work practices, and the extent of employee exposure or stress. This includes periodic monitoring of the workplace atmosphere.
- Procedures to ensure that atmosphere-supplying respirator equipment meet minimum requirements for breathable air.
- A guarantee that only MSHA or NIOSH approved or accepted respiratory protection devices will be used.

Respirator Selection

Selecting the right equipment involves several steps: determining what the hazard is and its extent, choosing equipment that is appropriate for the hazard, and assuring that the selected equipment is functioning as intended. A detailed list of selection criteria includes:

- Nature and concentration of contaminants present.
- Chemical and physical properties of contaminants.
- Toxicity of contaminants.
- Concentration of oxygen present.
- Characteristics of work activities.
- Location of the work area relative to the noncontaminated area with its breathable air.
- Period of time for which respiratory protection must be used.
- Level of activity of workers who must use respirators.
- Physical characteristics, functional capabilities, and limitations of available respirators.
- Respirator protection factors (PFs).
- Exposure limit and IDLH of known contaminants.

Respirator Approvals

Both OSHA (29 CFR 1910.134) and MSHA (30 CFR 11) require the use of approved respirators. Respirators are tested at the National Institute for Occupational Safety and Health (NIOSH) Testing Laboratory in Morgantown, West Virginia, and are jointly approved by MSHA and NIOSH if they pass the requirements of 30 CFR Part 11.

MSHA/NIOSH approval indicates that the respirator in use is identical to the one submitted for original approval. If a manufacturer changes any part of the respirator without resubmitting it for testing, the approval is invalid and will be rescinded. This provision is intended to protect the user from manufacturer changes that may affect the fit or the function of the respirator. Any unauthorized changes or hybridization of a respirator (as when parts of different respirators are interchanged) by the user also voids the respirator approval and all guarantees understood with the approval.

The approval number must be displayed on the respirator or its container. The entire number includes the prefix TC (Testing and Certification) and the schedule number, followed by the approval number. The approval label that is used also indicates the certifying agencies.

Periodically, NIOSH publishes a list of all approved respirators and respirator components. The current edition, issued in 1994, is entitled the *NIOSH Certified Equipment List as of September 30, 1993*. This document is used to determine if the respirator considered for use is approved for the anticipated work conditions, and if the respirator itself (mask and purifying elements) is an approved assembly. If either the use or the assembly is not approved, the respirator cannot be used.

An initial examination of the NIOSH approval schedule allows the user to determine the type of protection provided (Table 5–9).

On June 8, 1995, NIOSH published a final rule (42 CFR 84) changing certification requirements for particulate respirators (this action being the first in a series expected to revise approvals for all respirators). Also on June 8, 1995, MSHA transferred the requirements and authority for respirator approval from 30 CFR 11 to NIOSH. NIOSH and MSHA will continue to jointly review and approve respirators used for mine rescue and emergency response. Except for particulate respirators, most requirements for existing approvals under 30 CFR 11 remain unchanged. NIOSH is allowing three years for phasing out 30 CFR 11. OSHA is currently revising its respirator use standard (29 CFR 1910.134) and is allowing the continued use of particulate respirators approved under the old standard.

RESPIRATOR CARE AND MAINTENANCE

All users of APRs should receive instructions and practice on how the respirator should be worn, how to adjust it, how to determine if it fits properly, and how to perform positive and negative fit checks each time that it is worn. A program should be in place to ensure that all respirators are routinely inspected, cleaned, and properly stored.

Changing Cartridges/Canisters

Cartridges and canisters containing chemical sorbents should not be removed from protective packaging until needed. Once opened, they should be used immediately; efficiency and service life decrease because sorbents begin to absorb humidity and air contaminants when not in use. Cartridges should be changed regularly to prevent sorbent exhaustion and breakthrough. When not in use during the workday, as during breaks or intervals when a respirator is not required, the respirator, with air-purifying element attached, should be enclosed in a clean, sealed plastic bag.

The rule of D's is recommended for changing cartridges or canisters; that is, they should be changed at least *daily*, or more often if necessary, and *discarded* if *any* of the following conditions exists:

- Warning properties are *detected* by the wearer.
- It becomes *difficult* to breathe.
- The air-purifying element appears *dirty* or *damaged*.

Table 5-9. NIOSH approval schedule for respirators

Respirator type	Approval schedule
Self-contained breathing apparatus	TC-13F-
Gas masks (canister full-face APRs)	TC-14G-
Supplied-air respirators	TC-19C-
Dust, fume, and mist APRs	TC-21C-
Chemical cartridge APRs	TC-23C-
Vinyl chloride APRs	TC-11-

Used cartridges or canisters should be discarded immediately if any of these conditions exists at any time during respirator use. At the end of the workday, all air-purifying elements should be removed and discarded, prior to inspection, cleaning, and disinfecting of the facepiece.

Respirators should be inspected before use; make sure that all gaskets and O-rings are in place, and connectors are tight. Check the condition of the facepiece and all parts, including headbands, harnesses, connecting tubes, and hoses for pliability and signs of wear or deterioration.

Cleaning and Disinfecting

Clean and disinfect respirators after each use. Remove all air-purifying elements from the APR, and discard them; disconnect the facepiece and the hose from the SCBA regulator. Remove gaskets, seals, or O-rings, and set them aside. Remove and set aside the exhalation cover, the speaking diaphragm, and inhalation valves. Wash the facepiece and the breathing hose in a cleaner/sanitizer mixed with warm water. Wash the components separately. Remove all parts from the wash water, and rinse them twice in clean warm water. Air-dry all parts in a clean area. Wipe all pieces with a lint-free cloth to ensure dryness prior to reassembly. An antifog solution, which prevents exhaled breath from condensing inside the facepiece, should be applied to the inside lens of the facepiece after cleaning.

Most respirator manufacturers market their own cleaner/sanitizers as dry mixtures of a germicidal agent plus a mild detergent. Sanitizer mixtures are usually available in individual packets or in large, bulk packages. Disposable sanitizer towels are used to clean only the inside of the facepiece and the nosecup. Thorough cleaning and disinfection require disassembly and submersion of all respirator components in a sanitizing solution.

Maintenance and Storage

Repair and replacement of parts is acceptable as long as procedures are followed and parts used according to the manufacturers' recommendations. Never attempt to make unauthorized repairs or replacements.

Follow manufacturers' storage instructions. Respirator facepieces should be stored in a clean, sealed bag after cleaning; storage conditions should protect against dust, excess moisture, damaging chemicals, extreme temperatures, and direct sunlight. Storage conditions also should prevent distortion and damage of rubber or other flexible parts. Do not store respirators in clothes lockers, bench drawers, or tool boxes (Figure 5–11).

Inspection

Respirators should be inspected at regular intervals, even when not in use. A checklist often is helpful to ensure that inspections are completed. All respirators, regardless of type, should be inspected before and after each use, at least monthly when in storage, and every time they are cleaned or serviced.

Figure 5–11. An example of how *not* to store a respirator! The facepiece should be sanitized and placed in a protective bag or container. Cartridges should be discarded on a daily basis.

Specific checklists should be used for SCBAs, SARs, and APRs to ensure that all vital components are working properly:

Checklist for SCBAs:

- All connections for tightness.
- All straps and harness/belts for wear or deterioration.
- Regulators and valves for proper operation.
- Alarms for proper operation.
- Faceshield and lens for cracks and fogging.
- Manufacturer's expiration date of air bottle (if composite).
- Hydrotest date of air bottle.

Checklist for SARs and egress/escape bottles:

- Airlines for cracks, kinks, cuts, fraying, weak areas.
- Regulators and valves for proper operation.
- All connections for tightness.

- Alarms for proper operation.
- Faceshield and lens for cracks and fogging.
- Manufacturer's expiration date of egress/escape bottle.
- Hydrotest date of egress/escape bottle.

Checklist for APRs:

- Straps and harness for signs of deterioration or wear.
- Cleanliness of interior and exterior of mask.
- Faceshield and lens for cracks and fogging.
- Inhalation and exhalation flap valves.
- Cartridges for adequacy for intended use.
- Cartridges to see that they have not been previously opened or used.

6

Air Monitoring

Monitoring for air contamination is an essential component of hazardous waste site work. Air monitoring can provide information on the type and the relative amount of hazards present; and this information can be used to assess real or potential risk to site workers, the general public, and the environment. Results of air monitoring also can be used to help workers select appropriate personal protective equipment, delineate work zones or areas where protective equipment is required, identify site locations in need of remediation, and evaluate the effectiveness of contamination control and remediation procedures.

Air monitoring is an important aspect of the site characterization process required by the OSHA HAZWOPER rule covering hazardous materials site workers. Primary monitoring at uncontrolled sites must include detectors for ionizing radiation, flammable or explosive atmospheres, oxygen deficiency, and toxic substances. The results of initial monitoring will determine the type of personal protective equipment (PPE), including respiratory protection, that is appropriate for the particular site. An ongoing air monitoring program must be established to ensure continued proper selection of PPE, and to provide a means of assessing contamination controls and workplace practices.

As a minimum, periodic air monitoring must be performed when work begins at a different portion of the site, a different type of activity is begun (e.g., sampling drums as opposed to sampling soil), another type of contamination than that originally encountered is handled (e.g., vapors or gases emitted from a borehole), or workers are in an area of obvious contamination (e.g., lagoons or holding ponds, leaking drums, spill areas).

Finally, air monitoring results can eliminate questions or allegations of worker overexposure, as well as provide documentation of adherence to environmental and occupational regulations. To be so effective, however, the air monitoring be conducted by trained personnel who faithfully adhere to the requirements specified in the site air monitoring program. Workers operating monitoring equipment should be thoroughly trained in the use, operating characteristics, and limitations of each piece of equipment.

CHARACTERISTICS OF AIR MONITORING INSTRUMENTS

Instrument Design

The monitoring instruments selected for use in the field should be capable of generating reliable and useful information. They should be able to detect the contamination of interest and be sensitive at a useful concentration range; and they should be lightweight and portable, sturdy, compact, weather- and temperature-resistant, and simple to operate and maintain. An instrument should have a handle and a shoulder strap, an easily readable display, convenient controls, a variety of useful accessories, and audible and visible alarms. The ideal instrument should have a long battery life, a short warm-up time, and fast, easy procedures for calibration and sensor replacement. Finally, the instrument should be inherently safe to use in atmospheres containing flammable gases and vapors.

The components of a typical air monitoring instrument are represented in Figure 6–1. Many instruments are equipped with a battery-powered pump or a hand aspirator that draws sample air into the area where the sensors are located; these devices are called sample draw instruments. The alternative is diffusion instruments, which have internal or external sensors that must be physically present in the atmospheres being sampled.

Instrument Response

Every instrument is designed to respond to a particular type of hazard within a specific concentration range, which is often called the detection range or the operating range. For example, oxygen meters usually have a detection range between 0 and 25% oxygen by volume; carbon monoxide and hydrogen sulfide sensors are sensitive to ppm concentrations in air. The lower detection limit is the lowest concentration that the meter will respond to accurately; the upper detection limit is the concentration that saturates the sensor or the detector and elicits maximum meter response.

Response time is the interval required to obtain a sample and attain approximately 90% of the final response; direct reading instruments, which require little or no integration or interpolation, have response times of several seconds to several minutes. The lag time is the interval between introduction of the sample air and the first observable meter response. The recovery time is the time required for the instrument to return to zero or a normal ambient reading after removal from the sample atmosphere.

Sensitivity of an instrument is defined by the detection limit, or how small a concentration can be accurately and reliably detected. Sensitivity is important when very small changes in concentration may be hazardous. Selectivity determines what type of materials will be detected (e.g., combustible gases, oxygen, carbon monoxide, chlorine, organic vapors). Many instruments are not selective and require interpretation by the operator.

Precision is a measure of the reproducibility and the reliability of an instrument response to a known concentration. An instrument that repeatedly produces the same response or gives readings that are very close to each other is considered precise. Precision and accuracy are not the same thing (Figure 6–2). Accuracy is a measure of how close the instrument reading is to the actual concentration present. An instrument is considered accurate when

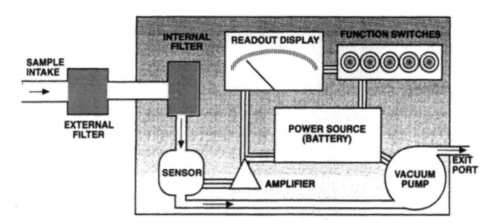

Figure 6–1. Schematic of a typical air monitoring instrument. (*Source:* Maslansky and Maslansky, 1993. Courtesy of Van Nostrand Reinhold.)

Figure 6–2. The difference between precision and accuracy can be demonstrated by using a bull's-eye. (*Source:* Maslansky and Maslansky, 1993. Courtesy of Van Nostrand Reinhold.)

the average of all readings falls within an acceptable, predetermined range about the true value (e.g., ±10–20%).

Inherent Safety

Instruments that require battery power or rely on a chemical reaction to obtain a response may be a source of ignition in the presence of a flammable atmosphere. Instruments should be constructed in such a way as to eliminate the possibility of igniting flammable gases or vapors. Any instrument operated in an area where flammable conditions are anticipated should be certified as intrinsically safe, explosion-proof, or purged. Explosion-proof instruments allow flammable gases to enter but have an enclosure designed to contain any explosion so that it does not spread. Intrinsically safe instruments are designed to be incapable of releasing sufficient energy to cause ignition of specific types of flammable gases and vapors. Pressurized or purged systems use a stream of inert gas to keep the flammable atmosphere from reaching the ignition source.

Standards for inherent safety in hazardous atmospheres are defined by the National Fire Protection Association (NFPA) in its National Electric Code (NEC). The code defines types of controls acceptable for use in hazardous atmospheres, locations where hazardous atmospheres may be generated, and the types of materials that can generate such atmospheres. Categories of hazardous atmospheres are defined by Class, Group, and Division.

Class describes the type of flammable material that produces the hazardous atmosphere. Class I represents flammable gases and vapors; Class II consists of combustible dusts such as flour, grain, and coal dust; and Class III is reserved for ignitable fibers.

Class I flammable gases and vapors are separated into Groups A through D (Table 6–1). Group A contains only one gas, acetylene; Group B contains gases and vapors with wide flammable ranges, such as butadiene, ethylene oxide, formaldehyde, hydrogen, and propylene oxide; Groups C and D contain a large number of gases and vapors.

Division is the term used to describe the location of the hazardous atmosphere. A Division 1 location is any area where a flammable atmosphere may be generated or released on a continuous, intermittent, or periodic basis, into an open or unconfined area; that is, flammable gases and vapors are anticipated to be present at any or all times. A propane filling station, a gasoline terminal, and a drum filling facility that handles flammable liquids would all be classified as Class I, Division 1 locations. A Division 2 location is an area where flammable gases or vapors are not anticipated under normal conditions; the generation or the release of a flammable atmosphere would occur only as a result of rupture, leak, or other failure of a closed system or container. An abandoned waste site containing intact drums of flammable liquids is a Division 2 location; if the drums are leaking, it is a Division 1 location.

Intrinsically safe instruments are approved for Division 1 locations, whereas nonincendive meters are approved only for Division 2 locations. Certification is awarded only for specific atmospheres; the instrument receiving it is not certified for use in atmospheres other than those indicated. Division 1 locations are more hazardous than Division 2; devices approved for Division 1 locations are also approved for Division 2. The highest level of inherent safety for flammable gases and vapors is Class I, Division 1, Groups A, B, C, D.

Instruments used in hazardous atmospheres should be certified as safe by Underwrit-

Table 6–1. Class I flammable atmospheres

GROUP A	GROUP D
Acetylene	Acetone
	Acrylonitrile
GROUP B	Ammonia
	Benzene
Butadiene	Butane
Ethylene oxide	Cyclohexane
Hydrogen	Ethyl acetate
Formaldehyde gas	Ethyl alcohol
Propylene oxide	Gasoline
Manufactured gases with more	Heptane
than 30% hydrogen	Hexane
	Isopropyl alcohol
GROUP C	Methane
	Methyl alcohol
Acetaldehyde	Methylethyl ketone (MEK)
Carbon monoxide	Natural gas
Crotonaldehyde	Petroleum naphtha
Cyclopropane	Pentane
Diethyl ether	Propane
Ethylene	Propylene
Hydrogen sulfide	Styrene
Methyl mercaptan	Toluene
	Xylene

ers Laboratories, Inc. (UL) or Factory Mutual Research Corp. (FM). The instrument should carry a permanent plate or other marking showing the logo of the testing laboratory as well as the approved class, division, and groups.

The Mine Safety and Health Administration (MSHA) tests and certifies electrical devices to be used in hazardous atmospheres associated with the mining industry. MSHA testing and certification are specific to only to atmospheres containing coal dust and methane gas.

The European Committee for Electrotechnical Standardization (CENELEC) requirements for construction and testing of instruments are similar to those of the NFPA/NEC. Certification that an instrument meets CENELEC standards may be awarded by a variety of European testing laboratories; certified instruments also carry an approval plate.

Instrument Calibration and Relative Response

Calibration is a process of adjusting the instrument response until the reading corresponds to the actual concentration present. Initial calibration is performed at the factory by using multiple concentrations of one particular vapor or gas, the calibration gas. After calibration, the instrument will respond accurately to the calibration gas within its detection range. The instrument will not, however, respond accurately to other gases and vapors. Instead, the meter will give a relative response reading, which may be higher or lower than the actual concentration present. When detecting gases and vapors other than the calibration gas, the meter reading is often expressed as calibration gas equivalents, response equivalents, equivalent units, or simply units.

For some types of instruments, the manufacturer may provide relative response curves or conversion factors for gases or vapors other than the calibrant gas. Such curves or factors can be used to determine the approximate actual concentration present.

In most cases, the instrument user does not actually calibrate the instrument but rather checks to ensure that it still maintains its factory calibration. The calibration check is performed with a single concentration of calibrant gas. If the instrument responds correctly to that concentration, the user can be reasonably assured that it will accurately respond to other concentrations within its detection range. The field calibration check is the *only* means of demonstrating that the instrument is working properly; if the meter does not respond appropriately to the calibration gas, it should be adjusted until an accurate reading is obtained. If an accurate reading cannot be obtained, the instrument should not be used. The calibration check should be performed prior to instrument use.

FACTORS THAT CAN AFFECT INSTRUMENT FUNCTION

A variety of factors can influence instrument function, and the user should always consider them during instrument operation and before interpreting meter responses.

Nature of the Hazard

One should use all reliable information about the nature of the hazard in order to make an informed decision about what instrument to use, as well as how, when, and where monitoring is conducted. The types of information available may include chemical/physical properties, estimated concentrations of contaminants, potential health and safety hazards, and how the instruments are likely to respond.

Interferences

Many hazardous gases and vapors can interfere with proper instrument function. Such interferences can result in changes in instrument sensitivity, which can lead to inaccurate (high or low) meter readings. For example, water vapor and relatively low concentrations of methane gas can interfere with photoionization detector readings; lead, silicates, and sulfates can desensitize the catalytic filament in %LEL combustible gas indicators. Methyl alcohol can elicit a false positive response from a carbon monoxide electrochemical sensor. Many gases and vapors can cause inaccurate responses in detector tubes.

A calibration check can verify that an instrument is responding properly to the calibration check gas; it does not, however, guarantee an accurate result during monitoring on-site. It is up to the user to determine if interfering compounds are present by comparing the response of one type of instrument to that of another.

Environmental Conditions

Environmental conditions may affect instrument function as well as the dispersion of hazardous materials. Humidity, temperature, barometric pressure, elevation, oxygen concentration, the presence of electromagnetic fields, static electricity, and the presence of interfering gases and vapors are among the common conditions known to affect instrument response. For example, some instruments are severely affected by high humidity; a photoionization detector will lose sensitivity because water vapor decreases ionization, and humidity interferes with the chemical reaction in some detector tubes. High voltage and static can affect the readout display on some instruments. Cold weather decreases battery life. It is important to review the instruction manual for conditions that affect instrument operation.

The Instrument Operator

Although someone should be responsible for overall instrument maintenance and upkeep, most personnel should be able to use each instrument, interpret readings, and troubleshoot problems in the field. Users often can choose between instruments that are comparable in terms of instrument response. In these cases, it is often better to get the simpler instrument that is easier to operate and interpret rather than a complex meter that can be used by only one or two individuals.

COMBUSTIBLE GAS INDICATORS (CGIs)

Combustible gas indicators, also called CGIs, explosion meters, or explosimeters, are used to test atmospheres that may contain a sufficient level of combustible gases or vapors to produce an explosion or support combustion. Three different scales or detection ranges are used on CGIs: ppm, %LEL, and %GAS.

%LEL Meters

The most common CGI is the %LEL meter, which has a detection range between 0 and 100% of the LEL. The LEL, or lower explosive limit, is the lowest concentration of a substance by volume in air that will explode or flash over in the presence of a source of ignition. Nearly all %LEL CGIs rely on catalytic combustion of gases on a filament.

The %LEL sensor contains two filaments. One filament is coated with a catalyst, which facilitates oxidation or burning of very low concentrations of combustible gases. The other filament has no catalyst and is called the compensating filament because it compensates for ambient conditions such as temperature and humidity. The catalytic filament and the compensating filament are incorporated into a Wheatstone bridge (Figure 6–3).

The battery supplies current to the Wheatstone bridge circuit and heats the two filaments; both filaments are heated to the same temperature, and they have the same resistance. Combustible gases burn on the catalytic filament but not on the compensating filament. This combustion causes the catalytic filament to increase in temperature, with a consequent increase in its resistance. The increase in resistance unbalances the Wheatstone bridge circuit, and the change is translated into a meter reading (Figure 6–4).

All CGI readings are relative to the calibration gas, which is usually either methane, pentane, or hexane. The meter readings correspond to the relative increase in resistance produced by the calibration gas when it burns on the catalytic filament. The meter reading, then, represents how hot the filament gets as the gas interacts with the catalyst and burns.

Some gases and vapors release more heat than the calibration gas releases during burn-

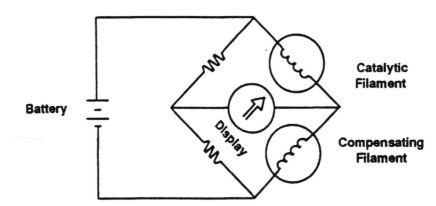

Figure 6-3. A Wheatstone bridge circuit. (*Source:* Maslansky and Maslansky, 1993. Courtesy of Van Nostrand Reinhold.)

ing; they are considered hot burning gases. When these materials are present, the filament becomes hotter at a lower concentration and gives a %LEL reading that is greater than their actual concentration. Cool burning gases, on the other hand, release less heat; a higher concentration is therefore required to heat the filament, so that the meter reading is less than their actual %LEL.

Many manufacturers provide response curves or conversion factors that indicate the meter response to individual gases or vapors throughout their LEL range (i.e., 0–100% LEL). The curves depicted in Figure 6–5 are response curves for the calibration gas and four other

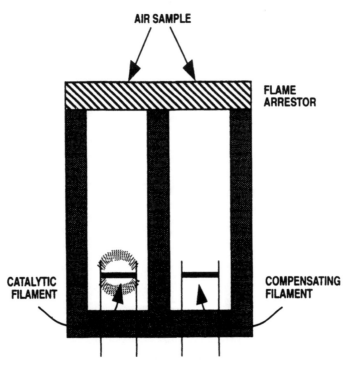

Figure 6-4. Line drawing of a %LEL combustible gas sensor. Combustion occurs only on the catalytic filament; there is no combustion on the compensating filament. The difference in temperature between the two filaments produces a meter reading. (*Source:* Maslansky and Maslansky, 1993. Courtesy of Van Nostrand Reinhold.)

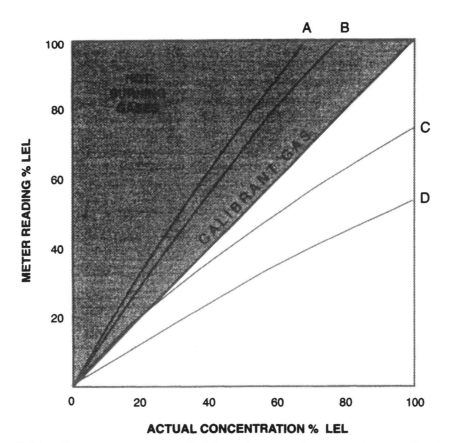

Figure 6-5. Response curves of different gases for a pentane-calibrated combustible gas indicator. Gases A and B are hot burning and lie above the calibrant gas curve. Gases C and D are cool burning and fall below the calibrant gas curve. (*Source:* Maslansky and Maslansky, 1993. Courtesy of Van Nostrand Reinhold.)

materials. If the identity of the gas or the vapor is known, response curves can be used to determine the approximate actual %LEL concentration present. For example, a meter reading of 50% LEL for Gas A represents an actual concentration of approximately 30% LEL.

A conversion factor is the number that best describes the meter response to a particular gas or vapor. Factors are usually carried out to one decimal place (e.g., 2.1, 1.6, 0.5). To obtain the approximate actual concentration, the user usually multiplies the meter reading by the conversion factor:

$$\text{Meter reading} \times \text{Conversion factor} = \text{Actual concentration}$$

Limitations of %LEL Meters

Oxygen is required for proper functioning of the %LEL sensor, as it is needed for combustion of the gas on the catalytic filament. The concentration of oxygen required varies by manufacturer and should be indicated in the instructions. Oxygen-deficient atmospheres will result in a lower reading or no reading at all; oxygen enrichment will result in inaccurately high readings and may damage the sensor.

The catalytic filament is susceptible to contaminants such as sulfur compounds, silicon compounds, and organic metals such as tetraethyl lead. These materials, when burned, form fumes that coat the filaments; and eventually the filaments no longer reach the proper temperature, or the catalyst becomes covered. When this happens, gases will no longer burn on the catalytic filament and the sensor will have to be replaced. Because the filament is so susceptible to damage, it is important to perform a calibration check before every use; it is also prudent to check the calibration after use as well, particularly if a meter is "going back on the shelf."

%GAS Meters

%GAS CGIs are used to measure the concentration of combustible gas present at concentrations greater than 100% LEL. These instruments measure concentration in percent by volume, up to 100%. Locations that may require a %GAS meter include gas pipelines, gas distribution facilities, tank farms, and landfills. It is not unusual to find methane concentrations of up to 40 to 60% by volume at landfills.

%GAS meters use a thermal conductivity (TC) filament sensor rather than the catalytic filament sensor employed in the %LEL meter. The TC sensor has two heated filaments, the TC filament and the compensating filament. Gases and vapors can only enter the TC filament side of the sensor; the compensating filament is sealed and remains at a fixed, constant temperature (Figure 6–6). Both filaments are incorporated into a Wheatstone bridge circuit. Incoming gases and vapors cool the TC filament; as the temperature of the filament decreases, the resistance also decreases, resulting in a meter response. In %GAS CGIs, the meter reading represents how cool the TC filament gets relative to the hot compensating filament. Because burning or oxidation is not involved, the %GAS CGI does not require oxygen for a valid reading.

Limitations of %GAS Meters

Gases and vapors differ in the capacity to cool or absorb heat from the TC filament. Calibration gases for %GAS CGIs, such as methane, propane, and natural gas, absorb heat relatively well and readily cool the filament. Some gases, however, will cool the filament more than the calibration gas will; the meter will then indicate a higher %GAS concentration than is actually present. On the other hand, some gases are poor heat absorbers and do not cool the filament as well as the calibration gas; these gases will indicate a lower %GAS concentration than is actually present. Major interfering gases, which will give a false positive reading on a %GAS CGI, include carbon dioxide and freon gases used as refrigerants. Nitrogen, often used as an inerting gas, does not affect the TC sensor and will not produce a meter response.

Low Concentration CGIs

The sensitivity of a CGI to detect and reliably measure ppm concentrations depends primarily on the type of sensor. Catalytic filament sensors are usually sensitive to concentrations as low as 0.5 to 1.0% LEL of the calibrant gas; or, for a pentane-calibrated CGI, approx-

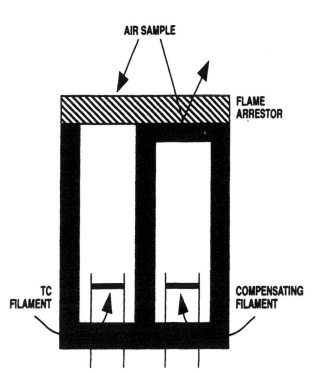

Figure 6-6. Line drawing of a %GAS sensor. Cooling occurs only on the thermal conductivity (TC) filament; the compensating filament is sealed and does not come in contact with the test atmosphere. The decrease in temperature produces a meter reading. (*Source:* Maslansky and Maslansky, 1993. Courtesy of Van Nostrand Reinhold.)

imately 75 to 150 ppm pentane. Sensitivity can be enhanced by modifying the catalyst or amplifying the meter response.

Semiconductor or solid state sensors, also known as metal-oxide sensors (MOS), employ a semiconductor material such as silica coated with a metallic oxide such as zinc oxide or aluminum oxide. These instruments are designed to respond to a wide range of gases and vapors at low concentrations. Many MOS instruments do not give a numerical reading but instead use an audible alarm, a bar graph, or warning lights to indicate a meter response. Solid state sensors can be sensitive to concentrations as low as 10 ppm of the calibrant gas.

ELECTROCHEMICAL SENSORS

Electrochemical sensors use an electrolyte solution to detect a specific gas of interest.* In the sensor, an electric current moves from one electrode, across the electrolyte solution, and into a second electrode. Gases such as oxygen, carbon monoxide, hydrogen sulfide, chlo-

---

Note: An electrolyte is a chemical substance that dissolves in water and can conduct an electric current (electric current is an organized flow of electrons, which have a negative charge, in one direction). The acid in a car battery is a good example of an electrolyte solution. An electrode is a material that can accept or donate electrons and thus complete an electric circuit; gold, mercury, silver, and copper are good electron acceptors, whereas lead, zinc, and aluminum are good electron donors.

rine, and sulfur dioxide diffuse into the sensor through a semipermeable membrane and dissolve in the electrolyte solution. The change in current across the electrolyte solution produces a meter reading, which indicates the amount of gas present.

Oxygen Sensors

Oxygen sensors contain an alkaline electrolyte solution and are calibrated to measure oxygen concentrations between 0 and 25% by volume. Normal air contains 20.9% oxygen. The meter readings are dependent on the partial pressure of oxygen in air, which forces the oxygen through the semipermeable membrane. At increasing elevations above sea level, the partial pressure decreases even though the concentration of oxygen in air remains the same (Table 6–2). An oxygen deficiency meter should be calibrated before use with clean ambient air at approximately the same elevation as that of the suspect atmosphere.

As nearly 80% of the air envelope contains gases other than oxygen, it takes approximately 5% of a displacing gas to decrease the oxygen concentration by 1%. In many cases, 5% (50,000 ppm) of a gas or a vapor is very hazardous. For this reason, combination meters that detect multiple hazards are often used.

Toxic Gas Sensors

Exposure to toxic gases such as carbon monoxide (CO) and hydrogen sulfide (H_2S) also are measured by electrochemical sensors. Sensors specific for either CO or H_2S use an acid electrolyte; the current generated is directly proportional to the concentration of CO or H_2S present. Electrochemical sensors are now available for numerous other organic and inorganic gases and vapors, including chlorine, freons, cyanide, nitrogen oxides, ozone, sulfur dioxide, trichloroethylene, and tetrachloroethylene.

Limitations of Electrochemical Sensors

The composition and the concentration of the electrolyte solution, the sensor design, and the composition of electrodes are modified to achieve sensor specificity within a limited detection range. Electrolyte solutions are either acid or alkaline and can be affected by

Table 6–2. Oxygen concentration readings at increasing elevations

Elevation above sea level	Oxygen reading, % by volume
0	20.9
500	20.4
1,000	20.1
2,000	19.3
4,000	18.0
6,000	17.3
8,000	15.4
10,000	14.3

materials that neutralize the solution. Oxygen sensors are gradually neutralized by normal atmospheric concentrations (330 ppm) of carbon dioxide, an acid gas, found in air. For this reason, the oxygen sensors are relatively short-lived; they last approximately one year under normal use conditions. The high concentrations of carbon dioxide found at landfills and in exhaled air will neutralize an oxygen sensor more quickly than usual. *Never exhale into an oxygen meter to determine if it is working properly.*

Interfering gases may inhibit the sensor response and produce an inaccurate, false negative result, or they may exaggerate the sensor response and produce a false positive. The manufacturer should provide a list of commonly encountered interfering gas for each chemical-specific sensor. Common interferents for carbon monoxide and hydrogen sulfide electrochemical sensors are acetylene, ethyl alcohol, methyl alcohol, isopropyl alcohol, hydrogen cyanide, propane, sulfur dioxide, and nitrogen dioxide.

Combination Meters

A combination meter has more than one sensor, for use in multiple hazard detection. The most commonly used combination meter is a %LEL CGI/oxygen deficiency meter. Meters that detect multiple hazards often have sensors that detect %LEL combustible gases, oxygen, CO, and/or H_2S. Such combination meters facilitate confined space entry when monitoring for combustible gases, oxygen, and toxic gases is required.

COLORIMETRIC DETECTOR TUBES

A detector tube is a sealed glass tube (Figure 6–7) containing an inert material impregnated with or mixed with one or more reagents that change color in the presence of specific types of air contaminants. The length of the color change or its intensity is proportional to the amount of contamination present.

Sample air is drawn into the tube by a pump, which may be a bellows, bulb, or piston type. One pump stroke pulls 100 cc of sample air through the tube. The number of pump strokes (ps) required to complete the tube protocol depends on the type of tube and the manufacturer. The amount of time required for 100 cc of air to pass through the tube is called the pump stroke interval; the interval may be as short as just a few seconds or as long as several minutes.

Many detector tubes contain a prelayer, or conditioning layer, that removes water vapor or other interfering gases; in some cases the conditioning layer reacts with the contaminant of interest to facilitate the color change produced in the indicating layer. Nearly all detector tubes are marked with an arrow that helps the user to properly place the tube in a sampling pump. The arrow indicates the direction of airflow and always points toward the pump.

The most common detector tube is the direct-reading tube, which has the calibration scale marked on it. After a specified volume of air is pulled through the tube, the length of the color change corresponds to the concentration on the scale. In some cases, the tube may have two scales, or the actual concentration present must be determined by multiplying the tube results by a conversion factor; other types of detector tubes rely on a conversion chart to convert millimeters of stain to a concentration. The color intensity tube also is provided

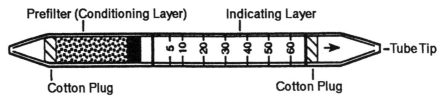

Prefilter (Conditioning Layer)   Indicating Layer

Cotton Plug   Cotton Plug
—Tube Tip

Figure 6-7. A colorimetric detector tube. (*Source:* Maslansky and Maslansky, 1993. Courtesy of Van Nostrand Reinhold.)

with a chart to determine the approximate concentration. Qualitative indicator tubes, also called poly tubes or poly test tubes, are used as yes–no tubes to alert the user to the presence of air contamination. These tubes usually have no scale.

Dosimeters and Badges

Dosimeters and badges are used to detect and measure specific contaminants in air over a period of time (e.g., over an 8-hour workday).

A long-term tube is a detector tube calibrated to a specific contaminant, such as benzene, carbon monoxide, or ammonia. The tube is placed within the breathing zone of the worker, usually clipped onto the collar. A small pump continuously pulls sample air

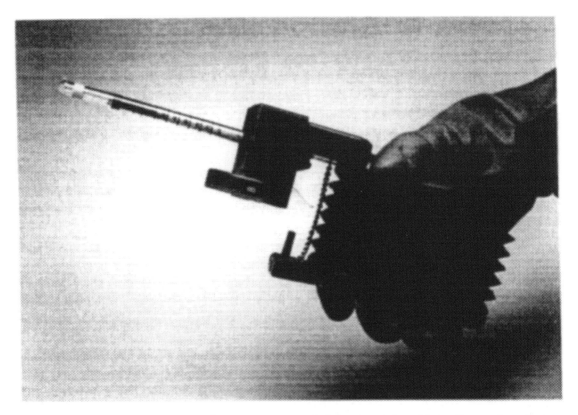

Figure 6–8. A Draeger bellows pump and detector tube. Note the color change extending down the tube. (*Source:* Maslansky and Maslansky, 1993. Courtesy of Van Nostrand Reinhold.)

through the tube at a fixed rate. The length of stain indicates the approximate actual concentration present.

A passive dosimeter tube does not require a pump but relies on diffusion to move contaminants into the tube. The dosimeter is worn in the breathing zone for a period of time; the length of stain indicates the approximate concentration present. Passive badges also rely on diffusion and permeation of gases through a membrane onto a tape designed to change color in the presence of certain type of contaminants. The badge color is compared to a chart, or an electronic reader scans the badge to determine the concentration present.

Some tubes and badges do not give a real-time response; they are designed to collect vapors and gases and must be sent to a laboratory for analysis.

Detector Tube Limitations

Although detector tubes, dosimeters, and badges are calibrated to one gas or vapor, they are not very specific and will respond to a broad class or category of contaminants, not just one material. For example, an ethyl alcohol tube will respond both to ethyl alcohol and to a variety of other alcohols. Other interfering gases and vapors, including water vapor or humidity, can affect tube readings by enhancing or inhibiting the chemical reaction responsible for the color change. When this occurs, the tube may give an inaccurate reading or no reading at all.

Other factors that can alter tube function include temperature extremes, which affect the chemical reaction; light, which can accelerate the chemical decomposition of tube reagents; and inadequate storage conditions, which also can facilitate reagent degradation. Time is another factor; all tubes have an expiration date, which represents the shelf life under optimal storage conditions as defined by the manufacturer and found on their box or in the instructions. The manufacturer's instructions should also describe the color change that indicates a positive reaction, and list any potential interferences that can cause inaccurate tube readings.

The accuracy of tube results, which may vary between different types of tubes and manufacturers, usually is indicated on the instruction sheet. The accuracy may be up to ±35% of the reading when it is within the calibration range of the tube. Finally, detector tube results are valid only for the time and the location of sampling; results may be markedly different at another sampling location just a few feet away. Detector tubes cannot replace instruments capable of continuous monitoring; but they can augment, enhance, or confirm results obtained from other instruments.

PHOTOIONIZATION DETECTORS

Photoionization detectors or PIDs are used as general survey instruments to detect low concentrations of air contaminants. Primarily used to indicate the presence of organic vapors and gases, they can also detect some inorganic materials. Most PIDs are equipped with a pump or a fan to draw the air sample into the meter; some of the smaller models rely on passive diffusion (Figure 6–9). PIDs are factory-calibrated to isobutylene or benzene.

Chemical compounds in sampled air form ions or are ionized as a result of being bom-

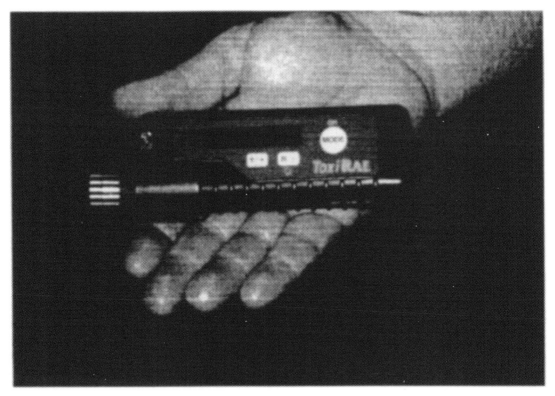

Figure 6-9. A PID small enough to clip to a lapel.

barded by high energy ultraviolet light, in a process called photoionization.* The ionization detector in a PID consists of a donating, or bias, electrode and a collecting electrode (Figure 6–10). When ions are formed, they are collected at the collecting electrode and produce an ion current. The ion current is amplified and then translated into a meter reading.

The energy required to remove the outmost electron from its orbit around the nucleus of an atom or molecule is called the ionization potential or IP. The IP of each molecule depends on its composition and is measured in electron volts (eV).

Ionization Potentials

Although all elements and chemical compounds can be ionized, they differ in the amount of energy required. Some materials lose electrons very easily, whereas others are difficult to ionize. Each element and chemical compound has its own IP; and the lower the IP is, the lower the amount of energy required for ionization.

Ionization potentials vary widely between different materials (Table 6–3); nitrogen, oxygen, and carbon dioxide, which are normally found in the atmosphere, have IPs greater

---

*Note: All materials are composed of atoms, each consisting of a nucleus and orbiting electrons, which are negatively charged. Atomic nuclei contain positively charged protons and neutrons, which have no charge. The positive charges of the protons keep the negatively charged electrons in their orbits. When an atom gains or loses an electron, it becomes an ion. Ions are either positively or negatively charged. The loss of an electron produces a positively charged ion.

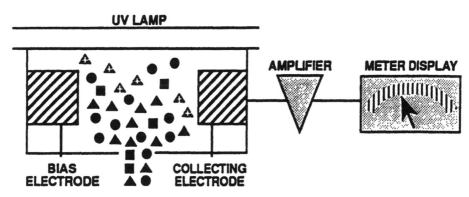

Figure 6–10. Schematic of an ionization chamber. Charged ions formed within the chamber are collected at an electrode, producing an ion current. Chemicals that cannot be ionized (and) will block the UV light from reaching ionizable materials ( ). (*Source:* Maslansky and Maslansky, 1993. Courtesy of Van Nostrand Reinhold.)

than 12.0 eV, whereas most organic compounds have IPs less than 12.0 eV. An important exception is methane gas, which has an IP of 12.98 and cannot be detected by PID.

UV Lamp Capacities and Limitations

The energy emitted by the UV lamp in PIDs is measured in electron volts (eV). The eV capacity of a UV lamp determines if a particular compound will be ionized: if the UV lamp energy is greater than the IP of the compound, ionization will occur; there will be no ionization if the UV lamp energy is less than the IP of the material. Most PID manufacturers offer UV lamps with capacities of 9.5, 10.2 or 10.6, and 11.7 eV.

The photoionization of chemical compounds depends on UV light; anything that interferes with the transmission of UV light will affect the PID response. Water vapor or dust in the ionization chamber scatters and deflects UV light, and gases that cannot be ionized will act in the same way. The meter response is dramatically reduced when nonionizable gases and high humidity are present. Workers at landfill sites and other locations where methane gas in present often use PIDs to detect low concentrations of nonmethane contaminants (see Figure 6–11). At high methane concentrations, however, PID responses to ionizable gases and vapors can be severely decreased or even absent. Considering all the limitations of PIDs, the absence of a PID reading should *never* be interpreted as the absence of contamination.

Sensitivity and Specificity of PIDs

When a PID responds to the calibration gas, the meter reading is equivalent to the actual concentration present. All other responses of the PID are relative to its response to the calibration gas. When gases and vapors other than the calibrant are present, the meter reading does *not* reflect the actual ppm concentration present. The reading indicates the ion current produced within the ionization chamber. A meter response of 20 indicates that the current is the same as when 20 ppm of the calibrant gas is present.

Table 6–3. Ionization potentials in electron volts (eV)

Chemical compound	eV
Elements	
Nitrogen	15.58
Oxygen	12.08
Chlorine	11.48
Simple Inorganics	
Carbon monoxide	14.01
Carbon dioxide	13.79
Hydrogen sulfide	10.46
Hydrogen cyanide	13.91
Nitrogen dioxide	9.70
Water	12.59
Saturated Hydrocarbons	
Methane	12.98
Ethane	11.65
Butane	10.63
Hexane	10.17
Octane	9.82
Unsaturated Hydrocarbons	
Acetylene	11.41
Butadiene	9.07
Cyclohexene	8.95
Ethylene	10.52
Propylene	9.73
Sulfur, Nitrogen Compounds	
Acrylonitrile	10.91
Ammonia	10.15
Carbon disulfide	10.08
Dimethyl sulfide	8.69
Methylamine	8.97
Methyl mercaptan	9.44
Chlorinated Hydrocarbons	
Carbon tetrachloride	11.47
Chloroform	11.42
Methylene chloride	11.35
Tetrachloroethylene	9.32
Trichloroethylene	9.45
Vinyl chloride	10.00
Aromatic Hydrocarbons	
Benzene	9.25
Naphthalene	8.12
Phenol	8.50
Toluene	8.82
Xylene	8.45

Table 6–3. *(continued)*

Chemical compound	eV
Aldehydes and Ketones	
Acetaldehyde	10.21
Acetone	9.69
Formaldehyde	10.87
Methyl ethyl ketone	9.53
Methyl isobutyl ketone	9.30
Oxygenated Compounds	
Acetic acid	10.37
Diethyl ether	9.53
Ethyl acetate	10.11
Ethyl alcohol	10.48
Ethylene oxide	10.56
Isopropyl alcohol	10.16
Methyl alcohol	10.85
Propylene oxide	10.22

PIDs are sensitive to low ppm concentrations of air contaminants. Under optimal conditions, concentrations of less than 1 ppm of the calibration gas can be detected. PIDs are nonspecific detection devices: because many different chemical compounds can be ionized, a PID cannot be used to measure a specific gas or vapor in a mixture, nor can a PID be used to identity contaminants in air or measure their actual concentration present.

The UV lamp eV capacity determines what compounds will be ionized; materials with IPs less than the lamp eV capacity will be detected. The lower the IP is relative to the lamp capacity, the greater the number of ions produced, and the greater the meter reading. The most commonly employed lamps are the 10.2 and 10.6 eV; 11.7 eV lamps can ionize more materials than the others, but they have a very short field life.

Materials with IPs greater than the lamp eV will not be ionized and will not be detected. Materials with IPs very close to the lamp eV capacity will be minimally ionized. The absence of a meter response does *not* mean the absence of contamination; it simply indicates that the instrument does not detect the presence of contaminants with IPs less than its lamp capacity. The lack of meter reading should always be reported as "no increase above background"; it should *never* be reported as "no contamination present."

FLAME IONIZATION DETECTORS

Flame ionization detectors or FIDs are low concentration instruments used in concert with PIDs as general survey instruments. FIDs are used to indicate the presence of organic compounds only; they cannot detect inorganic materials.

FIDs are similar to PIDs in that the meter reading is based upon ionization of contaminants in sample air. FIDs, however, create ions by burning the chemical contaminants in a hydrogen-fed flame. The ion current produced is then amplified and displayed as a meter reading. FIDs are calibrated to methane.

Figure 6–11. Workers in fully encapsulated suits use an HNU photoionization detector.

One advantage of FIDs over PIDs is that they are not restricted by the IP of the chemical contaminants. FIDs detect, with varying sensitivity, any material that can be burned—that is, any organic material, including light hydrocarbon gases such as methane, which has a very high IP. Water vapor is formed in the ionization/combustion chamber when hydrogen is burned; ambient humidity conditions do not affect FID efficiency or response. Some manufacturers are now offering combination meters, which combine both an FID and a PID in one instrument (Figure 6–12).

Limitations and Requirements of FIDs

The hydrogen-fed flame requires oxygen, which is provided in the incoming air sample. If the air sample is oxygen-deficient, the flame will be lower than normal, or it will be extinguished (flameout). A smaller flame results in less ionization and thus an erroneously low meter response.

The hydrogen fuel used to feed the flame must be free of contamination. Organic contamination in the hydrogen fuel will produce a high background and hinder efforts to detect very low concentrations of airborne contaminants. Water vapor in the fuel can facilitate corrosion of the fuel cylinder. Always use ultra-high pure (UHP) hydrogen, certified as 99.999% pure and containing less than 1 ppm total hydrocarbons (THC). Some FIDs require a nitrogen/hydrogen fuel; the UHP grade, with less than 1 ppm THC, should be used.

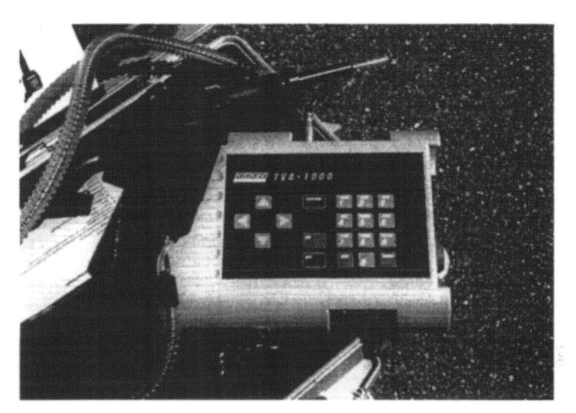

Figure 6-12. The Foxboro TVA 1000 offers a combination of an FID and a PID in the same meter. Readings from both detectors are displayed simultaneously.

Large concentrations of flammable gases in the sample air can act as an additional fuel source for the hydrogen flame. When this occurs in most FIDs, the hydrogen supply is momentarily shut off, and the flame is extinguished. In most situations, an intrinsically safe FID should be employed. Non-intrinsically safe FIDs can be a source of ignition and should be used only in situations where there is absolutely no possibility of a sudden or unexpected release of flammable gases or vapors.

Sensitivity and Specificity of FIDs

Except when responding to the calibration gas, FID readings are *never* equivalent to the actual concentration present. When gases and vapors other than the calibrant are present, the meter reading is relative to the calibration gas and does not reflect the actual ppm concentration present. The reading indicates the ion current produced within the combustion/ionization chamber. A meter response of 10 indicates that the ion current is the same as that produced when 10 ppm of the calibration gas is present.

FIDs are sensitive to very low ppm concentrations of air contaminants. Under optimal conditions, concentrations less than 1 ppm of the calibration gas can be detected. When used in the survey mode, however, FIDs are nonspecific detection devices and cannot be used to determine the actual concentration present, or to identify individual contaminants in a mixture.

The FID meter response varies with the amount of ion current generated. Moreover, the amount of ion current produced when an organic compound is burned is unique to that compound; the addition of an oxygen, a chlorine, a nitrogen, or a sulfur atom to the molecule affects the ionization efficiency and the amount of ion current produced. The absence of a meter response does *not* mean the absence of contamination; it simply indicates that the instrument does not detect the presence of organic contamination within the limits of its sensitivity. The lack of a meter reading should always be reported as "no increase above background"; it should *never* be reported as "no contamination present."

Survey vs. Chromatography

Some FIDs can operate in two different modes: survey and gas chromatography (see Figure 6–13). When operating in the survey mode, the FID continually draws air into the combustion/ionization chamber, where it is burned. The resulting meter response represents the total ion current generated in ppm calibration gas equivalents. It is not possible to distinguish between different individual hydrocarbons or classes of hydrocarbons.

When the instrument is operated in the gas chromatography or GC mode, sample air is injected into a GC column packed with an inert material that attracts hydrocarbon vapors. The vapors are pushed through the column by a carrier gas (usually hydrogen), which is not attracted to the column packing material. Individual contaminants become separated as they move through the column and into the combustion chamber (Figure 6–14).

The time it takes for a substance to pass through the column is known as its retention time. The retention time is a function of the composition of the substance, its affinity to the packing material, the carrier gas flow rate, and the column temperature. Lighter molecules, such as butane or propane, pass through the column more quickly than larger, more complex molecules, such as naphthalene or polyethylene.

When an FID is used in the GC mode, the meter response is usually displayed as peaks on a strip-chart recorder. FIDs with GC options are designed for use where there is a limited number of known contaminants present. A comparison of peak characteristics and retention times can be used to tentatively determine the number, identity, and approximate concentrations of known contaminants.

FIELD GAS CHROMATOGRAPHY

Portable gas chromatography (GC) units are available, which are small, portable, and usable under most field conditions. Usually equipped with flame ionization or photoionization detectors, portable GCs are used for air sampling at hazardous waste site remediations, for on-site soil and water sampling, and for in situ soil gas analysis.

A typical portable GC consists of a GC column, an injection system for introducing samples into the column, a detector, and a recording device (Figure 6–15). After the sample is introduced into the column, it is carried or pushed through it by a carrier gas; individual components are separated as they travel through the column. The individual components enter the detector and elicit responses that are depicted as peaks. The retention time is the length of time required for a material to travel the length of the column. Compounds can

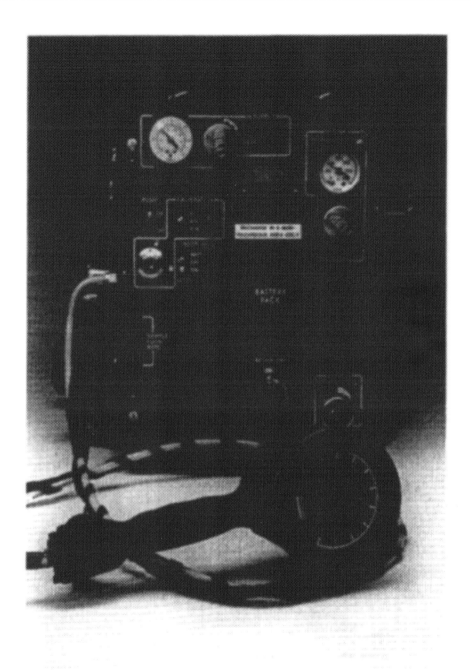

Figure 6-13. The Foxboro OVA Model 128 can be used in the survey or the GC mode. (*Source:* Maslansky and Maslansky, 1993. Courtesy of Van Nostrand Reinhold.)

be tentatively identified on the basis of their retention time and peak characteristics (Figure 6–16).

Backflushing is used to clean the column of residual contamination and materials with long retention times. Many GCs have a short precolumn to trap slow, heavy, less volatile materials with high retention times.

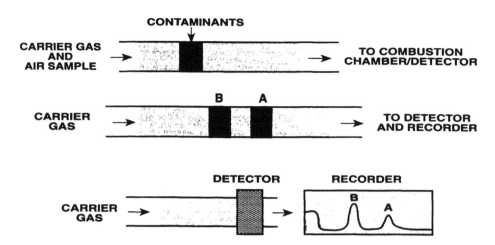

Figure 6-14. Separation of volatile contaminants in a gas chromatography column. (*Source:* Maslansky and Maslansky, 1993. Courtesy of Van Nostrand Reinhold.)

It is important to recognize that each peak may represent multiple contaminants with the same retention time. Individual contaminants do not necessarily have unique retention times. The presence of a peak at a particular time does not guarantee its identity. At sites where unknowns are present, the tentative identity of peaks should be confirmed by laboratory analysis.

Frequent calibration is required for portable GCs, to compensate for changes in detector output, sample flow, temperature, and column condition. Calibration with a gas mixture containing all known or suspected contaminants ensures the greatest accuracy. Surro-

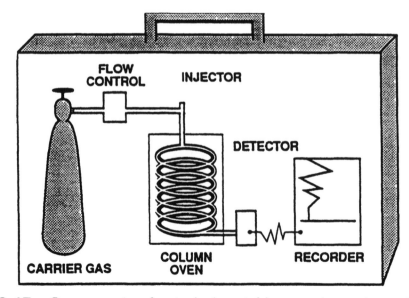

Figure 6-15. Components of a typical portable gas chromatograph. (*Source:* Maslansky and Maslansky, 1993. Courtesy of Van Nostrand Reinhold.)

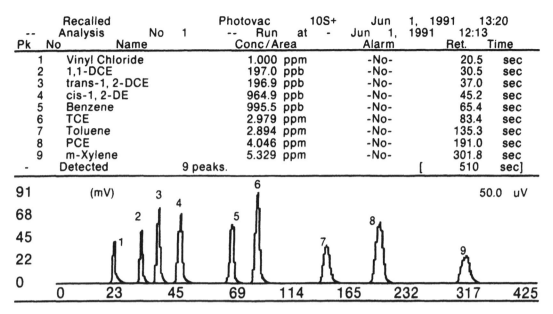

Pk No	Name	Conc/Area	Alarm	Ret.	Time
1	Vinyl Chloride	1.000 ppm	-No-	20.5	sec
2	1,1-DCE	197.0 ppb	-No-	30.5	sec
3	trans-1, 2-DCE	196.9 ppb	-No-	37.0	sec
4	cis-1, 2-DE	964.9 ppb	-No-	45.2	sec
5	Benzene	995.5 ppb	-No-	65.4	sec
6	TCE	2.979 ppm	-No-	83.4	sec
7	Toluene	2.894 ppm	-No-	135.3	sec
8	PCE	4.046 ppm	-No-	191.0	sec
9	m-Xylene	5.329 ppm	-No-	301.8	sec

Figure 6–16. A typical chromatogram, demonstrating relative peak heights of sample air contaminants separated in a gas chromatograph column. (*Source:* Maslansky and Maslansky, 1993. Courtesy of Van Nostrand Reinhold.)

gate calibration, using one or a few compounds, can be used to adjust the sensitivity and to ensure that the instrument is working properly under specific operating conditions.

Limitations of Portable GCs

The most often-cited limitation of portable GCs is the initial cost of purchase and setup. After acquisition of the GC unit, the carrier gases, columns, sample injection syringes, and carrier gases must be purchased. Calibration gas mixtures also must be ordered. Several workers must be trained to calibrate and use the instrument, and to troubleshoot problems in the field.

Most GCs are equipped with an onboard or separate computer that stores and retrieves calibration data. The user must provide data on measured response versus concentration and retention time for each compound of interest. During sample analysis, the computer compares detected peaks and retention times with stored data. The GC can identify or measure only those compounds for which it has calibration data.

GCs can be limited by temperature extremes; cold temperatures affect column separation and reduce the battery life, and high temperatures also can affect column separation. In many cases, more than one column is necessary to accommodate the variety of contaminants present on-site; new columns must be "conditioned" according to the manufacturer's instructions prior to use. GC columns must be handled gently; twisting or bending a column disrupts the interior packing and ruins its separation capabilities. Many manufacturers now offer GC columns that are protected inside a cartridge that snaps into the instrument, thus avoiding many of the problems associated with rough handling.

Figure 6–17. The Photovac 10S Plus field-portable gas chromatograph is relatively lightweight, can utilize a variety of carrier gases, and is equipped with a photoionization detector, computer, and modem.

PORTABLE INFRARED SPECTROPHOTOMETERS

Atoms or groups of atoms in a molecule are always in motion. The bonds that keep atoms together can be thought of as springs that also keep the molecule balanced. The motion or the vibration of each atom in a molecule is dependent on the atoms around it. Therefore, each unique chemical compound, such as the ethanol molecule (ethyl alcohol, Figure 6–18) has a vibration energy that is different from that of every other chemical.

When a chemical compound interacts with infrared (IR) radiation, it absorbs some of the IR energy. The pattern of energy absorption or the absorption spectrum (usually within a 2–15-micron wavelength range) represents the pattern of vibrations that is unique to that compound. This unique IR absorption spectrum can be considered a chemical fingerprint.

Each chemical substance demonstrates optimal absorption within a small portion of the IR spectrum. Optimal absorption bands for classes of chemicals have been defined, and they are usually quite small. For example, aromatic hydrocarbons such as benzene and toluene have an optimal absorption band between 3.25 and 3.35 microns, whereas ketones absorb best between 5.60 and 5.90 microns. It is possible, then, to use just a small portion of the IR spectrum and still measure the concentration of a chemical substance known to be present.

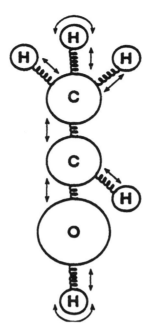

Figure 6–18. Bonding forces that hold a molecule of ethyl alcohol together can be visualized as springs. (*Source:* Maslansky and Maslansky, 1993. Courtesy of Van Nostrand Reinhold.)

An IR analyzer is a single beam spectrophotometer, and consists of a sample inlet, an IR radiation source, a filter to control the wavelength delivered, a sample cell, and a detector (Figure 6–19). IR absorption by a chemical over a given pathlength is proportional to its concentration. The longer at the pathlength is, the greater the sensitivity and the lower the detection limit. Mirrors reflect or fold the IR radiation within the sample cell to achieve a long pathlength. IR analyzers may have fixed or variable pathlengths.

IR analyzers are designed to measure the concentration of one or more compounds known to be present; they are not capable of identifying specific compounds in an atmosphere containing unknown contaminants.

Limitations of IR Analyzers

Like other direct-reading instruments, IR analyzers are susceptible to interfering gases and vapors that absorb at the same wavelength as that of the compound of interest. Optimal absorption bands are shared or overlapped by those of other chemicals; erroneously high readings will be obtained in the presence of multiple contaminants with the same absorption band. The most common interfering gas is water vapor, with an optimal absorption band between 5 and 8 microns. Carbon monoxide and carbon dioxide are other frequently encountered interfering gases.

Dust entering the sample cell can scatter the IR light and decrease the sensitivity. Water and gases can condense on mirrors in the sample cell, thereby decreasing the IR pathlength and the sensitivity. Also mirrors within the sample cell are susceptible to mechanical damage and faulty alignment.

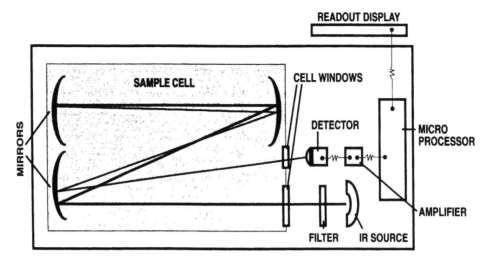

Figure 6–19. A simplified functional diagram of an infrared analyzer. Mirrors within the sample cell reflect the infrared radiation several times to achieve a long pathlength. (*Source:* Maslansky and Maslansky, 1993. Courtesy of Van Nostrand Reinhold.)

SAMPLE COLLECTION METHODS

Personal air samples may be collected by a variety of methods for subsequent analysis; whenever possible, the air samples should be collected within the breathing zone. NIOSH, OSHA, and ASTM have published manuals that describe specific collection procedures (NIOSH, 1994; OSHA, 1990; ASTM, 1987a). Exhaustive analysis to identify and measure all contaminants present is prohibitively expensive when a contract laboratory is used, and technically difficult and time-consuming for a field lab. The choice of analysis method and the number of chemical compounds quantified can often be optimized by estimating the site contamination on the basis of the site history, information collected during the initial site investigation, and the results of initial soil and water sampling.

Air sampling systems usually include a sampling pump that draws sample air into selected collection media. Special absorbents and sampling media can be used for specific types of contaminants. The type of sorbent used and the sample volume will vary according to the type and the concentration of substances known or anticipated on-site. Activated charcoal and porous polymers, such as Tenax, Chromosorb, and Poropak, can be used for a wide range of compounds. Polar sorbents, such as silica gel, will absorb materials that are not absorbed onto activated charcoal. Carbon molecular sieves, such as Carbosieve and Carboxen, are often used to collect volatile nonpolar organic compounds. PCBs are collected on glass fiber filters or fluorosil tubes.

The collected samples are usually capped and sent to a laboratory for analysis. When samples are taken for on-site field analysis, contaminants can be driven off the collection media by heat. A programmable thermal desorber extracts contaminants from charcoal and other collection media; the desorber can usually store the samples in a chamber and inject them directly into a gas chromatograph.

Specific information on sample collection and analysis can be obtained from the current *NIOSH Manual of Analytical Methods.*

RADIATION SURVEY METERS

It is not unusual to encounter ionizing radiation at hazardous waste sites, and high levels of such radiation can cause significant injury; so site workers should be able to determine if radioactivity is present at levels that pose a health hazard.

Radioactivity is the property of the nucleus of an atom to spontaneously emit energy in the form of radiation; radiation is excess nuclear energy emitted in the form of high energy electromagnetic particles or waves. Radiation can be broadly classified into two categories: ionizing and nonionizing.

Radioactive materials or radionuclides undergo spontaneous transformation or decay. As the decay takes place, the amount of radioactive material present per unit time decreases. The amount of time for half the atoms to undergo decay is unique for each radionuclide, and is called the half-life. (See also Chapter 1 discussion of this topic.)

Types of Ionizing Radiation

Three types of radiation are a major concern at hazardous waste site: alpha particles, beta particles, and gamma waves.

Alpha radiation is composed of particles ejected from the nucleus during nuclear disintegration. An alpha particle consists of two protons and two neutrons and has the same structure as the nucleus of a helium atom. Alpha particles are positively charged; because of their large mass and charge, they travel only a short distance.

Beta particles are negatively charged electrons emitted from radioactive atoms. Beta particles have a velocity very near the speed of light, and their range may be up to 60 feet. Beta particles are smaller, faster, and have a greater range than alpha particles. Skin burns can result from overexposure to beta radiation.

Gamma rays are high energy electromagnetic waves with a very short wavelength, which are emitted from the nucleus of the atom. Gamma radiation travels at the speed of light, has a very long range, and can penetrate several inches of steel. Because of its energy and penetration capabilities, gamma radiation is considered the most dangerous form of radiation.

Radiation Terminology

Ionizing radiation transfers energy to matter; in living tissue or organisms the amount of energy imparted per kilogram of irradiated material is defined as a unit called the gray (Gy). The more conventional unit of absorbed dose is the rad, or radiation absorbed dose; 1 Gy equals 100 rads, and 1 centigray (cGy) equals 1 rad.

The biological effect of radiation on living organisms differs according to the type and the energy of the radiation. The conventional unit is the rem, or roentgen equivalent in man. The roentgen (R) is a unit of exposure for gamma radiation; one R is equivalent to 1 rem. The dose in R, rem, or millirem (mR) is expressed on a per hour basis (mR/hr).

Naturally occurring or background radiation has always been present on earth. Background radiation levels vary, depending on the type of building materials or natural geologic formations. Typical natural radiation background levels are usually between 0.01 and 0.05 mR/hr.

Types of Radiation Detectors

Most radiation survey instruments operate on the same basic principle: the detector is filled with gas; radiation interacts with the gas and produces ions; the ions thus produced are counted electronically to give a meter reading.

Geiger-Mueller or GM counters are versatile, easy to operate, inexpensive, and fairly reliable. GM counters respond to gamma and beta radiation. A popular type of GM survey meter uses a detector tube inside a protective metal cylinder or casing, which is open on one side. A metal shield can be slid over the detector to absorb some of the incoming radiation. When the shield is removed, both beta and gamma radiation can reach the detector; when the shield covers the detector, only gamma radiation reaches the detector.

Ion detectors or ionization chamber detectors measure gamma rays and beta particles. If the covering over the ionization chamber is very thin, alpha particles can also be detected. Proportional counters are used primarily for detecting and measuring alpha radiation.

Scintillation detectors contain a solid, liquid, or gas phosphor or scintillator, which emits light when exposed to radiation. When radiation is present, the light signal is converted to an electrical pulse. Scintillation detectors are extremely sensitive, and are used for detecting very low levels of gamma radiation. Luminescent detectors contain a material that stores radiation energy. When processed, the stored energy is emitted as light, which is proportional to the amount of radiation received.

Radiation dosimeters are small detectors that measure the total radiation dose received over a period of time. Direct-reading pocket dosimeters are the size of a pencil and can be read immediately during or after use. Film badges resemble a credit card; the radiation dose received cannot be immediately determined by the wearer. Personal alarm dosimeters alert the user when the radiation dose exceeds a preset threshold.

PARTICULATE AND AEROSOL MONITORS

Potential health effects caused by inhalation of insoluble aerosols or particulates can be related to the size of the aerosol or the particle. Inspirable particulates can be inhaled through the nose or the mouth; they have a diameter of approximately 200 micons or less. Thoracic particulates are smaller, with a diameter of 25 microns or less; thoracic particulates can penetrate past the larynx but are captured before they reach the gas exchange region of the lung. Respirable particles are smaller than thoracic particulates, with a diameter smaller than 10 micons; these particles can reach the alveoli, the gas exchange area the lung.

Inspirable particulate and thoracic particulate samplers are available for personal as well as area monitoring. The most commonly used samplers, however, are those that sample for respirable particulates. Light scattering photometers offer the advantage of real-time monitoring for particulates. The RAM-1 is an example of a real-time photometer for field survey measurements. Particulate concentrations of up to 200 mg/m^3 are detected; the reading is updated at a selected measurement interval. The MINIRAM, which is a smaller version of the RAM-1 except that it does not have a sampling pump, is designed for personal sampling and should be worn on the lapel (within the breathing zone). The MINIRAM can be equipped with an adaptor that allows it to be used with a personal sampling pump.

Limitations and Requirements

Background drift can occur if large particles are deposited on the optical surfaces. These particles will be "recycled" and added to the particles in the sample air, but are too large to be evacuated from the chamber. Removing large particles at the inlet with a centrifugal separator or a cyclone can alleviate this problem. Particulate monitors should be designed so that humidity does not significantly affect the response; water droplets should be removed from the sample airstream at the inlet.

Field particulate monitors often use a filtered airstream to flush the optical surfaces; this airstream is dried with a chemical desiccant to protect the optics from water condensation. The filters and the desiccant must be inspected and replaced at regular intervals.

Some particles may be difficult to measure accurately by light scattering photometers; metal oxide fume particles, charged particles, or other particles that tend to condense into larger-sized particles or aggregates will be undercounted. Similarly, when the size composition of the particulates is varied, there will be some loss of sensitivity, as some of the particles will not reach the detection chamber.

Figure 6–20. Field conditions can make it difficult to get an accurate reading from a particulate monitor. Particulate concentrations are too high if particulates obscure objects within 5 feet of the monitor.

FIELD INSTRUMENTATION

It is essential that site workers understand how to use information obtained from air monitoring instruments. This requires an understanding of the operation, limitations, and proper application of each device employed.

Use the Appropriate Instrument

Instruments used on-site should be capable of assessing the hazards present or anticipated. For example, if workers are evaluating an underground storage tank that had been used to store gasoline, the anticipated hazards would include combustible gases, oxygen deficiency, and toxic gas buildup. It would be inappropriate to call for a radiation meter or a chlorine detector tube.

Although many instruments are versatile, it is not a good idea to get one instrument to serve many functions. For example, although a PID can detect hydrogen sulfide and many combustible gases and vapors, it cannot take the place of a CGI and a hydrogen sulfide detector. PIDs do not detect methane and many other combustible gases; the ionization efficiency for hydrogen sulfide varies with environmental conditions, and it is impossible to determine the actual concentration present.

The Absence of Evidence is Not Evidence of Absence

The lack of a meter reading does not automatically mean that no contamination is present. The instrument used may not be capable of detecting the type or the concentration of contaminants present. For this reason, multiple types of air monitoring instruments should be used to confirm the presence or the relative absence of contamination.

For example, suppose that there is sufficient oxygen in an underground storage tank to use a %LEL CGI. A reading of 0% LEL does not prove that no gasoline vapors are present; such a reading indicates that there are not enough gases or vapors present to elicit a reading above 0% LEL. To determine if low concentrations of gasoline vapors are present, it is necessary to use an instrument that measures in ppm-equivalents.

Never Assume Only One Hazard Is Present

Site workers may focus on what is perceived to be the primary hazard and forget that other hazards may also be present. Additional instruments can be used to rule out other potential hazards from air contamination. For example, an underground storage tank could be assessed for the presence of contaminants other than gasoline, such as hydrogen sulfide gas.

It is important to recognize, however, that the use of air monitoring instruments does not protect workers against other hazards commonly encountered on-site. The instrument response is only one of many different factors that must be integrated into the hazard and risk evaluation process.

Use One Instrument to Confirm Another

In most cases, there is at least one other type of detection device that can be used to confirm the response obtained from an instrument. For example, detector tubes can verify information collected by a CGI or an ionization detector. An FID can confirm readings obtained from a PID, a detector tube, or a ppm-equivalent CGI.

Interpret Readings in More than One Way

As instrument users become more familiar and comfortable with their instruments, they should be able to use the readings in different ways. For example, the ppm readings from a detector tube can be converted to %LEL and compared with a CGI response. A meter reading of 20.4% oxygen represents only a slight decrease in oxygen concentration but a rather significant concentration (up to 25,000 ppm or 2.5% by volume) of another material that is displacing the oxygen.

Establish Background Levels

Prior to on-site monitoring, noncontaminated areas surrounding the site should be assessed; such perimeter monitoring should be conducted before site activities begin, to determine the background readings for each instrument used. Background readings above zero are normal and expected, especially on low concentration instruments such as PIDs and FIDs. Such background readings should never be eliminated or zeroed out.

Whenever possible, background monitoring should encompass the same types of environments as those encountered on-site (e.g., blacktop, heavy vegetation, grassy areas, bare soil). Background readings should also be obtained from any equipment that will be used on-site; potential sources of volatile emissions that will affect instruments include vehicle engines, drilling rigs, compressors, pumps, and heaters.

Establish Action Levels

An action level is a meter reading or a response that triggers some action. The action may include evacuating unnecessary personnel, changing site work practices, watching meter readings more closely, upgrading levels of protective clothing or respiratory protection, or leaving the area. Action levels should be clearly stated in the site health and safety plan (HASP); the action taken should also be plainly explained and leave no room for misinterpretation.

Action levels that apply to uncontrolled sites include those for radiation (1 mR/hr), combustible gases (20–25% LEL meter reading), and oxygen deficiency (19.5%) or excess (23.5%). An action level of 10% LEL has been defined by OSHA for permit-required confined spaces; the action level for excavations is 20% LEL. Action levels for specific toxic gases, such as carbon monoxide, hydrogen sulfide, and sulfur dioxide, are usually one-half the 8-hour exposure limit value. For example, a commonly used action level for hydrogen sulfide is 5 ppm, which is one-half the OSHA exposure limit of 10 ppm.

The %LEL CGI is a safety meter, which is intended to inform the user if it is safe to be in the sampled atmosphere. A meter reading of 1% is safe in terms of flammability; but if the reading is from unknown contaminants, it may represent a significant health hazard to unprotected personnel. A CGI reading of 1% LEL usually indicates the presence of at least 100 ppm (and possibly more) of a flammable gas or vapor. Similarly, a 0.1% decrease in oxygen concentration may represent the presence of 5000 ppm of a displacing gas or vapor. When such readings are obtained at an uncharacterized site containing unknown contaminants, the area should be considered potentially hazardous, and appropriate respiratory and skin protection should be employed.

EPA advocates the use of PID or FID survey meter readings as action levels. A meter reading above background is the action level for the use of air-purifying respiratory protection; APRs may be used as long as meter readings remain less than 5 ppm-equivalents above background. Positive pressure supplied air respiratory protection is recommended when meter readings are 5 ppm-equivalents or greater.

EPA action levels for levels of protection are discussed in more detail in Chapter 8.

7

Personal Protective Clothing and Equipment

Site workers must be protected against potential hazards from vapors, gases, and particulates. Personal protective equipment (PPE) is used to decrease exposure to biological and chemical hazards and to shield against physical hazards. No single combination of protective equipment and clothing can protect against all possible hazards; PPE must be used in concert with safe work practices and, whenever possible, engineering controls.

Selection of PPE should take into consideration a variety of factors, including identification of hazards or suspected hazards, the nature of the hazard, concentrations of contaminants known or anticipated to be present, potential routes of exposure (inhalation, skin, eyes), ability of equipment to provide protection, equipment limitations, site conditions, and activities to be performed by the wearer.

Proper selection and use of PPE should ensure protection of the respiratory system, eyes, skin, face, hands, feet, body, and hearing.

PPE PROGRAM

Activities at a hazardous waste site are organized and coordinated under a site-specific health and safety plan or HASP. For sites that fall under the OSHA HAZWOPER rule, a comprehensive PPE program must be included in the HASP. The basic objective of the PPE program is to protect the wearer from safety and health hazards and prevent injury to the wearer from incorrect use and/or malfunction of PPE.

Basic elements of a PPE program, as described by OSHA [29CFR 1910.120(g)(5)], should include:

- Selection criteria based upon site hazards.
- Description of PPE use and limitations.
- Duration of work activities allowed for specific PPE ensembles.
- Procedures for regular inspection, cleaning, maintenance, and storage of PPE.
- Procedures for decontamination or disposal of contaminated PPE.

- Instruction and training on proper fitting, use, and limitations of PPE.
- Procedures for donning and doffing PPE.
- Evaluation of PPE program effectiveness.
- Limitations of PPE due to external or medical conditions.

STRUCTURE AND FUNCTION OF THE SKIN

The largest organ of the body, the skin has a surface area of approximately 20 square feet for potential contact with contaminants. It is a multipurpose organ that regulates body temperature, receives sensations, and repairs and replenishes itself. The skin is protective by design because of its thickness, resiliency, and capacity to inhibit the entrance of water and water-soluble chemicals. Its thickness and its elasticity protect underlying nerves, blood vessels, and muscle.

Skin Structure

The skin is composed of two parts, the epidermis and the dermis. The epidermis is the outermost portion of the skin, which is in contact with the environment. It is cemented to an inner, thicker dermis. Below the dermis lies a subcutaneous layer of tissues. (See Figure 7–1.)

The epidermis has an outer layer that consists of up to 30 rows of flat, dead skin cells containing keratin; this layer provides protection against light, heat, bacteria, particulates, and some chemicals. The lower layer of the epidermis consists of basal cells that multiply and push toward the surface, enter the outer cell layers, and are eventually shed as dead skin cells.

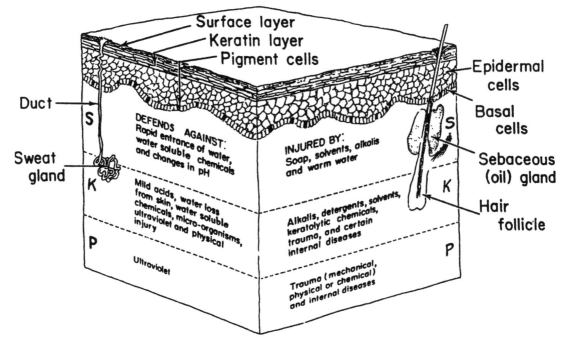

Figure 7–1. Diagram of the skin's protective layers. (*Source:* NIOSH, 1973.)

The dermis is thicker than the epidermis and contains sweat glands, hair follicles, oil glands, blood vessels, and nerves. The dermis also has elastic and connective tissues.

Functions of the Skin

Intact, normal skin serves many functions that are protective and regulatory in nature: The epidermis prevents absorption of many chemicals and is a physical barrier to bacteria and other microorganisms. The dermis contains nerves that make the skin sensitive to heat, cold, pain, and pressure. Sebaceous glands in the dermis secrete fatty acids that inhibit the growth of bacteria and fungi. The skin thickness and pigmentation provide some protection against UV radiation in sunlight. Sweat glands and blood vessels regulate body heat. The lymphatic system provides an immune response to foreign organisms and fights infection.

Skin Absorption of Chemicals

The ability of the skin to act as a barrier against chemicals is dependent on the chemical properties of the chemicals and the health of the skin. Absorption can be enhanced when the top portion or epidermis is abraded, lacerated, irritated, or otherwise physically damaged. Other factors that can lead to increased absorption of a chemical, include:

- Overhydrating the skin.
- Increasing the temperature of the skin; this in turn increases the secretion of sweat, which can solubilize chemicals on the surface of the skin.
- Increasing the blood flow to the skin surface.
- Altering the normal skin pH of 5.
- Increasing the skin contact time with the chemical.
- Increasing the surface area in contact with the chemical.
- Rubbing the chemical into the skin, as when clothing is wetted with the material and allowed to remain in contact with the skin.
- Adding agents that damage or defat the skin.
- Adding surface-active agents or other chemicals that can enhance absorption across the skin.

SKIN HAZARDS

Irritants

A variety of physical and chemical agents can damage the skin. Skin irritants act directly on normal skin at the site of contact; the quantity of irritant material, its concentration, and the duration of skin contact determine the extent of the injury. Acute irritants produce local, reversible inflammation in normal skin as a result of direct injury caused by a single application. Some substances cause defatting and drying of the skin; these materials are not acute irritants but can cause a cumulative irritation upon repeated or continued exposure. The result of such repeated exposure is often called chronic dermatitis.

Corrosives or absolute irritants are substances that cause disintegration and irreversible alteration of normal skin at the site of contact. Corrosive action is manifested by ulceration and subsequent necrosis (death of tissue); such damage is often described as a chemical burn. Because corrosive materials destroy the dermis and the epidermis, healing is achieved by scar formation.

Skin Sensitizers

Skin sensitizers are substances that cause an allergic skin reaction in sensitive individuals. The skin sensitizers or allergens most commonly encountered include poison ivy, poison oak, and poison sumac. Some chemicals also cause skin sensitization; these substances include the many organic isocyantes, alkaline dichromates, epoxy resins, and phenol-formaldehyde resins.

Other Causes of Skin Injury

The skin is susceptible to mechanical injury, which can result in lacerations, abrasions, friction burns, and blisters. Physical hazards, such as heat, cold, and UV and ionizing radiation, are also capable of injuring the skin. Exposure to high temperatures may cause heat stress and prickly heat or heat rash; extreme heat can cause burns. Cryogenic materials can produce frostbite-type injuries in exposed skin (cold burns). Exposure to low environmental temperatures also can cause frostbite. Biologic hazards, such as bacteria, fungi, parasites, and insects, can attack the skin and produce systemic disease or skin infections.

CHEMICAL PROTECTIVE CLOTHING

Chemical protective clothing (CPC) is an important part of the PPE ensemble. Such protective clothing is worn to prevent harmful chemicals from coming into contact with the skin. When used with respiratory protection, properly selected CPC can shield or isolate workers from the chemical and biologic hazards encountered on-site (Figure 7–2).

Classification of CPC

Protecting workers against skin exposure requires CPC that is resistant to chemicals known or anticipated to be present. Chemical protective clothing is classified by style, the protective material used, and whether the clothing is single-use (disposable) or reusable.

There are two basic styles of CPC: fully encapsulating and non-fully encapsulating. A fully encapsulating suit is a one-piece garment with a visor or a facepiece that completely encloses the wearer and provides chemical protection for the entire body when constructed of appropriate materials. Boots and gloves may be integral parts of the unit or separate. Some fully encapsulating suits, sometimes called cocoon suits, are designed to provide additional skin protection against solids and liquids, but they are not gas- or vaportight. Vaportight fully encapsulating suits prevent entry of gases and vapors and are considered to provide the highest level of protection to the wearer.

Figure 7-2. A worker wearing a chemical protective coverall with hood. Note that he is using an airline, but his facepiece also is equipped with air purifying cartridges. The airline hose has been sheathed in protective material to avoid excessive contamination.

Nonencapsulating suits or splash suits do not have a facepiece or a visor as an integral part of the suit. These suits are designed to provide protection against liquids and solids; they are *not* designed to protect against vapors and gases. Taping the ankles, wrists, and other susceptible areas can decrease skin exposure to gases and vapors. Splash suits come in a one-piece coverall style or a two-piece pants and coat style. Both styles may include a hood or other accessories.

Chemical protective clothing is also classified on the basis of the material from which it is made. Protective materials can be categorized as those that are made from elastomers and those that are not. Elastomers are polymeric (plastic-like) materials that return to their original shape after being stretched. Most CPC materials are elastomers. These materials include polyvinyl chloride, polyethylene, neoprene, nitrile rubber, polyvinyl alcohol, viton, Teflon, and butyl rubber. Nonelastomers are materials that are not elastic and cannot be stretched without ripping. Nonelastomers include Tyvek and coated-Tyvek fabrics.

Most of the CPC used today is considered disposable. The choice between disposable and reusable CPC is based on cost and ease of decontamination. In many situations, extensive decontamination procedures are not cost-effective, and disposable garments are used. In other cases, procedures required to put a reusable suit back in service, such as pressure-testing a vaportight fully encapsulating suit, are time-consuming and may not be feasible under field conditions. Wearing disposable or single-use garments over reusable clothing is a common practice employed to minimize wear and tear on the more expensive garment.

Performance Requirements for CPC

Performance requirements must also be considered in selecting the appropriate protective material. A variety of factors should be considered:

1. *Chemical resistance*, the ability of a material to withstand chemical and physical changes. The material's chemical resistance is the most important performance requirement. The chemical protective garment must maintain its structural integrity and protective qualities upon contact with a hazardous substance.

2. *Durability*, a measure of the material's inherent strength, which is manifested by the ability to withstand wear. Durable CPC should have the capability to resist punctures, abrasions, and tears.

3. *Flexibility*, which is extremely important for both glove and body-suit materials because it directly affects the wearer's mobility and agility. Flexible materials allow ease of movement and work.

4. *Temperature resistance*, is the material's ability to maintain its chemical resistance during temperature extremes (especially heat) and to remain flexible in cold weather. Temperature resistance can vary dramatically between different materials.

5. *Service life*, which is important for reusable garments and can be defined as the ability of a material to resist aging and deterioration. Exposure to chemicals,

temperature extremes, moisture, ultraviolet light, and oxidizing agents can decrease service life, as does inadequate decontamination of the garments after use. Proper cleaning, storage, and maintenance can help prevent aging.

6. *Shelf life,* which is important for disposable CPC because it may be stored for some time until needed. The shelf life can be extended by following the manufacturer's recommendations regarding storage conditions.

7. *Ease of decontamination,* a measure of the ability of a material to release chemical contamination on its surface. This property is especially important for reusable CPC. A few materials are very difficult to decontaminate, and in some cases a reusable garment can be covered by another material to prevent or minimize gross contamination.

8. *Suit design,* the way a suit is constructed or its compatibility with the anticipated type of respiratory protection that will be employed. Design also includes general and specific features of the suit and accessories. Considerations in determining the best suit design for a particular use include:

   • Whether it is fully encapsulating or nonencapsulating.
   • Whether it is one- or two-piece.
   • Availability of hoods, visors, gloves, and boots.
   • Location of zippers (front, side, back entry).
   • Location of pockets, straps, elbow/knee pads.
   • Seams and stitching.
   • Exhalation valves and ventilation ports.
   • The airline passthrough valve.

9. *Size,* is the actual dimensions of the CPC, a factor directly related to wearer comfort, dexterity, mobility, and efficiency. Ill-fitting clothing can be severely limiting to the wearer and increases the risk of accidents and falls. Manufacturers offer standard sizes in boots and gloves for men and women, but standard suit sizes for women are not yet available. Samples of CPC in different sizes should be tested for fit, flexibility, and utility prior to purchase.

10. *Color,* which can make it easier to maintain visual contact between workers. Color-coding garments (e.g., using different colors for different sizes of gloves or different types of glove materials) can also facilitate the selection of proper CPC and help ensure worker compliance to a PPE program. Dark-colored materials absorb radiant heat from the sun and other external sources and may contribute to heat stress.

11. *Cost,* which varies considerably and plays a significant role in CPC selection and frequency of use. Limited-use or disposable garments are often less expensive than multiple-use CPC, and do not require decontamination and cleaning. All CPC should be considered expendable; in some instances high quality, expensive clothing or equipment may have to be discarded after limited use.

Chemical Resistance

The effectiveness of a protective material against chemicals is based on its resistance to penetration, permeation, and degradation (Figure 7–3). Each of these properties should be evaluated in choosing CPC. When selecting a material for CPC, it is recommended that one remember these warnings:

- No protective material is truly impermeable.
- There is no single material that provides protection against every chemical.
- For certain chemicals or chemical mixtures, there is no material that will protect one for an unlimited period of time after initial contact with the substance.

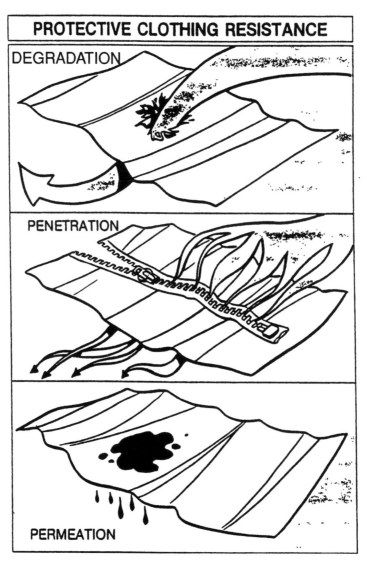

Figure 7–3. Protective clothing resistance to chemicals is described in terms of degradation, penetration, and permeation.

Penetration is the passage of a chemical through openings in a garment. A chemical may penetrate because of the suit design or imperfections. Stitched seams, buttonholes, pinholes, zippers, and woven fabrics can provide avenues for chemical penetration. Physical damage, such as rips, tears, punctures, or abrasions, will also allow chemical penetration. Suits with self-sealing zippers, seams with a tape overlay, flap closures, and nonwoven fabrics are resistant to chemical penetration.

Degradation is a breakdown of the protective material due to chemical action. Evidence of degradation can be detected by carefully examining CPC exposed to chemicals. The protective material may shrink or swell, become brittle or soft. Some changes may be manifested by slight or obvious discoloration or by a rough, spongy, or gummy surface, puckering, or cracks. Chemical degradation changes the properties of protective materials and may enhance permeation or allow contaminant penetration.

Degradation test data for specific chemicals or chemical classes are made available by manufacturers, suppliers, or other sources (Table 7–1). These data provide the user with a general degradation resistance rating, subjectively expressed as excellent, good, fair, or poor. Degradation data are important but must be supplemented with permeation data before garment selection. A protective material can have excellent degradation resistance for a chemical and poor permeation resistance. Permeation and degradation are not directly related and cannot be used interchangeably.

Permeation is a chemical action involving movement of chemical molecules through

Table 7–1. Degradation resistance of protective materials*

	Nitrile rubber	Neoprene	Natural rubber	Polyvinyl alcohol	Polyvinyl chloride
Acetic acid	G	E	E	NR	F
Acetone	NR	G	E	P	NR
Benzene	P	NR	NR	E	NR
Butyl alcohol	E	E	E	F	G
Carbon tetrachloride	G	NR	NR	E	F
Chloroform	NR	NR	NR	E	NR
50% Chromic acid	F	NR	NR	NR	G
Diethylamine	F	P	NR	NR	NR
Ethyl acetate	NR	F	G	F	NR
Formaldehyde	E	E	E	P	E
Gasoline	E	NR	NR	G	P
Hexane	E	E	NR	G	NR
Hydrochloric acid	E	E	G	NR	E
Kerosene	E	E	F	G	F
Methyl alcohol	E	E	E	NR	G
70% Nitric acid	NR	G	NR	NR	F
Nitrobenzene	NR	NR	F	G	NR
Pentachlorophenol	E	E	NR	E	F
Perchloroethylene	G	NR	NR	E	NR
Phenol	NR	E	NR	F	G
50% Sodium hydroxide	E	E	E	NR	G
Styrene	NR	NR	NR	G	NR
95% Sulfuric acid	NR	F	NR	NR	G
Toluene	F	NR	NR	G	NR
Trichloroethylene	NR	NR	NR	E	NR
Turpentine	E	NR	NR	G	P
Xylenes	G	NR	NR	E	NR

*E = excellent; G = good; F = fair; P = poor; NR = not recommended.

intact material. The permeation process involves sorption of the chemical on the outside surface, diffusion through the material, and desorption on the inside surface. A chemical gradient (high on the outside, low on the inside) is established. A high flow of chemical permeation is achieved and continues until the concentrations on the inside and the outside are approximately equal, and an equilibrium is achieved.

Permeation test data are measured and reported as rates. The permeation rate is the amount of chemical that will move through an area of protective material in a given interval, usually expressed in milligrams of chemical permeated per square meter per minute of exposure ($mg/m^2/min$). Other rates, such as $\mu g/cm^2/min$ or $mg/m^2/hr$, are also used.

A variety of factors can influence permeation, including the type of protective material and its thickness. Other factors to consider are the concentration of chemicals, contact time, temperature and humidity, and susceptibility of the protective material to chemical degradation.

Another measure of permeation resistance is the breakthrough time (Table 7–2), which is the elapsed time, usually in minutes, between the initial contact of a chemical with the outside surface of a protective material and detection of the chemical at the inside surface. The breakthrough time is chemical-specific for a particular material and is influenced by the same factors that affect permeation.

Breakthrough and permeation test data are available from manufacturers, which give specific rates and times. A given manufacturer's recommendations can serve as a guideline

Table 7–2. Permeation resistance of protective materials*

	Nitrile rubber	Neoprene	Natural rubber	Polyvinyl alcohol	Polyvinyl chloride
Acetic acid	270	420	135	—	180
Acetone	—	5	10	—	—
Benzene	—	—	—	7	—
Butyl alcohol	ND	240	15	30	180
Carbon tetrachloride	150	—	—	ND	25
Chloroform	—	—	—	ND	—
50% Chromic acid	240	—	—	—	ND
Diethylamine	45	—	—	—	—
Ethyl acetate	—	15	5	ND	—
Formaldehyde	ND	120	60	—	80
Gasoline	ND	—	—	ND	—
Hexane	ND	45	—	ND	—
Hydrochloric acid	ND	ND	>300	—	>300
Kerosene	ND	>300	30	ND	>360
Methyl alcohol	11	60	13	—	45
70% Nitric acid	—	140	—	—	345
Nitrobenzene	—	6	5	>300	—
Pentachlorophenol	ND	—	—	7	180
Perchloroethylene	300	180	—	300	—
Phenol	—	ND	60	30	75
50% Sodium hydroxide	ND	—	ND	—	ND
Styrene	—	180	—	ND	—
95% Sulfuric acid	—	—	—	—	220
Toluene	10	—	—	15	—
Tichloroethylene	—	—	—	30	—
Turpentine	30	—	—	300	—
Xylenes	75	—	—	ND	—

*As measured by breakthrough time in minutes; ND = not determined; > = greater than; — = poor no degradation resistance.

for proper selection of an appropriate protective material. Test data are usually produced according to Method F739 (ASTM 1985), a standard test method developed by ASTM for the measurement of protective clothing material permeation by liquids and gases. Method F739 is one of 15 test methods developed by ASTM for evaluation and selection of protective clothing and protective materials.

The best protective material for a specific chemical is one that has high degradation resistance, a low or negligible permeation rate, and a long breakthrough time. This optimal combination often is not obtainable. The data reflect test results for pure chemicals only; there are few or no data for chemical mixtures or low concentrations in aqueous solutions. In many cases, only chemicals that are considered skin hazards, such as corrosives, are subject to extensive testing.

In addition to manufacturers' chemical reference data, other sources of information on protective material performance are *Guidelines for Selection of Chemical Protective Clothing* (Schwope et al., 1987) and a handy pocket reference entitled *Quick Selection Guide to Chemical Protective Clothing* (Forsberg and Mansdorf, 1997). Both references are compilations of test data from manufacturers, vendors, and independent laboratories. Computerized data bases for protective clothing selection are also available.

Chemical Resistance Testing

Permeation resistance and breakthrough times are determined by using small swatches of the protective material. Its other aspects, such as penetration and the performance of zippers, visors, boots, and glove materials, are not considered or must be checked separately.

Most hazardous waste sites contain mixtures of chemicals, but chemical resistance data for multiple chemicals are not available. Furthermore, resistance data are collected under very controlled laboratory conditions that do not consider the effect of multiple chemical exposures. The effects of physical stress on the suit material caused by stretching, humidity, and temperature extremes also are not considered.

The breakthrough time is important, but the permeation rate is also of interest, as it indicates how "leaky" the material becomes when breakthrough occurs. For example, according to Table 7–3, acetone and ethyl acetate have similar breakthrough times for the protective material Chemtuff™, but the permeation rate for ethyl acetate is nearly seven times greater than that for acetone. Manufacturers that provide permeation only in qualitative terms (excellent, good, fair) should provide the criteria (e.g., breakthrough time and permeation rate) and test method used. If two protective materials have the same or similar breakthrough times, the material with the lower permeation rate should be selected.

All manufacturers should indicate which ASTM methods are used for testing CPC. ASTM D751 includes test methods to assess the strength of seams and suit closures, such as zippers (ASTM 1979). Vaportight, fully encapsulated suits should pass pressure testing requirements outlined in ASTM F1052 (ASTM 1987). The resistance to tearing, flammability, and temperature-induced flexibility changes should fall within acceptable ranges.

In addition to ASTM, NFPA has standards for the selection of chemical protective clothing. NFPA-1991 describes criteria for gas- and vaportight suits. For a suit to receive an NFPA-compliant label, it must resist breakthrough for each of 15 liquid chemicals, which represent different chemical classes. Resistance to chlorine and ammonia gases is also required by NFPA-1991.

Table 7-3. Chemical permeation rate for CHEMTUFF™

Chemical	Breakthrough time (minutes)	Permeation rate $(\mu/cm^2/hr)$
Acetone	20	8
Ethyl acetate	23	53
Hydrochloric acid	>480	NA
Methylene chloride	4	389
Nitric acid	>480	NA
Nitrobenzene	309	3
Sodium hydroxide	>480	NA
Styrene	16	1420
Sulfuric acid	>480	NA
Trichloroethylene	7	289

> = greater than; NA = not applicable.

NFPA-1992 describes a decision matrix for splash-protective garments. NFPA-1993 itemizes selection criteria for limited-use protective garments worn by support personnel or entry personnel once significant hazards have been removed or mitigated. NFPA standards also consider resistance to flame impingement and flexibility under cold temperatures.

Chemical Protective Materials—Elastomers

A wide range of protective materials are used for CPC, some of the more common of which are listed below. The ratings of these elastomers are based on resistance to degradation and permeation; but there are many exceptions within each chemical class, and whenever possible, the manufacturer should be contacted for specific chemical resistance data.

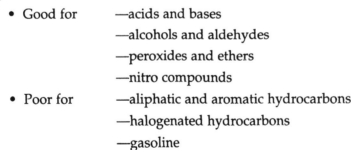

Butyl rubber:
- Good for —acids and bases

     —alcohols and aldehydes

     —peroxides and ethers

     —nitro compounds
- Poor for —aliphatic and aromatic hydrocarbons

     —halogenated hydrocarbons

     —gasoline

Neoprene (chloroprene):
- Good for —inorganic bases

     —dilute inorganic acids

     —organic acids

     —peroxides

     —alcohols

- Poor for   —aromatic hydrocarbons

                —halogenated hydrocarbons

                —nitro compounds

                —gasoline

                —concentrated organic and inorganic acids

Nitriles (acrylonitrile rubber, NBR, hycar):
- Good for   —alipathic hydrocarbons

                —gasoline and fuels

                —alcohols

                —organophosphates

                —acids and bases
- Poor for   —aromatic hydrocarbons

                —halogenated hydrocarbons

Polyvinyl alcohol (PVA):
- Good for   —almost all organics including:

                   aromatic and aliphatic hydrocarbons and

                   halogenated hydrocarbons

                —gasoline and fuels
- Poor for   —acids and bases

                —water-soluble organics such as:

                   alcohols, esters, ethers, acetates

                —water and solutions containing water

Polyvinyl chloride (PVC)
- Good for   —acids and bases
- Poor for   —aldehydes and ethers

                —halogenated hydrocarbons

                —aliphatic and aromatic hydrocarbons

Viton (fluorocarbon rubber, FPM)
- Good for   —acids

                —chlorinated hydrocarbons

                —aliphatic and aromatic hydrocarbons

                —gasoline and fuels

                —alcohols and peroxides
- Poor for   —aldehydes and ketones

                —amines, ethers, esters

Teflon (polytetrafluorethylene, PTFE)
- Good for   —chlorinated hydrocarbons

                  —aliphatic and aromatic hydrocarbons

                  —gasoline and fuels

                  —acids

Some manufacturers have developed techniques for layering protective materials. A suit can be designed with multiple layers that offer increased versatility and chemical resistance, but usually increase the thickness of the protective material and the weight of the entire garment. Examples of layered, nondisposable suits are Viton/butyl, Viton/neoprene, and butyl/neoprene. Other manufacturers have developed laminated protective materials for use in lightweight, disposable suits. In most cases, the garment material is Teflon or a similar material.

Silver Shield™ is made from a flexible elastomeric laminate that is available only in gloves. Silver Shield gloves display good chemical resistance to a wide range of organic and inorganic chemicals, including ketones, aromatic and aliphatic hydrocarbons, ethers, gasoline and fuels, and many chlorinated hydrocarbons. They are thin and lightweight but decrease wearer dexterity when used as an overglove. They are often used as a middle glove worn under a thicker, more rugged overglove.

Chemical Protective Materials— Nonelastomers

Nonelastomers are rated in the same way as elastomers:

Uncoated Tyvek (nonwoven polyethylene fibers)
- Good for   —dry particulates, dusts, asbestos

                  —minimizing gross contamination of CPC
- Poor for   —chemical vapors or liquids

Polyethylene (coated Tyvek)
- Good for   —acids and bases

                  —some alcohols and aldehydes

                  —low concentration vapors with no liquid splash

                  —minimizing gross contamination of underlying CPC, there by decreasing time and effort expended during CPC decontamination.

     Poor for   —halogenated hydrocarbons

                  —aliphatic and aromatic hydrocarbons

                  —gasoline and fuels

Saranex (Saran-coated Tyvek)
- Good for   —acids and bases

                  —some organics

> —low concentration vapors where there is low risk of splash or contact with liquids
>
> —minimizing gross contamination of underlying CPC, there by decreasing time and effort expended during CPC decontamination

- Poor for   —halogenated hydrocarbons

    —aliphatic and aromatic hydrocarbons

Thermal Protection

In general, CPC has poor thermal protective properties. A one-second flash of a flammable gas can destroy a coated Tyvek coverall, and a few seconds of flame impingement can cause complete destruction of a reusable fully encapsulated suit. For potential fires with unknown contaminants, fire-resistant coveralls made of Nomex or PBI/Kevlar should be worn underneath CPC.

In some situations, thermal protection is worn over CPC. Flash cover suits provide limited protection from flashbacks and are worn over fully encapsulated suits. These thermal protective overgarments may be made of aluminized Nomex or PBI/Kevlar. Proximity suits provide short-duration and close-proximity protection to radiant temperatures as high as 2000°F; these suits are not designed for entry into a total flame environment. Fire entry suits protect the wearer from fire impingement; these suits are utilized at petrochemical fires.

Selecting Chemical Protective Clothing (CPC)

Selecting the most effective CPC is difficult when there is little or no information about the number and the identify of contaminants on-site. As uncertainties about the substances present increase, selecting appropriate CPC becomes more complex. In some cases, knowledge about the contamination on-site is of minimal value because there often is insufficient information about the protective qualities of commonly used CPC against specific chemicals. In other cases, there is no single protective material available that offers optimal resistance against all the chemicals known or anticipated to be present.

When the identities of contaminants are known, the CPC selection process consists of:

- Identifying the chemicals involved and determining their physical, chemical, and toxicologic properties.
- Estimating the concentrations present on-site.
- Determining if there is a hazard to site workers who may come into contact with anticipated concentrations.
- Selecting the protective material that provides the best protection (least degradation, lowest permeation).
- Determining the appropriate style of CPC to use (fully encapsulating or nonencapsulating).
- Deciding if a fully encapsulated, vaportight suit is necessary.

Situations that could warrant wearing a fully encapsulating vaportight suit include:

- Presence of known contaminants that pose a dermal or inhalation hazard.
- Visible emissions of gases or vapors.
- Indications of airborne hazards based on direct-reading instrumentation.
- Presence of unknown chemicals that cannot be evaluated by using air monitoring instruments or other detection devices.
- Presence of container labels or vehicle placards that suggest the contents are very toxic or corrosive.
- Enclosed or poorly ventilated areas where toxic or corrosive gases could accumulate.
- Site activities that increase risk of exposure to high concentrations of toxic or corrosive materials.

Fully encapsulated clothing should be worn only when necessary, as nonencapsulating suits are less bulky, lighter, and easier to wear than fully encapsulating suits. Visibility is less impaired, and the wearer is less prone to accidents when wearing a nonencapsulating suit. Also it is more difficult to communicate while wearing a fully encapsulating suit. Heat and physical stress can be severe limitations of fully encapsulating suits.

Permeation and degradation data should be consulted to determine the optimal protective material. The best protective material to choose is one that is resistant to degradation and has a long breakthrough time and a low permeation rate. As explained earlier, permeation and degradation are not the same and cannot be used interchangeably. Penetration resistance is difficult to assess unless the entire garment has been tested.

There may be situations where the identity of the contaminants present is unknown, or there are no chemical resistance data for the contaminants known to be present. Then probably the best option would be to use a material that protects against a wide range of chemicals, such as Viton, Teflon, Silver Shield™ (for gloves only), or one of the laminated protective materials.

Other factors to consider in selecting CPC include:

- Durability and strength of the protective material and its resistance to tears, punctures, and abrasions.
- Flexibility of the material—an important consideration in selecting gloves.
- Effect of temperature on suit integrity and flexibility.
- Ease of decontamination and availability of decontamination procedures on-site.
- Compatibility of CPC with other equipment used on-site, such as hardhats or respirators.
- Duration of use (i.e., can the required task be accomplished before breakthrough or degradation occurs?).
- Presence of a fire/explosion hazard that requires the use of special protective equipment.

It must be noted that most ground-water contamination sites usually have low levels of chemical contamination. In working with such low levels (ppb to ppm) in water, protective clothing that is impermeable to water usually is sufficient. At sites with high levels of contamination, or during installation of monitoring or production wells into NAPL (nonaqueous phase liquids), workers must be careful to select the proper CPC and to follow contamination avoidance practices.

Physical Stress of Chemical Protective Clothing

Wearing CPC can increase the risk of personal injury because of decreased visibility, decreased audio and visual acuity, decreased mobility and dexterity, and increased physical exertion. The severity of these problems depends on the type and the style of CPC worn. Common accidents associated with wearing CPC include slips, trips, falls, and being struck by a vehicle. It is important to recognize that *health and safety are not mutually inclusive.* Minimizing a health risk by wearing protective equipment can increase a safety risk or another health risk.

The increase in physical exertion caused by working in CPC can increase fatigue and can lead to more serious conditions such as stroke or heart attack in susceptible workers. To minimize the effects of physical stress, workers wearing CPC should vary their work routine and include work and rest periods as appropriate. During periods of hot weather, projects should be scheduled for the cooler times of the day whenever possible. Fluid intake should be increased and body electrolytes replaced. The workers should be in good health, and they should be encouraged to maintain an optimum level of physical fitness and normal body weight.

Heat Stress and Chemical Protective Clothing

Heat stress is one of the most common illnesses encountered at hazardous waste sites, but regular monitoring and other preventive methods can reduce its incidence. Heat stress is caused by a number of factors, including environmental conditions, the use of CPC, the level of physical activity, and individual characteristics of the worker. Protective clothing interferes with the dissipation of body heat. Evaporation, the body's primary cooling mechanism, is reduced, as ambient air cannot come in contact with the skin's surface. Other heat exchange mechanisms (convection and radiation) are also impaired. Heat stress combined with dehydration from excessive perspiration can be debilitating. Indeed, workers wearing CPC will pour water out of their boots at the end of a work cycle. It is not unusual for workers wearing CPC to lose several pounds of body water weight in a single day.

Individuals vary in their susceptibility to heat stress. Factors that may predispose them to heat stress include:

- Lack of physical fitness.
- Obesity.
- Lack of acclimatization to working conditions.

- Age.
- Dehydration and loss of body salts (electrolytes).
- Alcohol and drug use.
- Sunburn (increases body heat).
- Diarrhea (increases dehydration/electrolyte loss).
- Chronic disease.

Avoidance of heat stress requires proper training and the use of preventive measures. Prevention is especially important because an episode of heat stroke or heat exhaustion may predispose a worker to additional heat stress injuries. To avoid heat stress, the following steps can be taken:

- Adjust work schedules to take advantage of cooler portions of the day.
- Rotate personnel to minimize overexertion at one activity or location.
- Modify work/rest schedules as needed.
- Provide shelter or shade to protect workers during rest periods.
- Maintain normal body fluid levels by providing cool liquids and encouraging workers to drink.
- Provide cooling devices such as field showers or hose-down areas to cool workers wearing protective clothing.
- Provide cooling jackets, suits, or vests.
- Train workers to recognize symptoms of heat stress.

Fluid replacement is necessary to avoid heat stress; the normal thirst mechanism is not sufficiently sensitive to ensure that the worker will drink enough liquids to replace the fluid lost through sweating. Workers should drink at least 16 ounces of fluid (water, diluted fruit juice, or diluted fluid-replacement beverage) before beginning work, and should then drink at least 8 ounces during each break. Each worker should consume at least one to two gallons of fluid per day.

Physiological Monitoring for Heat Stress

Site workers should be carefully monitored for signs of heat stress. Because the incidence of heat stress varies by individual and other factors, even workers not wearing PPE should be monitored. Monitoring methods include measuring the heart rate, the oral temperature, and body water loss.

If the heart rate exceeds 110 beats per minute at the beginning of the rest period, the next work cycle should be shortened by one-third, with the same rest period interval. If the heart rate still exceeds 110 beats per minute at the end of the shortened work cycle, the next work cycle should be shortened by one-third.

A clinical oral thermometer should be used to measure the worker's oral temperature at the end of the work period before any fluid consumption. If the oral temperature exceeds 99.6°F, the next work cycle should be shortened by one-third, with the same rest period

Figure 7-4. The carotid pulse of a worker is checked before he returns to work. A rapid heart rate is one sign of heat stress.

interval. If at the end of the next, shortened work period the oral temperature still exceeds 99.6°F, the following work cycle should be shortened by one-third. A worker with an oral temperature greater than 100.6°F should not be permitted to return to work.

Body water loss can be measured by weighing each worker at the beginning and the end of each workday to ensure that sufficient fluids are being taken to prevent dehydration. The scale should be accurate to within 0.25 pound; the worker should wear dry underwear or be nude. The daily body water loss should not exceed 1.5% total body weight; for a 170-pound worker, the body water loss should not exceed 2.5 pounds.

The frequency of monitoring and the work–rest regimen vary by temperature, type of PPE utilized, and worker activity levels. High worker activity and the use of impermeable CPC necessitate more frequent monitoring and rest. ACGIH recommends work–rest regimens that are temperature- and work load-dependent (Table 7-4). The ACGIH temperature TLV is for workers wearing permeable summer work uniforms. Correction factors are used for workers wearing other types of clothing, such as cotton coveralls or winter work uniforms. The ACGIH temperature TLV does not apply to workers wearing semipermeable or impermeable CPC. For these situations, workers should be monitored when the adjusted temperature exceeds 72.5°F (see Table 7-5). In some cases, the monitoring frequency must be increased to accommodate more frequent rest intervals, which may be required as the work load increases. Note that in both cases, the frequency of monitoring mandates the frequency of rest intervals for workers.

Both monitoring approaches presented above assume that workers have been acclimatized to working temperatures, are in good health, and are physically fit. Acclimatization

Table 7–4. ACGIH permissible heat exposure TLV

Hourly work–rest regimen	Worker Activity Level		
	Light	Moderate	Heavy
Continuous work	86*	80	77
45 minutes–15 minutes	87	82	78
30 minutes–30 minutes	89	85	82
15 minutes–45 minutes	90	88	86

*Wet bulb globe temperature in °F.
Source: ACGIH (1996).

to hot temperatures normally requires five to seven days. Workers should be assigned to activities that do not require PPE until they are fully acclimated.

OTHER TYPES OF PERSONAL PROTECTIVE EQUIPMENT

In addition to respiratory protection and protective clothing, other types of protective equipment, designed to protect the face, head, eyes, and extremities (Figure 7–5), must be provided and used whenever there is a potential for injury through physical contact or absorption. Protective equipment should be provided by the employer; but even if the employee owns the equipment, it is the employer's responsibility to ensure that it is adequate for its intended use and is adequately maintained. These requirements are part of the OSHA general industry standards and are not specific to HAZWOPER sites. The HAZWOPER rule states that all requirements of 29 CFR 1910 and 29 CFR 1926 (the construction industry standard) apply to hazardous waste and emergency response operations, whether covered by HAZWOPER or not.

Head Protection

Safety helmets or hard hats are needed at sites where there is a potential for head injury from falling objects or moving equipment. Most hard hats are molded plastic or fiberglass. Both materials resist water, oil, and electricity, and are impact-resistant; but fiberglass hard

Table 7–5. Frequency of physiological monitoring during work

Temperature * (in °F)	Normal work ensemble (cotton coveralls)	Chemical protective clothing
90 or above	every 45 minutes	every 15 minutes
87.5–90.0	every 60 minutes	every 30 minutes
82.5–87.5	every 90 minutes	every 60 minutes
77.5–82.5	every 120 minutes	every 90 minutes
72.5–77.5	every 150 minutes	every 120 minutes

*Temperature is adjusted by using equation: Adjusted temperature in °F = air temperature + (13 × % sunshine). Percent sunshine can be estimating what percent of time the sun is not covered by clouds thick enough to produce a shadow (e.g., 0% sunshine equals no shadows).
Source: NIOSH/OSHA/USCG/EPA (1985).

Body Part Protected	Type of Clothing	Type of Protection and Precautions
FULL BODY 	Fully-encapsulating suit (one-piece garment. Boots and gloves may be integral, attached and replaceable, or separate).	Protects against gases, dusts, vapors, and splashes. Does not allow body heat to escape. May contribute to heat stress in wearer.
	Non-encapsulating suit (jacket, hood, pants, or bib overalls, and one-piece coverall).	Protects against splashes, dust, and other materials but not against gases and vapors. Does not protect parts of head and neck. Do not use where gas-tight or pervasive splashing protection is required.
	Aprons, leggings, and sleeve protectors (may be integral or separate). Often worn over non-encapsulating suit.	Provides additional splash protection of chest, forearms, and legs. Useful for sampling, labeling, and analysis operations.
	Radiation-contamination protective suit.	Protects against alpha and beta particles. Does NOT protect against gamma radiation. Designed to prevent skin contamination.
	Flame/fire retardant coveralls (normally worn as an undergarment).	Provides protection from flash fires. May exacerbate heat stress.
HEAD 	Safety helmet (hard hat, made of hard plastic or rubber). May include a helmet liner to insulate against cold.	Protects the head from blows, must meet OSHA requirements at 29 CFR §1910.135.
	Hood (commonly worn over a helmet).	Protects against chemical splashes, particulates, and rain.
	Protective hair covering.	Protects against chemical contamination of hair, prevents hair from tangling in equipment, and keeps hair away from respiratory devices.
EYES & FACE 	Face shield (full-face coverage, eight-inch minimum) or splash hood.	Protects against chemical splashes, but does not protect adequately against projectiles. Provides limited eye protection.
	Safety glasses.	Protects eyes against large particles and projectiles.
	Goggles.	Depending on their construction, can protect against vaporized chemicals, splashes, large particles, and projectiles.

Figure 7-5. Types of personal protective equipment. (*Source:* NIOSH/OSHA/ USCG/EPA, 1985.)

hats usually have greater impact strength and higher dielectric strength (i.e., better electrical insulating properties) than plastic.

Type 1 safety helmets have a full brim, whereas Type 2 helmets are brimless with a peak in the front. Three classes of hard hats are recognized according to their intended use: Class A general service helmets are designed for protection against impact hazards and have only

Body Part Protected	Type of Clothing	Type of Protection and Precautions
EARS 	Ear plugs and muffs.	Protects against physiological damage from prolonged loud noise. Use of ear plugs should be reviewed by a health and safety officer because chemical contaminants could be introduced into the ear.
	Headphones (radio headset with throat microphone).	Provides some hearing protection while allowing communication.
HANDS & ARMS 	Gloves and sleeves (may be integral, attached, or separate from other protective clothing).	Protects hands and arms from chemical contact. Wearer should tape-seal gloves to sleeves to provide additional protection and to prevent liquids from entering sleeves. Disposable gloves should be used when possible to reduce decontamination needs.
FEET 	Chemical-resistant safety boots.	Protects feet from contact with chemicals.
	Steel-shank or steel-toe safety boots.	Protects feet from compression, crushing, or puncture by falling, moving, or sharp objects. Should provide good traction.
	Non-conductive or spark-resistant safety boots.	Protects the wearer against electrical hazards and prevents ignition of combustible gases or vapors.
	Disposable shoe or boot covers (slips over regular foot covering).	Protects safety boots from contamination and protects feet from contact with chemicals. Use of disposable covers reduces decontamination needs.

Figure 7.5 (Continued)

limited electrical resistance; they are used in mining, drilling, construction, lumbering, and manufacturing. Class B utility service helmets are intended for protection against impact and penetration, and from high voltage shock and burn; they are used by electrical and utility workers. Class C safety helmets are designed for lightweight comfort and impact protection; usually made of aluminum, they offer no electrical resistance.

The OSHA standard (29 CFR 1910.135) requires that safety helmets purchased after July 5, 1994 comply with *ANSI Z89.1–1986, Protective Headwear for Industrial Workers—Requirements*. Helmets purchased prior to July 5, 1994 must comply with *ANSI Z89-1–1969, Safety Requirements for Industrial Head Protection*. The ANSI head protection standard is periodically updated, and hard hats meeting later, more stringent requirements are acceptable for use under the OSHA rule. Hard hat users can identify the type of helmet they use by looking inside the shell for the ANSI designation and class; for example:

Manufacturer's Name

ANSI Z89.1–1989 (*or other date*)

Class A

Safety helmets can be fitted with a variety of auxiliary features, including liners for cold weather use, chin straps, brackets to support miners' lamps, ear muffs, goggles, and faceshields.

Face and Eye Protection

Face protection can be provided by faceshields made of clear plastic or a fine mesh made of nylon or steel. Plastic faceshields are designed to protect against impact from liquids and solids; nylon or steel mesh shields protect against solids only. Faceshields may be attached to a safety helmet, safety goggles, or a head harness.

In order to provide adequate protection, the faceshield should be long enough to cover the entire face (i.e., at least 9 to 12 inches long.) Many shields are sufficiently long to protect the neck. Shields vary and may include a wrap-around design, flared sides, or side extenders to protect the sides of the head and the ears. The OSHA rule (29 CFR 1910.133) requires that eye and face protection purchased after July 5, 1994 meet minimum standards defined by *ANSI Z87.1–1989, Occupational and Educational Eye and Face Protection.* Face protection devices purchased prior to July 5, 1994 may comply with the less rigorous requirements of *ANSI Z.87.1–1968.*

Although a faceshield is the first barrier against flying materials, it provides only secondary protection for the eyes. Primary eye protection is afforded by safety glasses or goggles. Faceshields do not protect against eye contact with vapors or gases.

Protective eyewear provides a barrier against flying particles, dusts, chemical splashes, and radiant light. Standard safety glasses offer protection against hazards that may strike from the front; the addition of side shields provides protection against hazards that may come from the side. Safety glasses with eyecup side shields provide protection against hazards that may come from the front, side, above, or below. Safety glasses protect against liquid splash and solids; they offer no protection against gases and vapors.

Goggles with direct ventilation fit snugly to the face and protect against particles that may strike from any angle. Goggles with indirect or hooded ventilation also block the entry of dusts and chemical splashes. Nonvented, gasproof or chemical goggles protect against gases, vapors, fumes, and impact from liquids and solids. Goggles with any type of ventilation allow gases and vapors to enter and contact the eye; only nonventilated goggles can protect against eye contact with gases and vapors.

Hand Protection

The potential for hand injury is increased whenever chemical, physical, or mechanical hazards are present. Hand protection in the form of gloves can prevent skin damage from chemicals and avoid their absorption into the body. Hand protection can also lessen skin damage from other causes, such as blisters or cuts from mechanical damage, burns from excessively hot or cold materials, allergic reactions from exposure to plants and insects, and infections from bacteria or fungi. The OSHA rule (29 CFR 1910.138) requires employers to evaluate hand protection devices with respect to the specific tasks involved and the identified hazard.

There are different types of hand and arm protection, designed for specific hazards. Gloves come in a variety of lengths, which extend to the wrist (8"–9"), above the wrist (12"), to the forearm (13"–15"), to the upper arm (18"), or to the shoulder (24"). Gloves that extend to the elbow or above are sometimes called gauntlet gloves or sleeved gloves. Protective sleeves or arm guards, made of the same material as protective gloves or a different material, can also be used for arm protection.

Chemical-resistant gloves and sleeves are made of the same protective materials as those used for CPC. Chemical resistance data are especially important for glove materials. Chemical protective gloves are often worn over an underglove of latex or vinyl, and covered by an overglove to protect against heat, cold, sharp objects, and abrasion. The use of the overglove guards chemical protective gloves against conditions that might affect their chemical resistance.

Overgloves are often made of leather, cotton, or other materials that have no chemical resistance. If contaminated, they should be discarded after use. When there is a high probability of chemical contamination, a second, chemically resistant overglove can be used.

Foot and Leg Protection

At hazardous waste sites, safety shoes should provide foot protection against falling objects, punctures, crushing injuries, and chemical exposure. Safety shoes come in a variety of styles and sizes, and can be comfortable if the proper size and width are worn. A typical safety shoe has a steel toe, a steel shank, and a cleated or chevron sole.

Most safety shoes or boots, however, have little or no chemical resistance. Leather safety shoes cannot be decontaminated and should not be worn on hazardous waste sites unless covered with a chemical-resistant overboot. Overboots are designed to slip over a safety shoe or a safety boot, and can be made of butyl rubber, nitrile rubber, vinyl, neoprene, PVC, or polyurethane materials. Overboots normally do not have aggressive soles and can increase the risk of slipping; lightweight, removable shoe chains or shoe spikes can be used for extra traction.

Chemical safety boots are over-the-sock boots that combine one or more safety shoe features with chemical resistance. These boots come in a wide choice of chemical protective materials, including blends such as PVC/urethane and PVC/nitrile rubber. Chemical safety boots are often covered with an overboot to increase chemical resistance and minimize contamination.

Foot guards which protect the entire foot (not just the toes), metatarsal guards which protect the area between the ankle and the toes, or shin guards can be added when there is a risk of injury to the areas they protect.

Safety footwear is classified according to its ability to meet minimum requirements for both compression and impact tests. OSHA (29 CFR 1910.136) requires that all safety footwear purchased after July 5, 1994 meet the minimum requirements described in *ANSI*

Figure 7–6. Leather safety shoes are not resistant to chemicals and cannot be decontaminated. This leather safety shoe shows excessive wear and should be discarded.

Z41–1991, Protective Footwear. Foot protection purchased prior to July 5, 1994 must meet the less rigorous requirements of *ANSI Z41–1967, Men's Safety-Toe Footwear.*

HEARING PROTECTION

A hearing conservation specialist should be consulted to recommend the most effective type of hearing protection for the noise encountered at specific hazardous waste sites. The best hearing protection is the one that is accepted and worn properly. Most properly fitted hearing protection devices can be worn continuously without significant discomfort.

There are two types of hearing protectors: those that are inserted into the ear, and those that cover the ear. Insert ear protectors or ear plugs are made of pliable rubber, plastic, or synthetic foam. Some pressure is required to fit ear plugs into the ear canal; a good seal cannot be obtained without some initial discomfort. Molded ear plugs are designed to fit an individual's ear canal. Molded plugs must be fitted by a trained professional, and each one will fit only one ear (they are not interchangeable).

Ear muffs cover the external ear. Muff effectiveness varies with the muffs' size, shape, and composition; also the size and the shape of the worker's ears and head should be considered. Most ear muff manufacturers offer only one size, but ear muffs need to make a perfect seal around the ear to be effective. Temple bars of eyeglasses, long sideburns, and long hair can affect their fit and effectiveness.

In situations that are extremely noisy, earplugs should be worn in addition to ear muffs. When used together, plugs and muffs change the nature of sounds; all sounds are reduced, but they are easier to hear and understand than they would be without the hearing protection.

OSHA (29 CFR 1910.95) requires that hearing-protective devices be made available to employees exposed to an 8-hour TWA of 85 decibels or greater. Hearing-protective devices should comply with *ANSI S3.19–1974, Measurement of Real-Ear Protection of Hearing Protectors and Physical Attenuation of Earmuffs.* The noise reduction rating (NRR) should also be checked to determine if it is adequate for the noise conditions. To determine the noise reduction capabilities in dBAs, subtract 7 from the NRR. Wearing muffs and plugs offers increased protection and a higher overall NRR, but the effect is not additive; the overall NRR will depend on the characteristics of the noise attenuated (frequency, impact vs. nonimpact) and the types of devices employed.

Inspection of Protective Equipment

An effective PPE inspection program should feature four different inspections:

1. Inspection and operational testing of PPE when it is first received.
2. Inspection of equipment when it is issued for use.
3. Inspection before and after use or training and prior to maintenance.
4. Periodic inspection of stored equipment.

Inspection checklists are often helpful in conducting inspections of chemical protective clothing (CPC). The following lists show what to look for:

1. Checklist for CPC before use:

   • Seams, zippers, other closures.
   • Color, surface irregularities, uneven coatings.
   • Cracks, swelling, stiffness.
   • Pinholes (hold clothing up to light; pressurize gloves).
   • Appropriate protective material for intended use.

- Appropriate size.
- Items certified clean and ready for use if reusable.

2. Checklist for CPC during use:

- Evidence of chemical or physical degradation.
- Seams, zippers, other closures.
- Tears, punctures, or abrasions.
- Gross contamination (see Figure 7–7).

Figure 7-7. Even the best personal protective equipment cannot protect against poor judgment. This worker is wearing chemical protective clothing that is contaminated. He removed his facepiece while waiting to go through decontamination. Note that the facepiece is resting against the contamination. A fellow worker supplied the cigarette.

8

Levels of Protection

Protective equipment should protect workers from specific hazards that may be encountered on-site. However, the selection of an appropriate PPE ensemble, or level of protection, can be a complex process. The more that is known about the site and its hazards, the easier it is to select the appropriate level of protection in terms of respiratory protection, CPC, and other accessory PPE.

The USEPA has defined four levels of protection for PPE ensembles (Level A through Level D; Figure 8–1); there are also recommended criteria for the use of each level of protection (Table 8–1). These criteria, defined in the EPA *Standard Operating Safety Guides* (SOSGs; USEPA, 1992a), are guidelines and should not be treated as hard and fast rules.

USEPA LEVELS OF PROTECTION

Level A

Level A protection is the highest level of respiratory, skin, and eye protection. The Level A ensemble consists of a one-piece, gas/vapor-tight, fully encapsulating chemical protective garment (Figure 8–2). Respiratory protection consists of a full-facepiece, positive pressure atmosphere supplying respirator (SCBA or airline with egress bottle).

If worker activities include handling contaminated materials, then inner chemical-resistant gloves are recommended. Chemical-resistant inner gloves provide an emergency barrier in case the outer gloves are inadvertently punctured or torn; the inner gloves also help reduce the potential for contamination while the garment is being removed. Optional equipment may include a hard hat, safety shoes or boots, inner coveralls, long cotton underwear, disposable outer gloves and overboots, and an internal cooling unit.

The EPA SOSGs provide guidance on when Level A should be employed. These are not strict rules but rather guidelines to assist in decision making. The EPA SOSGs list the following criteria that *may* warrant the use of Level A protection:

- Hazardous materials are known to be present that require the use of the highest level of protection for the skin, eyes, and respiratory system.
- The atmosphere contains less than 19.5% oxygen.
- Site activities involve a high potential for splash, immersion, or exposure to materials known to pose a skin contact hazard.
- Activities are being performed in a confined or poorly ventilated area, and the type and the amount of hazardous materials present cannot be determined.
- Direct-reading instruments indicate high levels of unidentified gases or vapors.

Many consider oxygen deficiency to be an insufficient criterion to warrant Level A. However, oxygen deficiency often is caused by the presence of another or displacing gas. An atmosphere that contains less than 19.5% oxygen should be considered an indicator that high concentrations of an unidentified, and potentially hazardous, gas or vapor are present.

Level B

Level B requires the same level of respiratory protection—full facepiece, positive pressure atmosphere supplying respirator—but less skin protection. In most cases, respiratory protection equipment is worn over CPC (Figure 8–3). Also recommended are inner and outer chemical-resistant gloves, chemical safety boots, and a hard hat. Coveralls or long cotton underwear may be worn under CPC, and disposable boot covers can be worn as overboots to minimize the contamination of reusable chemical safety boots. A faceshield attached to the hard hat decreases the risk of contamination of the facepiece.

The EPA SOSGs list the following criteria that warrant the use of Level B protection:

- The type and the amount of contamination require a high level of respiratory protection but less skin protection than for Level A.
- The atmosphere contains less than 19.5% oxygen.
- There are unknown contaminants present, but they are not suspected to be skin contact hazards.

Level C

Level C is defined by the use of air-purifying respiratory protection equipment (Figure 8–4). Level C air purifying respirators (APRs) may be full-face or half-face. Level C respiratory protection, as defined by the EPA, consists of a full-facepiece air purifying mask equipped with a canister. The EPA does not recognize the use of half-face or cartridge-equipped APRs for its own personnel; this does not preclude the use of such equipment for other workers.

Level C protection is required when the identity and the concentration of contaminants are known, and the criteria for using APRs can be met. A typical Level C ensemble also includes CPC, inner and outer chemical-resistant gloves, hard hat, escape mask, and chemical safety boots.

LEVEL A Protection

Totally encapsulating vaportight suit with full-facepiece
SCBA or supplied-air respirator.

LEVEL B Protection

Totally encapsulating suit
does not have to be vapor-tight.
Same level of respiratory protection
as Level A.

LEVEL C Protection

Full-face canister air
purifying respirator. Chemical
protective suit with full body
coverage.

LEVEL D Protection

Basic work uniform, i.e.,
longsleeve coveralls, gloves,
hardhat, boots, faceshield
or goggles.

Figure 8–1. Levels of protection are based on the type of respiratory protection employed. (*Source:* USEPA, 1992a)

Table 8-1 Criteria for the four levels of protection for PPE ensembles

LEVEL OF PROTECTION A			
Equipment	Protection Provided	Should Be Used When:	Limiting Criteria
RECOMMENDED: • Pressure-demand, full-facepiece SCBA or pressure-demand supplied-air respirator with escape SCBA. • Fully-encapsulating, chemical-resistant suit. • Inner chemical-resistant gloves. • Chemical-resistant safety boots/shoes. • Two-way radio communications. OPTIONAL: Hard hat. Coveralls. Cooling unit. Long cotton underwear. Disposable gloves and boot covers.	The highest available level of respiratory, skin, and eye protection.	1. The chemical substance has been identified and requires the highest level of protection for skin, eyes, and the respiratory system based on either: - measured (or potential for) high concentration of atmospheric vapors, gases, or particulates; or - site operations and work functions involving a high potential for splash, immersion, or exposure to unexpected vapors, gases, or particulates of materials that are harmful to skin or capable of being absorbed through the intact skin. 2. Substances with a high degree of hazard to the skin are known or suspected to be present, and skin contact is possible. 3. Operations must be conducted in confined, poorly ventilated areas until the absence of conditions requiring Level A protection is determined.	Fully encapsulating suit material must be compatible with the substances involved.

LEVEL OF PROTECTION B			
Equipment	Protection Provided	Should Be Used When	Limiting Criteria
RECOMMENDED: • Pressure-demand, full-facepiece SCBA or pressure-demand supplied-air respirator with escape SCBA. • Chemical-resistant clothing (overalls and long-sleeved jacket; hooded, one- or two-piece chemical splash suit; disposable chemical-resistant one-piece suit). • Inner and outer chemical-resistant gloves. • Chemical-resistant safety boots/shoes. • Hard hat. • Two-way radio communications. OPTIONAL: Coveralls. Face shield. Disposable boot covers. Long cotton underwear.	The same level of respiratory protection but less skin protection than Level A. It is the minimum level recommended for initial site entries until the hazards have been further identified.	1. The type and atmospheric concentration of substances have been identified and require a high level of respiratory protection, but less skin protection. This involves atmospheres: - with IDLH concentrations of specific substances that do not represent a skin hazard; or - that do not meet the criteria for use of air-purifying respirators. 2. Atmosphere contains less than 19.5 percent oxygen. 3. Presence of incompletely identified vapors or gases is indicated by direct-reading organic vapor detection instrument, but vapors and gases are not suspected of containing high levels of chemicals harmful to skin or capable of being absorbed through the intact skin.	Use only when the vapor or gases present are not suspected of containing high concentrations of chemicals that are harmful to skin or capable of being absorbed through the intact skin.

The EPA SOSGs list the following criteria that may warrant the use of Level C protection:

- Contaminants present do not pose a skin contact hazard.
- Contaminants have been identified, and their maximum concentrations do not exceed the IDLH.
- There is a cartridge or a canister available for the contaminants present.
- Oxygen concentrations are greater than 19.5% by volume.

Table 8-1. (Continued)

LEVEL OF PROTECTION C			
Equipment	Protection Provided	Should Be Used When:	Limiting Criteria
RECOMMENDED: • Full-facepiece, air-purifying, canister-equipped respirator. • Chemical-resistant clothing (overalls and long-sleeved jacket; hooded, one- or two-piece chemical splash suit; disposable chemical-resistant one-piece suit). • Inner and outer chemical-resistant gloves. • Chemical-resistant safety boots/shoes. • Hard hat. • Two-way radio communications. OPTIONAL: Coveralls. Disposable boot covers. Face shield. Long cotton underwear. Use of escape mask during initial entry is optional only after characterization (29 CFR 1910.120(c)(5)(ii)).	The same level of skin protection as Level B, but a lower level of respiratory protection.	1. The atmospheric contaminants, liquid splashes, or other direct contact will not adversely affect any exposed skin. 2. The types of air contaminants have been identified, concentrations measured, and a canister is available that can remove the contaminant. 3. All criteria for the use of air-purifying respirators are met.	Atmospheric concentration of chemicals must not exceed IDLH levels. The atmosphere must contain at least 19.5 percent oxygen.

LEVEL OF PROTECTION D			
Equipment	Protection Provided	Should Be Used When	Limiting Criteria
RECOMMENDED: • Coveralls. • Safety boots/shoes. • Safety glasses or chemical splash goggles. • Hard hat. OPTIONAL: Gloves. Escape mask. Face shield.	No respiratory protection. Minimal skin protection.	1. The atmosphere contains no known hazard. 2. Work functions preclude splashes, immersion, or the potential for unexpected inhalation of or contact with hazardous levels of any chemicals.	This level should not be worn in the Exclusion Zone. The atmosphere must contain at least 19.5 percent oxygen.

Source: USEPA (1992a).

• Site activities are not anticipated to increase the concentration of contaminants present in the work zone.

Level D

The Level D ensemble is a typical work uniform, consisting of coveralls, gloves, safety glasses or goggles, faceshield, and safety shoes or boots (Figure 8–5). Level D does not include respiratory protection. According to the EPA SOSGs, Level D is appropriate under the following conditions:

• There are no contaminants present.
• Maximum concentrations of contaminants known to be present do not exceed permissible exposure limits.

Figure 8–2. Two workers in Level A protection. Their suits are fully encapsulated and vaportight; SCBAs are worn under the suit. Note that they are wearing gauntlet overgloves.

Figure 8–3. These workers are wearing Level B protection. Their PPE ensemble also includes hard hat with faceshield, coated Tyveks with hood, and chemical-protective gloves. Unlike Level A protection, SCBAs are worn on the outside of the CPC.

Figure 8–4. A worker wearing Level C protection, consisting of full-facepiece air purifying respirator with cartridges. Additional PPE consists of coated Tyveks, hard hat, safety boots, and gloves.

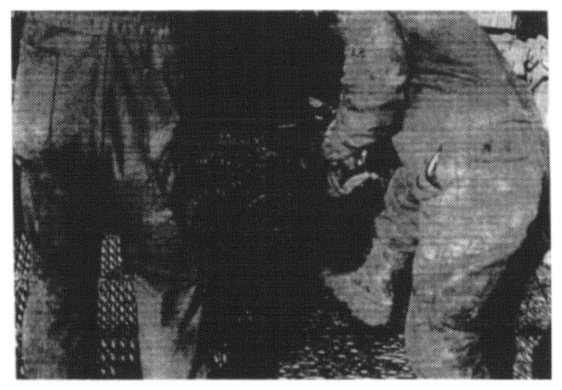

Figure 8-5. Level D protection does not employ respiratory protection. A typical Level D ensemble consists of coveralls, safety shoes or boots, safety glasses or faceshield, work gloves, and hard hat. This worker should be wearing eye protection.

- Site activities will not involve the potential for unexpected inhalation or contact with hazardous levels of any chemicals.

SELECTING LEVELS OF PROTECTION WITH UNKNOWN CONTAMINANTS

Many hazardous materials sites contain numerous substances that cannot be readily identified. Air monitoring can be employed to assess the location and relative concentrations of contaminants in the work zone; these concentrations can then be used to select the appropriate level of protection. The so-called total vapor concentration action levels are only for situations where the identity of contaminants is unknown. The PPE action levels should be used as a selection guide only until more information is available on the actual identity and concentration of contaminants present.

Limitations of PPE Action Levels

Air monitoring instrument readings must be carefully interpreted. Individual instruments differ in their ability to respond to various types of contamination. Most survey instruments such as PIDs and FIDs do not respond with the same sensitivity to several contaminants as they do to one contaminant; the relative concentration of unknowns detected is usually underestimated.

Air monitoring instruments cannot detect all potential hazards present. Infectious wastes, hazardous particulates, poison gases (arsine, phosphine, cyanides), and explosives will not be detected. Organic contaminants with very low exposure limits may also be present; air monitoring instruments not calibrated for these substances may indicate minimal readings that would not suggest an unsafe condition.

Air monitoring instruments do not detect all airborne contaminants. They do not respond identically to the same concentration of a detectable substance. Factors that may affect meter readings include the temperature, wind speed, water vapor or humidity, concentration of detectable substances present, and presence of interfering vapors and gases. It is imperative to have a thorough understanding of operating characteristics and limitations of each instrument before using meter readings as a guide to selecting a specific level of protection.

On-site activities should be evaluated before selection of a level of protection. Air monitoring readings may indicate that Level C protection is adequate, but the level of protection should be increased during higher-risk activities that increase the probability of encountering higher concentrations of the contaminants. Such higher-risk activities include drilling, digging test pits, excavation into wastes, moving or opening containers or drums, or bulking of wastes.

The following paragraphs give EPA SOSG recommendations for levels of protection for unknown contaminants based on air monitoring readings. These readings are used to assess the relative amount of contaminants present, and should be considered in concert with the other SOSG criteria for each special level of protection.

Level A (500–1000 ppm-equivalents)

Level A protection provides the highest degree of protection for the respiratory tract, skin, and eyes. The degree of protection afforded is based upon the resistance of the protective material to the contaminants present. Although Level A can provide protection against concentrations greater than 1000 ppm, the EPA recommends this restriction, which forces the wearer to:

- Evaluate the need to enter the unknown environment.
- Identify the specific contaminants contributing to the total concentration and their associated toxic properties, including the potential that one or more substances present may pose a severe skin hazard.
- Determine the concentration of each contaminant present.
- Evaluate the sensitivity of the instrumentation to the specific contaminants present.
- Evaluate the instrument response with respect to limiting factors such as wind, humidity, and temperature.

If such analysis is not possible, the operational range of 500 to 1000 ppm-equivalents is considered sufficiently conservative to provide a safe margin of protection if the readings are low because of instrument error or lack of sensitivity. The Level A operation range is also considered safe if higher concentrations are encountered, or if the substances present are toxic to the skin.

Work-zone concentrations in excess of 500 ppm-equivalents are not routinely encountered at hazardous waste sites, but such high concentrations may be found in confined spaces, closed buildings, or areas where contaminants have been spilled or released.

The decision to use Level A protection should also be balanced with its inherent limitations: limited mobility, visibility, and communications; increased risk of accidents such as slips, trips, and falls; and physical and temperature stress.

Level B (5–500 ppm-equivalents)

Level B is the minimum level of protection recommended for initial characterization of an unknown site (i.e., a site where the identity, number, and concentration of hazardous substances present are unknown). Level B protection affords a high degree of respiratory protection and can include good skin and eye protection as well.

Level B is recommended when meter readings greater than 5 units and less than 500 units are obtained. A limit of 500 ppm-equivalents on one or more air monitoring instruments is recommended as the upper limit for Level B. The EPA suggests a meter reading of 500 units as a decision point for evaluating the risks associated with work at higher concentrations. The EPA SOSGs recommend that the following factors be considered in selecting Level B protection:

- The need to enter an environment with unknown contaminants when meter readings indicate concentrations greater than 500 ppm-equivalents.
- The potential that one or more substances present pose a severe skin hazard.
- The probability that work activities will increase the concentration in air or increase the exposure hazard.
- The need for identification and quantification of contaminants present.
- The limitations of the air monitoring instruments used.
- The effect of environmental conditions on instrument sensitivity.

Level C (Background to 5 ppm-equivalents)

Level C is defined by the use of an air purifying respirator. The degree of skin and eye protection varies with the type of other protective equipment utilized. In dealing with unknown contaminants, an organic vapor/acid gas cartridge or canister with an HEPA filter should be employed. The EPA does not recognize the utility of cartridges, and it recommends using a canister.

The EPA recommends using Level C when meter readings above background up to 5 ppm-equivalents are obtained. When Level C is selected, it should be considered only temporary, until the contaminants responsible for the instrument readings are identified and quantified. If this is not possible, Level C protection is considered by the EPA to provide reasonable assurance that the respiratory tract is protected, provided that the *absence* of highly toxic substances has been confirmed.

In situations where increased concentrations may be encountered, or when high risk activities are undertaken (drilling, excavating, opening containers), a higher level of protection (Level B) should be used if engineering controls cannot be employed or guaranteed.

Level C and NIOSH Criteria

The EPA criteria for Level C protection are often misused and misinterpreted. The EPA criteria clearly state that the individual contaminants contributing to instrument readings should be identified. Once they are identified, toxic properties, the actual concentration present, and the optimal detection device for each substance can be determined. The appropriate level of protection can also be selected.

The EPA criteria also insist that the absence of highly toxic gases and vapors must be established before use of Level C. This requirement is often interpreted in terms of easily detectable gases such as hydrogen sulfide or carbon monoxide; the potential for other highly toxic air contaminants that may not be detectable by a single instrument or sensor is often overlooked.

A major concern is that EPA guidance documents are not compatible with the NIOSH criteria (and therefore OSHA regulations) for use of Level C respiratory protection. Air purifying respirators may be used only when the specific NIOSH criteria are satisfied for each contaminant present. If all criteria cannot be met, then Level C protection should not be used.

According to NIOSH criteria, air purifying respirators can be used only under *all* of the following conditions:

- The identity and the concentration of the contaminants in air are known.
- The oxygen content in air is at least 19.5%.
- The contaminants have sufficient warning properties when sorbent or filter capacity has been exceeded.
- There is periodic monitoring for contaminants in air.
- IDLH or LEL concentrations are *not* present.
- The respirator assembly is approved for protection against the specific contaminants present and their concentration levels.
- The type of respirator used has been successfully fit-tested by the user.

EPA-issued guidance documents are superseded by the OSHA HAZWOPER rule. The rule requires that direct-reading air monitoring equipment be used during site characterization to identify hazards and select personal protective equipment. An ongoing air monitoring program is also required after site characterization has established that the site is safe for initial start-up and subsequent operations. The OSHA rule makes no mention of EPA criteria for selecting levels of respiratory protection. Nevertheless, many site safety plans incorporate the EPA SOSG method for selecting levels of protection.

Level A at Hazardous Materials Sites

Level A protection is used primarily during emergency response to hazardous materials when personnel must enter highly toxic or corrosive atmospheres, such as a chlorine vapor cloud, to control a release. Level A is used infrequently at investigatory or remedial operations unless there is risk of an immediate release of poison gases, corrosives, or chemical warfare agents. Remediation activities that may warrant Level A include handling of deteriorated cylinders, pressurized drum removal, or recovery well installation for highly hazardous materials.

Level A protection is used when there is risk that contaminants in the ambient atmosphere may damage or impair respiratory protection, skin contact with contaminants may cause burns or incapacitation, or inhalation may cause incapacitation or death. A good rule of thumb is to ask, "Where is my next breath coming from if my respirator fails?" If severe burns, immediate illness, or death will result from taking just a few breaths of contaminated air, then Level A is warranted. If severe burns, immediate illness, or death will result from skin exposure to contaminated air or liquids, then Level A is warranted. Although it is not always possible, it is better to employ engineering controls than to put workers in Level A.

In most cases, however, Level B provides excellent protection for most investigatory and remedial operations. Level B employing an airline system with egress bottle is less stressful than Level A; most workers prefer Level B airline to Level B SCBA or even Level C.

Action Levels for Health and Safety

Protecting workers at hazardous waste sites can be difficult because the identity and the concentration of substances present are often unknown. Action levels that are easily applied to uncharacterized sites include those for radiation (1 mR/hr), combustible/flammable gas (20–25% LEL meter reading), oxygen deficiency (19.5%), and oxygen excess (23.5%). Action levels for specific gases and vapors are usually one-half their exposure limit value; the action level for hydrogen sulfide, for example, is 5 ppm, or one-half its OSHA exposure limit of 10 ppm. At uncontrolled or uncharacterized sites, action levels for specific contaminants do not apply because the identities of the materials present are unknown.

Direct-reading air monitoring instruments are inadequate to identify and quantify all airborne contaminants because not all gases and vapors are detectable. As it is impossible to guarantee that IDLH conditions are not present, Level B protection is usually required during initial investigations at uncharacterized hazardous waste sites, particularly where waste is exposed.

When there is an extensive site history, however, and previous soil and water analyses have identified contaminants and demonstrated that the level of contamination present is very low, Level C may be employed because the NIOSH criteria can be satisfied. Level C can be used as contingency protection on sites with dusts or unknown particulate hazards that may become airborne, where engineering controls and/or real-time dust monitoring is limited. At sites containing ppb levels of contamination, Level D should be considered. Air monitoring should be conducted continuously whenever Level C or Level D is used.

Table 8-2. PPE ensembles for Level X* site workers

Level X–	Uncoated Tyvek, latex gloves, hard hat, chemical-resistant safety shoes/boots
Level X	Coated Tyvek with hood, nitrile gloves, latex inner gloves, hard hat, faceshield, chemical-resistant safety shoes/boots
Level X+	Reusable CPC splash suit with hood, neoprene apron, nitrile gloves, latex inner gloves, gauntlet gloves, hard hat, faceshield, chemical-resistant safety shoes/boots, boot covers

*Level X = Level D, Level C, or Level B.

Tiered Levels of Protection

Table 8-1 presents EPA recommendations for PPE ensembles for Level A through D. However, for Levels B, C, and D, the level of protection defines only the type of respiratory protection employed; it has nothing to do with the type of CPC and other PPE used. Accordingly, the type of CPC and PPE utilized by site workers should be determined by site characteristics and the hazards associated with specific tasks. It is important to maintain some flexibility in defining specific PPE requirements within a particular level of protection; this allows workers to adapt to changes in site conditions.

Some organizations have adopted a tiered approach to the levels of protection used onsite (Table 8-2). For example, for a site that requires a particular level of respiratory protection, workers assigned to different tasks or locations may wear different PPE ensembles that are designated by a plus (+) or a minus (–).

Other organizations allow for even greater flexibility in delineating the type of body protection employed at a particular location or for a particular task. Subscripts are often used to indicate what PPE is required for the particular level of protection employed (Table 8-3). With this method, the level of protection (Level B, C, or D) is indicated, and the basic PPE ensemble is defined; the definition is further refined for each specific task using subscript nomenclature. For example, a typical Level C PPE ensemble may include full-facepiece air purifying respirator, Saran-coated Tyveks, hard hat, chemical-resistant safety shoes, and inner and outer chemical-resistant gloves. Workers at risk for chemical splash would require greater chemical protection; their assigned ensemble would include $C_{R,H,F,B,T}$ (Level C respiratory protection, basic PPE ensemble, coated Tyveks replaced by a *Reusable* CPC splash suit, *Hood*, *Faceshield*, *Boot* covers, *Taped* wrist and ankles). Drillers working

Table 8-3. PPE subscript nomenclature for level of protection

*X_T Taped wrists and ankles	X_H Hood
X_F Faceshield	X_A Apron
X_B Boot covers	X_G Gauntlet gloves
X_E Escape mask	X_R Reusable CPC

*X = Level D, Level C, or Level B.

with unknowns would require Level B protection (positive pressure airline with egress bottle); their assigned ensemble may include $B_{R,H,F,B,T,G}$ (Level *B* protection, basic PPE ensemble, with *R*eusable CPC, *H*ood, *F*aceshield, *B*oot covers, *T*aped wrist and ankles, *G*auntlet overgloves).

In the final analysis, it is the responsibility of the employer to ensure that the appropriate level of respiratory protection, as well as adequate PPE, is utilized for each site and task. This can only be determined after a thorough hazard versus risk analysis is performed. It is equally important, however, that workers understand why they are wearing a particular ensemble and the purpose of each item required.

9

Site Control

Site control, and the establishment of control zones, is an important aspect of the health and safety plan because it controls worker activities and movement of personnel and equipment on-site. Control zones help to reduce the potential for worker exposure and to minimize the spread of contamination to other areas. Site control also eliminates unnecessary personnel in contaminated areas and helps to reduce the risk of injury to unprotected support staff and unauthorized personnel.

A site control program should be developed during the planning stages of a hazardous waste site operation, and should be modified to reflect new information as it becomes available. Changes in site conditions resulting from ongoing activities, weather, or other unforeseen occurrences should be immediately incorporated into the site control program. Although site control measures are determined on a site-specific basis, all site control programs have the same basic components.

A site map (Figure 9–1) should be developed that includes the geographic layout and site hazards. The site map should be available before initial entry is made onto the site, and should include site drainage, natural topographic features (ponds, streams, culverts, ditches, hills), structures (buildings, tanks, impoundments, containers), location of buried utilities (electrical, water, gas, telephone), overhead hazards such as electrical wires, and pipes, fences, and other features that can affect site activities. Restricted areas must be clearly defined. Overlays or other mapping techniques should be used to reduce clutter and confusion.

WORK ZONES

A basic element of the site control program, work zones are used to designate the types of activities that will occur in each zone, the degree of hazard present, and specific areas that must be avoided by unauthorized or unprotected personnel. The purpose of work zones is to:

- Ensure good control of site activities and personnel.
- Reduce the unintentional spread of contamination, by workers or equipment, to clean areas.

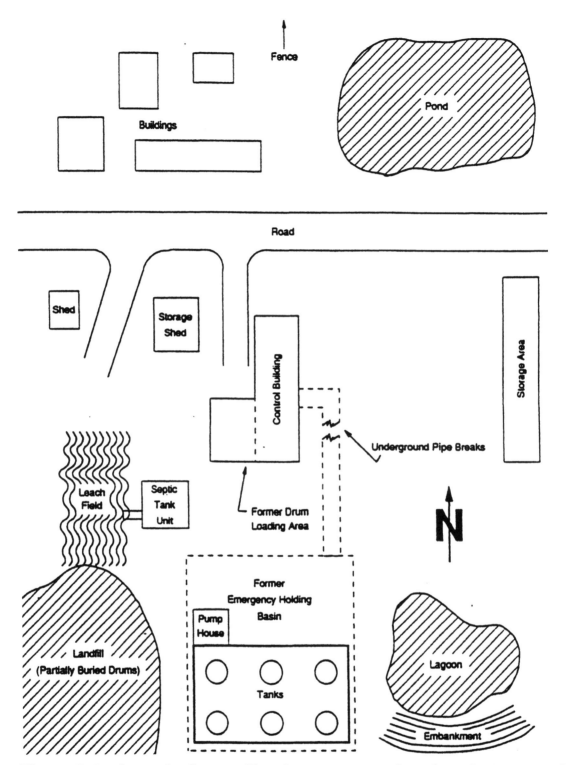

Figure 9–1. A sample site map. The site map represents an important source of information about the site, including the geographic layout and the location of hazards. (*Source:* USEPA, 1992a.)

- Confine site activities to specific areas, thus reducing the potential for accidental exposure.
- Facilitate the location and the evacuation of personnel in case of emergency.
- Restrict access of unauthorized entrants.
- Protect off-site receptors.

In establishing work zones, information collected on-site and off-site should be evaluated and compiled to determine the best location. The site map can be used to assemble collected data. All actual or potential hazards, locations of obvious soil or water contamination, and results of air and soil sample analyses should be plotted. Readings from real-time air monitoring instruments also should be considered; however, it is important to remember that the absence of readings *cannot* be considered evidence that the area is clean.

Although a site can be divided into as many zones as necessary, the three most frequently identified zones are the Exclusion Zone, the Contamination Reduction Zone, and the Support Zone. The movement of workers between these zones should be restricted to specific access control points to prevent cross-contamination and unauthorized entry (Figure 9–2).

The size and the shape of each zone must be based on hazards and conditions specific to the site. Each zone should be large enough to allow space for all necessary operations, provide an adequate distance to prevent the spread of contamination, and eliminate the possibility of injury due to fire or explosion. The following factors should be considered in establishing the dimensions of each zone:

- Physical and topographical features of the site.
- Weather conditions and prevailing wind.
- Results of laboratory analysis of field samples.
- Results of real-time air monitoring.
- Dispersion pattern for airborne contaminants.
- Dispersion pattern for surface and ground-water contaminants.
- Potential for explosion and flying debris.
- Potential for fire.
- Site hazards.
- Potential for exposure to chemical contaminants.
- Type of equipment and site activities required.
- Decontamination procedures required.
- Proximity to off-site receptors and sensitive areas.
- Availability of methods to monitor the spread of contamination during operations.
- Traffic patterns.

Figures 9–3 and 9–4 depict the site layout and work zones of actual sites.

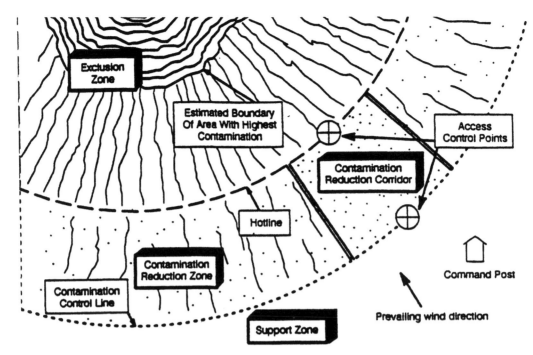

Note: Area dimensions not to scale. Distances between points may vary.

Figure 9-2. Illustration of work zones (not to scale). (*Source:* USEPA, 1992a.)

Exclusion Zone

The Exclusion Zone (also known as the Hot Zone, Red Zone, Zone A, Zone 1) is the area where contamination is known or anticipated to exist. The Hot Line separates the Exclusion Zone from the rest of the site; this line is a real or imaginary barrier determined by instrumentation, visual observation of hazards, or fragmentation distance. The distance to the Hot Line should be sufficient to prevent fire, explosion, or other emergency occurrences from affecting personnel outside the Hot Zone. Most commonly, however, the location of the Hot Line is determined by a geopolitical, institutional, or topographic boundary—usually a fence, road, or stream, some other preexisting natural or constructed barrier, or simply a property line.

An often-overlooked consideration in determining the location of the Hot Line or the size of the Exclusion Zone is the space requirements for conducting site activities. As a minimum around drilling equipment, the radius of the Exclusion Zone should be equivalent to the height of the boom, derrick, or mast. For excavation equipment, the minimum radius of the zone should be defined as the maximum extension of the boom plus 20 to 30 feet. The Hot Line should be physically secured with chains, rope, or fencing, or be clearly marked with hazard tape, signs, or placards. The Exclusion Zone can be separated into different subzones based upon known or anticipated hazards, site activities performed, site conditions or accessibility, or other factors. Areas with the highest level of contamination or greatest hazard can be indicated by additional signs or placards (Figure 9–5).

All persons who enter the Exclusion Zone must wear the appropriate level of protection for the degree and the type of hazards present. Different levels of protection may be

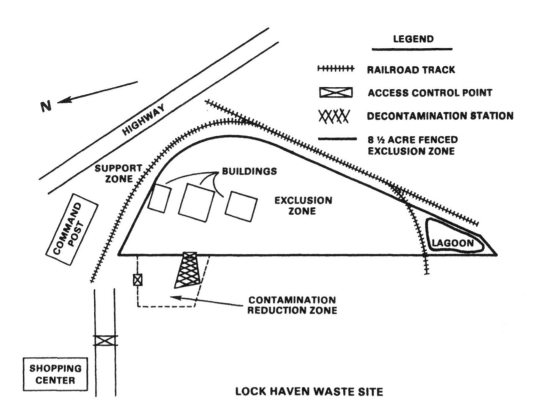

LEGEND

┠┼┼┼┼┼┨	**RAILROAD TRACK**
⊠	**ACCESS CONTROL POINT**
XXXX	**DECONTAMINATION STATION**
———	**8 ½ ACRE FENCED EXCLUSION ZONE**

LOCK HAVEN WASTE SITE

Figure 9–3. Site layout and work zones at an actual site. Note that railroad tracks are used to separate the Exclusion Zone from the Support Zone. The Contamination Reduction Zone is located in only one corner of the site. (*Source:* USEPA, 1988.)

used if the Exclusion Zone is subdivided, or if workers are engaged in different activities. The assignment of different levels of protection with the Exclusion Zone allows for a more flexible and less costly operation while still maintaining a high degree of safety.

Entry to and egress from the Exclusion Zone should be restricted to access control points at the Hot Line. The number of control points is determined by the size of the zone. Separate access points should be established for personnel and equipment.

Contamination Reduction Zone (CRZ)

The Contamination Reduction Zone (Warm Zone, Yellow Zone, Zone B, Zone 2) acts as a buffer between contaminated areas and clean areas. The purpose of the CRZ is to reduce the potential for clean areas to be contaminated or affected by worker activities or hazards in the Hot Zone. The barrier between the CRZ and the Hot Zone is the Hot Line. The degree of contamination decreases as workers and equipment move through the CRZ toward the noncontaminated Support Zone.

Decontamination of personnel and equipment is performed in a designated area within the CRZ, called the Contamination Reduction Corridor. Within the corridor, personnel are decontaminated at a Personnel Decontamination Station (PDS); and equipment is decontaminated in a separate Equipment Decontamination Station (EDS). A large opera-

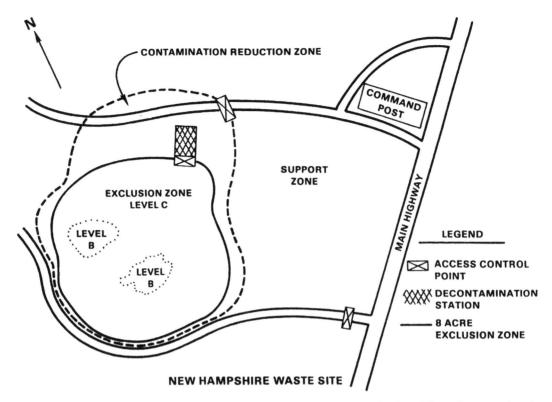

Figure 9-4. Site layout and work zones at an actual site. The Contamination Reduction Zone surrounds the Exclusion Zone. Note that two different levels of protection are utilized within the Exclusion Zone. (*Source:* USEPA, 1988.)

tion, or a site with multiple access control points from the Hot Zone, may require more than one Contamination Reduction Corridor.

The boundary between the CRZ and the noncontaminated Support Zone is the Contamination Control Line. Personnel within the CRZ must wear the prescribed level of protection; so to enter the noncontaminated Support Zone, personnel must remove protective equipment and exit through an access control point.

The dimensions of the CRZ are determined by individual site conditions and hazards; as a minimum, the zone should be large enough to comfortably contain both the PDS and the EDS. The CRZ should also be able to accommodate a protective equipment resupply area, sampling equipment, fire extinguishers, spill control supplies, first aid equipment, tools, a sample preparation area, a temporary rest area, and shower and toilet facilities. Although it is not always possible, every attempt should be made to orient the CRZ uphill and upwind from the Hot Zone.

Support Zone

The Support Zone (Cool Zone, Green Zone, Cold Zone, Zone C, Zone 3) is the outermost portion of the site, and is considered clean or noncontaminated. Workers in the Support Zone have a very low risk of being exposed to site hazards, and it is unlikely that they will be affected by activities in the Hot Zone.

Figure 9–5. Areas that are restricted because of structural, biologic, radiological, topographical, toxic, or other hazards of concern must be clearly defined.

The Support Zone contains administrative and technical support personnel, equipment, and personal hygiene facilities, as well as trailers, a staging area for equipment and personnel prior to site entry, and parking areas; traffic is restricted to authorized personnel. The location of support facilities and staging and parking areas is usually determined by the proximity to roads, site accessibility, and location of utilities. Because Level D protection (normal work ensemble without respiratory protection) is used within the Support Zone, potentially contaminated personnel, equipment, and samples must remain in the CRZ until they are decontaminated.

Strict use of control zones will minimize the spread of contamination into the Support Zone. Periodic monitoring of personnel and equipment at the Contamination Control Line will help determine if changes in decontamination procedures are required. A sampling and monitoring program for the Support Zone should be developed to ensure that it remains free of contamination.

SITE SECURITY

Site security is necessary to prevent inadvertent entry and exposure of unauthorized, unprotected individuals to site hazards. Although many unauthorized entries are indeed accidental, there is always the potential for vandalism, theft, and illegal dumping of other hazardous wastes on-site. The site and the equipment may represent an "attractive nui-

Figure 9-6. Contaminated dust from the Exclusion Zone is drifting toward the decontamination trailer in the CRZ (far left). The Support Zone should be relocated and dust control methods employed to minimize the spread of contamination.

sance" from which the general public must be protected. At "politically sensitive" sites, unauthorized individuals or groups may seek entry in order to demonstrate their disapproval of or interfere with site activities.

Site security is enhanced whenever a fence or another type of physical barrier is used to surround the site and limits or prevents entry under normal circumstances. If the site is not fenced, then barrier tape, rope, flagging, or some other type of barrier should be used around the perimeter. Regardless of the type of fence or barrier, warning signs advising people to keep out should be prominently displayed.

During nonworking hours, guards can be used to patrol the perimeter of the site. The guards, as a minimum, should receive right-to-know training regarding site hazards, and should know what action to take in case of an emergency. The guards should understand that they may not enter the site.

Public safety agencies, including police and fire departments, should be informed of site hazards. If possible, there should be arrangements for periodic police surveillance during off-work hours. Site security guards should be capable of calling for emergency assistance when necessary.

During working hours, one or more individuals should be responsible for security in the Support Zone and its access control points. An identification system should be established to identify authorized workers; unauthorized workers should be denied entry to the

CRZ. It is not unusual to have 24-hour security guards at sensitive remedial projects or at sites in vulnerable locations.

Site workers should check into the site at a single access control point. The project manager or the site administrator should know how many workers are on-site during the day. Workers should check in at access control points before entering the CRZ or the Hot Zone, and check out upon exiting. At the end of the workday, all workers should sign out. Equipment entering and leaving the site should be checked in and out at a separate access control point.

All visitors should be approved before entering the site. Someone in authority, such as the project manager, a supervisor, or a team leader should give such approvals. All visitors should sign in and out, and state their purpose for visiting the site. On small jobs, the daily log should record entry and exit times of all workers and visitors. Visitors should be briefed about site hazards and where they can and cannot go, and, if necessary, provided with protective equipment, such as hard hats or safety glasses. A trained site worker should accompany visitors at all times.

SITE COMMUNICATIONS

A system should be established on-site for both internal and external communications. Internal communications encompass communications between personnel on-site; external communications are communications between site workers and off-site personnel.

Figure 9-7. Site workers should enter the site at designated access points. These workers are receiving a last-minute safety briefing before entering the site.

Internal Communications

Internal communication is used to alert site workers to emergencies, advise workers about changes in site conditions or site activities, provide updated safety information, and maintain site control.

Commonly used devices for internal communication include radios or walkie-talkies (FM or citizens band); noisemakers such as bells, air horns, megaphones, sirens, and whistles; and visual signals, including flags, hand signals, lights, and body movements. All battery-operated equipment such as flashlights, megaphones, and radios used in the Exclusion Zone and CRZ should be certified as intrinsically safe for the intended use if flammable materials are known or expected to be present or if site conditions have not been characterized.

The simplest form of internal communication is one worker talking to another; such verbal communication can be difficult at a site because of background noise and the use of PPE. Prearranged nonverbal signals, or audible or visible cues, should be used when verbal communication is not feasible, or when other communication devices fail. Some Simple nonverbal gestures that can be used during normal field activities include:

- Thumbs up: everything OK, or I understand.
- Thumbs down: not OK, or I do not understand.

Figure 9–8. A worker preparing to don a Level A suit has a radio headset to allow internal communication between site workers and the support staff.

- One hand around throat: I am running out of air.
- Two hands around throat: I am having trouble breathing.
- Grabbing the wrist of another worker: leave the area at once.
- Grabbing around the waist of worker: leave the area at once.
- Both hands above or on top of the head: I need assistance.
- Pointing to top of wrist: watch the time, or out of time.
- Hand slicing across neck: shut down equipment; stop activity.

A secondary set of nonverbal signals should be reserved for emergency situations only. Signals that are often reserved for emergencies include sirens, piercing whistles, airhorns, and flares (flares used only in the Support Zone).

To expedite communication between workers, each individual's name or initials should be marked on his or her suit or hard hat. Long-distance identification of workers can be promoted by using colored tape, vests, numbers, symbols, or large initials.

It is important that equipment operators and site workers know and understand the hand signals before beginning operations. Site workers directing heavy equipment should know the standard signals used in the industry.

External Communications

An external communication system is necessary to make routine reports to management and the client; to maintain contact with organization personnel off-site; and, when needed, to call for medical assistance or other types of emergency response.

The primary means of external communication are the telephone and radio. If telephone lines are not installed on-site, a cellular telephone can be used. All site workers and off-hours security guards should know the location of the nearest telephone or radio. Emergency and nonemergency phone numbers or call letters should be readily available in the Support Zone or another easily accessible location. Whatever the means of external communication, they should be checked daily to make sure they are in good working order.

THE BUDDY SYSTEM

A buddy system is used to organize workers or teams of workers so that each worker is under the observation of at least one other worker. Using the buddy system is not the same as having a safety backup. A safety backup is a team of at least two individuals who stand by to assist another work team in case of difficulty or emergency. During initial site entry and characterization, when site hazards are unknown, safety backup teams usually are on hand to assist entry teams. During normal site operations, safety backup may not be considered necessary; the buddy system, however, should always be used on-site.

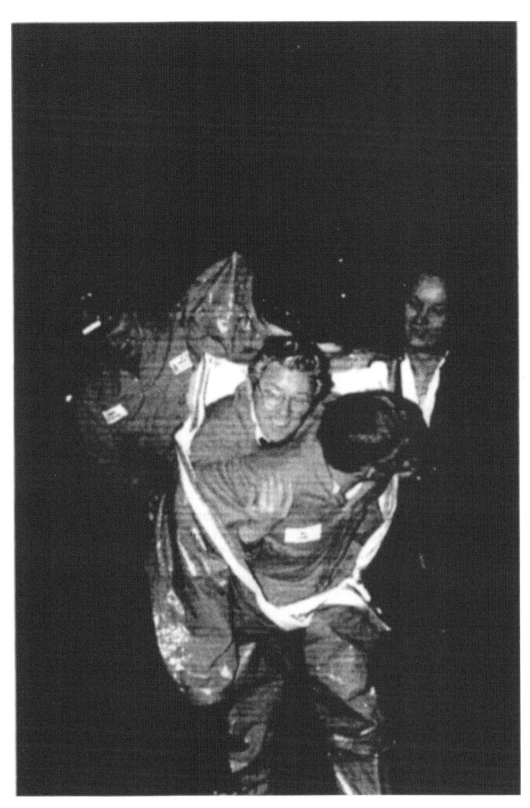

Figure 9–9. The buddy system should always be used at hazardous waste sites. It requires workers to be in visual or voice contact with each other; sharing the same CPC is strictly optional!

It is dangerous to allow one individual to work alone; as part of the buddy system, workers remain close together, maintain visible contact, and provide each other assistance in case of emergency. When needed, workers can use emergency signals.

The access control point into the Exclusion Zone or CRZ is the best location for enforcing the buddy system. Team leaders should assign buddies according to job assignments. The buddies need not be from the same organization as long as the workers are being observed by someone who can act in case of an emergency.

Workers using the buddy system should observe their partners for signs of chemical exposure, temperature stress, or fatigue. Also each worker can periodically check the integrity of the partner's PPE ensemble. Under this system, both buddies must enter and exit the zone together.

In case of emergency, the need for rapid assistance may not be readily recognized unless the workers remain in contact with their supervisor, the health and safety officer, or the project leader. Workers should be able to communicate with Support Zone personnel by line-of-sight, radio, or other means. If such communication is not feasible, a safety back-up team should be utilized. In areas where only one worker can work at a time, such as in a confined space, the buddy system still must be employed. This can be achieved by having direct-communication contact at all times, even when visual contact is not possible.

Figure 9-10. Unusual or unanticipated site conditions should be reported immediately. This drilling site was flooded after a heavy rain.

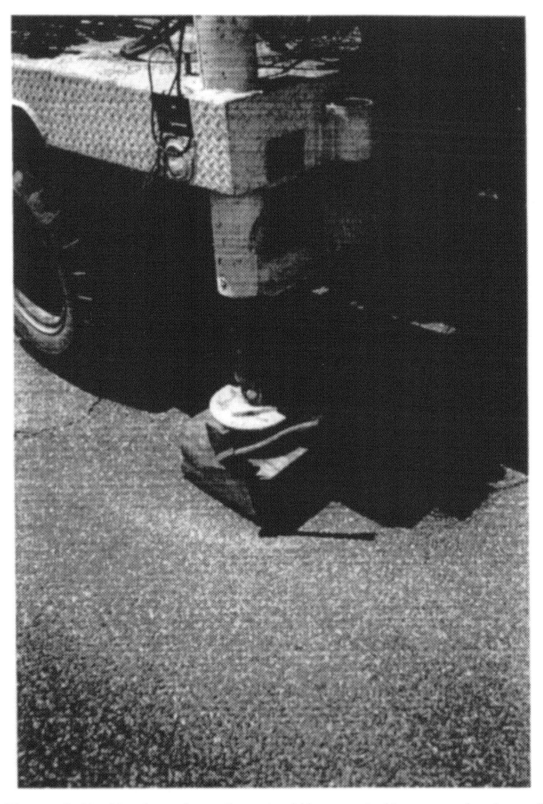

Figure 9–11. Unsafe work practices should be reported to a supervisor immediately.

STANDING ORDERS

Maintaining and enforcing safety awareness and safe work practices are facilitated by having a standard set of safe work rules or standing orders that *must* be followed at *all* times. Standing orders should be conspicuously posted and consistently enforced; they should be periodically reviewed at tailgate briefings and safety meetings. Copies of standard orders should be distributed to nonroutine workers and visitors.

Some commonly employed standard orders for the Hot Zone are:

- No smoking, eating, drinking, chewing of gum or tobacco, or application of cosmetics is allowed.
- The use of matches, lighters, or open flame is prohibited.
- Check in at the access control point prior to entry.
- Check out at the access control point prior to exiting the site.
- Use the buddy system at all times.
- Remove PPE and wash your face and hands prior to eating, drinking, or using toilet facilities.
- Report any unusual conditions or unanticipated hazards to your supervisor immediately (Figure 9–10).
- Report any unsafe work practices to the health and safety officer (Figure 9–11).
- Leave the area to report any malfunction of monitoring equipment or PPE.
- Avoid unnecessary contact with hazardous materials by staying clear of puddles, vapors, mud, discolored surfaces, containers, and site debris.
- Work only in areas where you are authorized to be.

10

Decontamination

At a hazardous materials site, workers' protective equipment is likely to come into contact with contaminated vapors and particulates, contaminated soil, drilling mud, and ground water. Driving on-site may contaminate vehicles and heavy equipment. During high risk activities, such as installation and testing of monitoring or recovery wells, digging of test pits, tank cleaning, lagoon remediation, and drum sampling or overpacking, workers may come into contact with relatively high concentrations of contaminants; these activities also involve contamination of heavy equipment and tools. All contaminated items must be properly decontaminated or containerized before they are removed from the site, or, in the case of reusable sampling equipment and tools, they must be thoroughly cleaned before the next use. The decontamination procedure used will depend on the nature and the concentration of contamination present, site characteristics and resources available, and the type of equipment being decontaminated.

OVERVIEW

Why Decontaminate?

Decontamination is a process that reduces the concentration of contaminants to some predetermined safe level, which is usually defined as normal background concentration. Decontamination, or decon, can be separated into three general categories: environmental decon, safety decon, and health decon.

Environmental decon is performed to protect some aspect of the environment (soil, water, or air) from low concentrations of pollutants that may have long-term environmental consequences. It is also done to prevent or minimize cross-contamination of soil, water, and equipment. This category includes contaminants present in very low concentrations, that is, parts per million (ppm) and parts per billion (ppb).

Safety decon is conducted when a material is not overly toxic or hazardous but presents safety problems, such as creating slippery walking or riding surfaces. Many mild alkalis, detergents, surfactants, oils, and diesel fuel fall into this category.

Health decon is performed when one or more contaminants present a health hazard to site workers or equipment. The hazardous substance may be toxic, flammable, reactive, or corrosive.

Decontamination Plan

A decontamination plan should be developed as part of the site-specific HASP, and should cover any workers or equipment that may be exposed to hazardous substances. The plan should be short, concise, and specific, and should include the following items for both personnel and equipment decontamination areas:

- The location of the decontamination area, usually within a contamination reduction corridor within the CRZ (see Chapter 9).
- The number of discrete decontamination areas required; large sites usually need multiple CRZs and decon areas.
- The layout of each decontamination station.
- Equipment needed for each decontamination station.
- Decontamination procedures for each station.
- Procedures to prevent contamination of clean areas and for monitoring of the CRZ.
- Procedures for disposal of contaminated clothing, equipment, and liquids.
- Procedures for placing decontaminated equipment back in service.
- Procedures for revising the plan as site activities, PPE, equipment, or site conditions change.
- An on-site training program for site workers that includes explicit instruction in decon procedures and sufficient practice for the workers to become proficient in their use.
- Procedures for ensuring worker compliance with the plan.

The initial decontamination plan is usually based on a worst-case situation in which all personnel and equipment leaving the Exclusion Zone will be grossly contaminated. During site characterization activities, decontamination stations should be available to wash and rinse all personnel and equipment coming out of the Exclusion Zone. As information on site hazards becomes available, the initial decontamination plan can be modified and unnecessary decon procedures eliminated.

Decontamination procedures can vary greatly, depending on the site and the nature of the hazards present. The following parameters should be considered when developing a decontamination plan:

- The *type of contamination*, in terms of toxicity, corrosivity, or other known or suspected hazard. If personnel are exposed to highly toxic, corrosive, or pathogenic materials, especially rigorous decon procedures will be required. Similarly, the physical nature of the contamination—sludge, vapor, particulate—

is also important; oily or greasy contaminants are difficult to remove and may pose a safety hazard even at very low concentrations.

- The *concentration of contaminants* present or anticipated on-site. Decontamination procedures should be developed for the highest concentrations anticipated and can be modified when more information regarding actual contaminant concentrations becomes available.

- *Site activities* and *worker assignments*. This factor is an important determinant of the potential for contamination. For example, workers installing recovery wells are more susceptible to gross contamination than workers conducting air monitoring.

- The *type and level of PPE* worn. This is a consideration in designing personnel decontamination stations. Disposable protective clothing requires decon procedures that allow the wearer to remove the garment safely; reusable clothing must be rigorously decontaminated, cleaned, and disinfected, and then inspected for signs of residual contamination or degradation. Decontamination methods must be developed for other PPE used, such as APRs, SCBA packs and air bottles, airline hoses, hard hats, and earmuffs.

- The *location and extent of contamination* on personnel and equipment. Contamination on the soles of boots requires less planning than possible whole body or upper body contamination. Unless removed, upper body contamination can pose a greater risk than contamination of the extremities, as off-gassing from upper body areas may generate hazardous concentrations within the breathing zone of the worker as well as decon personnel. Grossly contaminated workers have an increased probability of skin contact with the contaminant when removing protective garments.

- *Site resources and weather.* A decon plan developed for spring and summer may not be feasible in winter months when aqueous decon solutions will freeze and cause a severe slip/trip hazard. Lack of water, shade, and shelter is often a consideration on remote sites. The amount of space needed for decontamination presents a severe problem at some urban sites.

- The *size and duration of the project.* For activities that involve only a few people or are very brief, small decon kits may be sufficient. At larger sites involving many workers over a long period of time, more permanent arrangements, including personal hygiene trailers, may be warranted.

Contamination Avoidance

Avoiding contamination is the best way of reducing the amount of decontamination required. Equipment, personnel, and decon areas should be kept upwind of contamination whenever possible. Covering monitoring instruments, tools, and equipment with plastic, tarpaulins, or strippable coatings can minimize subsequent cleaning. Make sure that exhaust ports, sensors, intakes, and hot or moving parts are not covered. Porous items, such as wooden pallets, cloth-wrapped hoses, canvas tarpaulins, ropes, and wooden tool handles, cannot be decontaminated and should be considered expendable.

Workers should minimize contact with hazardous materials by walking around puddles, obviously contaminated soil, and visible vapors, mists or particulates in air (see Figure 10–1). They should be careful to avoid being splashed by liquids while sampling or opening containers. Contaminated instruments or equipment should not be used until decontaminated and put back into service. Unnecessary personnel should stay out of the Exclusion Zone to minimize the spread of contaminants into the CRZ.

The use of disposable overgarments can augment PPE and facilitate later decontamination. Overgarments can include disposable overboots to protect safety shoes, gauntlet gloves and aprons when front-only protection is needed, or a disposal coverall when whole body protection is desired. The grossly contaminated overgarment is removed and discarded prior to decontamination of the underlying PPE ensemble. The use of disposable overgarments also can extend the wear life of more expensive, reusable PPE.

DECONTAMINATION METHODS

Decontamination is achieved through a variety of processes, including removal, neutralization, absorption, chemical degradation, dilution, encapsulation (as in mixing into cement), isolation (as in burial), or weathering. Decon methods can be separated into three major categories: physical removal, chemical inactivation, or a combination of both physical and chemical means.

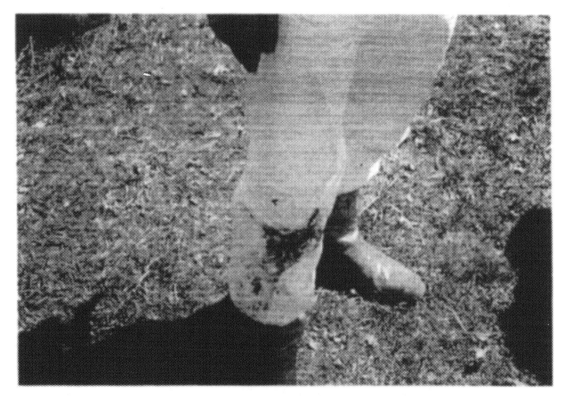

Figure 10–1. Avoiding contamination is the easiest way of reducing the amount of decontamination needed. This worker knelt on contaminated soil.

Physical Removal

Wiping, sweeping, scraping, blowing, vacuuming, rinsing with water, and steam cleaning are all examples of physical methods for removal of contaminants. In many cases, physical removal of gross contamination is an initial step in decontamination.

The physical removal method of choice depends on the type of contamination present. Loose contaminants, such as dirt, dust, or granular materials, often can be dislodged and brushed away (Figure 10–2). In most cases, rinsing with water or wash solutions is necessary to remove residual contamination. Adhering contaminants, such as muds, viscous liquids, sticky solids, and greases, often must be wiped or scraped off prior to washing and rinsing. Stubborn materials, such as sticky muds, glues, asphalt, adhesives, cements, and resins, may require special methods, such as heating, melting, freezing with dry ice, or adsorption onto dry pulverized clay, powdered lime, or kitty litter.

Volatile liquids may be removed from equipment by the natural process of evaporation, followed by washing and rinsing with warm water. Evaporation can be enhanced by steam cleaners, high pressure hot water spray cleaners, or high pressure air jets; however, workers must be careful not to inhale the vaporized contaminants. Steam cleaning or high pressure sprayers may remove necessary lubricants on heavy equipment; so the fluid reservoirs should be checked regularly. Wire ropes, bearings, and other vital components must be inspected and relubricated as necessary after decontamination.

Chemical Inactivation

Physical removal of gross contamination usually is not sufficient to reduce contaminant concentrations to background levels. In most cases, physical removal is followed by a wash and rinse process using cleaning solutions. These cleaning solutions often dissolve contaminants and keep them in solution until they can be rinsed away. Organic solvents, detergents, dilute acids, and water are the most commonly used of these dissolving agents. All dissolving agents used must be compatible with the equipment being cleaned. Most of the agents can be used on heavy equipment, drilling tools, and sampling equipment. Organic solvents can permeate and/or degrade personal protective clothing and are not recommended for decontamination under normal circumstances.

Organic solvents used for the decontamination of drilling tools and sampling equipment include acetone, *n*-hexane, and alcohols. These materials are often used in large quantities; it should be recognized that *in many cases solvents used for decontamination purposes present a greater flammability and toxicity hazard than on-site contaminants.* Care must be taken in selecting, using, and disposing of solvents used for decontamination. They should be used only when other cleaning agents will not remove the contaminant.

Solutions of trisodium phosphate or TSP have been used as all-purpose dissolving agents. TSP, however, has a high pH in solution; as a solid or in concentrated solution, it is a skin and eye irritant. In areas where phosphates are banned, nonhazardous decon wastewater containing TSP may have to be stored and disposed of elsewhere.

Low-sudsing detergents or surfactants are the most popular dissolving agents (Figure 10–3). These materials reduce the adhesive forces between contaminants and the surfaces being cleaned, and keep contaminants suspended in water. A liquid household detergent is

Figure 10–2. A decon worker brushes the boots of a site worker to remove dried soil and particulates. Note that the site worker is standing on a plastic bag; when the brushing is finished, the bag can be sealed and discarded.

Figure 10–3. Low-sudsing detergents are safe and effective dissolving agents for many types of contamination.

the best overall dissolving agent. Household liquid detergents are readily available, can be transported in concentrated form, are easily stored, inexpensive, and nontoxic, and go into solution easily. Detergent solutions work well against most forms of contamination. In some cases it may be necessary to add a small amount of solvent to effect a thorough decontamination.

Rinsing removes contaminants through the dual process of dilution and solubilization. Multiple rinses with clean water remove more contaminants than does a single rinse.

Neutralization using dilute acids and dilute bases can inactivate caustic contaminants as well as amines, hydrazines, phenols, thiols, and sulfonic compounds. Alternatively, a detergent solution and a water rinse can be used; and then used decon solutions can be neutralized prior to their disposal.

Disinfecting solutions are used as a means of inactivating infectious or pathogenic materials. The most commonly used disinfecting agents is dilute sodium hypochlorite or liquid household bleach. Alcohols and detergents also are effective against some pathogens.

Effectiveness of Decontamination

There is no single criterion for evaluating the efficacy of decontamination. Nevertheless, decon procedures must be assessed and revised as necessary. A variety of methods can be used, including visual observation, swipe testing, and analysis of decon solutions.

Visual observations of protective clothing in natural light may disclose discolorations, stains, residual dirt, or alterations in the fabric, which suggest that not all contaminants have been removed. Observing clothing and equipment under ultraviolet light may reveal chemical contaminants that fluoresce, such as polycylic aromatic hydrocarbons.

Swipe testing involves taking a swab, wetted cloth, or wetted filter paper and wiping the surface of a potentially contaminated object. The swipe is then analyzed for traces of contamination. The inner and outer surfaces of protective clothing should be tested.

Samples of the final decon rinse solution can be analyzed for contamination. Elevated concentrations in the final rinse suggest that residual contamination remains, and that additional decontamination is required. Samples of decontaminated protective clothing may also be analyzed for traces of contamination.

In some cases, methods to evaluate the effectiveness of decontamination are considered infeasible, and protective clothing and other equipment are not decontaminated but discarded. All PPE, tools, and other equipment should be collected, placed in containers, and labeled. All spent decon solutions and rinses must be collected and stored or discarded.

Selecting a Decon Procedure

The decontamination method should be selected after a consideration of the nature of the hazard, the physical state of the contamination (vapor, liquid, solid, sludge, fine particulate), the concentration known or anticipated to be present, and extent of worker contamination anticipated based on site activities (Figure 10–4).

Decontamination procedures and agents must be compatible with the contaminants being removed and the clothing and the equipment being cleaned, with the chemical and physical compatibility determined prior to their use. *Organic solvents should not be used on chemical protective clothing or other PPE unless recommended by the manufacturer.* Any method or decon agent that degrades or damages PPE or other equipment should not be used.

Similarly, procedures and decontamination agents should not pose a direct health hazard to workers. If a hazardous decon agent or method must be used, protective measures must be taken to protect all personnel involved.

General Decontamination Procedures

In general, wet contamination should be kept wet, and dry contamination should be left dry. Some dry compounds form solids, are water-reactive, or release heat when wetted. Decon operations should start with the simplest methods, such as a gentle brushing or spraying, followed by scrubbing of difficult areas if necessary.

Common sense must be used in selecting procedures. Gross contamination should be removed physically before washing with decon agents. Workers can use a shuffle pit filled with sand or boot scrapers to dislodge dirt, mud, sludge, and oily residues from boots.

Wet decon solutions and water rinses create unwanted runoff that spreads contamination; so it is generally a good practice to limit the amount of water used. Decon wash and rinse water can be contained in large tubs or children's wading pools. Tubs and pools should be large enough for a worker to stand in, and should have no drain or have a drain

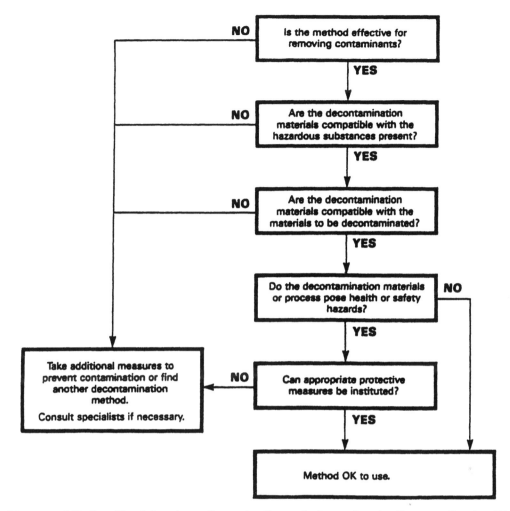

Figure 10–4. Decision tree for selection of decontamination methods. (*Source:* NIOSH/OSHA/USCG/EPA, 1985.)

connected to a collection tank or a disposal system. Alternatively, a small trench (4–6 inches deep) lined with plastic can be used to hold waste decon solutions. Small pneumatic garden sprayers help minimize the volume of water used; it is the pressure used, not the volume, that exerts the cleansing action.

Either steam cleaning or high pressure spraying using low volumes of water is the method of choice for heavy equipment and vehicles (Figure 10–5). Lower pressure units (90–120 psi) use decon agents, whereas high pressure units (up to 5000 psi) do not require decon solutions. Spray units operating at 500 to 800 psi with a flow rate of 3 to 5 gpm are satisfactory; 1000 to 3000 psi units using 0.5 to 2 gpm are ideal. Sand blasting or water blasting (up to 45,000 psi) of shovels, scoops, and buckets has also been used. In most cases, however, accessible parts are scrubbed with a detergent solution and rinsed under pressure or with steam.

Unfortunately, steam cleaners and high pressure sprayers produce aerosols that can

Figure 10–5. A high pressure spray unit for decontaminating drilling tools.

travel out of the decon area and spread contamination. The use of plastic sheeting, tarps, or other materials to curtain off the decon area can confine aerosols and minimize the spread of contamination; but aerosol confinement increases the likelihood that decon workers will be contaminated, and provisions must be made for their protection.

Equipment decontamination can generate significant amounts of water runoff. Decon may be performed on a wash pad constructed so that cleaning solutions and wash water can be recycled or collected for disposal (Figure 10–6). A raised graveled area lined with polyethylene can also be used. Alternatively, plastic-lined trenches can be used to divert the runoff into lined holding ponds. All parts of the equipment, including undercarriage, wheels or tracts, chassis, and cab, must be thoroughly cleaned. Air filters on equipment operating in the Exclusion Zone should be treated as highly contaminated; contaminated filters should be removed and replaced before the equipment leaves the site.

Contaminated respirator components can be soaked in cleaning solutions provided by the manufacturer, or soaked in hot soapy water and scrubbed with a brush. Used APR cartridges or canisters must be discarded. Tools with wooden handles should be kept on-site and handled only by workers wearing PPE. Such tools should be discarded at the end of the project or when they become grossly contaminated. Receptacles for decon fluids, contaminated soil, contaminated PPE, and tools should be clearly labeled. Workers should be careful when reopening containers holding contaminated material.

Reusable protective clothing, respirators, and noncontaminated work coveralls worn under protective clothing become soiled because of body oils and perspiration, and must

Figure 10–6. A decontamination pad for heavy equipment. Note that the pad is covered to minimize the spread of aerosols.

be sanitized after use. Respirators should be sanitized according to the manufacturer's directions, with recommended sanitizing solutions. Coveralls should be machine-washed and machine-dried. Reusable protective clothing, if not machine-washable, should be cleaned by hand. For some projects it may be cost-effective to have a local industrial uniform supplier pick up and deliver work clothes or coveralls to the job site.

Decontamination Equipment

Decon equipment and supplies usually are selected on the basis of cost, availability, and suitability (see Tables 10–1 and 10–2). Most decon equipment is disposed of when site activities are terminated; an exception may be large galvanized tubs or holding tanks.

Protecting Decon Personnel

Workers performing decontamination procedures must wear an appropriate level of protection to accomplish their task without exposure to contamination. The level of protection selected is based on the nature and the concentration of chemical contaminants, their toxi-

Table 10–1. Decontamination equipment for personnel and PPE

Containment Equipment	
Plastic sheeting	Pumps—hand or mechanical
Drums	Absorbent pads
Galvanized tubs	Buckets, 3–5-gallon
Stock tanks	Basins, 2–3-gallon
Wading pools	Lined trash cans
Plastic-lined boxes with sand or absorbent	Plastic bags

Cleaning Equipment	
Water	Long-handled bristle brushes
Scrub brushes	
Garden sprayers	Spray booths
Hoses	Paper or cloth towels
Washing and rinsing solutions	Scrapers brushes

Other Supplies
Storage space for clean equipment
Secure storage for personal items
Canopy for shade
Stools or benches
Personal hygiene facilities
Cloth towels
Shower facilities (as appropriate)
Change of clothes

Table 10–2. Decontamination equipment for heavy equipment

Containment Methods or Equipment	
Tanks or storage pools	Buckets
Pneumatic pumps	Plastic-lined trenches
Plastic sheeting	Absorbent-lined trenches
Drums, 55-gallon	Lined trash cans
Covered roll-offs	
Cleaning Equipment	
Long and short rods	Short-handled brushes and
Scrapers	brooms
Wire brushes	Steam cleaners
Sponges	High pressure sprayers
Wash and rinse solutions	Sand blasters
Long-handled brushes,	Water blasters
shovels, brooms	Absorbents

city and the route of exposure, the extent of contamination on personnel or equipment to be decontaminated, and the type of decon solutions used.

In the past, it was often assumed that workers performing decon should wear the same level of protection as the personnel being decontaminated. This may be appropriate some of the time, but not all of the time. For example, Level B protection is often used for workers during the initial investigatory activities at uncharacterized sites. If these workers encounter no contamination, Level B protection for the decon workers is not necessary. Similarly, where levels of protection of workers are often upgraded as a precautionary measure, upgrading the level of protection of the decon personnel is unwarranted.

Decon workers at personnel decontamination stations who come into contact with workers leaving the Exclusion Zone may require more protection than decon workers assigned to the last station. The dissolving of contaminants in wash and rinse solutions may enhance the absorption of these materials through the skin; also contaminants that are in solution can be aerosolized and inhaled. Therefore, decon workers who use sprayers or who come into contact with liquid decon solutions or rinses should wear additional splash protection, such as aprons and faceshields. Additionally, Level D protection is not recommended in the CRZ for decontamination workers; Level C protection with an HEPA filter can provide protection against aerosol inhalation during decontamination. A chemical cartridge should also be used if there is a risk of exposure to chemical vapors or gases. If site contaminants are known to be toxic and present at high concentrations, then Level B airline respiratory protection should be utilized. PPE for decon workers should be not be based on what Exclusion Zone workers may encounter, but rather on what they may bring back into the CRZ.

During equipment decon, workers using high or low pressure sprayers may be exposed to aerosolized contaminants in solution; steam cleaning enhances the evaporation of chemicals and increases the risk of inhalation exposure. In many cases, decon workers will encounter more contamination than will vehicle or heavy equipment operators; so equipment decon workers should wear good skin and eye protection, augmented with additional splash protection, such as a one- or two-piece splash suit, faceshield, and gauntlet gloves. The respiratory protection depends on the type of contaminants present on-site; a good rule of thumb is that equipment decon personnel should wear at least the same level of protection as personnel in the Exclusion Zone.

In some instances, organic solvents, acids, or bases are used in the decontamination of heavy equipment, drilling tools, and sampling equipment. It is important to recognize that these materials are hazardous, and decon workers must be adequately protected. Organic solvents can degrade CPC; care must be taken to outfit decon workers with splash suits and gloves that are solvent-resistant. Respiratory protection should also be employed to prevent overexposure. In many cases, organic solvents, such as *n*-hexane and acetone, or acids and bases used as neutralizers pose a greater hazard than that of the contaminants being removed.

DECONTAMINATION STATION DESIGN

The Contamination Reduction Corridor

Decon activities should be restricted to one or more designated areas within the CRZ; each area is known as a Contamination Reduction Corridor. A control access point controls access into and out of the Exclusion Zone; workers exiting the Exclusion Zone should immediately enter the Contamination Reduction Corridor. There should be a separate entry point for noncontaminated workers. The Contamination Reduction Corridor confines decon activities to a specific, limited area. The size of the corridor depends on the number of stations or the specific procedure involved in the decontamination process.

Each decon station represents one step in the decon process, such as glove wash, boot wash, or body wash. In cases where there is minimal contamination on site, several procedures may be combined into a single station. To prevent crowding and cross-contamination, there should be at least three to six feet between stations (Figure 10–7).

The decontamination process should proceed in a logical sequence. More heavily contaminated items, such as boots and gloves, should be decontaminated first at separate stations, followed by less contaminated items at later stations. The entire sequence of stations encompassing the entire process is called the decontamination line, decon line, or decon train (see Figure 10–8). Wherever possible, workers should approach from the downhill, downwind side. The least contaminated personnel should move through the decon line first.

Figure 10–7. Placing decon stations too close together results in crowding.

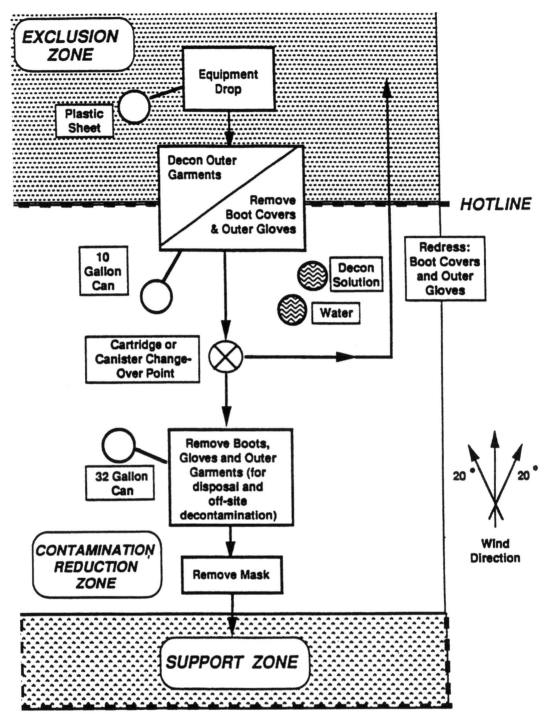

Figure 10–8. Decontamination layout for workers utilizing Level C protection. (*Source:* USEPA, 1992a.)

A cooling or a warming station is sometimes necessary within the decon line during periods of temperature extremes or inclement weather. This comfort station, which is covered to provide shade in hot weather, may include cool water hoses, ice packs, or cold towels. In inclement or cold weather, the station should be covered and have one or more sides to provide protection from the wind and the elements; a trailer or a building on-site is often used for this purpose (Figure 10–9).

A variety of decon lines have been described and depicted for various levels of protection. In some cases, decon lines with up to 19 stations have been described, with elaborate decon measures for washing and removing inner garments that were never even exposed! It is important to recognize that these procedures were originally prepared by the military for decontamination of equipment and personnel after exposure to nuclear, chemical, and biological warfare agents. Such procedures are unnecessary for most hazardous waste site work performed today. Decontamination planning must be on a site-by-site basis; each site is unique. It is a waste of time and resources to accept an elaborate decon plan when a shorter, simpler plan will suffice. There is no single decon plan that will fit every site.

Figure 10–9. Workers go through decon prior to entering a comfort trailer where they wash and change clothes.

Figure 10–10. Plastic sheeting can be slippery to walk on. Sheets of plywood have been placed over plastic to minimize slip rip hazards.

Regardless of the methods employed, all workers should practice decontamination procedures. Prior to entering the Exclusion Zone, site workers should be guided through the various decon station steps by workers assigned to the personnel decontamination station. A decon "walk-through" allows site workers wearing SCBA to estimate the amount of time required and ensures that sufficient time is allotted for workers to proceed through the decon line without running out of breathable air. This is especially important on sites where workers will perform self-decon.

11

Health and Safety Programs and Plans

The OSHA HAZWOPER rule requires that a written health and safety program be developed and implemented for hazardous waste operations at uncontrolled waste sites, at corrective actions, or at other applicable sites under 29 CFR 1910.120(b). The program should identify, evaluate, and control health and safety hazards on-site, and provide for emergency response during operations. The program must be maintained and updated by the employer, and be made available to:

- Workers or their representatives.
- Contractors and subcontractors.
- Any other individuals working for the employer who could be exposed to hazardous materials.
- OSHA personnel.
- Personnel from federal, state, or local agencies that have jurisdiction over the site.

As the health and safety program is intended to cover all employees, only one program should be developed, even if workers are engaged in different types of activities at more than one site. The program should define the organizational structure, describe general health and safety training and medical surveillance programs, and establish standard operating procedures for health and safety. The program must also develop a work plan and a Health and Safety Plan (HASP) for each site where hazardous materials operations are conducted.

COMPONENTS OF A HEALTH AND SAFETY PROGRAM

Organizational Structure

The section on organizational structure gives the specific chain of command and overall responsibilities of supervisors and employees in carrying out the health and safety program. It should identify the person responsible for the entire program, as well as the person responsible for all hazardous waste site operations. A list of health and safety supervisors for all sites and other personnel engaged in site activities or emergency response should be included. The organizational hierarchy should be delineated in terms of authority, communication, and coordination among managers, supervisors, and other personnel. It is necessary to review and update the organizational structure section on a regular basis to accommodate changes in personnel and site activities.

Health and Safety Training Program

The health and safety training program must establish health and safety training requirements for all site workers and supervisors. The HAZWOPER rule requires that on-site employees who are exposed, or potentially exposed, to hazardous substances, health hazards, or safety hazards receive training before they perform hazardous waste site activities. The training program must address the hazards found at hazardous materials sites, the use and limitations of personal protective equipment, work practices followed to minimize risks, the safe use of equipment, and medical surveillance requirements.

The OSHA standard specifies the hourly requirements for five different categories of site workers:

1. *General site workers performing activities that may result in exposure* to chemicals above their exposure limits or other health hazards are required to receive at least 40 hours of initial off-site training as well as a minimum of three days of actual field experience under the direct supervision of a trained, experienced supervisor.

2. *Employees who work in areas where there is minimal potential for atmospheric or health hazards* must receive a minimum of 24 hours initial off-site training and one day of field experience under direct supervision. These workers may not assume the duties of general site workers and may not use personal protective equipment until the additional 16 hours of initial training and two days of supervised field experience are obtained.

3. *Managers and supervisors* of employees described above must receive the same initial training and field experience as the employees they supervise. An additional 8 hours of specialized training is required, which must include their responsibilities under the health and safety program, the personal protective equipment program, the medical surveillance program, and the contingency or emergency response plan.

4. *Nonroutine site workers* who are on-site occasionally for specific tasks such as land surveying or ground-water monitoring, and who work in areas where there is minimal potential for atmospheric or health hazards, must receive a minimum of 24 hours initial off-site training and one day of field experience under direct supervision.

5. *Site employees with site emergency response duties* must be trained to a level of competency for duties that will be assumed (e.g., first aid or rescue) in addition to the above training requirements.

Annual refresher training is required for all site workers. The purpose of refresher training is to maintain the competencies essential for working safely on-site. Workshops, seminars, lectures, and other programs on health and safety at hazardous materials sites provide acceptable means of satisfying the annual refresher requirement. Organizational discussions or critiques of site activities are also acceptable. At least 8 hours should be spent on refresher training annually.

Documentation of initial and field-supervised training and of refresher activities should be maintained by the employer. Additional training in areas such as first aid, CPR, air monitoring instrumentation, field sampling methods, confined space entry, or emergency response should also be documented.

The OSHA rule was amended on August 22, 1994 (*Federal Register*, Vol. 59, No. 161, pp. 43268–43280). Appendix E of the amendment offers nonmandatory training curriculum guidelines, which include recommendations concerning student-to-instructor ratios, proficiency assessment methods, certificates, and recordkeeping.

There are no specific training requirements for visitors to a hazardous materials site, but they should be informed of potential site hazards and actions to take in the event of an emergency. Site visitors may not enter any hazardous areas of the site, and may not use personal protective equipment without proper training.

OSHA allows the employers to exempt employees from training requirements on certain low hazard sites or for low risk tasks. In many cases, however, client requirements and contractual prerequisites require training no matter how minimal the risk or the hazards.

Medical Surveillance Program

Site workers engage in activities that may expose them to health and safety hazards; so an ongoing medical program is necessary to assess and monitor their health, prior to placement as well as during and after hazardous waste site work. The goal of a medical surveillance program is to:

- Detect preexisting disease or medical conditions that may be exacerbated by site activities or may place a worker performing site activities at increased risk.
- Minimize worker exposures on-site.
- Assign workers to tasks that are commensurate with their capabilities, and maintain medical surveillance as they perform those tasks.

The medical surveillance program should be designed by an occupational health physician or a qualified occupational health consultant in conjunction with the organization's health and safety professionals. Medical examinations and monitoring procedures should be performed by or under the direction of a physician.

Employees who must be covered by the plan include:

- Workers who are, or may be, exposed to health hazards or hazardous substances at or above their exposure limits for 30 or more days per year.
- Workers who wear a respirator for 30 or more days per year.
- Members of organized hazardous materials emergency response teams.
- Workers who are injured because of an overexposure during a site emergency, or who show symptoms of illness that may have resulted from exposure to hazardous substances.

Under OSHA's proposed Respiratory Protection Rule (*Federal Register*, November 15, 1994, Vol. 59, No. 219, pp. 58883–58956), medical surveillance is required for workers who utilize respiratory protection for more than 5 hours per work week.

Site workers who participate in a medical surveillance program must undergo an initial, baseline medical examination prior to engaging in field activities. After the initial examination, an annual follow-up examination is required unless the physician considers another interval to be appropriate. The longest allowable interval between follow-up examinations is two years; more frequent examinations may be required upon consideration of the nature of site activities and the potential for exposure. Any worker who develops signs or symptoms of overexposure, is injured, or becomes ill from exposure to hazardous substances should be reexamined as soon as possible. All employees must be trained to recognize symptoms of overexposure and temperature stress.

Baseline medical screening determines a worker's fitness to use personal protective equipment and engage in site activities. The initial examination (Table 11-1) also provides baseline medical data for comparison with future medical data. Baseline medical examinations should focus on:

- The occupational and medical history, including prior chemical exposures, illness, diseases, and symptoms (shortness of breath, dizziness, palpitations).
- A complete physical examination.
- The ability to work while wearing protective equipment.

Workers who are unable to perform on-site activities because of their medical history (e.g., history of severe lung disease, heart disease, back or orthopedic disorders) should be disqualified. Any limitations on a worker's ability to wear protective equipment should be noted in writing. A written fit-for-duty statement should be provided by the physician.

The employer must provide the physician with a copy of the OSHA HAZWOPER rule and its appendixes to ensure that the physician understands the medical surveillance requirements. Substance-specific standards (e.g., benzene, formaldehyde, asbestos, lead) should also be provided if appropriate. The employer must also provide a complete

Table 11-1. Components of the baseline examination

Complete medical history	Pulmonary function
Complete occupational history	Electrocardiogram (EKG)
Physical examination by physician	Complete blood count
Visual acuity test	Routine urinalysis
Audiometry	Blood chemistry
Chest X ray	Cardiovascular stress test

Special Tests*

Cholinesterase	Urine and sputum cytology
Methemoglobin	Liver function
Heavy metal screen	Kidney function

*Performed when considered appropriate by physician.

description of the types of site activities and protective equipment use required of the worker.

The baseline and subsequent follow-up medical examinations and testing should be based upon the information provided by the employer regarding potential and actual exposure, respirator and protective equipment use, and site activities. Where worker duties are diverse and the potential for exposure is different among workers, several different medical monitoring protocols may be used. It is important to note that medical tests should be based upon anticipated exposures and not previous exposures.

At the end of their employment as hazardous waste site workers, employees should have a termination medical examination. A complete examination is necessary if any of the following criteria *cannot* be met:

- The last full medical examination was performed within the last 6 months.
- No exposure has occurred since the last examination.
- No symptoms associated with exposure have occurred since the last examination.

Medical records for site workers must be maintained for at least 30 years after their employment is terminated. Records should include:

- Names and social security numbers of the employees.
- Results of examinations and tests.
- Physicians' written opinions of fitness for duty.
- Recommended occupational limitations, if any.
- Employee medical complaints related to exposure.
- Injury and illness records.

The employer is responsible for retaining the records even if the company moves or goes out of business. All workers, or their authorized representatives, must have access to their medical records.

Respiratory Protection Program

Employers who provide respiratory protection equipment to employees are responsible for establishing an effective respiratory protection program; employees are responsible for wearing respirators and complying with the program. The program should be written, reviewed, and updated as needed.

The respiratory protection program should stress training to ensure that the workers understand the importance of using respirators. Workers should also learn how to properly fit and inspect respirators and how to maintain them in clean, serviceable condition. (See Figure 11–1.)

The respiratory protection program should include the following:

- Written procedures for selection and use of respirators.
- Selection criteria based on potential inhalation hazards.
- Instruction and training in the proper use and limitations of respirators.
- Procedures and schedules for respirator inspection, cleaning, maintenance, and storage.

Figure 11–1. A company's respiratory protection program should stress the importance of using respirators. It should also ensure that workers know how to wear them. Note that one worker is wearing his respirator on his head, rather than his face!!

- Medical examination requirements used to determine fitness for duty.
- Procedures for supervision of work areas where respirators are used to ensure that the proper level of protection is being used; this may include periodic or continual monitoring of the workplace atmosphere.
- Procedures to ensure that respiratory protective equipment meets minimum requirements and regulatory standards.
- Methods for evaluating the effectiveness of the program.

Personal Protective Equipment Program

The PPE program is designed to protect workers from health and safety hazards and to prevent injury to the wearer caused by incorrect use and/or malfunction of PPE. The program also makes good business sense, as it mandates an organized, logical approach to the selection of PPE before workers are on-site, which lessens the likelihood of making a last-minute, and often incorrect, decision regarding the type and the quantity of protective material needed.

Basic elements of a PPE program should include:

- Written criteria for selection of PPE.
- Description of PPE use and its limitations.
- Type and duration of activities allowed for specific PPE ensembles.
- Procedures and schedules for inspection, maintenance, cleaning, and storage of PPE.
- Procedures for decontamination or disposal of contaminated PPE.
- Instruction and training on proper fitting, use, and limitations of PPE.
- Procedures for donning and doffing PPE.
- Limitations of PPE due to external conditions or medical limitations.
- Supervision and monitoring of personnel wearing PPE; this may include monitoring for temperature stress.
- Methods for evaluating the effectiveness of the program.

Standard Operating Procedures

Procedures for routine, safe work practices should be clearly written and faithfully followed. Establishing such standard operating procedures or SOPs saves time, increases worker efficiency and productivity, and avoids unnecessary hazards and accidents.

SOPs may be administrative, technical, or management-oriented; all are intended to provide uniform instructions for accomplishing a specific task. SOPs are usually provided to cover a variety of site activities, including site entry and initial characterization, selection and use of PPE, respiratory protection, air monitoring instrumentation, drilling techniques, sampling procedures, drum opening/moving procedures, decontamination of equipment and personnel, and site control.

Figure 11–2. For company-owned equipment, SOPs should include instructions for inspecting equipment. This wire rope failed during on-site operations.

For SOPs to be effective, they must have certain characteristics:

- They must be written in advance. It is impossible for workers to develop and write safe, practical procedures while performing hazardous waste site activities.
- They must be based on the best available information, operational principles, and technical guidance.
- They must be field-tested, reviewed, and revised on a periodic basis or at appropriate times by a competent, experienced health and safety professional.
- They must be clearly written, understandable, and feasible.

All personnel involved in site activities must have copies of the SOPs and be briefed on their use. Site personnel must be trained and periodically retrained on SOPs, especially those pertaining to personal protection, health, and safety.

Site Safety Plans

There is much confusion regarding the site safety plan; just what exactly is its purpose? Whatever it is called—a health and safety plan (HASP), a contingency plan, or a safety and health emergency response plan (SHERP)—it should do more than just satisfy a regulatory requirement! A good plan satisfies two purposes:

1. It serves as an *active* accident-prevention plan.
2. It serves as a *reactive* contingency plan to mitigate the impact of Murphy's Law.

The OSHA HAZWOPER rule requires that a site-specific health and safety plan or HASP be developed for each site where workers are engaged in hazardous waste site operations. The purpose of the HASP is to address the health and safety hazards that may exist at each site and at each phase of operations, and to identify procedures for protecting employees.

It is not necessary to develop a new HASP when new hazards or tasks are identified on-site; the original HASP should be updated to reflect new information. Subcontractors working on-site must evaluate subcontracted activities, identify hazards, and prepare a HASP addressing these hazards. A subcontractor HASP should be submitted to the site manager or the health and safety coordinator, who will incorporate it into the general site HASP after it has been reviewed.

The development of the HASP utilizes information collected during site characterization. Generally there are three phases of site characterization:

1. Prior to site entry, a preliminary evaluation is conducted off-site to gather information about the site and to conduct reconnaissance from the perimeter of the site.

2. During initial site entry, a visual survey is taken, and preliminary air monitoring is conducted. During this phase, site entry is restricted to properly protected reconnaissance personnel.

3. Once actual or potential hazards have been identified, other activities may commence on-site. Monitoring must continue, to provide an uninterrupted source of information about site conditions.

The development of the HASP is a continuous process. At each phase of site characterization, the information obtained should be evaluated and incorporated into the HASP for the next phase of the work. As more accurate and thorough information becomes available, the HASP can be tailored to the actual site hazards that workers will encounter.

The first role of the health and safety program is that of an active accident prevention plan. This portion of the plan should describe:

- Site hazards.
- Associated risks to on-site personnel.
- How to monitor site hazards (visually and by instrumentation).

Figure 11-3. Safety should not be forgotten when workers leave the hazardous waste site. A welder is shown at work in a building designated as noncontaminated. Note the welder's precarious position, and the lack of fall protection and eye protection.

- How to protect personnel from such hazards (e.g., by PPE and by engineering controls).
- How to safely conduct normal job functions in the presence of such hazards.

The plan should actively address the actual hazards that will be encountered on-site. It is ludicrous for a plan to have page after page of data regarding the toxicity and the flammability of pure product chemicals, when it is known that the site contains only a few parts per million concentration in the soil or parts per billion in the ground water. The environmental contamination associated with such a site will have a different impact on residents from its effect on individuals working on-site for a short period of time.

Similarly, it is imperative for the plan to have specific information on other site hazards, particularly those associated with drilling or other types of heavy equipment (see Figure 11–4). The odds of physical injury are much greater than those of chemical overexposure. If you get hit by the bumper of a pickup truck or the bucket of a backhoe, your risk of injury from "heavy metal contamination" is much greater than that presented by a few parts per million of hexavalent chromium in the soil!

The second function of a site safety plan is that it should serve as a reactive contingency plan. This portion of the plan should tell workers what to do or what will be done when something goes wrong, such as a spill, a fire, or an emergency medical situation.

Figure 11–4. The HASP should describe the hazards associated with specific tasks. This drilling site has additional hazards because of poor housekeeping.

Many organizations divide their site safety planning between two or three separate documents. They have a generic HASP and a short site-specific plan that may take the form of a checklist or a fill-in-the-blanks document. Some organizations have off-the-shelf task-specific plans. These documents address the hazards of a particular activity, such as drum sampling, marine drilling, monitoring well installations at gas stations, or installing methane recovery systems at landfills; then they are augmented with a site-specific plan.

The components of a typical HASP can be summarized as follows:

- A list of key personnel and their alternates, and procedures to facilitate communications during routine operations as well as during emergencies.
- Health and safety risk analyses for each task and operation identified in the site-specific workplan.
- Site control measures specifying procedures used to minimize worker exposure to hazardous substances before and during site operations.
- Employee training requirements, including initial health and safety training, the annual refresher, supervisor training, site-specific training (if necessary), and other training such as first aid or CPR training.
- Medical monitoring procedures for personnel who are potentially exposed, or who must wear PPE. Medical monitoring must include initial, periodic, and termination medical examinations and medical recordkeeping.
- Criteria for selection and use of PPE, including a description of different PPE ensembles that will be used on-site. The HASP should include or refer to the organizational PPE program.
- Air monitoring techniques that will be used for on-site evaluation of potential exposure to air contaminants. Monitoring procedures must include methods for initial entry, periodic monitoring, and monitoring during high risk activities, such as drilling or excavation.
- Spill containment procedures for containment and isolation of hazardous materials spilled during site activities.
- Confined space entry procedures (if necessary).
- Decontamination procedures for personnel and equipment, including ways to minimize contact with hazardous substances.
- A contingency or emergency response plan, including ways that anticipated emergencies (fire, personal injury, spills) would be handled.

Table 11–2 presents a typical outline of a generic HASP for a firm that oversees site investigations at landfills. Subcontractors can adopt and utilize this plan (or portions of it) as their own; however, their plan should address the hazards specific to their activities (drilling, sampling, geophysical investigations, surveying, test pit excavation).

A site-specific HASP that calls for drilling and monitoring well installation adjacent to an old industrial tank farm is outlined in Table 11–3. No excavation or confined space entry is involved.

Table 11-2. Generic health and safety plan outline

Disclosure or Use of Information

Purpose and Policy

1.0 Introduction

2.0 Roles and Responsibilities
    2.1 Off-site Personnel
        2.1.1 Site Health and Safety Officer
        2.1.2 Project Manager
        2.1.3 Project Health and Safety Officer
    2.2 On-site Personnel
        2.2.1 Site Health and Safety Officer
        2.2.2 Field Team Supervisor
        2.2.3 Field Team Member
        2.2.4 Subcontractors
        2.2.5 Client
        2.2.6 Authorized Visitors
        2.2.7 Observers, Inspectors, Supervisors
        2.2.8 Use of the Buddy System

3.0 Elements of the Program and SOPs
    3.1 Site Characterization
    3.2 Risk Identification for Site Personnel
    3.3 Site Controls
    3.4 Selection of Personal Protective Equipment
        3.4.1 Company Use Policy
        3.4.2 Selection Criteria
        3.4.3 Respiratory Protection Decision Making
        3.4.4 Skin Protection
        3.4.5 Other Safety Equipment
        3.4.6 Authority to Make Changes
        3.4.7 Equipment Maintenance and Storage
    3.5 Site Monitoring
        3.5.1 Company Policy
        3.5.2 Equipment Availability
        3.5.3 Monitoring Practices and Procedures
        3.5.4 Action Level Determination
        3.5.5 Calibration and Maintenance Procedures
        3.5.6 Equipment Decontamination Procedures
        3.5.7 Real-time Monitoring Recordkeeping
    3.6 Decontamination
        3.6.1 Decon of Personnel and PPE
        3.6.2 Decon of Sampling Equipment
        3.6.3 Decon of Heavy Equipment
        3.6.4 HAZCOMM Program for Decon Agents
        3.6.5 Disposal of Contaminated Material
    3.7 Medical Surveillance Program
    3.8 Health and Safety Training Requirements

4.0 Site-Specific Health and Safety Plan
    4.1 Short Form/Checklist Plan
    4.2 Full Plan
    4.3 Tailgate Meeting Form
    4.4 Accident Prevention Plan
    4.5 Spill Prevention Plan
    4.6 Emergency Response and Contingencies

5.0 Respiratory Protection Program
    5.1 Purpose
    5.2 Selection Criteria
    5.3 Use and Limitations

Table 11-2. Generic health and safety plan outline

    5.4 Fit-Testing Procedures
    5.5 Maintenance and Repair Procedures
    5.6 Cleaning/Sanitizing Procedures
    5.7 Storage Procedures
    5.8 Inspection and Evaluation Procedures

6.0 Employee Right-to-Know
    6.1 Company Policy
    6.2 Training Requirements and Syllabus
    6.3 Material Safety Data Sheets (MSDS)
        6.3.1 For Lab Personnel
        6.3.2 For Field Personnel
    6.4 Container Labeling and Handling Procedures

7.0 Maintenance of Records
    7.1 Medical Certification
    7.2 Training Records
    7.3 Field Experience
    7.4 Respirator Training and Fit-Testing
    7.5 Prescription Safety Glasses Requirements
    7.6 Notification of Record Change Procedures

Appendixes
A. Checklist Health and Safety Form
B. Generic Health and Safety Plan (Example)
C. Medical Surveillance Protocols
D. Initial Health Questionnaire Form
E. Annual Health Questionnaire Form
F. Emergency Medical Information Form
G. Physician Fit-for-Duty Form
H. Training Certification Form
I. Respiratory Fit-Test Certification Form
J. OSHA HAZWOPER Standard, 29 CFR 1910.120
K. OSHA Excavation Standard, 29 CFR 1926.650
L. OSHA HAZCOMM Standard, 29 CFR 1910.1200
M. OSHA Respiratory Protection Standard, 29 CFR 1910.134
N. OSHA PPE Standard, 29 CFR 1910, Subpart I
O. Applicable State Requirements

Preparation of the HASP for a small job may require as much time and effort as that for a large one. Moreover, mistakes can occur if cost-cutting involves HASP preparation. In a number of incidents, workers have suffered temperature stress as a direct result of wearing PPE specified for only one time of year. The HASP should clearly indicate for which time period it is valid; active plans should always have a suspension or "tickle" date when the plan must be reevaluated and reapproved.

It is essential that the person who prepares the HASP not be the person who approves it! The final HASP approval should be given by someone in the company with the authority to assume financial responsibility for it. Also, it is advisable to have the plan reviewed by an in-house or an outside health and safety specialist who is familiar with the type of activities that will be undertaken on-site. It is important that the prime author of the plan visit the site; although this person may not be able to see the entire site, he or she should at least observe the perimeter. The plan's author needs to know the lay of the land, what obvious hazards might be present, and the availability of local agencies to respond in an emergency. (See Figure 11–5.).

Table 11–3. Site-specific health and safety plan outline

Preparer's Signature

1.0 Introduction
    1.1 Scope and Applicability
    1.2 Contractors
    1.3 Visitors

2.0 Organization and Coordination
    2.1 Key Personnel
    2.2 Site-Specific Health and Safety Personnel
    2.3 Organization Responsibility
    2.4 Employee Responsibility

3.0 Task/Operation Safety and Health Risk Analysis
    3.1 Historical Overview of Site
    3.2 Scope of Work
    3.3 Site Hazards
        3.3.1 Chemical Hazards
        3.3.2 Biological Hazards
        3.3.3 Physical Hazards
        3.3.4 Structural Hazards
        3.3.5 Mechanical Hazards
        3.3.6 Electrical Hazards
        3.3.7 Flammable Hazards
        3.3.8 Thermal Hazards
        3.3.9 High Pressure Hazards
        3.3.10 Active and Abandoned Utilities
        3.3.11 Traffic
        3.3.12 Noise
        3.3.13 Temperature Stress
        3.3.14 Weather
        3.3.15 Heavy Lifting
    3.4 Task-by-Task Safety and Health Risk Analysis

4.0 Site Requirements
    4.1 Preassignment and Refresher Training
    4.2 Site Supervisor Training
    4.3 Documentation

5.0 Medical Surveillance
    5.1 Preassignment Requirements
    5.2 Site-Specific Requirements
    5.3 Exposure/Injury/Medical Support

6.0 Air Monitoring
    6.1 Baseline Monitoring
    6.2 Rationale Monitoring
    6.3 Direct-Reading Instruments
    6.4 Instrument Use Guidelines
        6.4.1 Combustible Gas Indicators
        6.4.2 Oxygen Deficiency Meters
        6.4.3 Ionization Detectors
        6.4.4 Toxic Gas Monitors
        6.4.5 Dust Monitors
        6.4.6 Portable Gas Chromatograph
        6.4.7 Dosimeter Tubes
        6.4.8 Passive Colorimetric Detector tubes
        6.4.9 Detector Tubes
    6.5 Monitoring Frequency
        6.5.1 Borehole and Well Installation
        6.5.2 Test Pit Excavation and Sampling
        6.5.3 Monitoring Well Development and Sampling

Table 11-3. Site-specific health and safety plan outline

    6.6 Instrument Limitations
        6.6.1 Temperature
        6.6.2 Humidity and Water Vapor
        6.6.3 Water and Splash/Aerosol
        6.6.4 Dust and Particulates
    6.7 Action Levels and Levels of Protection
        6.7.1 Combustible Gas Indicator
        6.7.2 Oxygen Deficiency Meter
        6.7.3 Toxic Gas Monitors
        6.7.4 Ionization Detector
        6.7.5 Photovac Snapshot
        6.7.6 Dust Monitor

7.0 Personal Protective Equipment
    7.1 Rationale for Use
    7.2 Criteria for Selection
    7.3 Levels of Protection
        7.3.1 Basic Work Ensemble
        7.3.2 Chemical Protective Clothing Ensemble
        7.3.3 Level C
        7.3.4 Certification or Approvals
    7.4 Limitations of Protective Equipment
        7.4.1 Chemical Protective Material
        7.4.2 Physical Stress
        7.4.3 Heat Stress
    7.5 Field Upgrade/Downgrade Authority
    7.6 Storage

8.0 Site Control Measures
    8.1 Control Zones
        8.1.1 Exclusion Zone
        8.1.2 Contamination Reduction Zone
        8.1.3 Support Zone
        8.1.4 Access Control Points
    8.2 Buddy System
    8.3 Site Security
    8.4 Site Entry, Egress Procedures
    8.5 Hours of On-site Activities
    8.6 Site Communications
    8.7 Site Hygiene Practices
    8.8 Spill Control Procedures
    8.9 Fire Control Procedures
    8.10 Safe Work Practices
        8.10.1 Worker Safety Practices
        8.10.2 Personal Precautions

9.0 Decontamination Plan
    9.1 Rationale
    9.2 Personal Decontamination
    9.3 Sampling Equipment Decontamination
        9.3.1 Air Monitoring Equipment
        9.3.2 Water/Soil Sampling Equipment
    9.4 Heavy Equipment
    9.5 Protection for Decontamination Personnel
    9.6 Materials Handing and Storage
    9.7 Waste Disposition

10. Site Contingency Plan
    10.1 Pre-emergency Planning
    10.2 Personnel Roles and Lines of Authority
    10.3 On-site warning/Evacuation Procedures
    10.4 Emergency Contact/Notification Procedures

Table 11–3. Site-specific health and safety plan outline

10.5 Emergency Decontamination Procedures
10.6 Emergency Medical Procedures
    10.6.1 On-site First Aid
    10.6.2 Emergency Treatment
10.7 Hospital Transport Procedures
10.8 Spill Response Procedures
10.9 Fire/Explosion Response Procedures

11. Incident Reporting

Appendixes

Site Map
Directions and Map to Hospital
Correspondence with Fire Department
Correspondence with Hospital/EMS Provider
Company Forms: Employee Acknowledgment
    Daily Tailgate Safety Meeting
    Physician Certification
    Training Documentation
    Emergency Medical Information
    Hearing Conservation Program Participation
    Daily Safety Log
    Weekly Safety Report
    Incident/Accident Report
    Monthly Exposure and Injury Report
    Warning/Suspension Notice Record
    Incident Report Follow-up
    Air Monitoring Results Report

Site safety plans are more than required documents; they are necessary tools that allow workers to perform their jobs safely. They also tell us what to do in case something goes wrong. What is the test of a good HASP? A good plan informs you of what hazards can be found on-site, what hazards you or others may create; the plan is designed to protect all on-site workers. Ask these questions: What are the hazards associated with *my* job? How are the hazards going to be monitored? How will on-site job procedures, engineering controls, and PPE minimize my risk and protect me? Who will make sure everyone follows these safe work practices? What do I do if something goes wrong? What will happen, and who will take care of *me* if I get hurt?

Coordination Procedures

The health and safety program encompasses many elements that must be implemented within the organization as well as on a site-specific basis. The program must include procedures needed to coordinate organizational and site-specific health and safety activities. (See, e.g., Figure 11–6.) Such coordination procedures are necessary to ensure that all parts of the program function harmoniously.

Figure 11–5. The person who prepares the HASP should visit the site and confer with the author of the work plan. The position of this well would have been changed if each monitoring well location had been checked prior to the start of the project.

Figure 11–6. The HASP should address sanitation and personal hygiene facilities for on-site workers.

12

Contingency Planning

A contingency may be defined as an uncertain event, a chance, a possibility—something dependent on chance. Based on that definition, a contingency is not always an unfortunate event; winning the lottery, for instance, would be a very happy contingency. Contingency planning for hazardous waste sites, however, usually encompasses planning for unfavorable or unfortunate site emergencies such as worker accidents, chemical overexposures, spills, fires, or explosions. The site emergency response plan, designed to equip workers to deal with such emergencies, is an important part of the HASP and should be developed before initiation of on-site operations.

The contingency plan, as part of the HASP, is also prepared to protect local affected receptors (nearby homes or businesses, sensitive environmental areas) in the event of an accident or an emergency (Figure 12–2). It may incorporate an air monitoring plan, an accident prevention plan, and a spill control plan, if applicable.

ELEMENTS OF A CONTINGENCY PLAN

Personnel

The plan should include provisions for all potentially involved personnel and agencies, including on- and off-site personnel, nonroutine workers, subcontractors, visitors, and emergency responders (police, fire, medical). In all cases, a clear chain-of-command structure is required, and every individual should know his/her position and authority. The chain of command should be flexible to accommodate the absence of some individuals or multiple emergencies (those occurring at the same time).

In an emergency, one on-site individual should be designated to assume control of on-site operations. In this case, control means authorization to purchase supplies or contract for emergency services, authority to make decisions and resolve disputes concerning health and safety, and authority over the activities of all on-site personnel, including contractors and emergency responders. The person in charge may be the project leader, the health and safety officer, a site supervisor, or a team leader; this individual becomes the incident commander (IC) during the emergency. The designated IC should be identified in the plan, along with one or more alternates.

During an emergency, teams of two or more site workers provide greater efficiency and safety than individuals working alone. Teams comprised of site workers can be assigned emergency duties based on their normal work assignments, or they may receive additional training in first aid, CPR, rescue, or spill response.

Off-site personnel may include representatives from municipal or state agencies, public safety sectors, private organizations with specific expertise, utility companies, and cleanup and disposal contractors. In most cases, off-site personnel are from the local police, fire, and EMS departments; local or state health and environmental departments may also be involved. In many cases, when outside aid is summoned, the ranking police, fire, or agency official may be eligible, authorized, or required to assume the role of incident commander. As part of advance contingency planning, arrangements should be made with these agencies to determine their resources, response time, and willingness to assist with site emergencies. Information about site hazards and special precautions required must be made available to responding agencies.

The plan's author should verify that local responders will respond to a site emergency, that response personnel are certified as hazardous materials responders, and that the responders are properly equipped (Figure 12–3). In some cases, specialized training or equipment may be required before local responders are able to assist at on-site emergencies.

The contingency plan should list telephone numbers of appropriate local, state, and federal agencies, as well as national organizations that should be notified in case of a site emergency or a hazardous materials release. The plan should also list local and state spill reporting requirements and appropriate telephone numbers. Federal requirements under the National Oil and Hazardous Substances Pollution Contingency Plan (40 CFR 300) also must be followed.

Training

All site workers should be trained in how to respond in case of an emergency. A contingency training program should include:

- Site-specific hazards and anticipated emergency situations.
- Recognition of emergency situations or potentially dangerous conditions.
- Emergency signals; how to summon help.
- What to do, where to go, and whom to report to if an alarm or emergency signal is heard.
- Emergency procedures, including brief, frequent, and realistic drills or practice sessions.
- Primary and alternate evacuation routes, and safe assembly areas.

Although they should always be accompanied by a site worker, visitors must be briefed on basic emergency procedures including emergency signals and evacuation routes.

Figure 12–1. Accidents often happen because of carelessness or because of Murphy's Law. Contingency planning deals with accidents and other site emergencies.

Figure 12–2. Contingency planning is designed to protect local receptors in the event of an accident or an unexpected release.

On-site workers assigned to emergency response roles must have a thorough understanding of their duties and responsibilities. Training should be directed toward their specific emergency response duties, and should include instruction in the following:

- Emergency chain of command.
- Emergency communications and signals; how to call for help.
- Emergency equipment and its use.
- Emergency evacuation procedures.
- Recognition of chemical overexposure, physical injuries, and hot and cold stress.
- Rescue and treatment of injured personnel, including CPR.
- Decontamination of injured personnel.
- Confined space rescue.
- Using off-site emergency response resources.
- Brief but frequent practice sessions and drills, which should occasionally include off-site responders.

Off-site emergency responders, such as local police, fire, and EMS crews, could be exposed to hazards in case of an emergency. In some cases, local responders are less well-protected against hazards than on-site workers. If necessary, local responders should be

Figure 12-3. The willingness and the qualifications of local emergency responders should be verified before they are included in a contingency plan.

equipped to deal with site hazards (Figure 12–4). Local agency personnel may not have OSHA-required training or be qualified to respond to on-site emergencies. Site-specific training should be offered to local responders, which should include instruction in the following:

- On-site chain of command.
- On-site communication procedures.
- Communication procedures between off-site and on-site emergency response personnel.
- Site hazards and hazard recognition.
- Site access and restricted areas.
- Response techniques for specific hazards.
- Site emergency communications and signals.
- Site emergency evacuation procedures.
- Decontamination procedures for personnel and equipment.
- On-site emergency response resources.
- Anticipated role of off-site emergency responders.
- Practice sessions for specific procedures and occasional drills involving on-site responders.

Figure 12–4. It may be necessary to augment the equipment and the training of local responders before they are qualified to respond to on-site emergencies.

Communications

During an emergency, information must be transmitted quickly, easily, and accurately. On-site staff must be able to relay data regarding injuries, evacuation orders, and routes of entry and egress, as well as coordinate emergency response activities. To expedite on-site communication, a separate set of internal emergency signals should be developed and rehearsed on a regular basis.

Emergency signals should be different from routine signals, brief but obvious, and limited in number so that they are easily remembered. To prevent confusion, the signal for one type of emergency should be distinct and different from other emergency signals (e.g., a siren for evacuation, an airhorn for fire/explosion, a whistle for a medical emergency). An all-clear signal should also be used to indicate the end of the emergency.

A radio or another form of communication with Support Zone personnel should also be available to workers. Whenever possible, a separate channel should be designated for emergencies. All personnel should be trained in proper radio communications etiquette and procedures. Background noise can make speaking and listening difficult; so inexperienced users should practice using a radio while wearing PPE.

All site personnel should be familiar with the external communication system. Telephone and radio are the most commonly used communication devices. The notification protocol (telephone number and contact person) for specific emergencies (medical emergency, fire, spill) should be posted or otherwise readily available at all time.

In the unlikely event that there is no telephone available on-site, the location of the nearest public telephone should be posted, along with coins for the telephone (if required) and the necessary phone numbers. The public telephone must be readily accessible; transportation must be available if the phone is some distance away. Whatever the communication devices used, they should be checked daily to ensure that they are in good working order.

Site Mapping

Just as the site map is the basis for development of a site HASP, it is also an important tool in developing a site contingency plan (see Figure 9–1). The site map contains essential information on the location of site hazards, site terrain and topography, structures, routes of entry and egress, areas of worker activity, and locations of supplies, equipment, and resources needed during an emergency. Sensitive off-site receptors also should be noted, as well as site locations where emergencies are most likely to occur.

The site map should be made available to off-site emergency responders, who may wish to pre-plan for some types of emergencies and develop alternative response plans and strategies for placement of equipment and personnel.

Site Security

The importance of site security is most apparent during an emergency. It is imperative that the emergency IC know how many workers are on-site and where they are. Having workers check in and out at control access points allows this information to be readily available. During an emergency, workers who are recalled from the field must go through an access control point or assemble at a designated area. Once all workers are accounted for, only necessary rescue and response personnel are allowed on-site. (See Figure 12–5.)

Evacuation

During an emergency, normal evacuation routes through access control points may not be accessible or feasible. In such cases, alternate routes should be established during the initial contingency planning. Primary and alternate routes should be kept clear and accessible at all times. Evacuation routes, both on-site and outside the site boundaries, should be periodically checked to ensure that they are still usable. Road repair or construction can cause delays or reroute traffic through detours. Site activities can dramatically change the site terrain; if a once-flat area now contains trenches, pits, and steep slopes, it no longer qualifies as an escape route.

All site workers should know how to evacuate the site and where to assemble if it is not possible to exit through the access control points. Visitors should be informed of site hazards and the nearest evacuation route.

Figure 12–5. On-site personnel should not comment on their activities to the news media. Inadequate site security allowed this news crew onto the site.

A site emergency may be of such proportions that it threatens the health or the safety of the surrounding community. When this happens, the public must be so informed. Local authorities should be advised of the actual or the potential hazard, as it is up to the local authorities to determine whether the public should be evacuated. The site contingency plan should identify the public official who has the authority to order or recommend an evacuation. Evacuation plans should be coordinated with members of local public safety departments. For projects involving high hazards, or when a large number of residents may be affected, local television and radio stations, civil defense, Red Cross, and municipal transportation systems may be involved.

In some cases, depending on the nature of the hazard and the amount released, it may not be advisable for persons to go outdoors to evacuate. An alternative is protection in place, which calls for residents to stay indoors. It also requires residents to control all ignition sources, close all windows, and shut off all heating, air-conditioning, and ventilation equipment. This option is most feasible for hospitals, nursing homes, schools, and high-rise buildings where evacuation would be difficult, and the heavy construction and controlled interior environment offer protection to the occupants.

In planning for site emergencies, it is important to consider the local political climate and the local residents' perception of the site. For example, the on-site personnel may be regarded as "experts" by the local population. The decision to evacuate off-site receptors, however, should be made by the local authorities, in accordance with the preexisting evacuation plan.

Decontamination

Procedures for decontaminating accident or overexposure victims should be defined in the contingency plan and rehearsed by emergency responders. Decontamination procedures must address the needs of the victim and the safety of the responders. Decontamination should be performed if it does not interfere with medical treatment, but it should not be done if it delays life-saving treatment or aggravates a serious injury. Off-site emergency responders, ambulance crews, and the hospital must be aware that, in some cases, there may be contamination on the victim. Whenever possible, the contingency plan should include provisions for decontamination of the victim at the hospital.

If decontamination on-site is not possible, wrap the victim in blankets, plastic, or other wrapping to reduce the contamination of other personnel. Alert medical responders that the victim is contaminated, and provide information about the specific hazard, personal protective methods, and decontamination procedures.

Medical Treatment

Treatment of minor injuries, such as scrapes, cuts, and mild burns, should be easily handled by on-site workers, and these injuries should not be considered emergencies. In these cases, the medical treatment is fairly straightforward. On the other hand, toxic exposures or other situations that cause serious injuries and illnesses should be handled by a qualified physician. In some cases, however, medical assistance may not be immediately available, and on-site workers must administer life-saving techniques.

A group of general site workers should be trained in emergency medical treatment procedures, including, as a minimum, first aid and CPR. Emergency medical technicians (EMTs) are qualified to administer oxygen and perform more advanced treatment protocols. Site medical responders should also be trained in emergency decontamination procedures.

At long-term projects, on-site and off-site medical responders should plan a coordinated response to site emergencies and drill together on a regular basis. Emergency medical response procedures should be reviewed and approved by the organization's occupational physician, who should be on call at all times.

Local ambulance and hospital personnel should be educated about potential medical problems on-site, types of hazards and the consequences of exposure, and heat and cold stress. Procedures for handling medical emergencies and contaminated victims should be developed and practiced. The response time for off-site medical responders, the optimal route to the hospital, and the transit time to the hospital should be determined. If an ambulance is not available, an alternate means of transporting victims to the hospital should always be readily accessible. The route to the hospital should be checked regularly to ensure that it is still usable.

The local resources may be limited in terms of staff, space, and vehicles; the number of victims who can be transported to and treated at the local hospital must be determined. Additional medical facilities and other means of victim transportation may be necessary.

The following equipment and information should be readily available to on-site medical responders:

- Fully stocked first aid kit.
- Emergency eyewash and decontamination solutions.
- Sterile sheets and burn blankets.
- Stretchers and blankets.
- Oxygen, airways, and bag valve mask for resuscitation.
- Radios for on-site communication.
- Name and telephone number of on-call physician(s).
- Names and telephone numbers of medical specialists.
- Telephone numbers of receiving hospitals.
- Telephone numbers of ambulance services.
- Telephone numbers of police and fire departments.
- Telephone number of nearest poison control center.
- Name of person on-site (and alternates) to notify in case of medical emergency.
- Map and directions to each receiving hospital.

Small Projects and Common Sense

Much of the above discussion is oriented toward long-term projects of considerable size with a large number of workers. A small, mobile investigatory team (e.g., drill crew and geologist) cannot ignore the basic components of contingency planning. It is important that all members of the crew know how and whom to call in case of an emergency. The plan should cover spills or releases of hazardous substances (including hazardous decon agents), fires (incipient fires, drill rig fires, hazardous materials fires), and medical emergencies (involving a contaminated or a noncontaminated victim).

Each crew member must also understand his or her role during an emergency. In some cases, crew members may be prohibited from handling an emergency (e.g., a fire involving hazardous materials) because they lack proper training or equipment. Prompt recognition of these limitations will facilitate rapid notification of off-site emergency responders.

Even the best plans cannot prevent accidents (Figure 12–6), but common-sense planning is essential on a project of any size (Figure 12–7).

Items for a Preliminary Contingency Plan

The following is a list of items that could be included in a preliminary contingency plan:

- Name of on-site person responsible in the event of an emergency.
- Dates for meetings with: local community groups; local, state, and federal agencies involved; local emergency responders; personnel at local hospitals.

Figure 12-6. Even the best plan cannot prevent accidents. This crew hit a water line that was located 40 feet away from its marked position. The ruptured water line affected the sprinkler system to the entire facility.

Figure 12-7. For working in buildings, emergency lighting and alternate exits should be available.

- First aid and medical information, including names of site personnel trained as medical responders; map and directions to local medical facilities; emergency phone numbers for police, fire, EMS, and local hazmat teams, and state and federal reporting numbers.

- Air monitoring plan if site-specific hazards include risk of inhalation exposure or airborne dispersal; minimum requirements for on-site and perimeter monitoring.

- Spill control plan if there is a potential for spill and discharge: methods and equipment required to prevent contamination of soil, water, atmosphere, and uncontaminated structures; names of on-site personnel trained as spill responders; telephone numbers of hazmat teams or contractors trained to deal with hazardous waste spills.

- Decontamination plan for contaminated personnel, bystanders, or equipment.

13

Special Operations

Numerous hazards must be considered in working at a hazardous waste site, many of which were discussed in detail in earlier chapters. This chapter addresses site operational practices not yet discussed: handling drums and containers, drilling in flammable environments, and confined spaced entry. It also reviews information on chemical/physical properties, toxicity, and air monitoring, as an aid to workers using this information in special operations.

HANDLING DRUMS AND CONTAINERS

Size, Shape, and Composition of Containers

Many hazardous waste sites have within them a variety of containers, of different sizes and shapes, designed to hold liquid or solid materials. The size of a container usually is its capacity in gallons. Drums or barrels come in a variety of sizes, the 20-, 30-, and 55-gallon ones being the most common. Pails, cans, or buckets come with lids and vary in size from 1 to 10 gallons. Drums and pails may be made of steel, aluminum, other metals, plywood, fiber, or plastic. Jerricans have a rectangular or polygonal cross section and are usually made of steel or plastic (they derive their name from the German gasoline cans used in World War II). Boxes and bags come in a variety of sizes and shapes and are made of commonly used packaging materials. Boxes may be made of metal, wood, fiberboard, or plastic; bags may be made of plastic, textiles, or paper.

The design of the container often gives an indication of the physical state of the material it was designed to hold. For example, paper bags usually contain solids; jerricans usually hold liquids. Boxes typically contain solids, but lined boxes may also hold liquids. Drums and pails may hold liquids or solids. Pails and cans designed to hold liquids usually have a pour spout or another opening to facilitate removal of the liquid.

Closed head (nonremovable head) drums have a closed, sealed top with an opening or bung for filling or emptying them; they are designed to hold liquids (Figure 13–1). The closed head drums used to ship volatile liquids are usually equipped with a small pressure relief plug. Drum bungs and plugs are usually made of the same material as the drum itself; their gaskets are usually made of a synthetic material such as polyethylene or Teflon.

Open head (removable head) drums have a lid that can be removed; the lid or the cover of the drum is held in place by a retaining ring or a closure ring that is fastened by a nut and bolt or a lever locking device. Open head drums are designed to hold solids, but open head drums with bung openings can hold liquids or solids. The lid of an open head drum usually has the same gauge thickness as the body of the drum. The lid gasket may be rubber or a synthetic material.

Drums that have a 95- or 85-gallon capacity are overpack or salvage drums, designed to accept a damaged or leaking 55-gallon drum. Rolling hoops are thickened areas around the middle of the drum. Liners may be used to keep the contents away from the drum itself.

The chime of the drum is the seam or raised portion where the sides of the drum meet the bottom (and the top of a closed head drum). Because it is a seam where two pieces of metal have been joined, the chime is a weak point where leaks occur if the drum is stressed (Figure 13–1).

The design and the composition of a container may give an indication of the type of material it was designed to contain. For example, 16-gauge stainless steel drums are designed for strength, durability, and corrosion resistance; these drums are often used to ship corrosive, heavier-than-water materials (Figure 13–2). Drums made of other, so-called exotic metals, such as nickel or aluminum, often contain highly hazardous substances, as may older galvanized drums. Polyethylene drums are lightweight and rugged, and they too are used to transport corrosives.

Figure 13–1. A group of abandoned closed head drums. The larger bung opening is for filling the drum with liquid; the smaller bung is for venting.

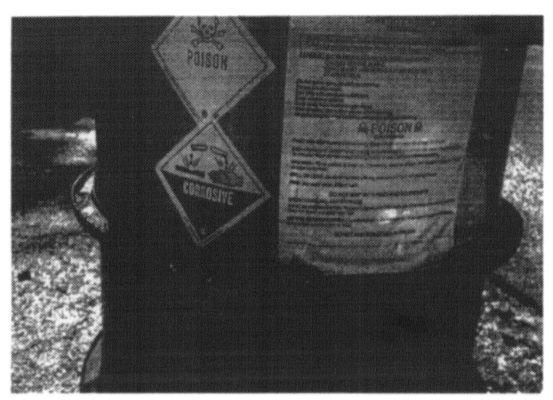

Figure 13–2. A stainless steel drum designed to hold corrosive liquids. This drum is equipped with rolling hoops.

Drum Specifications

Most drums is use today are made of 18-gauge carbon steel. New standards for container manufacture and performance were issued by the USDOT in 1990 as part of the changes to the DOT 49 CFR *Transport of Hazardous Materials Regulations,* known by their docket number HM-181. With HM-181, the USDOT decided to revise its cumbersome packaging regulations in favor of UN performance oriented packaging (POP). There are, however, millions of drums in storage or use that were manufactured under the old DOT specifications—most of the drums encountered at hazardous waste sites have been manufactured prior to 1990. It is helpful, then, to know the old DOT specifications as well as the new.

Each drum manufactured prior to HM-181 had a specification number stamped or embossed on its bottom (Table 13–1). The letters STC stood for single trip container; the

Table 13–1. Examples of DOT Specifications Prior to HM-181

Open head metal	17C, 17H
Closed head metal	17E, 17F
Closed head polyethylene	34
Open head fiber	21C
Aluminum composition	42B, 42D

container could not be reused unless it was reconditioned and certified for reuse. The gauge of metal used, volume capacity, and year of manufacture were also indicated. For example, 16-55-88 indicates a drum made of 16-gauge metal, with a 55-gallon capacity, that was manufactured in 1988; 16/18-55-88 indicates that 16-gauge metal was used for the body of the drum, and 18-gauge was used for the head or the cover. A worker in the field coming upon a drum lying on its side might see the following stamped on the bottom:

17E/17H

STC

16/18-55-73

These markings indicate that the drum is a 55-gallon, open head drum with bungs, manufactured in 1973; the cover is made of 18-gauge metal and the body of the drum of 16-gauge metal. Some manufacturers also put the month of manufacture and/or the company name or initials. Although the date of manufacture does not indicate when the drum was actually filled, it allows one to determine how old the drum is.

HM-181 requires that UNPOP tests be performed on packaging for hazardous materials. These tests include the drop test, leakage tests for liquid containers, hydrostatic pressure tests to ensure that the container prevents leakage of liquids under pressure, and a stacking test to ensure that packaging is uniform and remains stable in a stack of containers of similar type and size. UNPOP specifications must be clearly marked on each container by the manufacturer. A package identification code must be used, which indicates the type of container, its composition, and its category or subtype (Table 13–2). The letters UN, or a circle with the letters UN, should precede the identification code.

A 1A2 marking on a container indicates that it is a drum made of steel, with an open head; a 1A1 container is a closed head steel drum. A 4D1 container is an ordinary plywood box; a 3N1 container is a closed head metal jerrican. Composite packaging is indicated by two letters used in sequence; the first letter indicates the inner receptacle, and the second is for the outer container. A container labeled 6HA2 would have an inner plastic receptacle (H) inside a steel, closed head drum (A2).

Table 13–2. Examples of UN packaging identification codes (HM-181)

Type	Composition	Category
1: Drums	A: Steel	A, B, or H: Drums/Jerricans
2: Barrels	B: Aluminum	1—Closed head
3: Jerricans	C: Natural Wood	2—Open head
4: Boxes	D: Plywood	A or B: Boxes
5: Bags	G: Fiber	1—Ordinary; 2—Lined
6: Composite packaging	H: Plastic	C: Boxes
	N: Metal	1—Ordinary; 2—Sift-proof
7: Pressure receptacle	P: Glass, or porcelain	L: Bags
		1—ordinary; 2—Sift-proof

Each container must also meet specific criteria for Packing Group I, II, or III. Performance Standard Codes of X, Y, or Z are used. High hazards are assigned to Packing Group I, the lowest hazards to Packing Group III. Containers marked with an X meet standards for Packing Groups I, II, and III; a Y indicates that the container meets the criteria for Packing Groups II and III; containers marked with a Z meet standards only for Packing Group III.

A number designating either the weight that the container is designed to hold, in kilograms, or the maximum specific gravity should follow the performance standard code. Containers designed to carry liquids should be marked with the specific gravity for which the container has been tested. Markings are not required for containers designed to carry liquids with specific gravities that do not exceed 1.2. If the container is designed to hold liquids only, the hydrostatic test pressure should be indicated in kilopascals (rounded down to the nearest 10 kPa). A closed head drum marked Y1.8/300 meets the Performance Standard Code for Packing Groups II and III, is designed to carry liquids with specific gravities up to 1.8, and can withstand hydrostatic pressures up to 300 kPa or approximately 43 psi.

Packages intended to carry solids must be marked with the maximum gross weight, in kilograms (kg), for which the packaging has been tested. An S is used to indicate those packages that are intended for solids only. Thus an open head drum marked X280/S meets the performance criteria for Packing Groups I, II, and III, is designed to carry solids only, and can hold up to 280 kg or approximately 616 pounds.

Finally, the capacity of the container, in liters, as well as the year, date, and country of manufacture, should be clearly marked on the container. Put all these codes together, and they give a lot of information about the packaging, as well as the type of material inside. For example, a drum with the code 1A1/Y1.4/300/208/USA/96 is a closed head steel drum (1A1), meets Packing Group II and III criteria, and is designed to hold a liquid with a specific gravity of 1.4 or less (Y1.4); also, the drum is designed to withstand hydrostatic pressures up to 300 kPa or about 43 psi, has a capacity of 208 liters or 55 gallons, and was manufactured in the United States in 1996.

Drum Inspection

Any drum or container that is found on-site should be inspected for labels, manufacturer's markings, or other information that may indicate that the contents are hazardous. A drum that has a manufacturer's label may still have the factory seal on the bung; this is a good indication that the drum contains the original contents. Drum labels fade, or the drum may be improperly marked; one must not rely on drum labels or markings alone to identify hazards. Unlabeled drums should be considered to contain hazardous material until the contents are identified. If the contents of the drum are unknown, monitoring devices for flammability, radioactivity, and toxics should be employed. Inspecting for drum integrity requires close contact with the container; one should approach it cautiously.

The container also should be visually inspected for signs of deterioration, such as corrosion, rust, dents, or leaks. A drum that is dented on the top or the bottom will often leak at the chime. A drum that is bulging may also have a weakened or ruptured chime. Drums that bulge, or have been crushed or dropped, may also leak around the bung or the retaining ring. Crystallization or corrosion around the bung or the retaining ring suggests that vapors and perhaps liquids are escaping.

The container can be carefully thumped or tapped from top to bottom to determine if it is full, empty, or partially full. Observing condensation (on hot humid days) or feeling the drum with the back of a lightly gloved hand on both the sunny and the shaded side can help determine the liquid level. Drums containing volatile liquids will feel cooler than drums with aqueous materials, which will feel cooler than drums containing low vapor pressure oily materials.

Excessive pressure buildup within a drum will cause it to bulge. Some drums are designed to withstand fairly high pressures, whereas others are not. For example, heavy-gauge closed top drums intended to carry volatile liquids are designed to withstand pressures in excess of 300 kPa (43 psi); lighter-gauge drums are rated to only 100 kPa (14.5 psi). An increase of only 3 to 5 psi above normal pressure is sufficient to cause a light-gauge drum to bulge.

A bulging drum has *at some time* been subjected to excessive pressure buildup; not all bulging drums are actually under pressure. Drums may bulge because of a temperature increase, a chemical reaction, or biological decomposition. For example, the liquid in drums involved in a fire will be heated, and the vapor pressure will increase inside the drums. Some of the drums may rupture; others will bulge but not rupture. After the fire, the liquid in the drums will cool, and the pressure within the drums eventually will return to normal. The drums are no longer under pressure, but they continue to bulge; this condition is called plastic deformation (Figure 13–3). The top of a bulging drum that is not under pressure can be easily pushed down with one's hand; a drum under pressure will resist the pressure of the hand. Bulging drums that are not under pressure do not present a hazard because of excess pressure, but they have been weakened during the bulging process and should be handled accordingly. Metal drums under dynamic pressure typically show spalling of rust due to expansion and contraction.

Drums containing aqueous liquids, if allowed to freeze, will bulge at the bottom; they may also leak at the bottom chime. These drums will often tip to one side or another because the bottom of the drum is no longer flat. It is not unusual to find such drums inverted for stability reasons or to prevent leakage; the bungs will be on the bottom, and the bulge on top.

It is not unusual for drums stored outdoors to heat up and cool down during a normal temperature cycle. Closed top drums, especially those constructed of lighter-gauge metal, will expand and contract (undergo elastic deformation) as they heat up and cool down. As these drums expand and contract, they often give off telltale sounds; drums that "sing" usually contain volatile materials.

The design and the composition of the drum suggest what type of material it was designed to hold. Stainless steel, galvanized, aluminum, or plastic drums often contain corrosives or hazardous wastes; open top drums with liners often contain laboratory packs. Lab packs usually contain solid and liquid chemicals; but, in many cases, there is little evidence of the identity of the materials in a lab pack. It is not unusual to find radioactive, biological, and incompatible chemical waste materials in the same lab pack. Drums known or suspected to contain lab packs should be handled initially as if they contained shock-sensitive or explosive materials.

Figure 13-3. An example of plastic deformation. The bulging drum in the foreground was subjected to heat during a fire. Although no longer under pressure, the drum is permanently deformed.

Drums should be moved only when they must be. Unfortunately, drums are often stacked, buried, or crowded together in a manner that makes visual examination of each drum impossible. In these cases, the drums must be handled and moved to facilitate inspection. Empty drums are typically stacked horizontally on their sides, whereas full or partially full drums are stacked vertically, one atop the other. Drums that are filled tend to elongate and leak when stacked horizontally.

Moving and Handling Drums

Personnel responsible for moving drums should have experience in handling hazardous materials and be qualified to isolate, contain, and remediate a spill. They should wear proper PPE, including foot protection. Salvage or overpack drums and absorbents should be on hand before the drums are moved. Whenever possible, drums should be moved by mechanical equipment. A drum that is deteriorated or damaged may leak when moved, and it should never be rolled on its side. Moving a drum by tipping it on the bottom edge is asking for a back injury; also, on open top drums, the bolt on the retaining ring can catch the fingers!

Drum lift attachments can be used with forklifts, backhoes, or hoists to lift a drum onto a dolly or a pallet, or into a salvage container. Lift attachments are designed to lift, not to transport a drum any distance. Drum grabbers are designed to be used on backhoes or forklifts, and are used when a drum must be carried from one place to another. The grabber holds the drum securely around the middle; some grabbers can actually invert the drum if it needs to be emptied. In situations where mechanical equipment cannot be used, drum trucks can be employed; a strap or a chime hook should be used to keep the drum securely on the truck. Some drums that are corroded or have been buried may be in poor condition; these drums may leak or rupture if a drum lift or grabber is used. In these cases, the contents should be transferred before the drum is moved.

All equipment used for moving or handling drums should have sufficient capacity to handle the anticipated loads; equipment designed to handle a single drum at a time should be rated for 1000 pounds. Some drum-handling equipment is designed for warehouses or loading docks with smooth paved surfaces and will not perform well on the uneven surfaces encountered in the field.

Leaking Drums

Drums can be stressed to the point of leakage or rupture in a variety of ways. A drum leak can be mechanically induced, as when a drum is punctured by a forklift or a backhoe. Crushing the side of a drum can also cause a leak, either at the chime or where the drum is creased. Placing too heavy a load on the top or the side of a drum will cause leaking at the chime, the weakest point of the drum. Even drums that appear intact may leak at the bottom chime after being moved.

Whenever possible, the liquid in a drum that is deteriorated should be transferred to an

Figure 13–4. A buried drum is hoisted out of a trench with a backhoe.

Figure 13-5. The side of this drum was crushed by a forklift. Leaks occurred at the top and bottom chimes, not at the point of impact.

intact drum. Transferring the liquid is preferable to risking rupture during overpacking. Heavy-duty drum pumps are available to transfer flammable or corrosive materials. Pumps may be manual, electrical, or air-driven. Manual pumps used to transfer flammable liquids should be FM-approved and be equipped with a flash arrestor. Electric pumps should be intrinsically safe. Whenever flammable liquids are transferred from one drum to another, both drums should be grounded and bonded to prevent the ignition of vapors by static.

It is also possible to patch or repair a leak, but there is no such thing as a permanent patch. *Patching a leaking drum is a temporary measure to give workers additional time to safely overpack the drum or transfer the liquid.* A variety of patch kits are available; however, many are designed for hazmat teams and may not be suitable for on-site workers with only a few hours of formal training on spill control and containment techniques. Probably the easiest drum patch materials to use are the epoxy-based putties or pastes that plug relatively small holes or cracks in drums. The putty is usually kneaded by hand and then pressed into the hole; the paste is ready to use as it comes out of the tube container. Both begin to harden in minutes and work on all types of drums.

Drum "bandages" can be employed to stop liquid leaks. The bandage is usually a rectangular patch made of a chemical-resistant material; a heavy-duty ratchet and webbing hold the patch tightly over the leak. Some bandages employ a bladder to seal the leak; the bladder is filled with air, nitrogen, or carbon dioxide after it is slipped over the drum and positioned across the leak. Other patches are magnets capped with a chemical-resistant material. Drum bandages and patches vary in their resistance to chemicals; one should consult the manufacturers' chemical resistance guides to determine which patch material offers the greatest versatility.

Tapered plugs made of wood or rubber can be used to stop leaks in drums. A mallet is used to force the plug into the hole to create a liquidtight seal. Plugs come in all sizes, from as small as golf tees to up to 6 inches in diameter.

It is virtually impossible to patch a leaking drum without coming into contact with the contents. Personnel involved in chemical containment procedures should wear appropriate PPE. In cases of unknown or flammable liquids, air monitoring for flammable vapors should be conducted in the area. Tools used in drum handling and patching should be FM-approved and spark-resistant.

Overpacking Drums

Overpacking involves placing a damaged container inside a slightly larger, open top salvage container. A 95-gallon salvage container can hold one 55-gallon plastic or metal drum. An 85-gallon salvage container can hold a 55-gallon metal drum but cannot accommodate a 55-gallon plastic drum. The salvage container may have a chemical-resistant liner, or the space between the inner drum and the outer drum may be filled with an absorbent material. The salvage container may have a chemical-resistant inner liner. Salvage drums usually are made of steel or polyethylene.

Drums that have been patched or are deteriorated should be overpacked before they are moved to a staging or storage area. Overpacking the drums can be very hazardous; lifting the drum exerts additional stress that may cause it to rupture. The safest method of overpacking is to use a backhoe, hoist, or forklift equipped with a mechanical drum grabber or a drum lifter or sling; when this equipment is used, the drum can be lifted and placed into an overpack or salvage container. The operator should be supplied with appropriate PPE and protected with a heavy-duty splash shield. Other workers should use ropes or long poles to position the drum over the salvage container. Never stand directly under a raised drum; never use your hands to direct a lifted drum into an overpack container.

There may be instances when mechanical methods are not feasible, and overpacking must be performed manually. There are three basic methods (described below) that can be used, all of them somewhat hazardous. A 55-gallon drum may weigh in excess of 800 pounds; when such drums are mishandled, crushing injuries to the hands, fingers, legs, and feet are possible. Workers should be mindful of the lid ring bolt or the lever lock device on an open top drum, which can catch hands and fingers during rolling. Improper lifting of drums can cause back injury. Contamination from the leaking product is another possibility. The manual overpacking methods require a minimum of two workers, who must work as a team. Workers should practice these methods first with empty drums, and then with half-filled drums, before attempting to overpack drums containing hazardous materials in the field. Manual overpack methods should never be used on drums that contain explosive or unstable materials, or that are under pressure. Drums that may forcibly rupture during overpacking should be handled by a mechanical drum grabber or by a hoist.

The inverted overpack procedure is the simplest manual overpack method; it involves inverting the salvage container over the damaged drum, then tipping or rotating both drums into an upright position. This is the method of choice for dealing with a damaged drum that is upside-down because when the overpack is turned upright, the inside drum will be right-side-up and can be sampled. *Do not place a drum upside-down in a salvage con-*

Figure 13–6. The wrong way to overpack a drum. Workers should not use their hands to guide the drum into the overpack container.

tainer; the inner drum cannot be sampled, and its contents cannot be characterized. Drums with contents that cannot be characterized are often rejected by waste recovery facilities and returned to the sender.

A second method, the roller technique, uses small-diameter (1"–2") metal pipe. The damaged drum is placed on its side with the rollers underneath it. The drum is then rolled partway into the salvage container. The salvage container is manually turned upright, with the damaged drum right-side-up inside the outer container. Workers must pull the drum along the rollers while in a crouched position, which may result in back strain.

The third procedure, the angle roll method, requires the most teamwork; it can also be rather cumbersome and should be performed on level ground. The damaged drum and the salvage container are placed on their sides, the two of them end-to-end at an angle, with the bottom of the damaged drum facing the open end of the salvage drum. The drums are then rolled toward each other so that the damaged drum begins to roll into the salvage container. The drums roll toward each other until the inner drum jams against the wall of the salvage container; then they are repositioned and rolled in the opposite direction. Eventually, the salvage drum can be lifted into an upright position with the damaged drum right-side-up inside.

When necessary, the drums can be turned upright with a drum upender, which is equipped with a long handle to give the user sufficient leverage to lift the drum manually. Most drum upenders were designed for 55-gallon drums; usually the effort of two workers is needed to put an 85- or 95-gallon salvage drum in the upright position.

Drum Sampling

Sampling the contents of drums is necessary to characterize them. In most cases, drums with unknown contents should be segregated and sampled separately. The drums to be sampled should be placed upright on the ground, in rows no deeper than two drums each. It is dangerous to attempt to sample stacked drums; do not climb over or stand on drums during sampling. Use tools and equipment designed to open drums. Air monitoring for combustible gases and radiation should be conducted during drum opening and sampling.

Each drum must be opened for sampling, and a variety of methods can be used for this. The simplest method for closed top drums is to open the bung manually with a bung wrench. The smaller bung, if there is one, should be opened first; turn the bung slowly to release any pressure in the drum. A nonsparking bung wrench usually is employed with drums containing flammables or unknowns; however, the wrench cannot prevent sparking between the metal bung and the drum itself. Do not attempt to open a bung if it is coated with a crystalline material; crystal formation around the bung suggests the presence of peroxides, which can be shock-sensitive.

A deheader is essentially a giant can opener used to remove the top of a closed head drum. Deheaders are used only if the drum cannot be opened at the bung. The initial cut into the drum should be made very slowly to allow the pressure within the drum to dissipate. Deheaders should not be used on bulging drums that are under pressure, or on nonbulging drums if the tops cannot be easily depressed by a hand. Because of the potential for sparking, deheaders should not be used to open drums containing flammable liquids.

In some instances it may be advisable to open drums remotely. There are a variety of remotely controlled means of opening drums, a fairly low-tech one being the spike method.

A bronze spike is attached to the arm of a backhoe, and is punched through the top of the drum, creating a hole several inches in diameter. A more sophisticated variation is the hydraulic backhoe drum plunger, which employs a drum grabber and a hydraulic stainless steel plunger. The backhoe is equipped with an "explosion-proof" shield to protect the operator from splash and debris. These methods are usually employed when a large number of drums must be sampled. Also, a conveyer belt system for remote hydraulic puncturing of drums has been described (USEPA, 1992a).

Other remote opening devices include the hydraulic drum drill and the pneumatic bung opener; but both must be positioned on each drum opened, a fairly time-consuming procedure. The pneumatic bung opener is often used, however, when drums are known or suspected to contain shock-sensitive materials.

Extreme care should be taken in opening an overpacked drum. In many cases, the salvage or overpacked container already holds a damaged, deteriorated, leaking, or chemically reactive drum. Moving the drum may have shifted or mixed materials and produced a chemical reaction. For many years, salvage container lids did not have a bung for venting; newer salvage drums are now equipped with frangible discs or vent bungs designed to release pressure in excess of 5 psi. Most salvage containers in service today are of the older type and do not have pressure relief mechanisms. Overpack drums are typically heavier-gauge steel, which can take pressure of several psi before a bulge appears. Opening the lid by removal of the retaining ring of a pressurized salvage drum has resulted in serious injury and death. An increase of just 3 psi on an 85-gallon overpack lid can produce over 1700 pounds of thrust. Never assume that because a drum is not bulging it is not under pressure!

Excavating Drums

A geophysical survey should be conducted to determine the approximate location and depth of buried drums. The soil above buried drums should be removed carefully to a level of 6 to 12 inches above the drums. Backhoe buckets should be fitted with bars to reduce the risk of drum puncture, or the soil can be excavated by hand. Each drum should be completely uncovered to facilitate inspection. Drums that appear damaged or deteriorated should be emptied before excavation. A drum grappler or grabber is preferred; alternatively, drum lifters or slings may be used. Even a drum that appears intact may rupture when lifted; workers must keep well away whenever a drum is being moved. Bulging drums under pressure should not be handled until the pressure can be relieved. Extreme care should be used in excavating or moving buried or abandoned drums, as they may no loner meet minimum design specifications.

DRILLING IN FLAMMABLE ENVIRONMENTS

Not very long ago, it was common practice for test boring contractors and monitoring well installers to preflash drill holes while engaged in landfill investigations. The preflashing resulted in "controlled" methane fires and explosions, and led to some very tall tales about launching rods and augers, Without a doubt, preflashing was a very unsafe procedure; many drillers can attest to burns and lost drill rigs. Flammable conditions also may be

Figure 13-7. A number of 55-gallon drums have been exposed in this excavation. Note the lid of a drum at the bottom of of the trench. In this situation, the trench should be backfilled and the upper tier of drums removed before attempts are made to excavate drums at the lower levels.

encountered in drilling at spill sites or at locations with leaking underground storage tanks (LUSTs).

The hazards of monitoring in flammable environments must be addressed and appropriate measures taken to protect the health and safety of on-site personnel.

Properties of Flammable Gases and Vapors

It is important for workers to review certain properties of flammable liquids and gases in order to predict and monitor their presence. (See Chapter 2.) The most important parameter in defining whether a substance is flammable or not is its flash point, which is the temperature to which a liquid or a solid must be raised to generate sufficient vapors to form an ignitable mixture with air. At the flash point, the concentration of vapors present will ignite or flash over in the presence of an ignition source, such as an open flame, a cigarette, an electrical spark, a static discharge, a frictional spark (from metal tools or quartz and metal), and so forth. The lower the flash point is relative to ambient temperatures, the greater the hazard.

The flash point of gasoline is actually a narrow range of temperatures because it is a mixture of many chemicals. Fresh gasoline has a flash point range typically between –38°F and –50°F. Diesel fuel (heating fuel oil #2), also a mixture, has a flash point range between 126°F and 204°F. Fresh gasoline, then, can be ignited in most drilling situations, whereas diesel fuel will not be. As gasoline ages, the flash point changes; older, weathered fuels have much higher flash points than fresher fuels.

Substances that are gases at normal temperature and pressure do not have flash points, as the flash point is the temperature to which a liquid or a solid must be raised to release sufficient vapors for it to burn. No heating is necessary to release gases, as they are already in a vapor state. Methane does not have a flash point because it is a gas under normal temperatures and pressures.

The concentration in air of a gas or a vapor that will burn or support combustion is known as the flammable range. Within this range, the gas or the vapor mixes with oxygen in the air and burns; outside it, the mixture is too lean or too rich to burn. The flammable range is defined in terms of the percent by volume of gas or vapor in air. Every chemical capable of burning has a flammable range.

The lower explosive limit (LEL) or lower flammable limit (LFL) is the minimum concentration of a gas or a vapor in air that will ignite in the presence of a source of ignition; concentrations less than the LEL are too lean to burn and will not ignite. Liquids cannot generate an LEL condition unless they have been raised to the flash point temperature. The upper explosive limit (UEL) or upper flammable limit (UFL) is the maximum concentration of a gas or a vapor in air that will ignite; higher concentrations are too rich to burn.

In dealing with flammable vapors, one should remember that a lean mixture does not have to get richer; but one that is too rich must eventually lean out, putting the resulting concentration in the explosive range. Also, the presence of oxidizers or an increased oxygen content changes the flammable range. The wider the flammable range is, the more hazardous the material. Gasoline typically has a range of 1.2 to 7.6%; diesel fuel, 1.3 to 6.0%; and methane, 5 to 15%. Each of these ranges is quite narrow compared to that of a material such as acetylene gas, which has a range of 2.5 to 81%. Many more fires and explosions would occur at drilling sites if the common flammable gases and vapors present on-site had wider explosive ranges.

In most cases, liquids must be raised to their flash point temperature to generate an LEL condition. However, the rapid volatilization or aerosolization of a flammable liquid may result in a flashover regardless of the temperature of the liquid. This phenomenon may occur when flammable liquids in tanks are subject to compression, air sparging, or rapid agitation.

A flashover may occur in the absence of a source of ignition if the vapors or the gases reach their autoignition or self-ignition temperature. At its autoignition temperature, the gas or the vapor is so hot that it ignites itself; and once the autoignition temperature is reached, self-sustaining combustion can occur. The autoignition temperature range for the various grades of gasoline is 536°F to 853°F; for diesel fuel, 494°F; for methane, 999°F. Hot manifolds and catalytic converters on vehicles certainly can reach these temperatures.

In working with potentially flammable substances, one should understand vapor pressure, vapor density, specific gravity, and solubility (which are discussed in detail in Chapter 2). Vapor pressure indicates the ability of a liquid or a solid to evaporate into the air; substances with high vapor pressures are highly volatile and quickly evaporate into the air. Vapor pressure is dependent on temperature; as the temperature increases, the vapor pressure increases. Vapor pressures for liquids and solids are usually expressed in millimeters of mercury (mm Hg) at a reference temperature of 68°F (20°C). Substances that exist as gases at room temperatures have vapor pressures defined in pounds per square inch (psi) or atmospheres (atm). Normal atmospheric pressure at sea level is 760 mm Hg, 14.7 psi, or 1 atm. Atmospheric pressure compresses liquids and solids and inhibits their evaporation. Materials that exist as gases at room temperature have pressures greater than 1 atm and are able to overcome the compressive force of normal atmospheric pressure.

Temperature dramatically affects vapor pressure. Water, for example, has a vapor pressure of 18 mm Hg at 68°F. When heated to its boiling point (212°F), water has a vapor pressure of 760 mm Hg and quickly evaporates. A puddle of water evaporates quickly on a hot sunny day, but it may linger for several days during cold weather.

Most mobile liquids (liquids that readily pour at room temperature) have vapor pressures greater than 1 mm Hg. Compounds with vapor pressures between 1 and 100 mm Hg will release significant vapors into the atmosphere; materials with vapor pressures greater than 100 mm Hg are highly volatile.

As discussed in Chapter 2, it is possible to roughly calculate how much vapor will be given off by a liquid or a solid on the basis of its vapor pressure. There is a simple equation, known as the 1300 rule, used to estimate the parts per million (ppm) concentration of vapors that collect in the headspace of a container at equilibrium at 68°F (20°C):

$$\text{Vapor pressure (mm Hg)} \times 1300 = \text{Headspace concentration (ppm)}$$

This equation estimates the actual vapor concentration within the container. The 1300 rule can be used only for materials that are solids or liquids at normal temperature and pressure; gases such as methane are already in the vapor or gaseous state, so the rule does not apply to them. Temperature affects vapor pressure; the calculated value could be greater or less than the actual concentration, depending on the actual ambient temperature. Despite these caveats, the 1300 rule offers a quick indication of vapor potential in a confined space (tank, borehole, casing, drum). For example, the vapor pressure of gasoline is about 190 mm Hg at 68°F; utilizing the 1300 rule, a tank containing gasoline would have a headspace concentration of approximately 247,000 ppm (1300 ppm/mm Hg × 190 mm Hg). A tank containing diesel fuel, with a vapor pressure of only 2 mm Hg, would have a headspace concentration of 2600 ppm at 68°F.

Vapor density is the weight of a vapor or a gas relative to an equal volume of air at a comparable temperature and pressure. The vapor density of air is 1; all other gases are compared to air. Gases and vapors with densities greater than 1.0 are heavier than air; gases and vapors with densities less than 1.0 are lighter than air. Some references also report the weight of the vapor or the gas as its specific gravity$_{[gas]}$, in grams per liter (g/l); this information is of little value unless one knows the weight of air. Air weighs 1.29 g/l at 32°F. So a gas or a vapor that weighs 1.80 g/l is heavier than air, whereas a vapor or a gas weighing 1.10 g/l is lighter than air.

Vapor density can also be calculated by taking the molecular weight of the compound of interest and dividing it by the molecular weight of air, which is 29. For example, the molecular weight of methane is 16, and dividing 16 by 29 gives a vapor density of 0.55; therefore, methane is lighter than air. Gasoline, depending on its age and its octane value, has vapor densities of between 3 and 4, and diesel has similar densities; thus vapors from gasoline and diesel are considerably heavier than air.

Lighter-than-air gases can act as heavier-than-air gases if they are cooler than ambient temperatures; and heavier-than-air gases can act as lighter-than-air gases if released from a confining or hotter than ambient environment, as found at many landfills.

Specific gravity is the ratio of the density or the weight of a substance to that of a reference material. In working with liquids and solids, the density of the material in grams per cubic centimeter (g/cc) or grams per milliliter (g/ml) is compared to that of water, the density of water being 1 g/cc. Liquids or solids with specific gravities less than 1.0 will float on water, whereas materials with specific gravities greater than 1.0 will sink in water. The specific gravity of gasoline is 0.7–0.8, and that of diesel fuel is 0.88; both will float on water.

Of critical importance, when there is drilling in aqueous environments, is the solubility of contaminants in water. A material's water solubility is the amount of the material that will mix or dissolve completely in water. Materials that are miscible in water will mix freely regardless of the quantities present. Some chemical references consider a material to have negligible solubility or to be insoluble in water if its solubility is less than 1 part in 1000 parts of water, or 0.1%; in other references "practically insoluble" may be used to indicate the same thing. In many cases, the "field solubility" is higher than that predicted or measured in a laboratory—often because of adsorption or dispersal of the material in the presence of other dissolved or solid organic compounds in water, or because of selective sorption on fine-grained mineral matter. Gasoline has a solubility on the order of 0.01% to 0.06% (100–600 mg/l). Methane has a water solubility of 3.5%.

It is also important to consider the toxicity of the materials one may encounter while drilling. Exposure to toxic materials may occur through inhalation, skin and eye contact, ingestion, or injection. The water solubility of a vapor or a gas is important in determining how much of the inhaled material actually reaches the lungs. Highly soluble gases such as ammonia dissolve readily in the upper respiratory tract, whereas less soluble gases such as nitrogen dioxide readily reach the lung.

The acute toxicity of a chemical refers to its ability to produce adverse health effects as a result of a one-time exposure of short duration. This exposure may be as reversible as a headache or may be a life-threatening emergency. Chronic toxicity is the ability of a chemical to produce systemic damage as a result of repeated exposure. Some chronic effects have a very long latency interval and develop gradually, making it difficult to establish a cause-and-effect relationship. Factors that influence the toxicity of substances include the amount and the duration of exposure, the route of exposure, and the susceptibility of the individual. Individual susceptibility is determined by many factors, including age, sex, diet, inherited traits, overall physical health, and use of alcohol, tobacco products, medications, and drugs. Exposure limits are used to control employee inhalation exposure to specific substances in the workplace.

There are several organizations that recommend exposure levels, including the American Conference of Governmental Industrial Hygienists (ACGIH) and the National Institute for Occupational Safety and Health (NIOSH). Employers are required by law to comply with Occupational Safety and Health Administration (OSHA) exposure limits. Permissible Exposure Limits (PELs) were established by OSHA in 1971, and few have been changed since that date. A PEL is the average concentration of a substance to which a typical worker can be exposed during an 8-hour day over a 40-hour work week. ACGIH recommends Threshold Limit Values (TLVs), the most common TLV being the 8-hour time-weighted average (TWA) value, which has a similar definition to that of the OSHA PEL. Short term exposure limits (STELs) also exist for the 8-hour average values for many substances. NIOSH recommends Recommended Exposure Limits or RELs, which are typically lower than OSHA values. Concentration levels above which a substance is considered to be immediately dangerous to life and health (IDLH) are also recommended by NIOSH. ACGIH recommends a 300 ppm TWA and a 500 ppm 15 min STEL for gasoline vapors, but there are no recommended exposure values for diesel fuel vapors; methane is treated as a simple asphyxiant.

The odor threshold concentration can be useful in determining if a material is present. Great care must be taken in utilizing the sense of smell for exposure monitoring. Some substances have no perceived odor or are perceived at concentrations exceeding their IDLHs. Still other compounds can be smelled at low concentrations but produce olfactory fatigue with time or at higher concentrations. The odor threshold for gasoline is 0.005 to 10 ppm, with 0.25 ppm the common value reported; but one gets used to the odor fairly rapidly. Diesel fuel has an odor threshold of 0.08 to 0.11 ppm, but again one gets used to the odor. Methane is an odorless gas, but it typically carries sulfur compounds when encountered at landfills.

Atmospheric Hazard Testing

It is absolutely necessary, from the standpoints of both fire safety and worker exposure, to monitor the air during drilling operations. Atmospheric hazards that cannot be seen or smelled may exist around the drill rig. The environment around the rig (including that underneath it) should be tested for combustible gases, oxygen deficiency or excess, and toxic gases and vapors. Direct-reading instruments are portable, battery-powered meters that measure the amount of certain types of gases or vapors in air. Many direct-reading instruments are capable of continuously monitoring the atmosphere.

Combustible gas indicators (CGIs) measure the concentration of flammable or combustible vapors and gases in percent of the LEL. The CGI readings range from 0% LEL (no flammable gas detected) to 100% of the LEL (the gas or the vapor may ignite if a source of ignition is present). CGIs can also read relatively low concentrations (ppm range) or very high ranges (0 to 100% gas by volume) of gases or vapors.

Electrochemical sensors measure specific gases, such as oxygen, carbon monoxide, hydrogen sulfide, ammonia, and chlorine. Oxygen meters measure the concentration over a range of 0 to 25% oxygen by volume in air. Toxic gas monitors measure the concentration of a specific gas of interest in ppm. Many manufacturers offer combination instruments, which can measure oxygen, combustible gases, and toxic gases such as carbon monoxide and hydrogen sulfide simultaneously.

A detector tube contains a reactive mixture of chemicals designed to change color in the presence of certain types of contaminants in air. The tips of the tube are broken off, and a known volume of air is drawn through it with a manually operated pump. The length of the color change in the detector tube can be used to determine the approximate concentration of contaminant present.

All air monitoring devices have limitations. High relative humidity (i.e., 95–100%) can cause a decrease in sensitivity or erratic readings in electrochemical sensors and detector tubes. Electrochemical sensors have a limited life span (approximately one year), and they may give erroneous readings when exposed to corrosive gases. The LEL sensor in a CGI may be damaged by leaded gasoline, silicone vapors, or exposure to high concentrations of halogenated hydrocarbons. CGIs also need oxygen to work properly. Chemical-specific devices such as detector tubes or electrochemical sensors may be subject to cross-sensitivity; this means that the sensor can give the same response or similar responses to more than one chemical, not just the one it is designed to detect. Other devices are susceptible to interfering gases, which can cause inaccurate readings. Many devices are affected by temperature extremes and electromagnetic fields.

Instruments must be checked and calibrated, according to the manufacturer's recommendation, each day prior to use. All workers who enter confined spaces should be able to use, and understand the limitations of, each instrument; to interpret readings; and to troubleshoot problems.

Never operate an instrument or a probe while augers are turning. Because gases and vapors may stratify within a "nonproducing" borehole, various levels of the hole should be tested. An extension hose or a probe can be used to reach into the bore. The oxygen content should be tested first to ensure that there is enough oxygen available to support the use of a CGI. If an extension hose or a sampling probe is used, the instrument user must wait for a response and adjust the sampling interval to the length of the extension hose and the detector response. (Air monitoring in confined spaces is discussed further below, under "Testing and Control of Atmospheric Hazards.")

Drilling Techniques

The selection of a drilling method is a complex issue, with the choice based on site geology, potential impact on sample integrity, required hole size and depth, equipment availability, and cost, just to name a few factors. However, certain drilling methods are inherently safer than others for drilling in flammable environments.

Typically, the critical zone for an explosion or a flash to occur extends one foot into the hole and one foot above the hole, or the top of the casing or the hollow stem auger. At depths greater than one foot, the gas concentrations tend to become too rich, and/or the oxygen concentration is too low for combustion. Conditions one foot above the hole, assuming that the site has normal ventilation, usually produce rapid dispersion of lighter-than-air gases and either dispersion or a cascading of heavier-than-air gases. The authors have measured methane concentrations at dozens of wells in excess of 60% (600,000 ppm) immediately inside the bore or the casing annulus. These concentrations were found to be reduced an order of magnitude within one foot in every direction, and to ppm levels within three feet in every direction. This testing was done under various wind speed conditions, including no wind.

A dangerous condition can exist if heavier-than-air gases settle into low or confined areas. For example, gasoline vapors have collected under vehicles at gasoline station tank pulls and then exploded upon vehicle start-up. Similarly, landfills can release carbon dioxide that may cascade into low areas.

Common drilling techniques have certain advantages as well as disadvantages. Most drilling equipment presents potential sources of ignition from frictional sparking and power trains. The venting of aerosolized hydraulic fluids during sonic operations can increase the fire potential, should a flashover occur. Although the use of an air rotary rig utilizing a compressor may reduce some sources of ignition, it can blow vapors out of the hole, aerosolize liquids and blow them from the hole, and produce static charges. Some of these problems have been lessened through the use of drilling foams. Auger rigs allow for the placement of dry ice around the auger base, so that the dry ice is covered with spoil; the heavier-than-air carbon dioxide then fills the voids near the top of the hole and produces a cascading effect down the hole. The use of hollow stem augers facilitates monitoring as well as the introduction of inerting gases. Mud rotary techniques are used to suppress vapors and ignition sources, but the fluids present can bring nonaqueous flammable liquids to the surface. Cable tools generate frictional sparking hazards and can also displace or cause upward migration of vapors, as the tools are placed into and removed from the hole. This problem is of particular concern when a casing is not utilized.

Inerting and Purging

Inerting is the introduction of a nonflammable gas (argon, nitrogen, helium, or carbon dioxide) into a space to displace oxygen and thus make combustion less likely. Purging is the introduction of a nonflammable gas or fresh air into a space to remove flammable gases or vapors. Inerting is typically a down-hole process, whereas purging takes place around the hole and underneath vehicles.

Rig Positioning

On some jobs, particularly those of short duration (where the rig is moved every day), it may be beneficial to take advantage of the prevailing wind direction. The rig position depends upon the drilling technique, where the driller stands or sits, if a helper is present while tools are being advanced, whether the bore is "producing," and if a heavier-than-air or a lighter-than-air material is present. In dealing with heavier-than-air gases and vapors (i.e., gasoline), the rig should be positioned so that gases are blown away from it and do not accumulate underneath the rig (with the cab facing into the wind). In cases of lighter-than-air gases and vapors (i.e., methane), the cab is typically downwind. For unknowns or long trailered rigs, a perpendicular position is sometimes employed.

Use of Blowers

Portable ventilation devices have been utilized at many drilling sites. In particular, they have been used successfully at landfill investigations and during the installation of gas extraction/recovery wells. Ventilation systems utilize vaneaxial, tubeaxial, centrifugal, and venturi (air horn) types of configurations. There are explosion-proof electric-powered units, gasoline-powered units, and pneumatic and steam-powered units. Blowers can be used with and without duct work and typically move between 1000 and 5000 cubic feet per minute (cfm). Because the production of static is of great concern, many units have built-in grounding and bonding systems or come equipped with a grounding lug. Venturi-style or pneumatic systems with a low noise muffler and compressor are the most commonly employed. Ventilators and blowers can be easily rented from safety supply companies or locally from construction equipment rental companies. Blowing across the hole to push vapors and gases away from the rig is considered positive ventilation. Pulling gases and vapors into duct work and discharging away from the rig is considered negative ventilation.

Other Considerations

All electrical equipment should be intrinsically safe (explosion-proof), and all equipment should be properly grounded. For work with heavier-than-air materials, vertical exhausts and spark arrestors should be employed.

Smoking must not be permitted within an established Exclusion Zone. For most jobs this zone can be considered to be within a "shadow of the mast" distance from the rig. Welding and cutting should be done away from the hole whenever that is possible. If welding or cutting tasks must be performed at the hole, they should be initiated only after flammable levels of gases have been monitored, and spaces must be inerted or purged before hot work is begun.

Extra care must be exercised in working in topographic lows and at the toes of landfills, and in employing wind screens and barriers. Heavier-than-air gases and vapors may collect at such locations.

The importance of good housekeeping cannot be overlooked. The presence of gasoline cans near the hole and excessive oil and grease on rig components will only compound the problem, should a hole flash.

CONFINED SPACE ENTRY

Hazardous waste site investigations are by their very nature inherently dangerous. Working around heavy equipment and above- and below-ground utilities, often in extreme temperature and weather conditions and in difficult terrain, is commonplace. Site investigators

Figure 13–8. A worker using a disposable mouth-held combustible gas indicator, also known as a cigarette.

and drillers usually associate severe injury and death with being crushed by the equipment or inadvertently hitting a utility line. However, there is another, often overlooked, hazard to consider: confined spaces.

What Is a Confined Space?

Ask hazardous waste site workers or drillers what a confined space is, and most will mention a structure such as a tank, a silo, or even a sewer manhole (Figure 13–9). Few would consider a well, a cistern, or a pump shed a confined space. Most victims of fatal confined space incidents would not have entered the space had they thought it would kill them; they simply failed to recognize the hazard.

Although many organizations, including the National Institute of Occupational Safety and Health (NIOSH) and the American National Standards Institute (ANSI), have issued confined space guidance documents, and several states have confined space regulations, there was until recently no national standard available to protect the approximately 1.6 million workers who enter confined spaces each year. After a 17-year effort, OSHA published its Permit-Required Confined Space Rule (29 CFR 1910.146) on January 14, 1993.

The OSHA rule makes a distinction between a confined space and a permit-required confined space. A *confined space* is an area or space that satisfies these criteria:

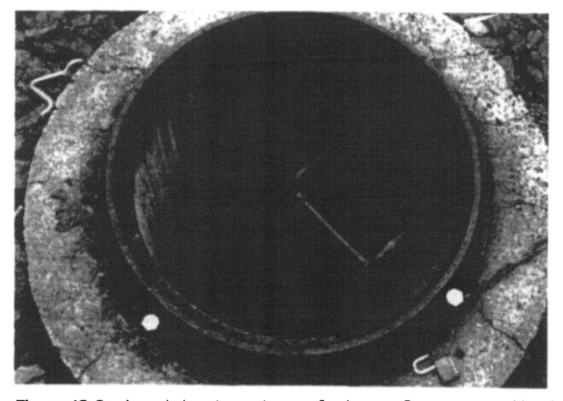

Figure 13–9. A manhole entrance to a confined space. Sewers are considered permit-required confined spaces because they have the potential to contain hazardous atmospheres, and they may contain water or other materials that can engulf and drown a person. Two untrained and unsupervised workers died in this space.

1. It is large enough for a person to bodily enter it and perform work. (The space must be large enough to accommodate the worker's entire body.)
2. It has limited or restricted means of entry or exit. (For example, the hatch or the entryway is smaller than usual, or a person must step up or bend down to go through the opening, or the worker must ascend or descend into the space via a ladder.)
3. It is not designed for continuous human occupancy.

Spaces that appear to satisfy all three criteria but have mechanical ventilation installed to allow human occupancy for a period of time are not confined spaces. Thus a submarine, even though it satisfies the first two conditions, would not be considered a confined space because it contains mechanical ventilation and is designed for human occupancy.

Doorways or other entryways through which a person can walk are not considered to be limited means for entry or exit, but a space containing a door or a portal may be considered to have limited or restricted access if the ability to exit or escape in a emergency could be hindered. For example, any obstruction that blocks the door, prevents the door from opening, or limits the use of the door, would satisfy the limited access criterion. A confined space entry is considered to occur if any part of the body breaks the plane of the opening to a confined space. This includes poking one's head into the space, reaching into the space with an arm and a hand, or dangling one's feet into the space.

A *permit-required confined space* is a confined space that is dangerous for one or more of the following reasons:

1. It contains, or has the potential to contain, a hazardous atmosphere.
2. It contains material with the potential for engulfment, such as water, sawdust, soil, sludge, or any other flowable material that can cause death by its aspiration into the respiratory system or by exerting a constricting or crushing force on the body.
3. It has an internal configuration that could allow a worker to become trapped or asphyxiated by inwardly sloping or converging walls, or it has a floor that slopes or tapers into a smaller cross section.
4. It contains any other recognized serious safety or health hazard.

Confined Space Hazards

Clearly a permit-required confined space is more dangerous than other confined space. Hazards that make a space dangerous can be placed in several categories: atmospheric hazards, physical/mechanical hazards, engulfment/entrapment hazards, and other hazards.

Atmospheric Hazards

The most dangerous and the most frequently ignored hazards, these hazards cause the greatest number of fatalities and injuries. Atmospheric hazards include oxygen deficiency (less than 19.5% oxygen) or excess (greater than 23.5% oxygen), concentrations of flammable gases or vapors in excess of 10% of their lower explosive limit (LEL), a contaminant concentration of any toxic material above its OSHA permissible exposure limit (PEL), the presence of airborne combustible dust that obscures one's vision at a distance of 5 feet or less, or any atmosphere containing contaminant concentrations above the NIOSH Immediately Dangerous to Life and Health (IDLH) value.

Oxygen-deficient air contains less than 19.5% oxygen by volume. Normal air contains 20.9% oxygen. Although the physiological effects of oxygen deprivation are readily apparent when oxygen concentrations drop below 16%, the presence of an oxygen-deficient atmosphere is not detectable until it is too late. Oxygen deficiency may occur when air is displaced by another gas. For example, naturally occurring gases such as methane and carbon dioxide can displace the air in trenches, pits, wells, sumps, and cisterns; and heavier-than-air gases and vapors found on hazardous waste sites may seep into low-lying areas and replace air. Oxygen deficiency also may occur when the oxygen in air is consumed. Such oxygen consumption may occur via a chemical reaction, by combustion during a fire or welding, or by oxidation of metals or rusting. Biological oxygen consumption by algae or bacteria during decomposition of organic matter is another frequently encountered mechanism of oxygen consumption.

Oxygen-enriched atmospheres contain more than 23.5% oxygen by volume. The enriched atmospheres are dangerous because materials are more flammable in them, and there is a greater potential for a flash fire or explosion than in normal air.

Flammable gases and vapors in excess of 10% of their LEL are considered hazardous in confined space situations. Combustible dust that obscures one's vision at a distance of 5 feet is also considered a hazardous condition.

It is important to realize that a confined space may be tested and found to have acceptable oxygen and combustible gas levels, but may still present a toxic gas hazard. Many toxic gases and vapors can cause serious effects at relatively low concentrations. For example, hydrogen sulfide (H_2S) has an LEL of 4.5% by volume in air, or 45,000 ppm; but the IDLH for H_2S is 300 ppm, which is less than 1% of the LEL. It is important, then, to evaluate the confined space for other atmospheric hazards, such as toxic gases and vapors.

Carbon monoxide, hydrogen sulfide, and nitrogen dioxide are toxic gases commonly encountered in confined spaces. Other frequently encountered vapors and gases are gasoline, petroleum solvents, chlorine and hydrochloric acid, ammonia, and chlorinated hydrocarbons. These materials can cause acute illness or incapacitate a worker at concentrations above the OHSA permissible exposure limit (PEL). The NIOSH IDLH often defines the minimum concentration required to incapacitate a worker or to impair the worker's ability for self-rescue.

Physical/Mechanical Hazards

Confined spaces may contain such physical hazards as unstable slopes and sharp debris. Poorly stabilized ladders, slippery surfaces, and falling objects are other physical hazards, which can be prevented by good housekeeping and safe work practices. Mechanical hazards are created whenever there are moving parts that can catch, cut, shear, or crush. It is important to identify any potential mechanical hazards present and deactivate the mechanical system to prevent inadvertent injury.

Engulfment/Entrapment Hazards

Engulfment occurs when a worker is surrounded by a liquid or solid substance. Death is caused by asphyxiation, either by aspiration of the material into the respiratory system or by compression of the torso that physically prevents the victim from exhaling. Entrapment hazards exist if the confined space has a funnel-shaped or sloping configuration that tapers to a small cross section, which could trap or asphyxiate a worker.

Other Hazards

Other confined space hazards include: biological hazards, such as molds, spores, bacteria, insects, and rodents that may be found in dark, damp spaces such as sewers, cisterns, and sumps; corrosive chemicals, such as those used to disinfect wells, which may cause severe skin, eye, and respiratory irritation; excessive noise generated within the space by welding, drilling, or mechanical ventilation; and electrical hazards, as presented by electric-powered tools and equipment unless they are equipped with ground-fault interrupters (GFI) or low voltage transformers. Poor lighting in a confined space is an often overlooked hazard. Adequate lighting must be provided to allow workers to safely enter, perform their jobs, and exit the confined space.

Testing and Control of Atmospheric Hazards

Atmospheric hazards that can neither be seen nor smelled may exist in a confined space; so air monitoring equipment must be used to test the air within the space before entry. The space should be tested for combustible gases, oxygen deficiency or excess, and toxic gases and vapors. Many direct-reading instruments can be used for continuous monitoring of the atmosphere within a confined space.

All air monitoring devices have limitations, as discussed above under "Atmospheric Hazard Testing." High relative humidity (i.e., 95–100%) can cause a decrease in sensitivity or erratic readings in electrochemical sensors and detector tubes. Electrochemical sensors have a limited life span (approximately one year) and they may give erroneous readings when exposed to corrosive gases. The LEL sensor in a CGI may be damaged by leaded gasoline, silicone vapors, or exposure to high concentrations of halogenated hydrocarbons. CGIs also need oxygen to work properly.

Chemical-specific devices such as detector tubes or electrochemical sensors may be subject to cross-sensitivity (i.e., the sensor can give the same or a similar response to more than one chemical, not just the one it is designed to detect); or they may be susceptible to interfering gases, which can cause inaccurate readings, or to temperature extremes or electromagnetic fields.

Instruments must be checked and calibrated, according to the manufacturer's recommendation, each day before use. All workers who enter confined spaces must be adept at using each instrument, interpreting readings, and troubleshooting problems.

As gases and vapors stratify within a confined space, *all levels (top, middle, bottom) must be tested before entry;* an extension hose should be used to reach all levels and remote areas for testing. The oxygen content should be tested first to ensure that there is sufficient oxygen available to support CGI use. If an extension hose or a sampling probe is used, the instrument user must wait for a response and adjust the sampling interval to the length of the extension hose and the detector response.

Ventilation is an effective means of controlling atmospheric contaminants and minimizing the risk of fire or explosion. Natural ventilation is achieved by allowing natural air currents to remove gases and vapors. The configuration of some confined spaces, as well as the unpredictability of wind currents and thermal effects, makes natural ventilation unreliable as a primary control method. Mechanical dilution ventilation uses fans or blowers to push fresh air into a confined space; this is usually the method of choice. Localized exhaust ventilation is often used to capture contaminants at or near the point where they are generated, so that they can be discharged outside the space. Exhaust ventilation is especially effective for welding, cutting, and braising operations where fumes or dusts are generated.

Purging with an inert gas such as nitrogen or carbon dioxide is often used to remove flammable atmospheres from a confined space. After purging, mechanical ventilation is used to introduce fresh air into the space prior to atmospheric retesting and entry.

Permit-Required Confined Space Entry

Because permit-required confined spaces (PRCS or permit spaces) are more dangerous than other confined spaces, and there is a greater potential for injury or death in the PRCS, only trained workers may enter or supervise others who enter such spaces. Employers of workers who enter such spaces must develop a permit-required confined space program and a permit system. Entry into any PRCS requires a specially trained and equipped team, which consists of one or more authorized entrants, an attendant, and an entry supervisor.

Each member of the team must be able to recognize the hazards associated with permit spaces. They also must be able to use the equipment and monitoring devices required for entry, and be able to assist in evacuating the space if an emergency exists. In many cases, the attendant, with the help of a tripod and a mechanical retrieval device, can remove an injured or incapacitated entrant without actually entering the space. If nonentry rescue is not possible because of the configuration of the space or for lack of equipment, then rescue personnel must be available in the event of an emergency.

If a nonentry rescue or retrieval system cannot be used to remove an incapacitated entrant, then the employer must provide other workers trained and equipped as rescuers, or off-site rescue services must be utilized. Rescuers must be trained in the proper use of protective equipment and retrieval devices for making rescues from permit spaces. They must also be trained in basic first aid and CPR, and must practice a permit space entry and rescue every year.

A permit space cannot be entered until the supervisor fills out a written entry permit. The permit must show the purpose of the entry, authorized entrants and attendants, entry supervisor, hazards of the space, methods required to monitor and/or control the hazards, rescue services available, communications procedures, and special equipment, such as respiratory protection, protective clothing, or retrieval devices, required for entry. After work in the space is completed and all entrants have exited, the supervisor is responsible for canceling the permit. The same worker can act as both attendant and entry supervisor if trained for both positions.

Table 13–3. OSHA action levels for permit-required confined spaces

Oxygen deficiency	less than 19.5% oxygen
Oxygen excess	greater than 23.5% oxygen
Combustible gas	greater than 10% LEL
Combustible dust	obscures vision at 5 feet
Toxic gas or vapor	greater than OSHA PEL, ACGIH TWA
	greater than NIOSH IDLH
	greater than other regulatory or recommended limits

Every permit must define acceptable entry conditions and the results of atmospheric monitoring. OSHA has defined criteria for acceptable entry conditions and has provided specific action levels for atmospheric hazards in permit spaces (Table 13–3). An action level is a meter reading at or above which a certain action must be taken. In this case, the action is to *not* allow entry into the space until:

- The space is ventilated and readings become more normalized; or
- The entrants don respiratory protection prior to entering the space.

Reclassifying a Permit-Required Confined Space

A confined space must be initially classified as a permit space because of the presence one or more hazards (i.e., atmospheric, engulfment, entrapment, mechanical). In many cases, however, only the presence of an atmospheric hazard makes the confined space a permit space. Under these conditions, if the atmospheric hazard can be eliminated, the space may be reclassified as a non-permit space, and alternate entry procedures can be used (Table 13–4).

When alternate entry procedures are used, a formal written program is not required, nor must there be a written permit, an attendant, a supervisor, or rescue/retrieval equipment available. Although an entry permit is not required, a written certification form must be completed prior to entry. Many provisions of the OSHA rule are no longer followed when alternate entry procedures are used, but training for confined space entry still is required.

Table 13–4. Requirements for confined space entry

OSHA requirements	Permit space entry	Alternate entry
Written program	Yes	No
Permit system	Yes	No
Attendant	Yes	No
Supervisor	Yes	No
Rescue provisions	Yes	No
Training	Yes	Yes
Atmospheric monitoring	Yes	Yes
Other hazards allowed	Yes	No

Entry into the space is prohibited until it can be demonstrated that no atmospheric hazards are present; air monitoring must be conducted during entry to ensure that the space remains safe. If continuous ventilation is required to maintain safe levels, the space cannot be reclassified. Also, if work in the space may introduce a hazard (e.g., welding, cleaning with corrosives or solvents, brazing), the space cannot be reclassified.

It must be noted that if entry into a space is required to eliminate an atmospheric hazard or for initial atmospheric monitoring, this entry must be conducted as a permit-required confined space entry.

Training Requirements

A permit-required confined space training program is required for workers who may enter a permit space. Training must cover a variety of topics, including: confined space hazards; components of the employer's written PRCS program; components of the entry permit system; how to fill out a permit; air monitoring equipment use and limitations; atmospheric testing protocols; methods for control or elimination of atmospheric hazards; action levels and procedures to follow when an action level is reached; use of entry equipment; limitations and use of personal protective equipment; use of rescue/retrieval devices (see Figure 13–10); duties of the entrant, the attendant, and the supervisor; basic first aid and CPR procedures.

Workers entering confined spaces using alternate procedures require similar training but are not required to use rescue/retrieval equipment or a permit system. However, they

Figure 13–10. A worker prepares to enter a confined space. Note that he is attached to a tripod, which would facilitate his rescue in case of an emergency.

must be trained to document the lack of atmospheric and other hazards on a written certification form.

A permit-required confined space recognition form is a useful tool. If a "yes" answer is given to any question in Part II, the confined space is a permit space. Note that the use of alternate entry procedures or reclassification to a non-permit space may be possible.

CONFINED SPACE AND PERMIT-REQUIRED CONFINED SPACE
RECOGNITION FORM

Part I

1. Is the space large enough for a person to enter it and perform work?
2. Does the space have limited/restricted means for entry or exit?
3. Is the space designed for occupancy?

Part II

1. Does the space contain a hazardous atmosphere? Is there a potential for the space to contain a hazardous atmosphere?
2. Does the space contain any chemicals or chemical residues?
3. Does the space contain any flammable/combustible materials?
4. Does the space contain or potentially contain any decomposing organic matter?
5. Does the space have any pipes that bring chemicals into it?
6. Does the space have any materials that can trap or potentially trap, engulf, or drown an entrant?
7. Is the worker's vision obscured by dust at 5 feet or less?
8. Does the space contain any mechanical equipment?
9. Does the space have converging walls, sloped floors, or a tapered floor to a smaller cross section that could trap an entrant?
10. Does the space contain rusted interior surfaces?
11. Does the space have thermal (extreme heat or cold) hazards?
12. Does the space have excessive noise levels that could interfere with communication with an attendant?
13. Does the space present any slip, trip, or fall hazards?
14. Are any operations conducted near the space opening that could present a hazard to entrants?
15. Are there any hazards from falling objects?
16. Are there lines under pressure servicing the space?
17. Are cleaning solvents, paints, or acids going to be used in the space?

18. Is welding, cutting, brazing, riveting, scraping, or sanding going to be performed in the space?

19. Is electrical equipment located in or required to be used in the space?

20. Does the space have poor natural ventilation that would allow an atmospheric hazard to develop?

21. Are there any corrosives in the space that could irritate the eyes?

22. Are there any conditions that could prevent any entrant's self-rescue from the space?

23. Are any substances used in the space that have acute hazards?

24. Is mechanical ventilation needed to maintain a safe environment?

25. Is air monitoring necessary to ensure that the space is safe for entry, because of a potentially hazardous atmosphere?

26. Will entry be made into a diked area where the dike is more than 5 feet in height?

27. Are residues going to be scraped off the interior surfaces of the space?

28. Are nonsparking tools required to remove residues?

29. Does the space restrict mobility to the extent that it could trap an entrant?

30. Is respiratory protection required because of a hazardous atmosphere?

31. Does the space present any hazards other than those noted above?

APPENDIXES

Appendix A

Generic Health and Safety Plan

Whatever it is called, a health and safety plan (HASP), a safety plan, a contingency plan, or a safety and health emergency response plan (SHERP) should do more than just satisfy a regulatory or contractual requirement. A good plan serves in two capacities: (1) as an active accident prevention plan and (2) as a reactive contingency plan to mitigate the impact of Murphy's Law.

The accident prevention part of the plan should delineate:

- Site hazards (most site hazards are not chemical in nature).
- Risks to drilling and technical personnel.
- How to monitor for the hazards.
- How to protect personnel against hazards.
- How to safely conduct site activities.

For personnel on-site for a limited period of time, the hazards associated with site activities will be very different from those of nearby residents or environmentally sensitive areas. It is ludicrous for a plan to include page after page of data discussing all the chemicals found on-site from toxicity and flammability standpoints when only a few parts per billion of the chemicals have been found in the ground water or soil. The same plan may not even mention the many, more significant hazards associated with drilling, heavy equipment operation, or trenching.

Many organizations divide their site safety planning responsibilities among two or three separate documents, which may include a generic HASP and a much shorter site-specific plan in the form of a checklist or a fill-in-the-blanks document. Other organizations prepare task-specific plans that address the hazards of a particular activity, such as monitoring well installations at gasoline stations, marine drilling, or methane recovery well installation at landfills. The task-specific documents are then augmented with a site-specific plan.

It is important for a HASP to be flexible in some areas. For example, although specific guidance is necessary regarding the use of permeable or impermeable protective clothing; the HASP may be somewhat flexible in the areas of taping, use of hoods, and use of accessory protective equipment, such as aprons or gauntlet gloves. Flexibility should also be possible in the extent of the decontamination procedures, according to the level of contamination encountered.

The following generic health and safety plan provides for both accident prevention and contingency planning.

HEALTH AND SAFETY PLAN
FOR THE
SAWMILL HILLS INDUSTRIES, INC. SITE

Prepared by
Anything for a Buck Consultants and Contractors, Inc.

APPROVED   APPROVED

_____   _____

I. M. Bigshott   U. R. Safenow
Principal   Manager, Health & Safety
June 18, 1998   June 18, 1998

NOTE: This plan is valid between June 18, 1998 and October 15, 1998. The corporate health and safety manager must revise and reissue this plan if the described project activities are delayed or anticipated to extend beyond October 15, 1998.

HEALTH AND SAFETY PLAN
FOR THE
SAWMILL HILLS INDUSTRIES, INC. SITE

CONTENTS

HASP APPENDIX

AFBCCI Forms:

Form 1. Employee Acknowledgment
Form 2. Daily Tailgate Safety Meeting
Form 3. Physician's Certification
Form 4. Training Documentation
Form 5. Emergency Medical Information
Form 6. Hearing Conservation Program Participation
Form 7. Daily Safety Log
Form 8. Weekly Safety Report
Form 9. Incident/Accident Report
Form 10. Monthly Exposure and Injury Report
Form 11. Warning/Suspension Notice Record
Form 12. Incident Report Follow-up
Form 13. Air Monitoring Results Report

HEALTH AND SAFETY PLAN
FOR THE
SAWMILL HILLS INDUSTRIES, INC. SITE

1.0 INTRODUCTION AND PURPOSE

Anything for a Buck Consultants and Contractors, Inc. (AFBCCI) employees may be exposed to actual, potential, or perceived hazards during field activities. It is AFBCCI corporate policy to minimize the potential for site-related injuries and illnesses through a program of qualified site supervision, health and safety training, medical surveillance, the use of personnel and engineering controls, and, where appropriate, the wearing of personal protective equipment. AFBCCI has an established health and safety program to implement this corporate policy and requires that this policy be adhered to in a manner that protects its employees to the maximum reasonable extent. The *Corporate Health and Safety Program Manual* is issued as a separate document to all employees and is updated periodically through amendments, attachments, and directives. This site-specific health and safety plan has been designed to protect site personnel and the surrounding community.

1.1 Scope and Applicability

This site-specific health and safety plan (HASP) applies to all AFBCCI employees and subcontractors being supervised directly by AFBCCI personnel working at the Sawmill Hills Industries, Inc. (SHII) site, where operations involve employee exposure or the reasonable possibility of employee exposure to safety or health hazards. Governmental Oversight Agency (GOA) personnel also must comply with the provisions of this plan. This plan incorporates, where applicable, the Occupational Health and Safety Administration (OSHA) requirements set forth in 29 CFR 1910 and 1926. A copy of this HASP will be present at the site during all on-site activities conducted by AFBCCI employees and their subcontractors.

1.2 Contractors

This HASP is also intended as information and guidance to other contractors working on-site who have been retained directly by the Client. This HASP will be made available to all parties; however, AFBCCI has no control over the actions of any other parties, and all parties will enter the site with this understanding.

1.3 Visitors

The Client will be responsible for any and all visitors (except GOA personnel) to the site, a category including but not limited to local officials, regulatory personnel, public interest groups, and the media.

2.0 ORGANIZATION AND COORDINATION

2.1 Key Personnel

The following AFBCCI employees are designated and required to carry out stated position responsibilities as related to the SHII Site project. These responsibilities are delineated in detail in the AFBCCI *Health and Safety Program Manual*, and are outlined below in Section 2.3.

> Principal-in-Charge: *I. M. Bigshott, P.E.*
> Project Manager: *M. Bissen*
> Health and Safety Manager: *U. R. Safenow*

2.2 Site-Specific Health and Safety Personnel

> Site Manager: *O. Kraft*
> Site Safety Office (SSO): *G. Disorders*
> Assistant Safety Officer (ASO): *C. McCall*

2.3 Organizational Responsibility

The Principal-in-Charge has corporate responsibility for the SHII Site project and will approve all personnel assigned to this project. The Project Manager is responsible for overall project administration and management and for implementation of the HASP. The Health and Safety Manager is responsible for approving all AFBCCI HASPs and for periodic inspection and evaluation of all HASPs. The Site Manager will assume all on-site activities of the Project Manager and on small projects is authorized to act as the Site Safety Officer. The Site Safety Officer and the Assistant Safety Officer are responsible for overseeing daily safety issues. In the event of any dispute concerning implementation of this plan, the Project Manager will be consulted to resolve said dispute. Nevertheless, the Site Safety Officer or the Assistant Safety Officer has the authority to suspend any or all site work conducted by AFBCCI employees or subcontractors.

2.4 Employee Responsibilities

All on-site AFBCCI employees and AFBCCI subcontractors who participate in site activities are responsible for adhering to this HASP.

Any person who observes any health or safety hazards or potential hazards that have not been discussed in the HASP of at daily safety briefings should immediately report them to the site safety officer or other appropriate key personnel.

3.0 TASK OPERATION SAFETY AND HEALTH RISK ANALYSIS

3.1 Historical Overview of Site

The Sawmill Hills Industries, Inc. (SHII) site is a 12-acre tract of land situated in the Town of Sawmill Hills, Westchester County, New York. The site is situated south of Dana Road, approximately 500 feet from State Route 9A (Saw Mill River Road).

In 1961 the site was purchased by Mr. U. Chucka, who founded Sawmill Hills Industries, Inc. This company formulated furniture strippers sold under the brand names of Outdamspot™ and Shelac-No-Mor™. The company went out of business in 1971 after a fire destroyed the production building. Several tenants leased part of the site from SHII, and they occupied a remaining warehouse. Past tenants included: the Dwight Makeswrite Corp., manufacturers of fluorescent, glow-in-the-dark writing instruments for the military (1972–1974); Toan, Deff, and Flatt, manufacturers of musical instruments (1974–1977); and DOGMEET, pet food wholesalers (1977–1981). Mr. Chucka died in 1981, and the property was later sold to LANDSHARCO, which wants to develop a shopping center on the site.

Over the past 10 years LANDSHARCO has voluntarily conducted remedial investigations and feasibility studies at the site, with oversight from both the State of New York (NYSDEC) and the USEPA. Surface debris and approximately 60 drums of solvent waste were removed in 1989, and a 5000-gallon-capacity underground storage tank (UST) was removed in March 1996. The tank contained 3000 gallons of contaminated water. Laboratory analysis of the water revealed the following contaminant concentrations: tetrachloroethylene, 190 mg/l; toluene, 400 mg/l; isopropyl alcohol, 8300 mg/l.

The site contains five monitoring wells, which were installed in 1989. Two wells (MW-1 and MW-2) are upgradient and have shown no contamination after five sampling events. One of three downgradient wells (MW-3) has consistently shown tetrachloroethylene (8–50 µg/l) and vinyl chloride (35–60 µg/l) over five sampling events; the other two downgradient wells (MW-4 and MW-5) have shown no detectable levels of contamination.

None of the site occupants used radioactive materials in the manufacture of its products. The site was screened for radioactivity by a radiation consulting firm, RadCheck, in 1974, and was checked again in 1989 by the NYS Health Department. No above-background radiation readings were obtained anywhere on the site.

For additional details refer to the Site Work Plan.

3.2 Scope of Work

Work covered under this HASP includes the completion of 16 test borings and co-located monitoring wells, well development, the first round of ground-water sampling, and surveying of the completed monitoring wells. Previously installed monitoring wells (MW-1 through MW-5) also will be sampled. Shallow test pits, 4 feet or less in depth, will be dug in the vicinity of the old (removed) underground storage tank. Test pits, test borings, monitoring well installation, and ground-water sampling procedures are discussed in detail in the Site Work Plan.

3.3   Site Hazards

3.3.1   Chemical Hazard

Site activities will involve handling materials that are known or suspected to be contaminated. Levels of ground-water contamination generally are very low, but site personnel must be aware that well installation areas and the test pit area may contain higher than usual contaminant concentrations in the ground water and the soil. Contaminates known or suspected to be on-site include:

> Tetrachloroethylene (perchloroethylene), CAS 127-18-4
>
> Vinyl chloride, CAS 75-01-4
>
> Toluene, CAS 108-88-3
>
> Isopropyl alcohol (isopropanol), CAS 67-63-0

Low concentrations of vinyl chloride (a breakdown product of tetrachloroethylene) and tetrachloroethylene have been found in two monitoring wells. Tetrachloroethylene, toluene, and isopropyl alcohol were found in contaminated wastewater in the UST. Toluene and isopropyl alcohol have not been found in monitoring well samples.

Please see the the individual data sheets on each contaminant (Data Sheets 1 through 4) for information on chemical/physical properties, exposure limits, and toxicity for contaminants as pure products. On the basis of previous sampling data, site personnel are not expected to encounter high concentrations of these contaminants. Data on pure products are provided for informational purposes only. Exposure to contaminants may occur through one or more of the following pathways:

Inhalation. Substances that pose an inhalation hazard are volatile at ambient temperatures. All contaminants found on-site are considered volatile. Inhalation hazards may exist when contaminated soil particles become airborne, when contaminated water is aerosolized, or when contaminants in soil or water volatilize into the air.

Based on previous sampling data, the level of contamination in soil and ground water is not expected to present a significant inhalation hazard to unprotected personnel. No information is available concerning the level of soil contamination in the area where shallow test pits will be dug. To prevent inadvertent inhalation exposure, air monitoring will be conducted continually during any site activity in areas known or suspected to be contaminated.

Skin/Eye Contact. Given the levels of contamination found in the ground water, significant skin and eye absorption is not expected to occur upon direct contact with the contaminated ground water. No information is available concerning the level of soil contamination in the area where shallow test pits will be dug. To prevent inadvertent absorption of contaminants through the skin, skin protection will be utilized during soil excavation and well installation activities.

Significant eye or skin irritation is not expected to occur from direct contact with the contaminants as vapors, dissolved in ground water, or absorbed onto soil. Physical irritation may occur upon direct eye contact with soil or other particulates.

Data Sheet 1. Isopropyl alcohol

CHEMICAL NAME: Isopropyl Alcohol		CAS 67-63-0
EXPOSURE LIMITS (in ppm)		NIOSH REL 400
ACGIH TWA 250	OSHA PEL 400	
SKIN NOTATION? No		NIOSH IDLH 12,000
CARCINOGEN? (list agency) No		
ORGAN TOX?		
OTHER		

PHYSICAL DESCRIPTION (PURE PRODUCT) Colorless, non-viscous liquid		
CHEMICAL/PHYSICAL PROPERTIES (PURE PRODUCT)		
BP 181 °F	VP 33 mm Hg	VD 2.1
SpG 0.79	H20 Solubility: Miscible	
FLAMMABILITY DATA		
FP 53 °F	LEL 2.0%	UEL 12.7%
OTHER INFORMATION		IT 750°F
ODOR Like rubbing alcohol		OT 7.5-200 ppm
IP 10.10 eV	RESP organic vapor cartridges	

MAXIMUM CONCENTRATIONS FOUND ON SITE		
WATER 8300 mg/l	SOIL not measured	AIR not measured
COMMENTS found in UST that was removed from site		

BP=boiling point; VP=vapor pressure; VD=vapor density; SpG= specific gravity; FP=flash point; LEL=lower explosive limit; UEL=upper explosive limit; IT=ignition temperature; OT=odor threshold; IP=ionization potential; RESP=recommended type of respiratory protection. Print! Indicate if °F or °C are used.

Data Sheet 2. Toluene

CHEMICAL NAME Toluene	CAS 108-88-3
EXPOSURE LIMITS (in ppm)	NIOSH REL 100
ACGIH TWA 50 OSHA PEL 100	
SKIN NOTATION? Yes	NIOSH IDLH 2,000
CARCINOGEN? (list agency) No	
ORGAN TOX? Liver, kidney, CNS	
OTHER	

PHYSICAL DESCRIPTION (PURE PRODUCT) Colorless, non-viscous liquid		
CHEMICAL/PHYSICAL PROPERTIES (PURE PRODUCT)		
BP 232 °F	VP 20 mm Hg	VD 3.1
SpG 0.87	H20 Solubility 0.05%	
FLAMMABILITY DATA		
FP 40 °F	LEL 1.2%	UEL 7.1%
OTHER INFORMATION		IT 896°F
ODOR Sweet, pungent		OT 0.17-40 ppm
IP 8.92 eV	RESP organic vapor cartridges	

MAXIMUM CONCENTRATIONS FOUND ON SITE		
WATER 400 mg/l	SOIL not measured	AIR not measured
COMMENTS found in UST that was removed from site.		

BP=boiling point; VP=vapor pressure; VD=vapor density; SpG=
specific gravity; FP=flash point; LEL=lower explosive limit;
UEL=upper explosive limit; IT=ignition temperature; OT=odor
threshold; IP=ionization potential; RESP=recommended type of
respiratory protection. Print! Indicate if °F or °C are used.

Data Sheet 3. Tetrachloroethylene

CHEMICAL NAME Tetrachloroethylene	CAS 127-18-4
EXPOSURE LIMITS (in ppm)	NIOSH REL carcinogen lowest feasible concentration
ACGIH TWA 50 OSHA PEL 25	
SKIN NOTATION? No	NIOSH IDLH 500
CARCINOGEN? (list agency) Yes - NIOSH, IARC, NTP	
ORGAN TOX? Liver, kidney, CNS	
OTHER NIOSH REL:lowest feasible concentration	

PHYSICAL DESCRIPTION (PURE PRODUCT) Colorless, non-viscous liquid		
CHEMICAL/PHYSICAL PROPERTIES (PURE PRODUCT)		
BP 250 °F	VP 14 mm Hg	VD 5.8
SpG 1.62	H20 Solubility 0.02%	
FLAMMABILITY DATA Non-combustible liquid		
FP NA	LEL NA	UEL NA
OTHER INFORMATION		IT Unknown
ODOR Mild, chloroform-like		OT 4.68-50 ppm
IP 9.32	RESP organic vapor/supplied air	

MAXIMUM CONCENTRATIONS FOUND ON SITE		
WATER 190 mg/l	SOIL not measured	AIR not measured
COMMENTS 190 mg/l in former UST; 50 ug/l in MW-3.		

BP=boiling point; VP=vapor pressure; VD=vapor density; SpG= specific gravity; FP=flash point; LEL=lower explosive limit; UEL=upper explosive limit; IT=ignition temperature; OT=odor threshold; IP=ionization potential; RESP=recommended type of respiratory protection. Print! Indicate if °F or °C are used.

Data Sheet 4. Vinyl chloride

CHEMICAL NAME Vinyl chloride	CAS 75-01-4
EXPOSURE LIMITS (in ppm)	NISOH REL: carcinogen lowest feasible concentration
ACGIH TWA 5 | OSHA PEL 1	
SKIN NOTATION? No	NIOSH IDLH 3600
CARCINOGEN? (list agency) Yes - NTP, NIOSH, OSHA, IARC	
ORGAN TOX? Liver, CNS, blood, circulatory system	
OTHER	

PHYSICAL DESCRIPTION (PURE PRODUCT)
Colorless gas; colorless liquid below 56°F.

CHEMICAL/PHYSICAL PROPERTIES (PURE PRODUCT)		
BP 7 °F	VP > 1 atm	VD 2.2
SpG Gas (0.91)	H20 Solubility 0.1%	

FLAMMABILITY DATA		
FP Gas	LEL 3.6%	UEL 33.0%

OTHER INFORMATION	
ODOR pleasant	OT 260-3000
IP 9.99 eV	RESP supplied air

MAXIMUM CONCENTRATIONS FOUND ON SITE		
WATER 60 ug/l	SOIL not measured	AIR not measured
COMMENTS decomposition product of chlorinated solvents; found in MW-3.		

BP=boiling point; VP=vapor pressure; VD=vapor density; SpG= specific gravity; FP=flash point; LEL=lower explosive limit; UEL=upper explosive limit; IT=ignition temperature; OT=odor threshold; IP=ionization potential; RESP=recommended type of respiratory protection. Print! Indicate if °F or °C are used.

Ingestion. Based on previous sampling data, the level of contamination in soil and ground water is not expected to present a significant hazard due to ingestion of contaminants. Ingestion may occur when personnel work in situations in which soil becomes airborne, or when personnel perform hand-to-mouth activities with contaminated hands or in contaminated work areas.

Injection. There is minimal hazard from injection of contaminants, which may be injected into the bloodstream when the skin is punctured by foreign objects, such as nails, spikes, and glass or wood slivers.

3.2.2 Biological Hazards

Site activities are anticipated to occur during the months of July, August, and September 1998. Biologically hazardous plants and animals may be present on-site during this interval. One hazardous plant known to grow on-site is poison ivy, which has been observed along the fence line and at the southwest corner of the abandoned warehouse. Poison ivy can cause an allergic skin reaction in sensitive individuals.

Potentially hazardous insects found on-site include yellow jackets and paper wasps. Paper wasp nests have been observed in the eaves of the abandoned warehouse; one or more yellow jacket nests may be present in the former production building site. Yellow jacket nests may also be present in the ground, and these insects have been known to nest in monitoring well casings. The sting from a yellow jacket or a paper wasp can be extremely painful; moreover, these insects can inflict multiple stings. A sting may produce an allergic reaction in sensitive individuals; and an extreme allergic reaction may result in anaphylactic reaction, which is marked by difficult breathing, hives, and intense itching. Anaphylactic reactions may cause asphyxia and death.

Dog ticks and deer ticks have been found in high grass and scrub areas of the site. Deer ticks can transmit Lyme Disease, which is endemic in the area.

Any individual who is sensitive to poison ivy or insect stings should so inform the site safety officer.

3.3.3 Physical Hazards

Structural debris and other material may be encountered on the ground surface, and during excavations and drilling. Pieces of lumber, broken glass, nails, broken bottles, crushed cans, and other loose debris have been observed around the entire site. Loose debris on the ground presents a trip hazard; sharp edges may cut or puncture the skin or protective clothing. Loose surface debris that may pose a hazard should be carefully removed as necessary to ensure safe working conditions. To avoid foot injury from surface debris, all workers will wear foot protection, consisting of steel-toe, steel shank safety boots. No worker will be allowed on-site without foot protection.

Slipping on muddy or wet surfaces may occur at any time during site activities. Slip hazards are increased during wet weather, especially during excavation and well installation activities. When necessary, boards, wooden slats, dry deck panels, coarse gravel, sand, or other materials can be placed over muddy or slippery surfaces to improve traction. Boots with aggressive soles should be worn in areas with poor footing.

Test pits or excavations present a hazard to workers who must collect samples or perform inspections within the pit. No worker will enter an excavation that is more than 4 feet deep. Excavations will be cut back, with a slope of 1.5 on 1.0. No workers may stand on the loaded side of any excavation. Whenever possible, pits will be ramped to allow easy access; at other times, a ladder will be used for entering the pit.

Deenergized overhead electric lines and telephone lines to the warehouse and the former production building site may present a physical hazard during raising of the drill rig mast and hoisting of drilling tools.

3.3.4 Structural Hazards

The production building was destroyed by fire, and the remaining building debris is structurally unsafe. The warehouse has been abandoned since 1981; there is no information regarding its structural integrity. These areas should not be entered at any time during site activities.

3.3.5 Mechanical Hazards

Site activities require the use of heavy equipment, including drill rigs and backhoes. Rigs and backhoes have moving parts that can catch or crush all or part of the body and are hazardous. All moving parts must be guarded whenever possible; no device may be used if its machine guarding is inoperative. Emergency shutoff switches must be tested daily; no device may be used unless the emergency shutoff is functioning properly. All workers must know the location of the emergency shutoff and how to use it.

A backhoe can run over or back over personnel working nearby; the bucket can strike workers within its reach. Safety precautions that should be strictly observed in working around a backhoe include:

- Always stay within the line-of-sight of the operator.
- Observe backhoe operations from a distance of at least 5 feet more than the reach of the bucket.
- Never walk behind, to the side of, or within the reach of heavy equipment without first alerting the operator to your presence.
- Never remain within the test pit while the backhoe continues to excavate.

Drill rigs have many moving parts, including the auger, cathead, and hammer. Safety precautions that should be strictly followed in working around a drill rig include:

- Never wear loose clothing around the drill rig.
- Never move the rig with the mast raised.
- Always lower jacks or outriggers before raising the mast.
- Whenever possible, work on a level surface; never operate on a steep slope.
- Always chock the wheels after stopping.

High pressure hoses, such as hydraulic hoses, can present a severe mechanical hazard upon rupture. Pinhole leaks of fluids under excessive pressure can inject liquids deep into human tissue. Never attempt to use a bare or a gloved hand to stop a high pressure leak;

instead, shut off the equipment, and then repair or replace the hose. If a high pressure injection injury occurs, seek medical assistance immediately.

Head injuries may be caused by falling equipment or by personnel walking into equipment. Eye injuries also may occur. All workers will wear head protection and safety glasses while on-site. No worker will be allowed on-site without head and eye protection.

3.3.6 Electrical Hazards

There are no active electrical utility lines on or under the site. A small gasoline-powered generator will be used when portable electrical equipment is used. Electrocution or electric shock is a hazard whenever electrical equipment is used. Ground-fault interrupters and watertight connecting cables will be used on all electric equipment.

3.3.7 Flammable Hazards

Small quantities of gasoline will be used on-site to fuel a small generator and a hot water sprayer. Gasoline will be stored in an flammable liquid container meeting current NFPA specifications. Gasoline-powered equipment will be kept upright to prevent spillage. The 5-gallon gasoline container will be stored in a portable flammable liquid storage cabinet adjacent to the tool locker.

3.3.8 Thermal Hazards

The hot water sprayer used for decontamination will heat water to a temperature of 180°F, and contact with the hot water can cause skin or eye burns. Thermal burns also may occur if the hot spray wand or the hot water reservoir contacts unprotected skin.

3.3.9 High Pressure Hazards

The hot water sprayer used for decontamination utilizes a high pressure (2000 psi) hot water spray. Also, heavy equipment used on-site has high pressure hydraulic lines. Severe injury may occur if liquid is injected into the skin under high pressure.

3.3.10 Active and Abandoned Utilities

The only active utilities on-site are water lines to two fire hydrants; no other active utilities exist. Piping to the former UST was removed with the tank. Aboveground electrical lines to the warehouse and the vicinity of the former production building still exist but are not energized. These lines may present a physical hazard during raising of the drill rig mast and hoisting of drilling tools.

3.3.11 Traffic

Heavy equipment and vehicles present a hazard to pedestrians and to each other. Vehicle operators must take care when entering and exiting the site via Dana Road. They must always signal the intent to turn, and use flashers when traveling at less than normal speeds.

No personal or company vehicles are permitted past the parking lot, with the exception of the backhoe and the drill rig. All vehicles will yield to pedestrians in the parking lot.

The entrance road and the parking lot may be slippery in wet or muddy conditions. The entry area by Dana Road will be swept daily to prevent any buildup of dirt or mud.

3.3.12 Noise

Exposure to excessive continuous noise is anticipated for personnel working around heavy equipment, and excessive continuous and impact noise is anticipated for personnel working around the drill rig. If normal speech cannot be heard, and workers must raise their voices to be heard, the noise levels probably are excessive, and hearing protection is needed. Hearing protection, in the form of ear muffs, ear plugs, or both, should be used by personnel working around heavy equipment.

3.3.13 Temperature Stress

Site activities will be performed during the hottest period of the year (July through September). It is anticipated that site workers will experience some form of heat stress during levels of intense physical activity and high temperatures. Personnel wearing chemical protective clothing are particularly prone to heat stress because evaporative cooling cannot occur.

Heat stress can be prevented or minimized by increasing the fluid and electrolyte intake, acclimatization of workers, modifying work schedules, and heat stress monitoring.

All site workers should drink at least 16 ounces of water or diluted fruit juices before the start of work and at every rest break. Cool water or juice and disposable cups will be available throughout the day. Trees around the parking lot should provide adequate shade during rest breaks. Chairs or benches will be provided under the trees. Every effort will be made to avoid the hottest period of the day by scheduling work for early morning and late afternoon. Workers will be periodically rotated to less demanding jobs to reduce physical exertion and minimize heat stress. Workers who have difficulty adjusting to hot temperatures will be allowed additional time for acclimatization.

Heat stress monitoring will be performed for persons working in protective equipment; the heart rate and the oral temperature will be taken before donning of protective equipment, at rest breaks, and after equipment is doffed. Specific criteria for monitoring are given in Section 5.2.

3.3.14 Weather

Severe weather may happen at any time during site activities; torrential rain, lightning, and high winds may occur during summer thunderstorms. In the event of severe weather, site activities will stop, and all equipment will be secured. Workers will remain in company or personal vehicles until the weather has passed. If the forecast is for continued bad weather for the remainder of the day, workers will leave the site and meet at the LANDSHARCO maintenance garage, approximately one-half mile south of the site on Route 9A.

Site activities will continue during periods of light or moderate rainfall, at the discretion of the Site Safety Officer. Activities that may be affected by slippery or muddy conditions may be suspended after consultation with the Site Safety Officer or the Assistant Safety Officer. No site activities may be performed during periods of heavy rain, fog, or other conditions that impair visibility.

3.3.15 Lifting Hazards

Lifting and carrying of heavy objects may be necessary during site activities. Heavy objects should be lifted with proper techniques, such as using the legs and keeping the back straight.

3.4   Task-by-Task Safety and Health Risk Analysis

The various site activities present different risks to workers performing or observing the activities (risk being the likelihood that harm may be caused by a particular hazard). Some site activities carry higher risks because they increase the potential for encountering a hazard. A risk analysis matrix (Table A-3.4) for site activities describes the potential health and safety risks for workers performing specific tasks.

4.0   SITE REQUIREMENTS

4.1   Preassignment and Refresher Training

All AFBCCI personnel working on-site will have completed the 40-hour initial training and three days of supervised field training required for workers at hazardous waste sites. Surveyors under contract to AFBCCI must have completed a 24-hour initial training program and have received 8 hours of supervised field training.

All site workers must have received 8 hours of refresher training annually after their initial training. No workers will be allowed on-site unless their refresher training is current.

Table A-3.4.   Risk analysis matrix for SHII site

Description of site activity	Exposure hazard	Other hazards
1) Excavate test pits to a depth of no more than 4 feet using a back hoe. 2) Collect soil samples from trench or trench spoils.	Potential for inhalation, and skin exposure to, gases, vapors, and, contaminated soil.	Physical and mechanical hazards and excessive noise associated with heavy equipment. Slip/trip/fall hazards for workers entering trench.
3) Advance test borings with hollowstem auger. 4) Collect soil samples from each test boring.	Potential for inhalation and skin exposure to gases, vapors, aerosols, and contaminated soil.	Physical and mechanical hazards associated with heavy equipment and drilling. Excessive impact/continuous noise.
5) Install monitoring wells.	Potential for inhalation and skin exposure to gases, vapors, aerosols, and contaminated soil and water.	Physical and mechanical hazards associated with heavy equipment. Lifting hazards. Excessive impact/continuous noise.
6) Well development.	Potential for inhalation and skin exposure to gases, vapors, aerosols, and contaminated water.	Electrical hazards. Slip/trip/fall hazards. Lifting hazards.
7) Collect water samples from new and preexisting monitoring wells. 8) Measure water levels.	Potential for inhalation and skin exposure to gases, vapors, aerosols, and contaminated water.	Slip/trip/fall hazards. Lifting hazards.
9) Survey monitoring wells.	Minimal potential for exposure.	Slip/trip/fall and Lifting hazards. Physical hazards from site debris. Biological hazards.
10) Heavy equipment decontamination.	Potential for inhalation and skin exposure to gases, vapors, aerosols, and contaminated water.	Mechanical hazards. Slip/trip/fall hazards. Heavy equipment hazards. Thermal hazards. Noise hazards.

4.2   Site Supervisor Training

All AFBCCI site-specific health and safety personnel must have completed 8 hours of supervisor training in addition to the training required of all site workers.

4.3   Site-Specific Training/Tailgate Meetings

All AFBCCI workers will attend a site-specific health and safety meeting conducted by the Site Supervisor on the first day of site activities. Tailgate safety meeting will be held in the parking lot at the site each morning before site activities begin.

4.4   Documentation

All subcontractors must provide documentation of training before initiating any activities on-site. Documentation of completed training will be submitted to the Site Safety Officer and kept on file with a copy of the site HASP at AFBCCI.

5.0   MEDICAL SURVEILLANCE

5.1   Preassignment Requirements

All AFBCCI personnel must participate in a medical surveillance program. All AFBCCI workers must have received and passed an initial baseline medical examination, or a follow-up medical exam within one year from the initial baseline exam, as described in the AFBCCI *Health and Safety Program Manual*. An occupational physician must certify AFBCCI workers as medically qualified to work at hazardous materials sites and wear protective equipment, including a respirator.

AFBCCI subcontract personnel must participate in a medical surveillance program. Subcontract personnel will not be allowed on-site unless they have passed a medical examination and have been certified by a physician to be fit-for-duty and capable of using respiratory protection.

5.2   Site-Specific Requirements

Heat stress monitoring will be conducted for personnel wearing chemical protective clothing and/or respirators when the temperature exceeds 70°F. Monitoring will consist of measuring the heart rate and the oral temperature. The frequency of monitoring is based on the ambient temperature (see Table A-5.2).

If the heart rate exceeds 120 beats per minute at the beginning of the rest period, the next work cycle will be shortened by 25%. If the heart rate still exceeds 120 beats per minute at the next rest period, the following work cycle will be shortened by 25%. Rest periods will be extended until the heart slows to the rate measured before the beginning of work.

If a worker exhibits one or more signs of heat stress (profuse sweating, hot and flushed skin, nausea, muscle spasms, pale and cool skin, dizziness) but has a heart rate of less than

120 beats per minute, the oral temperature should be measured. If the oral temperature exceeds 99.6°F at the beginning of the rest period (before drinking), the next work cycle will be shortened by 25%. If the oral temperature still exceeds 99.6°F at the next rest period, the following work cycle will be shortened by 25%. Rest intervals will be extended until the oral temperature falls below 99.0°F.

No worker will be allowed to wear chemical protective clothing or respiratory protection devices with an oral temperature above 101.6°F or a heart rate above 120 beats per minute.

5.3   Exposure/Injury/Medical Support

Any site worker who is injured or develops signs or symptoms of heat stress, overexposure, or any other type of physical distress will be evaluated by a physician as soon as possible. The examination content will be determined by the physician and will be based on the circumstances of the incident.

On-site emergency medical support procedures for exposure or injury are described in Section 10.6.

6.0   AIR MONITORING

6.1   Baseline Monitoring

Before site activities are initiated on day 1, air monitoring will be conducted at the outside perimeter of the site, adjacent to the fence line, to determine background readings for each instrument. Background readings should be taken at three intervals (A.M., midday, P.M.) to see if readings fluctuate during the day.

Baseline readings will be obtained from the area around the heavy equipment that will be used on-site. Variability in readings when the equipment is operating or turned off will be determined.

Instrument readings will also be obtained from the areas of invasive site activities (monitoring well installation, test pits, test borings). Readings will be taken at the soil surface and within small boreholes in the soil, made with a hand auger, to a depth of at least 18 inches, before invasive activities are begun.

Table A-5.2.   Monitoring of workers wearing protective equipment

Temperature in °F	Frequency of heat stress monitoring
> 90	Every 15 minutes of work
85–90	Every 30 minutes
80–85	Every 60 minutes
75–80	Every 90 minutes
70–75	Every 120 minutes

6.2 Rationale for Monitoring

Periodic or continuous real-time monitoring is conducted throughout remedial investigation activities to help verify that the level of protection for site workers is appropriate, and to assure compliance with OSHA and other published exposure limits. Air monitoring information also can be used to establish work zones, work practices, or engineering controls.

Real-time air monitoring using direct-reading instruments will be performed when site invasive activities begin, when obviously contaminated materials are encountered, and when containers that may contain contaminants are handled. Air monitoring will also be conducted periodically during decontamination procedures when heavy equipment has encountered action-level concentrations of contamination.

Air monitoring will focus primarily on those workers engaged in site activities that carry a risk of personal exposure to contaminants in air. (These tasks were identified in Section 3.4.) Whenever possible, monitoring around personnel will be performed in the breathing zone, defined as an area within a 24-inch radius of the head. Ground-level monitoring will also be performed as appropriate.

6.3 Direct-Reading Instruments

Instrument calibration and operation are described in the *Health and Safety Program Manual* and are performed in accordance with manufacturers' recommended procedures. AFBCCI also has backup instruments available in the event of instrument failure. Direct-reading instruments that will be available for use on-site are indicated in Table A-6.3.

6.4 Instrument Use Guidelines

6.4.1 Combustible Gas Indicators

A CGI will be used under the following conditions:

- Before any entry into a test pit.
- During invasive activity when the level of contamination is unknown.
- When other instrument readings indicate that high concentrations of contaminants are present.

Table A-6.3. Instruments available for use on-site

Combustible gas indicator	MSA Model 360; MSA Model 361
Oxygen deficiency meter	MSA Model 360; MSA Model 361
Flame ionization detector	Foxboro OVA Model 128
Photoionization detector	Photovac MicroTIP Model 2000
Portable gas chromatograph	Photovac Snapshot
H_2S, CO monitors	MSA Model 360; MSA Model 361
Dust monitor	MIE Miniram
Detector tubes and pumps	MSA, Draeger
Passive dosimeter tubes	MSA, Draeger
Personal monitoring pumps and charcoal tubes	MSA, Draeger

- Prior to handling of containers that may contain contaminants.
- Anytime, at the discretion of the SSO or the ASO.

6.4.2 Oxygen Deficiency Meters

An oxygen deficiency meter will be used under the following conditions:

- Before any entry into a test pit.
- Whenever a CGI is used.
- When other instrument readings suggest increased or decreased oxygen concentrations.
- Anytime, at the discretion of the SSO or the ASO.

6.4.3 Ionization Detectors

Flame ionization detectors (FIDs) and photoionization detectors (PIDs) are able to detect relatively low concentrations of contaminants and will be used under the following conditions:

- During invasive activities when a combustible gas reading is obtained.
- During invasive activities when an oxygen reading below 20.8% is obtained.
- When obviously contaminated materials are encountered.
- Prior to handling of containers that may contain contaminants.
- Anytime, at the discretion of the SSO or the ASO.

6.4.4 Toxic Gas Monitors

Hydrogen sulfide and carbon monoxide monitors will be used under the following conditions:

- Before any entry into a test pit.
- Anytime, at the discretion of the SSO or the ASO.

6.4.5 Dust Monitors

A dust monitor will be used under the following conditions:

- If dust is produced, and control measures appear inadequate.
- Anytime, at the discretion of the SSO or the ASO.

6.4.6 Portable Gas Chromatograph

A portable gas chromatograph (Photovac SnapShot with an ethylene-vinyl chloride applications module) will be used to determine if vinyl chloride is a contaminant in the breathing zone. The SnapShot will be used under the following conditions:

- When FID or PID readings consistently demonstrate readings at or above 3 ppm-equivalents above background.
- Anytime, at the discretion of the SSO or the ASO.

6.4.7 Dosimeter Tubes

Personal sampling pumps and dosimeter tubes (charcoal tubes) will be attached to clothing within the breathing zone of workers under the following conditions:

- When the Photovac SnapShot consistently demonstrates vinyl chloride readings above the action level (0.5 ppm).
- Anytime, at the discretion of the SSO or the ASO.

6.4.8 Passive Colorimetric Dosimeter Tubes

Passive dosimeter tubes for carbon monoxide (CO) and hydrogen sulfide (H$_2$S) will be attached to clothing within the breathing zone of workers under the following conditions:

- When toxic gas sensors detect CO or H$_2$S at concentrations above the action level but less than the TWA.
- Anytime, at the discretion of the SSO or the ASO.

6.4.8 Detector Tubes

Detector tubes for toluene, chlorinated hydrocarbons, and alcohols will be used to assist in identifying the source of FID, PID, or LEL readings and at the discretion of the SSO or the ASO.

6.5 Monitoring Frequency

Air monitoring instruments will be functioning and available for use whenever invasive activities are conducted on-site. It is not feasible to require continuous monitoring of the breathing zone during some activities because monitoring will interrupt and can interfere with some procedures. Continuous monitoring will occur in the immediate vicinity of workers performing invasive activities in Level D or Level C protection.

6.5.1 Borehole and Well Installation

Readings will be obtained in the breathing zone and 6 inches inside the annulus of the auger each time that the split spoon is removed.

6.5.2 Test Pit Excavation and Sampling

Continuous instrument readings will be obtained as the test pit is excavated. Readings will be obtained from within the pit prior to and during workers' entry to collect soil samples.

6.5.3 Monitoring Well Development and Sampling

Readings will be obtained from the headspace of each well prior to its development or sampling. If readings are obtained, the well will be allowed to off-gas.

6.6 Instrument Limitations

6.6.1 Temperature

High temperatures, possibly in excess of 100°F, are anticipated during the summer months when site activities will be conducted. Most instruments that will be used on-site have a maximum operating temperature of 104°F.

To avoid erroneous readings caused by instrument overheating, the instruments will be shaded and will not be left in the sun. Background readings will be carefully monitored when ambient temperatures are high, and the instruments will be shut off if spurious increases in background readings are noted.

6.6.2 Humidity and Water Vapor

Ambient humidity is typically high during summer months; high concentrations of water vapor are encountered during monitoring well headspace monitoring. High water vapor concentrations may also be encountered during borehole advancement, test pit excavation, and monitoring well installation.

Water vapor can affect the ionization efficiency of PIDs, decreasing it by up to 90%, depending on the water vapor concentration as well as the model and the manufacturer. The Photovac MicroTIP is less susceptible to the effects of humidity than other PIDs and will be the primary PID used on-site.

Other instruments used on-site are not susceptible to high ambient humidity. Whenever possible, PID and FID readings will be taken concurrently and compared. The higher of these readings will be used in determining whether a breathing zone action level has been reached.

6.6.3 Water Splash/Aerosol

All direct-reading air monitoring instruments can be damaged if water or water aerosols are drawn into them. Water can clog filters, block flash arrestors, and damage sensors. Water traps or filters will be used whenever possible. Care will be taken to avoid immersing the sample intake in water.

6.6.4 Dust and Particulates

All direct-reading air monitoring instruments can be affected by dust or particulates in the air. All instruments used on-site will be equipped with a dust filter. Clogged dust filters decrease the sample flow and decrease meter efficiency. When sites are dusty, the filters will be inspected daily and changed as necessary.

6.7 Action Levels and Levels of Protection

Instrument readings will be used to determine if the level of protection for site workers is appropriate and to assure compliance with OSHA and other published exposure limits and guidelines. Invasive site activities will be initiated in the basic Level D PPE ensemble described in Section 7.3.2. This level of protection will remain unchanged unless an action

Table A-6.7a. Breathing zone action level summary

Type of reading	Action level
Combustible gas	20% LEL
Oxygen deficiency	19.5% O_2
Oxygen excess	23.5% O_2
Carbon monoxide	18 ppm
Hydrogen sulfide	5 ppm
FID and/or PID	3 units
Vinyl chloride	0.5 ppm

level is reached. Action levels are summarized in Table A-6.7a; the actions taken based on meter readings are summarized in Table A-6.7b.

6.7.1 Combustible Gas Indicator

The health action level is 1% LEL. Site workers encountering readings in excess of 1% LEL should immediately determine if the flammable gas is methane by using a charcoal tube. If the flammable gas is determined to be methane, work may proceed with continuous monitoring up to the action level of 20% LEL. If the source of the %LEL reading is not methane, Level C protection should be used and SnapShot readings obtained to determine if vinyl chloride is present. Charcoal dosimeters should be worn if SnapShot readings indicate the presence of vinyl chloride.

A-6.7b. Actions taken when action levels reached

Action level	Action taken*	Protection level
1% LEL	Charcoal tube Detector tubes SnapShot analysis	Level C
20% LEL	Discontinue work; call H & S Manager.	Leave proximate worksite area.
19.5% O_2 23.5% O_2	Discontinue work; call H & S Manager.	Leave proximate worksite area.
5 ppm H_2S	Passive H_2S dosimeters	Level D
18 ppm Co	Passive CO dosimeters	Level D
≥ 3 units on FID or PID	SnapShot analysis	Level C
≥ 5 units on FID or PID	Detector tubes	Level C
0.5–10 ppm vinyl chloride	Charcoal dosimeters SnapShot analysis	Level C
5 mg/m$_3$ dust	Engineering controls	Level C

*Includes continued monitoring.

The flammable action level is 20% LEL. This action level is based upon the flammability hazard only; it does not reflect other hazards that the flammable vapor or gas may present. Work will be discontinued and the Health and Safety Manager consulted if the flammable action level is reached.

6.7.2 Oxygen Deficiency Meter

The action level for oxygen excess is 23.5%; for oxygen deficiency, 19.5%. The oxygen deficiency action level applies only to situations in which the displacing gas is known to be nontoxic and nonflammable. Work will be discontinued and the Health and Safety Manager consulted if an oxygen action level is reached.

6.7.3 Toxic Gas Sensors

The action level for hydrogen sulfide (H_2S) is one-half the TWA or 5 ppm. The action level for carbon monoxide (CO) is one-half the TWA or 18 ppm. Work may continue if the action level for CO or H_2S is reached. Continuous monitoring will be conducted, and workers will wear long-term passive dosimeters for the toxic gas detected. The dosimeters will be inspected by the SSO or the ASO at the end of each workday. Work will cease and the Health and Safety Manager will be consulted if toxic gas sensors detect concentrations above the TWA.

6.7.4 Ionization Detectors

The action level for SnapShot sampling is consistent readings of 3 ppm-equivalents or more above background on the FID and/or the PID.

The action level for a possible upgrade to Level C protection is consistent readings of 5 ppm-equivalents or more on the FID and/or the PID. Detector tubes (alcohol, toluene, chlorinated hydrocarbons) will be used to help quantify the concentrations present. If detector tube readings are not obtained or cannot be interpreted, and the source of the readings cannot be identified, Level C protection will be used.

FID readings in the absence of PID readings will be checked with a charcoal filter to determine if the source of the reading is methane. If the source is not methane gas, the action levels described above will apply.

6.7.5 Photovac SnapShot

The Photovac SnapShot will be employed when any %LEL readings are obtained, or when PID or FID readings of 3 units or more are observed. Also, the Photovac SnapShot will be used to determine if vinyl chloride is present. If vinyl chloride is present at concentrations between 0.5 and 10 ppm, Level C protection will be used, and charcoal dosimeters will be worn.

If vinyl chloride concentrations greater than 10 ppm are obtained, work will cease, and the Health and Safety Manager will be notified.

6.7.6 Dust Monitor

The action level for the Miniram is one-half the TWA for nuisance dust or 5 mg/m^3. If the action level is reached, engineering controls such as wetting the soil will be employed. If dust cannot be adequately controlled and concentrations exceed the action level, Level C protection (HEPA filter only) will be used.

7.0 PERSONAL PROTECTIVE EQUIPMENT

7.1 Rationale for Use

Site workers engaged in monitoring well installation, test pit excavation and sampling, advancement of the borehole, and borehole sampling have the potential for inhalation exposure to, and skin and eye contact with, vapors, gases, aerosols, and contaminated soil particles.

Existing data suggest that the concentrations of contaminants in ground water are low. Nevertheless, the possibility of encountering significant concentrations in soil or water cannot be excluded.

7.2 Criteria for Selection

All AFBCCI personnel on-site will wear a basic Level D work ensemble. Hearing protection will be available for use at all times.

Workers engaged in invasive activities at locations known or suspected to be contaminated will wear a chemical protective clothing (CPC) ensemble consisting of personal protective equipment designed to protect them from inhalation, skin, and eye contact with airborne contaminants or contaminated soil.

Additional protection against splash will be used during borehole advancement, well installation, and well development.

Hearing protection will be used for work around heavy equipment or in other situations where noise levels are too high for normal conversation.

Respiratory protection will be employed when air monitoring instruments indicate that a Level C action level has been reached. Level C respiratory protection can also be used at any time, at the discretion of the SSO or the ASO.

7.3 Levels of Protection

7.3.1 Basic Work Ensemble

The basic work ensemble will consist of AFBCCI flame-resistant cotton coveralls, steel toe–steel shank safety boots, hard hat, safety glasses with side shields, and cotton work-gloves. All AFBCCI employees will have ear muffs and disposable ear plugs available for use at all times.

7.3.2 Chemical Protective Clothing Ensemble

The CPC ensemble will consist of flame-resistant cotton coveralls, steel toe–steel shank safety foots, hard hat with faceshield, goggles, disposable latex or vinyl undergloves, disposable nitrile or neoprene overgloves, disposable nitrile or neoprene overboots, and disposable uncoated Tyvek coveralls. When necessary, lightweight workgloves may be worn over chemical-resistant gloves. All AFBCCI employees will have ear muffs and disposable ear plugs available for use at all times.

Additional protection against splash will be provided by the use of saran-coated, one-piece, Tyvek coveralls. Rubber or neoprene aprons and oversleeves, or a one-piece apron with sleeves, may be substituted for saran-coated coveralls at the discretion of the SSO or the ASO.

The use of duct tape and hoods will be at the discretion of the SSO or the ASO. Decisions about the use of tape and/or hoods will be based on temperature, meter readings, and the potential for splash or particulate contamination.

7.3.3 Level C

Level C is defined by the use of a full-facepiece air-purifying respirator with organic vapor and/or HEPA filter cartridges. The basic work ensemble or the CPC ensemble may be used with Level C respiratory protection.

Organic vapor cartridges will be utilized as protection against volatile organic contaminants in air. HEPA filter cartridges will be used as protection against dust or particulates. In situations where both hazards are present, a combination organic vapor/HEPA filter cartridge will be used.

7.3.4 Certifications or Approvals

Air-purifying respirators and cartridges must carry current NIOSH approvals and must have been listed in the NIOSH Certified Equipment List as of September, 1993. Respirators will be used and maintained according to ANSI guidelines as specified in the American National Standard for Respiratory Protection Z88-1992, and described in detail in the AFBCCI *Health and Safety Program Manual.*

Eye and face protection devices, including faceshields, goggles, and safety glasses with side shields, must meet current requirements as specified in ANSI Occupational and Educational Eye and Face Protection, Z87.1-1989.

Safety footwear for men and women must meet the minimum requirements described in ANSI Men's Safety-Toe Footwear, Z41.1-1983. Hard hats or other head protection devices must meet the standards specified in ANSI Requirements for Industrial Head Protection, Z89.1-1989.

7.4   Limitations of Protective Equipment

7.4.1   Chemical Protective Material

No single type of chemical protective material is resistant to all chemicals. Pure-product concentrations are not anticipated on this project. Chemical contaminants are found in water or adsorbed onto soil particles. The chemical protective clothing (CPC) selected is more than adequate to protect personnel from vapors, gases, aerosols, and spray. All chemical protective equipment (gloves, clothing, overboots) should be inspected prior to and during use for tears, rips, seam openings, pinholes, abrasions, and other avenues of penetration.

7.4.2 Physical Stress

Wearing CPC decreases mobility and dexterity, makes it difficult to see and hear, and is physically demanding. Increased physical exertion can lead to fatigue and heat stress. Wearing CPC also increases the risk of an accident. To minimize the effects of physical stress, workers will be able to vary their work schedule, at the discretion of the SSO or the ASO, to include shorter work and longer rest periods. Tasks requiring excessive physical exertion will be scheduled for the cooler periods of the day whenever possible.

7.4.3 Heat Stress

Protective clothing interferes with the body's normal evaporative cooling mechanism, which dissipates body heat. Heat stress may occur during site activities, which will be conducted during the hottest period of the year. To avoid heat stress and excessive physical fatigue, activities for workers wearing chemical protective clothing will consist of work–rest cycles as indicated below. Heat stress monitoring, as described in Section 5.2, will be conducted during each rest cycle, as summarized in Table A-5.2.

All workers must be familiar with the three types of heat stress illness:

1. *Heat cramps:* muscle pain or spasms in arms, legs, and abdomen; faintness, dizziness, weakness, fatigue; normal body temperature; cool and moist skin.
2. *Heat exhaustion:* pale, moist, clammy skin; dilated pupils; weakness, dizziness, nausea, headache; normal or even low body temperature.
3. *Heat stroke:* hot, dry, red, or flushed skin with no sweating; elevated temperature; strong, rapid pulse; constricted pupils; faintness, weakness. The person may be incoherent, and muscle twitching and unconsciousness may occur.

7.5   Field Upgrade/Downgrade Authority

Personal protection ensembles and levels of protection may be modified, upgraded, or downgraded to accommodate site activities and conditions, and upon the specific approval of the SSO or the ASO. The Site Manager will be notified of all changes in ensembles or levels of protection. Documentation of conditions and criteria for such changes will be added to the daily log; a copy of the rationale will be sent to the AFBCCI Project Manager and the Health and Safety Manager.

7.6   Storage

Protective clothing will be stored at AFBCCI and transported to the site on a weekly basis by the SSO or the ASO. Protective clothing will be stored in the tool locker. Respirators will be assigned to individual workers for the duration of the project. Workers will keep the respirators clean and store them in a cool, dry place when they are not in use.

8.0 SITE CONTROL MEASURES

8.1 Control Zones

8.1.1 Exclusion Zone

Control zones will be used to minimize the potential for the spread of contamination. The Exclusion Zone is defined as the area of invasive activities where contamination is known or suspected to occur.

For test pit excavations, the radius of the Exclusion Zone will be defined as the maximum extension of the backhoe plus an additional 20 feet.

For advancement of test borings and monitoring well installation, the radius of the Exclusion Zone is defined as the height of the mast.

For open test pits, the Exclusion Zone will include the trench and the area around excavated soil known or suspected to be contaminated.

Air monitoring instruments will be used to verify that airborne contamination does not extend beyond the Exclusion Zone. If necessary, the radius of the Exclusion Zone can be increased; background readings should be obtained at the Hot Line, which separates the Exclusion Zone from the rest of the site.

Each Exclusion Zone will be marked off by orange cones and barrier tape.

8.1.2 Contamination Reduction Zone

The remainder of the site, with the exception of Exclusion Zones, will serve as the Contamination Reduction Zone (CRZ). Decontamination of personnel and equipment will take place in separate areas adjacent to each Exclusion Zone. Heavy equipment and personnel may not leave the site after working in the Exclusion Zone without first undergoing decontamination procedures.

8.1.3 Support Zone

The Support Zone will be located in the parking lot adjacent to Dana Road. No personnel or equipment may enter the Support Zone after working in the Exclusion Zone without first undergoing decontamination procedures. The SSO or the ASO will inspect and monitor the Support Zone on a daily basis to ensure that it remains free of contamination.

8.1.4 Access Control Points

A plywood table will be placed at each of the two access control points, located at the eastern and western corners of the parking lot. Workers will sign in and out of the CRZ at an access control point. No workers may be present on-site unless they have signed in.

8.2 Buddy System

The buddy system will be employed to ensure that quick assistance is available in case of an emergency. Workers will be assigned buddies from their work team; if a team consists of only two or three workers, all team members are buddies. Each worker on a team must be under the observation of at least one other worker. Buddies must enter and leave the CRZ together at an access control point.

No one may continue work activities in the absence of a buddy. If a buddy is unavailable, do not enter the site; consult the site safety officer or the assistant safety officer.

8.3   Site Security

A vehicle gate is located at the entrance to the site. This gate will be locked at all times unless ABFCCI personnel are present on-site. Keep-out signs will be place on the gate and at 100-foot intervals on the surrounding fence, reading: AUTHORIZED PERSONNEL ONLY, NO TRESPASSING. FOR INFORMATION CALL 914-555-1234.

The back gate will be locked at all times unless AFBCCI personnel are present on-site. The SSO or the ASO will be responsible for unlocking and locking both gates.

8.4   Site Entry, Egress Procedures

The site will be accessed through the front gate off Dana Road. All personal and company vehicles will remain in the parking lot. The SSO or the ASO will record the presence of all workers and subcontractors on-site in the daily log book. Visitors will be instructed to remain in the parking area and will also be logged in by the SSO or the ASO.

Workers will access the CRZ by one of two access control points. Members of each work team, consisting of two or more workers, will enter the site together at the access control point. All team members will sign in and out in the daily log whenever they enter or leave the CRZ.

8.5   Hours of On-site Activities

Work will commence no earlier than 30 minutes after sunrise and will cease no later than 30 minutes before sunset. No worker may remain on-site longer than 12 hours.

8.6   Site Communications

Worker-to-worker communications on-site will be normal, direct oral communication or hand signals. Hand signals will be used as indicated below:

> Hands around throat: having difficulty breathing.
>
> Pointing to top of wrist: out of time; watch your time.
>
> Hands around buddy's waist or wrist: leave area immediately.
>
> Cutting motion across throat: stop engine; stop activity.
>
> Both arms above head: need help.
>
> Thumbs up: yes; OK; I understand; I am all right.
>
> Thumbs down: no; not OK; I don't understand; I am not all right.

Cellular phones will be available in the vehicles of the Site Manager and the Site Safety Officer. An airhorn and a whistle will be available for emergencies.

The Site Manager will contact the Project Manager or the AFBCCI office at least once each day. In the absence of the Site Manager, the SSO will contact the Project Manager or the AFBCCI office at least once each day.

8.7   Site Hygiene Practices

A portable lavatory with potable water for washing and drinking will be available in the parking lot. Water will be provided by Poland Springs in sealed 5-gallon plastic containers, to be delivered daily.

Meals and breaks may be taken under the trees by the parking lot. Folding chairs will be available. The trash will be put into two covered galvanized garbage cans, which will be emptied daily by LANDSHARCO employees.

8.8   Spill Control Procedures

Sand will be stockpiled on-site, adjacent to the parking lot, for use in monitoring well installation, and it can be used for emergency spill control. Buckets of sand will be used for small spills; the backhoe will be utilized for large spills.

8.9   Fire Control Procedures

There will be no smoking or fires in any area on-site, with the exception of the immediate vicinity of the portable lavatory. A sign designating the smoking area will be displayed. No smoking will be allowed inside the lavatory or in the meal/break area.

Dry brush or debris that may present a fire hazard will be cleared from around heavy equipment. Two 20-pound ABC DriChem fire extinguishers with a minimum rating of 2A:10B:10C will be available on-site. One extinguisher will be positioned by the drill rig; the other will be kept outside the portable tool trailer in the parking lot.

8.10   Safe Work Procedures

All site workers will adhere to established safety procedures and practices for their respective specialties.

8.10.1   Worker Safety Practices

To enhance site safety, general worker safety procedures have been established:

- Always use the buddy system; never enter the site without a buddy.
- Always sign in and out of the site at one of the access control points.
- Take breaks at specified intervals; drink plenty of fluids.
- Avoid contact with known or potential contaminants. Do not inhale chemical vapors. Do not expose skin or eyes to vapors, particulates, or sprays. Do not walk through discolored soil or puddles that may be contaminated.
- Use the appropriate protective ensemble.

- Always carry your hearing protection device with you at all times.
- Discard respirator cartridges, and clean and sanitize your respirator, at the end of each day's use.
- Perform the established decontamination procedures for all personnel and equipment leaving the Exclusion Zone.
- Inspect equipment on a daily basis; do not use equipment that is in need of repair or replacement.
- Know and understand all aspects of this site-specific HASP.

8.10.2 Personal Precautions

- Workers should wash their face and hands before eating, drinking, smoking, or other hand-to-mouth activities, and wash their hands before and after using the toilet.
- Jewelry, including rings, bracelets, chains, and watches, must be removed before any site activity is conducted.
- Smoking is prohibited except in the designated area in the Support Zone.
- Eating is prohibited except in designated areas in the Support Zone.
- Contact lenses are not allowed on-site; instead workers should use prescription safety glasses with side shields.
- Off-hour use of alcohol or caffeine may cause dehydration and increase workers' susceptibility to heat stress.
- Lack of sleep during off-hours may affect workers' alertness to potential hazards.
- Prescribed drugs may affect workers' alertness or interfere with their safe performance of some work activities; some drugs may also increase their susceptibility to heat stress.

9.0 DECONTAMINATION PLAN

9.1 Rationale

Contaminants on equipment or personnel must be removed when they leave the Exclusion Zone. Decontamination protects workers, limits the spread of contamination to clean areas, and prevents contamination from leaving the site. Equipment that cannot be completely decontaminated must be safely discarded after leaving the Exclusion Zone.

Contaminant concentrations in soil and water are expected to be low. The decontamination practices described herein are applicable for low-level soil and water contamination.

9.2 Personal Decontamination

Self-decontamination of personnel will be performed adjacent to the Exclusion Zone. Work team members will assist each other with decontamination as necessary.

Workers who do not contact hazardous materials, and who do not observe evidence of contamination on their protective ensemble, may use a dry decontamination procedure. Outer gloves, boot covers, and disposable outer clothing will be rolled down with the insides out, from the top down, while the wearer stands in an open plastic bag within a galvanized tub.

Bandage scissors will be available to cut away disposable outer boots if necessary. The wearer will then step out of the bag and shut it securely using a twist tie. After the bag is secured, the respirator may be removed. Inner gloves may be removed and discarded in other dedicated receptacles within the CRZ.

A decontamination solution of Liquid Tide detergent in water will be available in garden sprayers and 5-gallon buckets. Decontamination solution will be available in small tubs for washing the outer gloves. Decontamination solution will then be sprayed on, or long- or short-handled brushes will be used, to remove excessive gross contamination from disposable garments while the wearer stands in a galvanized tub or a plastic kiddie pool. After the wearer steps into another tub or pool, clean water in a garden sprayer and a 5-gallon bucket will be used as a rinse. The wearer will then step into a collapsed plastic bag within a clean, dry, galvanized tub; disposable garments will be removed as described above.

(A schematic of the minimum requirements for a personal decontamination station may be found in Appendix B)

9.3 Sampling Equipment Decontamination

9.3.1 Air Monitoring Equipment

Monitoring equipment will be wiped with a damp disposable cloth or sponge before leaving the CRZ. Sensors, hoses, and internal components will be purged by allowing the instruments to run for at least 15 minutes in the Support Zone.

9.3.2 Sampling Equipment

Monitoring well sampling will be conducted with dedicated, disposable bailers and lines. All sampling equipment will be discarded in dedicated receptacles after each sampling event.

Split spoons will be washed after each use with a solution of Liquid Tide in water; a garden sprayer containing clean water will be used for rinsing. Spoons will be air-dried; or, if necessary, the spoons will be manually dried with clean paper towels.

9.4 Heavy Equipment

An equipment decontamination station (EDS) will be set up on the asphalt road, approximately 50 feet north of and uphill from the dirt pile. A 3-inch asphalt curb will be added to retain wastewater (details of construction may be found in the work plan).

Drill rig and backhoe decontamination procedures will consist of a high pressure (2000 psi), hot water (180°F), low volume (1 gpm) spray. The drill rig will be driven to the EDS and decontaminated after the completion of each borehole.

The backhoe will be decontaminated on an undisturbed area, immediately adjacent to each completed test pit; it will be decontaminated after completion of each test pit. After completion of all test pits, the backhoe will be driven to the EDS for decontamination prior to leaving the CRZ.

Heavy equipment components (wire ropes, lubricated surfaces) must be inspected after decontamination or prior to moving to the next borehole or test pit location.

9.5   Protection for Decontamination Personnel

Personnel performing heavy equipment decontamination will wear the chemical protection clothing ensemble with splash protection described in Section 7.3.2. The operator of the high pressure, hot water sprayer will also wear thermal-protective outer gloves. Personnel should attempt to stay upwind of steam, spray, or aerosols generated during decontamination.

Heavy equipment decontamination procedures will be monitored by the SSO or the ASO on a daily basis. The SSO or the ASO may upgrade personnel protection to Level C (organic vapor cartridge with HEPA filter, or HEPA filter only) if steam, spray, or aerosol concentrations are considered excessive, or if action level concentrations are obtained as a result of air monitoring of the breathing zone of personnel performing decontamination.

No worker may remain in the EDS or other areas where heavy equipment decontamination procedures are conducted without the appropriate protective ensemble.

9.6   Materials Handling and Storage

Materials required for the conduct of this project will be brought to the site on a daily or weekly basis by the SSO and the Site Manager. The SSO or the ASO will issue cotton coveralls and disposable clothing and equipment to all workers. Fresh coveralls will be worn each day. Cotton coveralls will be will be picked up for cleaning and delivered twice a week by Uniforms-R-Us.

No fuels will be stored on-site, by order of the local fire department. Fuel for heavy equipment will be provided by the Fill It Up Fuel Company. A fuel delivery truck will be allowed onto the road adjacent to the parking lot before the start of site activities each morning. The backhoe and the drill rig will be parked adjacent to the road for fuel transfer. At no time will the fuel delivery truck be allowed off the road or past the parking lot.

9.7   Waste Disposition

All drill rig cuttings, split spoon contents remaining after sampling, and contained water will be placed into 55-gallon DOT-approved drums and left on-site adjacent to each finished monitoring well until receipt of TCLP analysis. All drums will be marked with well number, project name, and AFBCCI phone number.

Wash water from heavy equipment decontamination performed at the EDS will be collected with a manual transfer pump and stored in portable, covered tanks adjacent to the EDS until receipt of the TCLP analysis. All tanks will be marked with contents, project name, and AFBCCI phone number.

Wash and rinse water from personal decontamination will also be collected with small manual pumps and transferred from galvanized tubs into 55-gallon drums until receipt of

TCLP analysis. All drums will be marked with contents, project name, and AFBCCI phone number.

At this site, it is unlikely that hazardous concentrations of residual contamination will be present on protective clothing that initially had no visible contamination, or on clothing that had all visible contamination removed before it was doffed. Such protective equipment will not meet hazardous waste criteria and can be discarded as noncontaminated solid waste.

Items of disposable equipment suspected of containing significant levels of contamination should be placed in plastic bags. Each bag should be secured with a twist tie and marked with contents, project name, and AFBCCI phone number. Bags will then be placed in drums to await analysis and final disposition.

10.0 SITE CONTINGENCY PLAN

10.1 Pre-emergency Planning

Copies of the site safety plan have been distributed to the local fire department (Sawmill Hills Fire Department), police department (Sawmill Hills Police Department), and emergency medical service (Sawmill Hills Volunteer Ambulance Corps or SHVAC) for informational purposes. Fire and emergency medical agencies have trained and OSHA-certified personnel available to respond to the site in cases of emergency. Letters stating their willingness to do so are attached to this plan (see Exhibit A). The local hospital (Sawmill Hills Community Hospital) has been contacted and is willing to accept site workers; it also has the capability to perform decontamination procedures. Police department members have no hazardous materials training and will not be allowed past the parking area.

10.2 Personnel Roles and Lines of Authority

The SSO or the ASO will be in charge in the event of a site emergency. In case of fire, upon the arrival of the local fire department, the senior fire official will be in command. During medical emergencies, the SHVAC responding with the highest level of emergency medical training will be in command. Site personnel will provide technical or other assistance only if requested to do so.

10.3 On-site Warning/Evacuation Procedures

In case of a site emergency that requires the evacuation of site workers (e.g., a site fire), the SSO, the ASO, or the Site Manager will notify other site workers. Notification of evacuation will be made orally (person-to-person) or by sounding three blasts on an air horn. Air horns will be available at the EDS and in the tool locker in the parking lot.

Any site workers who see an imminent hazard will leave the area immediately with their buddies and notify the SSO, the ASO, or the Site Manager. At the signal to evacuate, all personnel will quickly assemble in the northwest corner of the parking lot. Equipment operators will shut the engines off and leave their equipment. If smoke or fumes are blowing toward the parking area, personnel will move upwind, exit the parking lot, and assemble on the other side of Dana Road.

10.4 Emergency Contact/Notification Procedures

Emergency calls will be made from on-site cellular telephones available in the company cars of the SSO and the Site Manager. Both cellular phones will be checked daily to ensure their proper function. A pay phone is located at the intersection of Route 9A and Dana Road. Change is not required for emergency calls. The pay phone will be checked each morning by the SSO or ASO to ensure that it is working properly. Emergency numbers are listed below:

Site Emergency Notifications

Local Responders

Fire	911
Police	911
Emergency Medical	911

Spill Emergencies

State Spill Hotline (all releases)	1-800-457-7362
National Response Center (releases into the river)	1-800-424-8802

Other notifications will be made as appropriate:

Organization	*Telephone*	*Contact Person*
LANDSHARCO (client)	914-555-1234	S. Minella
NYSDEC	914-555-5787	L. Stream
USEPA	212-555-5858	I. Greene
Fire Dept.	914-555-8399	Chief S. Robinson
Police Dept.	914-555-2343	Lt. D. Duncan
Ambulance Corps	914-555-4961	Cpt. B. Tylar
Hospital	914-555-7437	Dr. M. McCann
AFBCCI Physician	203-555-1001 or	Dr. M. Lane
(beeper)	203-555-9540	

All numbers will be posted on the door of the tool locker.

10.5 Emergency Decontamination Procedures

In the event that a site worker encounters pure chemical product (gasoline, diesel fuel, chemical contamination), emergency decontamination will be performed by using the nearest available supply of clean water. Potential water supplies include the garden hose attached to the hydrant adjacent to the abandoned warehouse, and clean rinse water located at decontamination areas in the CRZ.

The affected worker will immediately strip off his/her clothing and wash him/herself thoroughly with water. After washing, the worker will don clean Tyvek coveralls and overboots and immediately report to the SSO or the ASO. If necessary, decontamination may be continued by using a garden hose attached to a hydrant. The worker will then report to the hospital for examination, additional decontamination, and possible treatment.

If a worker is injured during site activities, other workers will help the injured worker remove his/her protective ensemble. The outer protective garments will be decontaminated by using the available clean water supplies only if the garments are grossly contaminated. Decontamination will not be performed if it could endanger the welfare of the injured worker by delaying treatment.

10.6 Emergency Medical Procedures

Treatment for minor physical injuries may be handled on-site by the SSO or the ASO. For serious injuries, emergency medical care will be provided by the Sawmill Hills Volunteer Ambulance Corps.

Any signs or symptoms of potential overexposure or heat stress will be evaluated by the SSO. Signs or symptoms that should be reported include skin or eye irritation, headache, nausea, disorientation, dizziness, lightheadedness, fatigue, weakness, loss of appetite, and muscle cramps.

10.6.1 On-site First Aid

An AFBCCI physician-approved first aid kit will be positioned next to the fire extinguisher at the tool locker during working hours. First aid may be administered on-site for certain injuries, including the following:

- *Insect sting:* Remove wasp or bee stinger if present by scraping it off to the side. Keep the affected part below the level of the heart. Apply ice or a cold pack if swelling occurs. Apply an anesthetic spray or a soothing lotion. Get immediate emergency medical attention if redness and swelling extend beyond the sting site, or if the victim's breathing becomes difficult.

- *Minor burns:* Apply cold water until the pain decreases. Apply an anesthetic spray, and cover the burn area with a clean dry gauze dressing. Apply an ice pack. Do not break blisters. If pain continues, seek medical attention from the hospital or the AFBCCI physician.

- *Minor cuts:* Apply pressure with a clean gauze dressing or pad. Apply anesthetic and antiseptic spray. Cover the cut with a band-aid or a bandage.

- *Objective in eye:* Flush the eyes will running water for several minutes until particulates are washed out. Get medical attention if irritation persists, or the object cannot be removed.

- *Minor sprains:* Keep the affected part elevated, and apply ice or cold packs. Seek medical attention from the hospital or the AFBCCI physician if pain or swelling persists.

- *Heat stress:* Comply with the site-specific requirements outlined in Section 5.2 for minor heat stress symptoms.

10.6.2 Emergency Treatment

Emergency treatment for injuries or illnesses will be administered by the SSO, the ASO, or other workers who have received the AFBCCI 8-hour first aid training. Emergency treatment is designed to stabilize or maintain the victim until emergency medical treatment and transport arrive.

- *Heat stroke:* Cool the victim immediately by removing any excess clothing; move the victim to shade. Apply cool water over the victim's body or ice packs under the arms, against the groin, and to both sides of neck. Get immediate emergency medical attention.
- *Fainting:* Keep the victim lying down with the feet elevated. Loosen any tight clothing. Bathe the victim's face with cool water, or wipe it with a cool towel. Get immediate emergency medical attention.
- *Unconsciousness:* Keep the victim flat with the feet elevated. Maintain an open airway. Check the victim for breathing and a pulse. Give artificial respiration if the victim is not breathing. Give CPR if the victim has no pulse. Get immediate emergency medical attention.
- *Difficulty in breathing:* Keep the victim sitting up if possible, encouraging him/her to breathe slowly. Try to keep the victim calm. Get immediate emergency medical attention.
- *Chest pain (possible heart attack):* Keep the victim calm; let the victim assume his/her most comfortable position. Do not allow the victim to walk; if necessary, carry him/her to a more suitable location. Get immediate emergency medical attention.
- *Fractures:* Do not move the victim unless it is absolutely necessary to do so. Do not move the affected part. Get immediate emergency medical attention.
- *Chemical burns to skin:* Remove the victim's clothing if it is soaked with a chemical. Flush the skin with copious amounts of water until all traces of the material are removed. If the chemical or contaminant is dry, brush as much of it as possible from the skin before flushing it with water. Apply a sterile dressing. Get immediate emergency medical attention.
- *Chemical in eyes:* Flood the eyes with water for at least 15 minutes. Cover both eyes, and get immediate emergency medical attention.
- *Bleeding:* Apply pressure with a sterile dressing or a clean cloth. Elevate the affected part if direct pressure is not effective. Apply a pressure bandage. Get immediate medical attention.
- *Vomiting:* Do not give anything by mouth. Allow the victim to assume a comfortable position; keep the victim calm. Get immediate medical attetion.

10.7 Hospital Transport Procedures

After emergency medical services and transport are requested by calling 911, a site worker will stand by the Dana Road entrance to direct the ambulance into the site. The ambulance will not proceed past the parking area. Ambulance crew members, assisted as necessary by site workers, will remove the patient from the site. In an extreme emergency (i.e., cardiac arrest or severe bodily injury), the ambulance will be allowed to enter the site to expedite patient treatment.

If an ambulance is not available for transport to the hospital, the victim will be taken to the hospital by site workers in a personal or company vehicle. All site workers will become familiar with the route to the hospital. A map to the hospital will be posted on the inside door of the tool locker, along with emergency and contact telephone numbers. (A copy of the map to the hospital can be found in the Appendix of this HASP.)

10.8   Spill Response Procedures

If a spill of a chemical, fuel, or contained water occurs, sand and/or clean dirt will be used for diking and as an absorbent. All spills will be prevented from entering catch basins in the parking lot or going into the river.

All spills will be immediately reported to the NYSDEC Spill Hotline and the AFBCCI. When chemicals or fuels are involved, an AFBCCI or subcontractor crew will be dispatched to clean up the spill. If decon wastewater is spilled, site workers will shovel up the material, absorbed onto dirt or sand, and place it in 55-gallon drums.

10.9   Fire/Explosion Response Procedures

In the event of a fire or an explosion, the local fire and police departments will be notified at once by calling 911. Sound three short blasts on an air horn and evacuate all personnel.

Fire extinguishers will be used only on small incipient fires. All personnel not engaged in fire-fighting activities or evacuation will assemble at the designated area.

11.0   INCIDENT REPORTING

The Site Safety Officer will prepare an incident report of any unusual events, including accidents, injuries, spills, and fires. The SSO and the Site Manager will review the incident and determine if a change in work practices is warranted. Changes in safety procedures or workplace practices will be noted in the incident report. The corporate Health and Safety Manager and the Project Manager will receive a copy of all incident reports.

HASP APPENDIX

This appendix includes Exhibits A through D and Forms 1 through 13, which are examples of the correspondence, maps, and reports used to guide and document a comprehensive health and safety plan.

Exhibit A. Letter to Sawmill Hills Fire Department

AFBCCI

4 LANDFILL ROAD
FORT DRUM, NY 13456
315-555-7788

April 14, 1998

Chief S. Robinson
Town of Sawmill Hills Fire Department
51 Sawmill River Road
Sawmill Hills, NY 10593

Re: SHII Site

Dear Chief Robinson:

As a follow-up to our conversation of April 9, 1998, this letter serves to inform you of work planned for the Sawmill Hills Industries, Inc. Site (SHII). AFBCCI has been contracted by the present owners (LANDSHARCO) to perform investigatory activities at the site. These activities include the installation of monitoring wells and the cutting of shallow (<4 ft) trenches. A small backhoe and truck-mounted drill rig will be utilized on site. To date, only minor concentrations of isopropyl alcohol, vinyl chloride, tetrachloroethylene, and toluene have been found.

We are required by OSHA to prepare an emergency response plan as part of an overall site safety program. This letter is to formally notify you of our upcoming work at the site. We plan to commence work in early June and should be off the site by September 15th. I will contact your office prior to our start-up.

Enclosed is a draft copy of our Health and Safety Plan for your review and comments. It is my understanding that you have fire and emergency medical personnel trained in accordance with 29 CFR 1910.120(q) and will respond to the site for emergencies, and you will call the County's Hazardous Materials Response Team for assistance if required. Your signature on the enclosed copy of this letter signifies your receipt of the site hazard information and your willingness to respond.

We appreciate your assistance and look forward to any comments or concerns you might have. If you have any questions, do not hesitate to contact me or Mr. Disorderi at the above number.

Very truly yours,

U. R. Safenow

U.R. Safenow
Manager, Health and Safety

URS/ms
Enclosures

S. Robinson

Chief S. Robinson, for the Sawmill Hills FD

Exhibit B. Site map

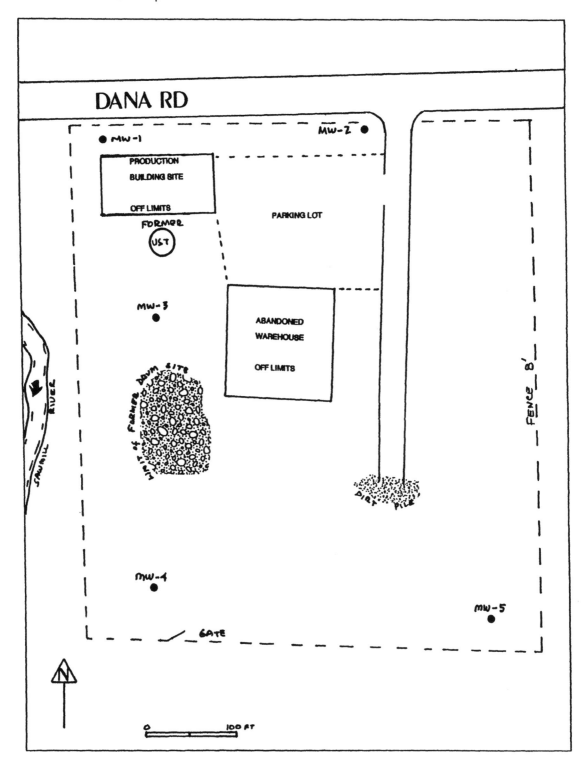

Exhibit C. Map of area surrounding site

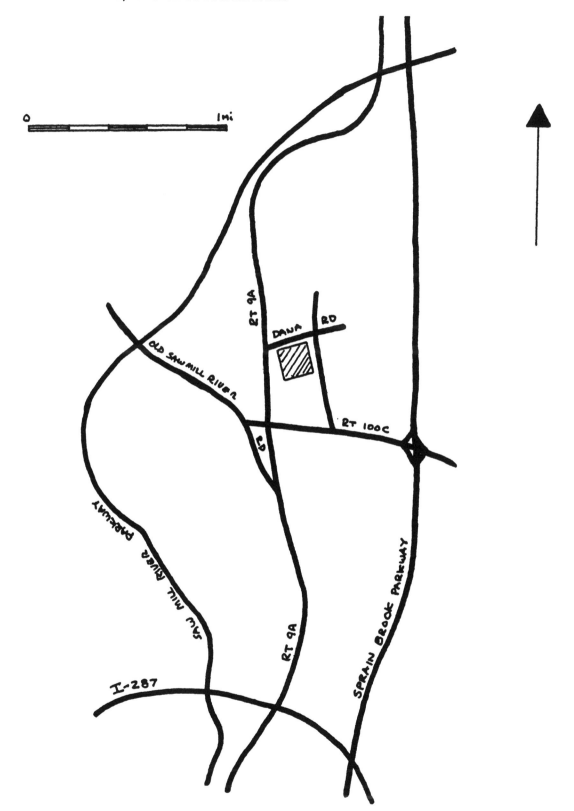

Exhibit D. Directions and map to hospital

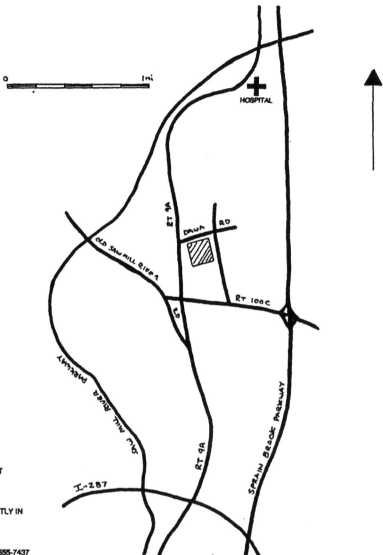

DIRECTIONS TO HOSPITAL

UPON LEAVING SITE TURN LEFT
GO TO BOTTOM OF HILL AND TURN RIGHT
PROCEED NORTH ON RT 9A 2 MILES
HOSPITAL IS ON THE LEFT SIDE
EMERGENCY ROOM ENTRANCE IS DIRECTLY IN
FRONT

EMERGENCY ROOM CONTACT NUMBER 555-7437

Form 1. Employee acknowledgment

SITE HEALTH AND SAFETY PLAN

AFBCCI EMPLOYEE ACKNOWLEDGEMENT FORM

I, _____, have read, reviewed,
     (PRINT your name)

and have been briefed as to the contents of the SHII

Site Health and Safety Plan. I understand the plan and

hereby agree to comply with those requirements and

directions described therein as provided to me by:

                    _____
                         (Site Safety Officer)

EMPLOYEE SIGNATURE _____

           DATE   _____

Form 2. Daily tailgate safety meeting

DAILY TAILGATE SAFETY MEETING

Date_____ Time_____ Job Number_____

Client_____

Site Location_____

SAFETY TOPIC PRESENTED

Planned site activities for the day_____
_____
_____

Protective clothing/equipment_____
_____

Chemical hazards_____
_____

Physical hazards_____
_____

Heat stress_____
_____

Equipment demonstrated/discussed_____
_____

Other_____
_____

Emergency procedures_____
_____

Worker questions/comments_____
_____

Special training_____
_____

ATTENDEES

Print names below   Print names below
_____   _____
_____   _____
_____   _____
_____   _____
_____   _____
_____   _____

Meeting conducted by:

_____   _____
      (Print name)   (Signature)

Form 3. Physician's certification

```
+---------------------------------------------------------------+
|            PHYSICIAN'S CERTIFICATION FORM                     |
| (Required medical surveillance record-OSHA 29 CFR 1910.120)   |
|                                                               |
| Last Name_____First_____Initial___     |
|                                                               |
| S.S.#____-____-_____ Age_____Sex_____ D.O.B._____      |
|                                                               |
| Exam Date_____ Type of Exam (Check one):            |
|                                                               |
|  [ ] Pre-Employment Physical                                  |
|                                                               |
|  [ ] Post-Employment Physical    [ ] Annual Physical Exam     |
|                                                               |
|  [ ] Post-Exposure Physical      [ ] Other _____      |
+---------------------------------------------------------------+
| In my professional opinion, the examination of the above-     |
| named employee:                                               |
|                                                               |
| [ ] Showed no evidence of a detected medical condition that   |
|     would place the employee at increased risk of material    |
|     impairment of his/her health from working in hazardous    |
|     waste operations or emergency response.                   |
|                                                               |
| [ ] Showed the employee is physically able to perform         |
|     work while using respiratory protective equipment.        |
|                                                               |
| [ ] Revealed a detected medical condition that would place    |
|     the employee at increased risk from working in            |
|     hazardous waste operations or emergency responses         |
|     unless special precautions or measures  are taken to      |
|     limit his/her involvement. These special precautions      |
|     are described below under recommended limitations.        |
|                                                               |
| [ ] Has been informed of the results of the exam and of any   |
|     medical conditions that require further treatment.        |
|                                                               |
| The following are recommended limitations on the employee's   |
| assigned work (e.g. prescription lenses, allergies to wasp    |
| or bee sting)_____        |
| _____         |
| _____         |
|                                                               |
| Physician's Name _____ Date_____   |
|                      (Please print)                           |
|                                                               |
| Signature _____Telephone_____   |
+---------------------------------------------------------------+
```

Form 4. Training documentation

TRAINING DOCUMENTATION FORM

Last Name_____First_____Initial___

S.S.#_____-_____-_____ Age_____Sex_____ D.O.B._____

Date of Initial Training _____(24 hr)_____(40 hr)

ON-SITE EXPERIENCE RECORD

LEVEL D:

SITE	ACTIVITY	SUPERVISOR	DATE	HOURS

LEVEL C:

SITE	ACTIVITY	SUPERVISOR	DATE	HOURS

LEVEL B:

SITE	ACTIVITY	SUPERVISOR	DATE	HOURS

LEVEL A:

SITE	ACTIVITY	SUPERVISOR	DATE	HOURS

REFRESHER TRAINING

DATE	TOPICS DISCUSSED	INSTRUCTOR

Form 5. Emergency medical information

```
┌─────────────────────────────────────────────────────────────────┐
│                 EMERGENCY MEDICAL INFORMATION                     │
│ Last Name_____First_____Initial___      │
│ S.S.#_____-_____-_____ Age_____Sex_____ D.O.B._____         │
│ Height_____ Weight _____ Blood Type_____           │
├─────────────────────────────────────────────────────────────────┤
│ TO WHOM IT MAY CONCERN:                                           │
│                                                                   │
│ The above-referenced employee completed a physical exam on        │
│ _____,  _____. Our records show the following:             │
│                                                                   │
│ HEALTH RESTRICTIONS:                                              │
│                                                                   │
│                                                                   │
│ CURRENT MEDICATIONS:                                              │
│                                                                   │
│                                                                   │
│ CURRENT IMMUNIZATIONS:                                            │
│                                                                   │
│                                                                   │
│ OTHER SIGNIFICANT MEDICAL INFORMATION:                            │
│                                                                   │
├─────────────────────────────────────────────────────────────────┤
│ In case of emergency, please notify the following family          │
│ members:                                                          │
│                                                                   │
│ _____ Telephone_____             │
│ (Print name and give relationship)                                │
│                                                                   │
│ _____ Telephone_____             │
│ (Print name and give relationship)                                │
├─────────────────────────────────────────────────────────────────┤
│ Physician's Name _____ Date_____              │
│                      (Please print)                               │
│                                                                   │
│ Signature _____Telephone_____         │
│                                                                   │
│ Address _____               │
└─────────────────────────────────────────────────────────────────┘
```

Form 6. Hearing conservation program participation

HEARING CONSERVATION PROGRAM PARTICIPATION

Last Name_____First_____Initial____

S.S.#_____-_____-_____ Age_____Sex_____ D.O.B._____

I, _____ am participating in the
       (print your name)
hearing conservation program on the _____ project.

	Initials
1. I have been informed about the health hazards associated with exposure to excessive noise and the effects of noise on hearing.	
2. I have been informed about the types of work that may result in exposure to excesive noise levels, and the necessary protective steps to prevent exposure, which include engineering controls and safe work practices.	
3. I understand the purpose for, proper use and limitations of hearing protective devices, and I have received instructions on selection, fitting, use and care of such devices.	
4. I will wear company-issued equipment when instructed to do so by my supervisor or when noise levels do not permit normal conversation.	
5. I have been informed about the monitoring and medical surveillance programs, including information on audiometric testing and have received an explanation of the test procedures.	
6. I have been provided access to copies of the applicable regulations governing occupational exposure to excessive noise, including 29 CFR 1910.95.	

Employee Signature   Date

Form 7. Daily safety log

```
                    DAILY SAFETY LOG

Date_____ Time In_____ Time Out_____

Weather Conditions_____

Site Location_____Job Number_____

Summary of Day's Work Activities_____
_____

Equipment Used by Site Safety Officer_____
_____

Protective Clothing/Equipment (by task)_____
_____
_____

Physical Condition of Workers (temperature stress, medical
problems)_____
_____

Incidents or Breach of Procedures_____
_____
_____

Description of Air Monitoring or Air Samples Taken_____
_____
_____

Decontamination Procedures:
Heavy Equipment_____
_____
Personnel_____
_____
```

```
All personnel accounted for at end of day _____
Gates locked at _____(give time)


_____        _____
        Print Name                          Title

Signature _____
```

Form 8. Weekly safety report

```
                    WEEKLY SAFETY REPORT

Site Location_____Job Number_____

Site Safety Officer_____Week Ending_____

Describe work performed during reporting period:
_____
_____
_____
_____

Summarize air monitoring data; include air sample analyses,
action levels exceeded, and actions taken:
_____
_____
_____

Indicate if there were incidents involving the following:
_____non-use of protective devices in assigned areas
_____non-use of specific assigned protective clothing
_____disregard of buddy system
_____violation of personal hygiene requirements (eating,
      smoking in prohibited areas, not washing, etc.)
_____misuse of standard operating procedures
_____instances of job-related injury of illness
_____spills or releases
_____fires or other emergencies

Briefly describe incident, including employee name and date
(use additional pages as needed)._____
_____
_____
_____
_____
_____

Attach all necessary documentation.
Comments_____
_____
_____


_____    _____
       Print Name                       Title

Date received by H&S Manager _____
```

Form 9. Incident/accident report

```
INCIDENT/ACCIDENT REPORT
(Page 1 of 3)
PERSONAL INFORMATION              DATE_____

Name_____ Age____ Sex____DOB_____

Home Address_____

Home Telephone_____ Title_____

SITE INFORMATION
Location_____ Job Number_____

Hazardous materials on site_____
_____

Activity on site_____
_____

INCIDENT INFORMATION

Date of incident_____ Time_____
Check one: Exposure_____ Possible Exposure_____
(or more)  Physical Injury_____ Illness_____
           Property Damage_____ Near Miss_____
           Fire _____ Motor Vehicle_____
           Other_____

Severity of illness or injury:
           Non-disabling_____ Disabling_____
           Medical treatment_____ Fatality_____
Estimated number of days away from job_____

Nature of injury (check one or more as applies):
_____Fracture          _____Thermal burn      _____Cold Stress
_____Dislocation       _____Sunburn           _____Frostbite
_____Sprain            _____Radiation burn    _____Heat Stress
_____Abrasion          _____Bruise(s)         _____Heat Stroke
_____Laceration        _____Blister(s)        _____Concussion
_____Puncture          _____Resp Exposure     _____Faint/Dizzy
_____Animal bite       _____Skin Exposure     _____Resp Allergy
_____Insect sting      _____Ingestion         _____Skin Allergy

Body part(s) affected_____
```

Form 9. (*continued*)

INCIDENT/ACCIDENT REPORT
(Page 2 of 3)

Briefly describe injury/exposure and area affected _____
_____

Where on-site did the incident occur?_____
_____

What caused the incident (object, substance, machinery)? __
_____

Additional contributing factors (weather, fatigue, stress,
improper attitude, lack of knowledge or training, slow
reaction time, etc.)_____
_____
_____
_____

Names of other individuals involved/exposed_____
_____

Witnesses_____

Was employee(s) in compliance with protective clothing and
equipment as defined in the site Health & Safety Plan?
_____Yes_____ No. If No, how and why did protection differ?
_____
_____

Did the incident require improper removal of protective
equipment in an area of potential exposure? Describe _____
_____
_____
_____

RESPONSE TO INCIDENT

Persons informed of incident_____
_____

Was medical attention received? Yes_____ No_____. If Yes,
briefly describe attention received_____
_____
_____

Form 9. (*continued*)

INCIDENT/ACCIDENT REPORT
(Page 3 of 3)

Where was medical attention received? _____

Attending physician's name and address_____

_____

_____

Preventive measures to avert recurrence of same type of
incident (machine modificiation, workplace practice change,
mechanical guards, training, etc.) _____

_____

_____

_____

_____

_____

PROVIDE ADDITIONAL INFORMATION ON BACK OF THIS FORM

_____   _____
Site Health & Safety Officer   Date

_____   _____
AFBCCI Health & Safety Manager   Date

_____   _____
AFBCCI Project Manager   Date

Form 10. Monthly exposure and injury

MONTHLY EXPOSURE AND INJURY REPORT

Last Name_____First_____Initial____

S.S.#_____-_____-_____ Age_____Sex_____ D.O.B._____

During the month of _____, 19_____, I have
worked at sites involving the following hazardous materials

   List chemicals if known   Job Number   Dates

_____
_____
_____
_____
_____
_____

The following personal protective equipment was used (e.g.
tyveks, respirators, etc.)

   Equipment Used   Job Number   Dates Used

_____
_____
_____
_____
_____

To the best of my knowledge, I have/have not (circle one)
received a significant exposure to hazardous materials
requiring the filling out of an Incident/Accident Report.

If a significant exposure was/is suspected, an Incident/
Accident Report was filed with the Site Safety Officer on
_____, 19_____.

I have/have not (circle one) been injured on the job during
the past month.

If injured on the job, an Incident/Accident Report was
filed with the Site Safety Officer on _____, _____.

_____   _____
    Employee Signature   Date

_____   _____
AFBCCI Health & Safety Manager   Date

Form 11.   Warning/suspension notice record

WARNING/SUSPENSION NOTICE RECORD

Any violation of the safety rules issued for the SHII

Project Site will result in the following disciplinary

actions to be taken:

1) First Offense - verbal warning with documentation to
                    personnel file.

2) Second Offense - written warning with documentation to
                    personnel file.

3) Third Offense - one (1) working day suspension from the
                    site.

4) Fourth Offense - immediate dismissal from the site.

Name _____ S.S.# _____

Type of Warning: Verbal/Written (circle one)

Date of one-day suspension _____, _____

Date of dismissal from site _____, _____

Reason and comments_____
_____
_____
_____

_____   _____
     Employee Signature   Date

_____   _____
 AFBCCI Health & Safety Manager   Date

_____   _____
   AFBCCI Project Manager   Date

_____   _____
 AFBCCI Human Resources Manager   Date

Form 12. Incident report follow-up

```
                    INCIDENT REPORT FOLLOW-UP

Date of Incident_____

Name of Employee_____

Site Location_____ Job Number_____

Outcome of Incident_____
_____
_____
_____
_____
_____
_____

Physician Recommendations_____
_____
_____
_____
_____
_____
_____

Date returned to work_____


  ATTACH ANY ADDITIONAL PERTINENT INFORMATION TO THIS FORM


_____      _____
AFBCCI Health & Safety Manager                 Date

_____      _____
  AFBCCI Project Manager                       Date

_____      _____
AFBCCI Human Resources Manager                 Date
```

Form 13. Air monitoring results report

```
AIR MONITORING RESULTS REPORT

Date_____

Site Location_____ Job Number_____

Duration of Monitoring_____

Work Location and Task_____

List instrument used:


Reading      Time      Reading      Time      Reading      Time
_____      _____      _____
_____      _____      _____
_____      _____      _____
_____      _____      _____
_____      _____      _____
                       _____      _____


Calibration:


Instrument   Time      Instrument   Time      Instrument   Time

Perimeter samples collected_____
_____
_____

Perimeter and Personal Sample results from previous date's
sampling events (record date of sampling event and results)

Results      Date      Results      Date      Results      Date
_____      _____      _____
_____      _____      _____


Comments_____
_____
_____


_____      _____
   Site Safety Officer            Date Completed

_____
   Signature of SSO
```

Appendix B

Generic Emergency Response Plan and SOPs

The following pages give an example of an emergency response plan (ERP) and standard operating procedures (SOPs) for a midsize, hazardous waste consulting and contracting company that provides on-site emergency technical consulting and remedial services to a large chemical shipping and recycling company. The services include responding to large spills to provide mitigative services, including soil excavation and ground-water recovery and treatment, while an emergency may still be declared. The spill may be at a client facility or involve a client-owned transport vehicle.

The ERP is pertinent for employees covered under Section (q) of the HAZWOPER rule (emergency responders, emergency cleanup contractors), who respond to hazardous materials incidents without regard to location. The ERP is not designed for employees engaged in operations at uncontrolled hazardous waste sites or RCRA facilities; at these sites, a site-specific health and safety plan (HASP; see Appendix A) would be utilized.

ANYTHING FOR A BUCK CONSULTANTS AND CONTRACTORS, INC. SITE HAZARDOUS INCIDENT RESPONSE TEAM
SITE EMERGENCY RESPONSE PLAN

APPROVED   APPROVED

_____   _____

I. M. Bigshott   U. R. Safenow
Principal   Manager, Health & Safety
June 18, 1998   June 18, 1998

ANYTHING FOR A BUCK CONSULTANTS AND CONTRACTORS, INC.
SITE HAZARDOUS INCIDENT RESPONSE TEAM
SITE EMERGENCY RESPONSE PLAN

CONTENTS

I. PLANNING BASIS

A. Purpose

The purpose of the Anything for a Buck Consultants and Contractors Inc. (ABCCI) Site Emergency Response Plan (hereinafter called the Plan) is to establish a strategy to mitigate hazardous materials incidents as required by client authorization.

B. Objectives

1. To describe operational concepts, organization, and support agencies required to implement the Plan.
2. To identify authority, responsibilities, and actions of federal, state, local, and private industry agencies necessary to minimize damage to human health, the environment, and property, and to aid in mitigating the hazard.
3. To establish an operational structure for the ABCCI Site Hazardous Incident Response Team (SHIRT) that has the ability to function within a geographic area defined by ABCCI.
4. To utilize SHIRT members who have been trained to handle hazardous materials incidents.
5. To establish lines of authority and management for a hazardous materials incident.

II. ADMINISTRATION

A. Scope

1. Geographic limitations: This plan addresses the response to hazardous materials incidents that occur within the United States of America, its territories, or its possessions. Geographic limitations will be mutually determined by ABCCI and its client.
2. Hazards: Hazards will include actual or potential fires, spills, leaks, rupture of containers, contamination, and any threat to life, safety, and the environment, involving hazardous materials.
3. Hazardous materials: Hazardous material may include explosives, flammables, combustibles, compressed gases, corrosives, cryogenics, poisons, toxics, etiological agents, reactive and oxidizing agents, radioactive materials, or any combination of these materials.
4. Incident: The Plan is designed to accommodate any hazardous materials incident associated with any mode of transportation, industrial processing and/or storage sites, waste disposal procedures, and illegal usage and disposal.

B. Authority

Authority is established in accordance with the contract dated 20 January 1998 between Southern Pacific Industrial Lagoons, Ltd. (SPILL) and Anything for a Buck Consultants and Contractors, Inc. (ABCCI).

C. References

1. *Occupational Safety and Health Guidance Manual for Hazardous Waste Site Activities,* DHHS (NIOSH) Publication No. 85–115.
2. USEPA, 1992. *Standard Operating Safety Guides.*
3. ABCCI, August 1987. General Site Safety Plan.
4. ABCCI, September 1987. Respiratory Protection Plan.
5. ABCCI, July 1988. Hazard Communication Program.
6. 29 CFR 1926 Subpart P, originally published in 54 FR p. 45959.
7. 29 CFR 1910.120, originally published in 54 FR p. 9294.
8. NFPA, 1992. *Hazardous Materials Response Handbook* (NFPA 471, 472).
9. ABCCI, January 1990. Excavation Safety Program.
10. ABCCI, February 1991. Health and Safety Operating Procedures.
11. ABCCI, April 1996. Permit-Required Confined Space Program.
12. 29 CFR 1910.146, originally published in 58 FR p. 4462.

III. HAZARDOUS MATERIALS INCIDENT CLASSIFICATION

There are four levels of hazardous materials classifications. The bases used for determining the level of a hazardous materials incident are:

1. Level of technical expertise required to mitigate the incident.
2. Extent of participation or involvement by municipal, county, state, and federal agencies.
3. Extent of evacuation required.
4. Extent of injuries and/or deaths.
5. Extent and type of decontamination procedures required.

A. Level I Incident

1. A Level I Hazardous Materials Incident (Level I HMI) is any spill, leak, rupture, and/or fire involving hazardous materials that can be controlled, contained, extinguished, and/or mitigated by utilizing equipment, supplies, and resources immediately available to the local agency, the department having jurisdiction, or the industrial response organization that has been authorized to act on behalf of the agency or department having jurisdiction; and
2. A Level I HMI does not require evacuation of civilians.

B. Level II Incident

The Incident Commander (IC) may upgrade a Level I HMI to Level II at any time.

1. A Level II HMI is an incident at which the IC desires the presence of a SHIRT supervisor or his/her designee, for the purpose of obtaining information, guidance, or other forms of assistance.
2. The SHIRT representative at the scene, upon consultation with the IC, may upgrade a Level II HMI to Level III.

C. Level III Incident

1. A Level III HMI is an incident that can only be identified, tested, sampled, controlled, contained, extinguished, and/or abated by utilizing resources of the SHIRT; it is a hazardous materials incident that requires the use of specialized equipment, including chemical-protective clothing.
2. A Level III HMI requires evacuation of civilians within the area of the department having jurisdiction; and/or
3. A Level III HMI may have fires involving hazardous materials that are permitted to burn for a controlled period of time, or are allowed to consume themselves.

D. Level IV Incident

A supervisor of the SHIRT, or the incident commander, can upgrade a Level III HMI to Level IV.

1. A Level IV HMI is any spill, leak, and/or rupture involving hazardous materials that can be controlled, contained, and/or mitigated by utilizing highly specialized equipment, supplies, and resources available to the SHIRT; and/or
2. A Level IV HMI has fires involving hazardous materials that are allowed to burn because of: the ineffectiveness or dangers of using extinguishing agents, or the nonavailability of sufficient water, and/or the threat of large container failure, and/or an explosion, detonation, BLEVE, or container failure that has occurred; and/or
3. A Level IV HMI requires evacuation of civilians extending across jurisdictional boundaries, and/or there are serious civilian injuries and/or deaths as a result of the incident; and/or
4. A Level IV HMI requires multiagency involvement of large proportions.

IV. INCIDENT COMMAND AND SCENE MANAGEMENT

A. Incident Commander

The Incident Commander (IC) shall be the designated fire, police, or site representative responsible for mitigating the hazards at the scene of a hazardous materials incident. Upon arrival the IC shall secure and maintain immediate control until the situation has been mitigated.

1. The local authority having jurisdiction shall accept and provide the position of Incident Commander for the scene of all hazardous materials incidents. The chief(s) of the local police, fire, or other department(s) shall coordinate and direct all department activities within their jurisdiction and have responsibility for, but not limited to, rescue and first aid, scene stabilization and management, suppression activities, protection of exposures, agency notification, scene isolation, and evacuation.

2. Anything for a Buck Consultants and Contractors, Inc. and the SHIRT shall maintain a list, for each client site, of employees or facility representatives who can act as Incident Commander, and/or provide technical assistance at a hazardous materials incident.

3. The senior representative of the SHIRT present shall report to and function through the Incident Commander.

V. ABCCI SITE HAZARDOUS INCIDENT RESPONSE TEAM (SHIRT)

A. Purpose

1. Anything for a Buck Consultants and Contractors, Inc. shall maintain a specially trained team for the specific purpose of responding to hazardous materials emergencies at client remediation sites. The SHIRT can provide expertise and equipment especially developed to help control and mitigate a hazardous materials incident.

2. The equipment, instruments, protective clothing, tools, and any other equipment and supplies of the SHIRT shall not be loaned or used by any other individual or agency without the express consent of the SHIRT supervisor on scene.

B. Responsibilities of the SHIRT Supervisor

1. The SHIRT supervisor shall assist in determining what local, county, state, and federal agencies should be notified of the incident, and if representatives

of local, county, state, and federal agencies, or private support representatives should be present at the scene, and shall convey this information to the Incident Commander.

2. The SHIRT supervisor shall upgrade a Level III hazardous materials incident (HMI) to Level IV through proper notification procedures when:

   a. The incident is beyond the capabilities of the SHIRT.
   b. The SHIRT supervisor requires additional SHIRT personnel or equipment to control the incident.
   c. The SHIRT supervisor requires subcontractor support personnel or equipment to control the incident.

3. The SHIRT supervisor shall work with, and be subordinate to, the designated client representative and/or the Incident Commander of the agency or department having jurisdiction.

VI.   CONTROL ZONES

The SHIRT supervisor shall assess zones already established, and recommend establishment or redefinition of zones if appropriate. Emergency response control zones may overlap or be independent of already existing control zones established to protect general site workers engaged in site investigation or remediation activities. Entry into and exit from specific control zones shall be through access control points.

A.   Limited Access Zone

1. The Limited Access Zone (Yellow Zone, Warm Zone) is defined as an area where a potential or real danger exists to the public or the environment.
2. Identification of a Limited Access Zone shall be implemented by general site workers, SHIRT members, or first-arriving personnel of an emergency response agency.
3. Access to the Limited Access Zone shall be limited to those members of agencies on scene who are appropriately protected and directly engaged in rescue, control, and preliminary stabilization.

B.   Restricted Zone

1. The Restricted Zone (Red Zone, Hot Zone) shall be established as necessary to identify and define an area of exceptional danger, including extreme threat to health, safety, and life.
2. The SHIRT may assist in identification and establishment of the Restricted Zone.

3. Access to the Restricted Zone shall be coordinated by the SHIRT. Only qualified SHIRT personnel and other designated personnel shall be allowed entry into the Restricted Zone.

C. Decontamination Zone

1. One or more Decontamination Zones (within Contamination Reduction Corridors) shall be established as necessary to facilitate decontamination of SHIRT personnel, other responders, civilians, and equipment in an effort to reduce or prevent the spread of suspected contaminants.

2. Identification and setup of the Decontamination Zone may be performed by the SHIRT at the request of the IC.

   a. Access into the Decontamination Zone shall be coordinated with the SHIRT. Only a SHIRT member, upon direction of an SHIRT supervisor, may allow personnel to exit the Decontamination Zone.

   b. Personnel entering the Decontamination Zone to assist the SHIRT shall be under the direction of SHIRT members, and shall be appropriately protected.

   c. Procedures that involve decontamination of SHIRT personnel and/or equipment shall be conducted only by SHIRT personnel.

D. Staging Area or Base

1. The location of the staging area or base shall be determined by the IC.

2. A SHIRT staging area within the base shall be established by the SHIRT supervisor for SHIRT personnel.

ANNEX A
THE OPERATIONAL PLAN

I. ACTIVATION

This plan will become operational when the SHIRT receives notification of a hazardous materials incident.

II. NOTIFICATION

A. SHIRT

1. The SHIRT is activated by the on-call SHIRT manager after receiving a request from the client.
2. The SHIRT shall respond to any request determined to be a site emergency by the requesting client. The SHIRT shall also respond to standby and informational requests.
3. All SHIRT members shall have Skynet pagers on their person or nearby at all times when they are on call.
4. Response to any site hazardous materials incident shall be in accordance with local vehicular and traffic regulations unless the responding unit is escorted by a police agency.

B. Dispatch

1. All SHIRT vehicles shall communicate with corporate headquarters, branch offices, field personnel, and client representatives by cellular telephone.
2. Vehicle radios shall be used for emergency communications only while enroute to the scene of a site emergency.
3. SHIRT members at the scene shall use portable radios to communicate with SHIRT personnel and the SHIRT supervisor.
4. SHIRT radio traffic at the scene shall be coordinated by the SHIRT supervisor or his/her designee.

C. Agency Call Lists

1. ABCCI and the SHIRT shall maintain a checklist, for each client site, of the necessary and required agencies that can provide technical assistance or that may become involved in a response to a hazardous materials incident.
2. ABCCI and the SHIRT shall make notifications to local, state, federal, industrial, and private agencies and mutual aid requests *only* as requested by the client.

III. IMMEDIATE ON-SCENE ACTIVITY

A. First on Scene

1. Identification

   a. The fire, police, other agency, or client on-scene representative will attempt to identify:

      (1) The type of material involved.
      (2) The quantity of material involved.
      (3) The potential for/possibility of contamination.
      (4) The immediate exposure potential.

   b. This information is to be transmitted to the local authority having jurisdiction and to the SHIRT as soon as possible.
   c. The IC or client representative on the scene shall inform the SHIRT of the initial evaluation and actions taken, and shall direct SHIRT members regarding which access routes should be utilized when approaching the incident.

2. Command Post

   a. The first SHIRT supervisor at the scene shall establish a Command Post and ensure that it is in a safe location.
   b. The client or local authority having jurisdiction shall assume the position of Incident Commander or appoint a designee to the position.

3. Base or Staging Area

   a. A base or staging area shall be established as necessary outside the anticipated hazard area in a removed and safe location. This area, and its resources, shall be coordinated by assigned personnel.
   b. The base location shall be transmitted to ABCCI as soon as possible, and all responding SHIRT units dispatched to the incident shall be directed to report to the staging area.
   c. A SHIRT staging area within the base shall be established for standby SHIRT members. All responding SHIRT personnel shall respond to the SHIRT staging area unless otherwise directed.

4. Communication

   a. On-scene SHIRT communication to ABCCI will be maintained by using a cellular telephone.

    b. Incident Command communications will be established on the scene by the client.

5. Incidents on Roads and Streets

    a. If the hazardous materials incident is on a highway, state road, or city or municipal street, the appropriate law enforcement agency shall be notified and summoned to the scene. The law enforcement agency will establish communication with the Incident Commander and the SHIRT.

    b. Successful mitigation of a hazardous materials incident requires that good communications be established and maintained between the Incident Commander, the SHIRT supervisor, and all other agencies involved.

6. Site Hazardous Incident Response Team Operation

    a. It shall be the responsibility of the SHIRT supervisor:

        (1) To evaluate the problem.
        (2) To determine the potential consequences of the incident.
        (3) To determine possible courses of action that could be undertaken to mitigate the incident and the potential effects of such action.
        (4) To relay this information to the Incident Commander.

    b. The Incident Commander shall evaluate all information and determine the best course of action to mitigate the hazard. The Incident Commander shall then convey his/her decision to the SHIRT supervisor and other affected agencies.

    c. The SHIRT supervisor shall inform the Incident Commander of the need for additional personnel or resources. The Incident Commander or his/her designee shall then inform the appropriate agencies and/or authorize the use of additional SHIRT or contract personnel or resources.

7. Evacuation

    a. The Incident Commander shall notify appropriate agencies if an evacuation is necessary. If requested to do so, the SHIRT supervisor may assist the IC in determining the need for and the extent of such an evacuation.

    b. The law enforcement agency having jurisdiction, assisted by other appropriate agencies, shall plan and conduct an orderly evacuation within a specified geographic area.

    c. The SHIRT shall not participate in evacuations unless specifically requested to do so by the client; SHIRT personnel assisting in evacuations shall operate under the supervision or the authority of the law enforcement or other agency having jurisdiction.

8. Termination

    a. When an actual, potential, or perceived hazard is no longer present, SHIRT members shall initiate termination procedures and leave the scene as determined by the SHIRT supervisor.

    b. The SHIRT shall remain on standby at the scene only at the request of the IC and with the approval of the client when a representative of the client does not act as IC.

B. Additional Resources

1. The SHIRT supervisor shall promptly notify the client when additional SHIRT personnel and/or equipment is required at the scene of an incident.

2. The SHIRT supervisor shall promptly notify the client when additional outside resources, in the form of personnel or materiel, are required.

ATTACHMENT ONE

I.   RESPONSIBILITIES OF AGENCIES

A.   Municipal Agencies

1. A local fire or police department or facility official shall assume the role of Incident Commander (IC) at the scene of a hazardous materials incident. The ———— . department (*add appropriate agencies here—fire, police, EMS, or facility department*) shall coordinate and effect rescue efforts, first aid, containment, and hazard reduction activities, within the capabilities of available personnel, resources, training, and equipment.

2. The police department or the site security staff shall have responsibility for crowd control, traffic control and relocation, and scene security, and shall coordinate and control excavation activities.

3. The public works agency, site security personnel, site support staff, or facility grounds or maintenance department shall be responsible for necessary surface road closures and detours, and for establishing traffic control zones. The public works agency shall also assist the fire department in appropriate scene stabilization and mitigation of hazards for incidents on surface roads. The public works agency can also provide sand and barricades.

4. Water and sewer departments or appropriate facility personnel shall be responsible for maintaining clear and flowing water distribution systems and monitoring remedial actions when a hazardous material may affect water sources and distribution systems.

B.   County Agencies (where appropriate)

1. The County Health Department shall provide assistance and information regarding environmental health dangers and can provide information regarding cleanup and disposal procedures.

2. The Department of Public Safety or the Sheriff's Office may provide an official for the position of Incident Commander for hazardous materials incidents.

3. The County Office of Disaster and Emergency Services or Civil Defense may be used to coordinate procurement of required resources at hazardous materials incidents and to coordinate efforts to acquire the services of other agencies.

C.   State Government (where appropriate)

1. The State Emergency Management Office (SEMO) is responsible for notification and coordination of all state agencies that may become involved in a response to a hazardous materials incident. SEMO shall coordinate state radi-

ological monitoring of areas, personnel, and equipment in support of local authority.

2. The State Police have the responsibility for traffic supervision and control on all state roads, state-owned bridges, and highways within certain unincorporated areas. The State Police shall provide for traffic control, traffic routing, road closure, and prevention of unauthorized entry into restricted and limited access areas, and shall assist local authorities as requested. They shall function as overall Incident Command for any hazardous materials incident occurring on highways and roads within their primary responsibility. The State Police may also dispatch personnel for enforcement of federal DOT regulations.

3. The Department of Transportation (DOT) is responsible for maintaining the state highway system. DOT as requested can assist in the identification, containment, and cleanup of hazardous materials on the state highways, and as requested can provide for highway closure, traffic management, and restoration of orderly traffic flow. DOT may assist as requested in monitoring contamination, and shall repair and restore contaminated highways. DOT shall coordinate its on-scene activities through the Incident Commander or his/her designee.

4. The State Environmental Agency (SEA) is responsible for protecting the state's natural life and wildlife resources and habitats. SEA can provide recommendations and guidelines when a hazardous material has contaminated or may contaminate streams or waterways. SEA shall coordinate its on-scene activities through the Incident Commander or his/her designee. Whenever there is a kill of wildlife or a serious threat to it, SEA should be notified; it may elect to send a representative to the scene.

5. The Department of Health (DOH) is responsible for protecting public health from low-level radioactivity and hazardous materials. DOH is also responsible for protecting food and water supplies from the effects of hazardous materials incidents, and for designating a location for the disposal of hazardous waste.

D. Federal Government

1. The U.S. Environmental Protection Agency (USEPA) is responsible for ensuring the protection of the environment from all types of contamination, and it must be notified of incidents involving hazardous materials that result in contamination. The National Contingency Plan specifies that the federal on-scene coordinator for inland waters and ground water is the USEPA.

2. The Department of Energy (DOE) has the responsibility for and capability of assisting and providing technical information in the handling and the disposal of radiological and nuclear materials.

3. The U.S. Department of Transportation (USDOT) is responsible for regulating the transportation of hazardous materials.

4. The U.S. Coast Guard (USCG) is responsible for the nation's coastline and for major navigable waterways within the state. The USCG can provide for decontamination and cleanup of any material that enters and affects the waters. The National Contingency Plan specifies that the federal on-scene Incident Coordinator for coastal waters is the USCG.

ANYTHING FOR A BUCK CONSULTANTS AND CONTRACTORS, INC.
SITE HAZARDOUS INCIDENT RESPONSE TEAM
STANDARD OPERATING PRACTICE AND PROCEDURES

APPROVED

APPROVED

I. M. Bigshott
Principal
June 18, 1998

U. R. Safenow
Manager, Health & Safety
June 18, 1998

ANYTHING FOR A BUCK CONSULTANTS AND CONTRACTORS, INC.
SITE HAZARDOUS INCIDENT RESPONSE TEAM
STANDARD OPERATING PRACTICES AND PROCEDURES

CONTENTS

ATTACHMENTS

Purpose: to define the practices and procedures of the Site Hazardous Incident Response Team (SHIRT).

I. INCIDENT SIZE-UP

Upon arrival at the scene, the SHIRT Supervisor (or his/her designee) will report to the Incident Commander (IC) and will do the following:

A. Perform an initial size-up of the situation.

   1. Determine if the hazardous material has been identified.
   2. Determine immediate risks to health and safety if hazardous materials are involved.
   3. Conduct a rapid evaluation for effects of wind, topography, location of incident and receptors, extent of spill or release, and fire condition or potential, in order to identify the nature of the incident and the severity of immediate hazards.
   4. Attempt to quickly confirm the identity of the materials involved, and to determine their primary hazards.
   5. Assume command of the SHIRT.

B. Report on conditions and requirements for additional resources to the on-scene IC and the client if not the IC.

C. Establish or redefine control zones.

   1. If possible, establish a Restricted Zone to define the area of greatest hazard.
   2. If it is not already designated, suggest that a Limited Access Zone be established to designate the area where a potential danger exists.
   3. If it is not already designated, establish an operational perimeter in order to keep unnecessary people away from the scene and SHIRT operations.
   4. Use barrier tape, traffic cones, traffic barricades, sawhorses, fire hoses, or other impediments to vehicle and foot traffic to delineate control zones.

D. Inform the IC, and the client if not the IC, of which agencies should be notified.

E. Initiate/continue the identification of materials involved and their hazards.

   1. It is imperative that the SHIRT Supervisor determine the identity and the volume of the hazardous materials involved prior to initiating mitigation activities.
   2. No SHIRT member should enter the scene of an incident to make an identification unless all other attempts have been exhausted. Sources of information include:

      a. Placarding and/or labeling.

    b. Paperwork/shipping papers/MSDS associated with materials being transported or stored.

    c. Persons directly related to the incident (i.e., site workers, facility employees, plant manager, driver, owner, etc.).

    d. References that are readily available.

3. Action taken prior to determining the hazardous material involved may worsen the situation. *A decision based on inadequate information can be worse than no decision at all.*

II. INITIAL OPERATIONS

Upon determining the need, the SHIRT Supervisor (or designated members) will conduct initial operations:

A.   Establish the SHIRT Command and Operations Area and inform the IC of its location.

    1. Appoint other SHIRT personnel to operational positions as necessary.

    2. Utilize a status board to track personnel, assignments, and resources.

B.   Continue to obtain technical information pertaining to the hazardous materials involved, using sources of information delineated in Section I.E.2 above and the following:

    1. ChemTrec (phone 800-424-9300).

    2. The shipper or the manufacturer (ChemTrec can provide assistance or act as an intermediary).

    3. Computer data bases.

    4. Other sources of information as appropriate.

C.   Oversee the use of air monitoring instruments and other means to help define the nature and the extent of the hazards involved, and verify that site control procedures are limiting or preventing the spread of contamination. Monitoring may be done with:

    1. Combination combustible gas indicators, designed to detect flammable vapors and gases, oxygen deficiency or excess, and/or toxic gases such as carbon monoxide or hydrogen sulfide.

    2. Combustible gas indicators, to detect flammable vapors and gases.

    3. Oxygen deficiency indicators, to detect oxygen deficiency or excess.

    4. Toxic gas monitors, for toxic gases such as hydrogen sulfide or carbon monoxide.

    5. Radiation meters.

6. Hazardous gas leak detectors, to detect the presence of low ppm concentrations of hydrogen, silane, and poison A gases such as arsine, phosphine, and diborane.

7. Survey instruments such as photoionization detectors and flame ionization detectors, to detect ppm-equivalent concentrations of organic vapors or gases.

8. Colorimetric detector tubes, to determine the presence and the concentration of specific types of air contaminants.

9. pH paper, to detect the presence of acids and bases.

10. Specific kits, to detect the presence of chlorinated hydrocarbons in liquid or soil.

D. Identify objectives, as follows:

1. Objectives must be based upon priorities identified according to the following criteria:

   a. The type and the magnitude of the life hazard involved.
   b. The type and the magnitude of the hazardous materials involved.

2. The objectives should focus on containment and/or control of the materials involved in a manner designed to prevent or to minimize unnecessary exposure of on-scene personnel as well as nearby receptors. Protection of uninvolved property and the environment should also be considered.

3. The SHIRT Supervisor shall identify all possible objectives to the IC. The IC, or the client if not the IC, shall approve the objectives of the SHIRT.

4. Objectives must be clearly understood and communicated to all levels of on-scene personnel. Close cooperation and coordination are essential. The SHIRT Supervisor must verify that all SHIRT personnel understand the objectives of any action undertaken.

E. Develop an Action Plan.

1. Upon identification of clearly defined objectives, the SHIRT Supervisor shall develop an Action Plan and present it for approval to the IC, and the client if not the IC.

2. The Action Plan should be centered around the following:

   a. Protection of life.
   b. Containment and control of the material and its effects on humans, property, and the environment.

3. The action plan must be understood by all personnel involved at the scene. The SHIRT Supervisor must verify that all involved SHIRT personnel understand the Action Plan before it is initiated.

F. Initiate and monitor progress of the Action Plan.

1. The SHIRT Supervisor shall monitor progress of the Action Plan to ensure that the objectives are accomplished.

2. The Action Plan may be modified to meet the requirements of the incident, and it must be sufficiently flexible to allow such modification.

3. The SHIRT Supervisor shall keep the IC, and the client if not the IC, informed regarding SHIRT activities and the progress of the Action Plan.

III. SAFETY

All SHIRT activities must always be undertaken with safety as a primary consideration. It is the duty of the SHIRT Supervisor and all SHIRT members to ensure that all activities are conducted according to the basic priorities of personal safety, safety of others, the environment, and property.

A. Basic Guidelines

The following basic guidelines should be observed:

1. Move people away from the incident, and keep them away.

2. Do not walk into, touch, or come in contact with known or suspected hazardous material.

3. Avoid inhalation of all gases, fumes, vapors, smoke, or mists at an incident, even if the material involved is not considered hazardous.

4. Wear the minimum protective clothing ensemble within the operational perimeter.

5. Never make an entry into an incident without backup; if a backup team is not ready or is unavailable, do not make the entry.

6. Never make an entry into an incident unless the need for decontamination has been assessed and addressed. If personnel decontamination is required, do not make the entry until the decontamination station is set up and operational.

7. Never make an entry into an incident unless the need for medical/ambulance backup has been assessed and addressed.

8. If the presence of an ambulance or additional medical personnel is required, do not make an entry until such equipment and personnel are available.

9. Do not make an entry unless all SHIRT members know and understand the planned activity, and their role.

10. Do not eat, drink, chew gum, or engage in any other hand-to-mouth activities in the Restricted Zone, Limited Access Zone, or Decontamination Area.

11. Do not smoke at any time when inside the SHIRT operational perimeter; this includes the Restricted Zone, Limited Access Zone, Contamination Reduction Corridor, and Decontamination Area, and any other area where there is a potential for contamination or contact with a known or suspected hazardous material.

12. Any SHIRT member who observes an unsafe or potentially unsafe act performed by another SHIRT member should report it to the Supervisor or the Safety Officer as soon as possible.

B. Safety Officer

The SHIRT Supervisor shall appoint a member to act as Safety Officer at the scene of an incident.

1. The Safety Officer will assist the SHIRT Supervisor in monitoring the progress of the Action Plan, and will report to the SHIRT Supervisor or his/her designee.

2. The Safety Officer is empowered to stop any action deemed unsafe or potentially unsafe for the health and safety of SHIRT members. The SHIRT Supervisor will be informed when activities are stopped because of health and safety concerns.

3. The Safety Officer may report to the SHIRT Supervisor or the IC any unsafe or potentially unsafe actions performed by other agency members.

4. The Safety Officer will ensure that protocols and procedures are followed as they pertain to the health, safety, and welfare of SHIRT members.

5. The Safety Officer will inform the SHIRT Supervisor of potential signs and symptoms of overexposure to the hazardous materials involved.

6. The Safety Officer will tell the SHIRT Supervisor when the SHIRT activities may be initiated, and/or if additional resources are required.

7. The Safety Officer shall keep a record or log of SHIRT activities as they relate to potential exposure, time on supplied air, and other actions that may affect health and safety.

C. Medical Officer

The Safety Officer shall appoint, as appropriate, a SHIRT member as Medical Officer.

1. The Medical Officer shall determine the potential signs and symptoms of overexposure to the hazardous materials involved.

2. The Medical Officer must verify that all involved SHIRT personnel understand the potential signs and symptoms of overexposure before SHIRT activities are initiated.

3. The Medical Officer shall inform the Safety Officer when SHIRT members have been cleared for activity.

4. The Medical Officer shall inform the Safety Officer if any SHIRT member displays signs or symptoms of overexposure. Members who have been exposed to a hazardous material or who are displaying signs or symptoms of overexposure shall receive immediate medical treatment.

5. The Medical Officer is empowered to remove any member from activity because of medical considerations. The Medical Officer shall inform the Safety Officer of any decisions to remove SHIRT members from activity.

6. The Medical Officer shall inform the Safety Officer if additional medical personnel or ambulances are required.

7. The Medical Officer shall inform responding ambulance or rescue personnel of the potential signs and symptoms of overexposure, and coordinate SHIRT activities with such personnel if so required.

8. The Medical Officer shall document all medical monitoring and decision-making activities of a medical nature.

9. Whenever possible, the Medical Officer should be a certified Emergency Medical Technician.

10. The Medical Officer may appoint additional SHIRT members to assist him/her, or utilize ambulance or rescue personnel, as necessary.

11. The Medical Officer will confer with the Decontamination Officer regarding the location of medical monitoring activities for decontaminated SHIRT personnel.

D. Decontamination Officer

The Safety Officer shall appoint, as appropriate, a SHIRT member as Decontamination Officer.

1. The Decontamination Officer shall determine the need for decontamination, and if required, the optimal procedure to utilize for the hazardous materials involved. These decisions will be made upon consultation with the Safety Officer.

2. The Decontamination Officer shall set up and make operational a decontamination station.

3. The Decontamination Officer may appoint additional SHIRT personnel to assist him/her. Fire department or facility personnel may be also utilized during initial setup.

4. The Decontamination Officer may request additional resources through the Safety Officer.

5. Requests for fire department, facility, or other personnel to assist with station setup must be made through the IC.

6. The Safety Officer will inform the Decontamination Officer regarding the following:

   a. Number of personnel who will require decontamination.
   b. Type and extent of contamination anticipated.

c. Personnel protective equipment required for personnel conducting decontamination activities.

d. Potential signs and symptoms of personnel overexposure to the hazardous materials involved.

e. Disposition of decontamination water and decontaminated equipment and clothing.

f. Termination of decontamination activities.

7. The Decontamination Officer shall confer with the Medical Officer regarding medical monitoring of decontaminated personnel.

8. The Decontamination Officer shall ensure that SHIRT members know and understand the decontamination station setup and procedures that will be utilized.

9. The Decontamination Officer shall report to the Safety Officer when the decontamination station is operational.

E. Equipment Coordinator

The Equipment Coordinator shall be responsible for maintenance of the response trailer and the equipment of the SHIRT.

1. The Equipment Coordinator shall ensure that all equipment is in good working order.

2. The Equipment Coordinator shall keep a record of all equipment used during a hazardous materials incident.

3. The Equipment Coordinator shall remain with the trailer and the equipment during a hazardous materials response.

4. The Equipment Coordinator shall ensure that all equipment is properly stowed upon termination of a response. The Equipment Coordinator shall document any equipment that has been damaged or lost, or requires replacement.

5. The Equipment Coordinator shall report any equipment deficiencies to the SHIRT Supervisor.

6. The Equipment Coordinator may designate other SHIRT members to assist him/her.

IV. SHIRT MEMBER RESPONSE

A. No SHIRT member shall respond to an incident unless officially requested to do so by the SHIRT.

B. SHIRT members shall respond to a location designated by the Supervisor.

1. SHIRT members shall report to the Supervisor or other designated personnel already at the scene.

2. SHIRT members shall document the time of arrival on a roster sheet.
3. SHIRT members shall inform the Supervisor or the Safety Officer when they leave the scene prior to termination.
4. SHIRT members shall document the time of departure on the SHIRT roster sheet.

V. COMMUNICATIONS

A. SHIRT Notification

1. SHIRT members are notified by individual pagers or telephone.
2. SHIRT notification codes are as follows:

   a. Code A: all SHIRT members contact on-call Supervisor.
   b. Code B: SHIRT on-call supervisor contacts on-call Project Manager.
   c. Code C: SHIRT Equipment Coordinator contacts on-call Project Manager.

B. Radio Communications

1. Radio communications between response vehicles and the SHIRT Supervisor shall be conducted on channel 1.
2. Radio communications between responders and the SHIRT Supervisor shall be conducted on channel 2.
3. Radio communications between responders shall be conducted on channel 3.
4. Radio communications between the SHIRT Supervisor and designated field supervisors shall be conducted on channel 4.
5. Periodic progress reports shall be transmitted by the SHIRT Supervisor to ABCCI by cellular telephone.

C. Communications at an HMI

1. The most accurate communication is person-to-person.
2. Radio communications must be clear and concise.
3. Cellular telephone communications may also be utilized.
4. Runners may be utilized if radio or cellular telephone communication is not possible, or radio frequencies are overcrowded. When runners are utilized, use written notes to augment verbal messages to ensure the accuracy of the information transmitted.
5. Communication at an incident must be two-way in nature. It is imperative that SHIRT members in charge of safety, medical assistance, decontamination, instrumentation, and references provide up-to-date information. Clear directions and coordination must flow down from the IC and the SHIRT Supervisor.

6. If communication of directions and coordination between elements of the SHIRT are unclear, and the lack of such communication jeopardizes the safety of SHIRT members, the SHIRT Supervisor or the Safety Officer shall direct that SHIRT activities cease until the problem is remedied.

D. Other Methods of Communication

1. One long blast of an air horn is the emergency signal for all SHIRT members to leave the Restricted Zone *immediately.*

2. The following hand signals shall be used when radios are not available:

   a. Hands gripping throat: out of air, can't breathe.
   b. Gripping partner's wrist: leave area immediately.
   c. Hands above head: I need assistance.
   d. Thumbs up: OK; I am all right; I understand.
   e. Thumbs down: no; not OK; I do not understand.

VI. PERSONAL PROTECTIVE CLOTHING

A. The basic minimum protective clothing ensemble shall be worn by all SHIRT members within the operational perimeter.

1. The basic ensemble shall consist of the following:

   a. Nomex or PBI-Kevlar coveralls or turnouts.
   b. Chemically resistant boots or boot covers.
   c. Hard hat.
   d. Safety glasses with side shields.

B. The minimum protective clothing ensemble may be altered to accommodate specific SHIRT activities or hazards at the incident. The Safety Officer, upon consultation with the SHIRT Supervisor and chemical compatibility guides, shall determine the appropriate types of chemical-protective clothing to utilize. The type of protective clothing used may be changed to accommodate changes in the Action Plan or the status of the incident.

1. Tyvek coveralls, oil-resistant boots or boot covers, undergloves, and nitrile overgloves must be worn at oil spill incidents when there is no danger of fire.

   a. Cotton or cotton and leather overgloves must not be worn unless worn over the chemically resistant glove ensemble defined above. Cotton or cotton and leather overgloves must be discarded after use.
   b. No SHIRT personnel shall participate in spill mitigation activities in leather shoes, sneakers, or other types of street footwear.

    c. Fire-fighting turnout gear cannot be easily decontaminated and should not be worn at oil spill incidents when there is no danger of fire.

2. At incidents where there is an actual or a potential threat of fire or explosion, full fire-fighting turnout gear must be worn within the Restricted and Limited Access Zones, including:

    a. Bunker pants, coat, and boots, and
    b. Fire-fighting helmet with face shield, and
    c. Fire-fighting gloves, and
    d. Nomex or PBI-Kevlar inner gloves and hood.

3. At incidents where there is an actual or a potential threat of chemical exposure, chemically resistant clothing must be worn within the Restricted and Limited Access Zones (a–f worn at all times):

    a. Inner Nomex or PBI-Kevlar coveralls, and
    b. Chemically resistant coveralls with hood, and
    c. Chemically resistant boots or overboots, and
    d. Chemically resistant gloves, and
    e. Latex or vinyl inner gloves, and
    f. Hard hat with faceshield.
    g. Thin cotton undergloves, which may be worn under inner gloves in hot weather.
    h. Cotton undergloves, which may be worn under outer gloves in cold weather.
    i. Gauntlet gloves, which may be worn over outer gloves to provide additional hand and arm protection.
    j. Cotton work gloves, which may be worn over chemically resistant outer gloves to provide additional hand protection or improve dexterity.
Note that cotton or cotton and leather overgloves must be discarded after use.

4. At incidents where there is an actual or a potential threat of chemical exposure *and* fire, chemically resistant clothing with fire-fighting turnout gear shall be worn within the Restricted and Limited Access Zones, including:

    a. Fire-fighting bunker pants, jacket, and boots, and
    b. Chemically resistant overboots, and
    c. Chemically resistant outer coveralls with hood, and
    d. Nomex or PBI-Kevlar hood and inner gloves, and
    e. Vinyl or latex inner gloves, and
    f. Chemically resistant outer gloves, and
    g. Fire-fighting helmet with face shield.

h. Fire-fighting gloves, which may be added to the ensemble at the discretion of the Safety Officer. Fire-fighting gloves shall be used whenever fire is present, or hot surfaces are anticipated.

5. At incidents where the risk of chemical exposure is extreme, a fully encapsulated, vaportight protective clothing ensemble shall be worn within the Restricted Zone:

   a. Nomex or PBI-Kevlar inner coveralls, and
   b. Latex or nitrile undergloves, and
   c. Chemically resistant, vaportight protective outer garment, including chemically resistant outer gloves and boots, and
   d. Inner hard hat, if applicable.
   e. Thin cotton undergloves, which may be worn under inner gloves in hot weather.
   f. Cotton undergloves, which may be worn under inner gloves in cold weather.
   g. Gauntlet gloves, Silver Shield™ gloves, or another type of disposable, chemically resistant glove, which may be worn over outer gloves to provide additional hand protection.
   h. Cotton work gloves, which may be worn over outer gloves to provide additional hand protection or improve dexterity; cotton work gloves must be discarded after use.

6. At incidents where the risk of chemical exposure is extreme, and there exists a real or a potential threat from flashover and/or fire, a fully encapsulated, vaportight protective clothing ensemble with vaportight flashover protection shall be worn within the Restricted Zone, including:

   a. Nomex or PBI-Kevlar coveralls, and
   b. Nomex or PBI-Kevlar hood and inner gloves, and
   c. Vinyl or latex inner gloves, and
   d. A chemically resistant, vaportight protective outer garment, including chemically resistant outer gloves and boots, and
   e. A flashover-resistant protective outer garment, including outer gloves and boot covers, and
   f. An inner hard hat, if applicable.

7. At incidents involving unknown or uncharacterized hazardous materials, where there is evidence of a potential fire or explosion threat (based on previous events or instrument readings), chemically resistant clothing with fire-fighting turnout gear shall be worn within the Restricted and Limited Access Zones, including:

   a. Fire-fighting bunker pants, jacket, and boots, and
   b. Chemically resistant overboots, and

    c. Chemically resistant coveralls with hood, and

    d. Nomex or PBI-Kevlar hood and inner gloves, and

    e. Vinyl or latex inner gloves, and

    f. Chemically resistant outer gloves, and

    g. A fire-fighting helmet with face shield.

    h. Fire-fighting gloves, which be added to the ensemble at the discretion of the Safety Officer. Fire-fighting gloves shall be used whenever fire is present, or hot surfaces are anticipated.

8. At incidents involving unknown or uncharacterized hazardous materials, where there is no evidence of a potential fire or explosion threat (based on instrument readings), chemically resistant clothing shall be worn within the Restricted and Limited Access Zones (a–f worn at all times):

    a. Nomex or PBI-Kevlar coverall, and

    b. Chemically resistant coveralls with hood, and

    c. Chemically resistant boots or overboots, and

    d. Chemically resistant gloves, and

    e. Latex or vinyl inner gloves, and

    f. A hard hat with faceshield.

    g. Thin cotton undergloves, which may be worn under inner gloves in hot weather.

    h. Cotton undergloves, which may be worn under outer gloves in cold weather.

    i. Gauntlet gloves, which may be worn over outer gloves to provide additional hand and arm protection.

    j. Cotton work gloves, which may be worn over outer gloves to provide additional hand protection or improve dexterity.
Note that cotton or cotton and leather overgloves must be discarded after use.

9. SHIRT members reporting to the SHIRT staging area within the base area shall, whenever possible, wear company coveralls and a baseball cap with logo. These articles of clothing are useful in facilitating identification of the SHIRT members on scene.

C. The SHIRT has the following types of protective clothing available for use:

1. Disposable Saran-coated Tyvek coverall with hood.

2. Reusable PVC coverall with hood.

3. Disposable Chemrel coverall with hood.

4. Fully encapsulated, vaportight Fyrepel butyl rubber ensemble with chemically resistant boots and gloves.

5. Fully encapsulated, vaportight MSA Viton/neoprene ensemble with chemically resistant boots and gloves.

6. Fully encapsulated, vaportight Chemrel Max GT ensemble with chemically resistant boots and gloves, and Flash Max radiant heat protection.

7. Aluminized Nomex or PBI-Kevlar Fyrepel flashover ensemble with gloves and boot covers.

8. Neoprene oversleeves and aprons.

9. Fire-fighting turnout coats, boots, and helmets.

10. Nomex or PBI-Kevlar gloves and hoods.

11. Hard hats with face shields.

12. Chemically resistant gloves:

   a. Nitrile.
   b. Viton.
   c. Butyl.
   d. Silver Shield™.
   e. Neoprene.

13. Nonchemically resistant gloves:

   a. Latex or vinyl surgical (inner) gloves.
   b. Cotton undergloves.
   c. Cotton work gloves.
   d. Fire-fighting gloves.

14. Chemically resistant footwear:

   a. Butyl overboots.
   b. Neoprene overboots.
   c. PVC/urethane blend safety boots.
   d. Neoprene safety boots.

D. Selection of the most appropriate type of chemically resistant protective material shall be based upon manufacturer-supplied chemical resistance charts and other chemical-protective clothing selection guides.

VII. RESPIRATORY PROTECTION

A. The SHIRT Supervisor, upon consultation with the Safety officer, shall determine the type of respiratory protection required within the Restricted Zone and the Limited Access Zone.

1. At incidents where there is an actual or a potential threat from fire or explosion, full fire-fighting turnout gear must be worn within the Restricted Zone and the Limited Access Zone, and:

    a. A positive pressure self-contained breathing apparatus (SCBA) or an air-line with egress bottle shall be utilized in the Restricted Zone.

    b. SCBA shall be used, or donned and ready for use, in the the Limited Access Zone.

2. At incidents where there is an actual or a potential threat from chemical exposure to high concentrations of known hazardous materials, or exposure to unknown or uncharacterized materials:

    a. A positive pressure self-contained breathing apparatus (SCBA) or an air-line with egress bottle shall be utilized in the Restricted Zone.

    b. SCBA shall be used, or donned and ready for use, in the Limited Access Zone.

3. At incidents where there is an actual or a potential threat from chemical exposure *and* fire:

    a. A positive pressure self-contained breathing apparatus (SCBA) or an air-line with egress bottle shall be utilized in the Restricted Zone.

    b. SCBA shall be used, or donned and ready for use, in the Limited Access Zone.

4. At incidents where a fully encapsulated, vaportight protective clothing ensemble is required:

    a. A positive pressure self-contained breathing apparatus (SCBA) or an air-line with egress bottle shall be used.

5. At incidents where the oxygen concentration in the workzone is less than 19.5%:

    a. A positive pressure self-contained breathing apparatus (SCBA) or an air-line with egress bottle shall be utilized.

6. At incidents where there is a real or a potential threat of low-level radioactivity in the form of alpha and beta particles *and* chemical exposure:

    a. A positive pressure self-contained breathing apparatus (SCBA) or an air-line with egress bottle shall be utilized in the Restricted Zone.

    b. SCBA shall be used, or donned and ready for use, in the Limited Access Zone.

7. At incidents where there is a real or a potential threat of low-level radioactivity in the form of alpha and beta particles and *no* threat from chemical exposure:

   a. A positive pressure self-contained breathing apparatus (SCBA) or airline with egress bottle shall be used in the Restricted Zone, or

   b. An air-purifying full-faced respirator equipped with HEPA cartridges and escape bottle shall be utilized in the Restricted Zone.

   c. SCBA or APR with HEPA cartridges and escape bottle shall be used, or donned and ready for use, in the Limited Access Zone.

8. At incidents where there are low concentrations of known materials that can be quantified by using instrumentation:

   a. A positive pressure self-contained breathing apparatus (SCBA) or an air-line with egress bottle shall be used in the Restricted Zone, or

   b. An air-purifying full-faced respirator (APR) equipped with organic vapor/acid gas cartridges (OV/AG) and an escape bottle shall be utilized in the Restricted Zone.

   c. SCBA or APR with OV/AG cartridges and escape bottle shall be used, or donned and ready for use, in the Limited Access Zone.

9. Restrictions on the use of APRs are as follows:

   a. APRs shall be used only by personnel who have successfully completed the ABCCI Respiratory Protection Employee Training Program.

   b. APRs shall be used only by personnel who have demonstrated a proper fit test within the previous six months.

   c. Personnel may only use the specific make(s), model(s), and size(s) of APRs for which a satisfactory fit test has been demonstrated with the previous six months.

   d. APRs shall be utilized only when:

     (1) The identity and the concentration of the contaminant are known.
     (2) Oxygen levels are greater than or equal to 19.5%.
     (3) The contaminant has adequate warning properties.
     (4) Approved cartridges are available.
     (5) The contaminant concentration does not exceed the capacity of the cartridge.
     (6) The contaminant concentration does not exceed the IDLH.

   e. APRs shall be full-face; half-face APRs shall not be used.

   f. APRs shall be utilized only if there is an approved cartridge for the chemical hazards present.

   g. APRs shall *not* be utilized when:

     (1) The concentration of oxygen in the workzone is less than 19.5%; or
     (2) The contaminant has inadequate warning properties; or
     (3) The contaminant concentration present exceeds the capacity of the cartridge; or

    (4) The contaminant concentration present exceeds 50 times the PEL or the TWA, for those materials that have a recognized OSHA or ACGIH exposure limit; or

    (5) The contaminant concentration present exceeds the IDLH for those materials that have an assigned IDLH; or

    (6) There is no approved cartridge for the hazard present.

  h. Cartridges shall be promptly discarded and replaced when:

    (1) The user detects a *warning property,* such as odor, eye, or nasal irritation; or

    (2) The user experiences *difficulty in breathing* or shortness of breath, or inhaled air becomes hot; or

    (3) The *cartridges* appear *defective or damaged.*

  i. At termination, all used cartridges shall be discarded.

10. The qualitative fit-test procedure followed by the SHIRT is defined in the ABCCI Respiratory Protection Program.

11. All male SHIRT members who utilize respiratory protection must be clean-shaven within 24 hours.

12. All SHIRT members who utilize respiratory protection or other forms of personal protective equipment shall be examined by an occupational physician and certified as medically competent to use respiratory protection equipment.

  a. The physician shall be given a copy of the pertinent sections of the OSHA rule 29 CFR 1910.120, as well as a copy of Chapter 5 of the *Occupational Safety and Health Guidance Manual for Hazardous Waste Site Activities,* NIOSH Publication No. 85–115.

  b. SHIRT members deemed medically unfit for duty shall not be permitted entry into the Restricted Zone or the Limited Access Zone.

  c. SHIRT members deemed medically unfit for duty shall not be permitted to use respiratory protection.

13. All SHIRT personnel who utilize respiratory protection equipment shall participate in the ABCCI Medical Monitoring Program.

VIII. DECONTAMINATION

A. The objective of decontamination is to remove contaminants from SHIRT and/or other affected personnel, their clothing, and equipment; and to prevent the spread of contamination.

B. The decontamination area shall provide a corridor between the Limited Access Zone and the noncontaminated portion within the operational perimeter.

  1. Whenever possible, the decontamination area should be positioned:

    a. In an area accessible to the Restricted Zone.
    b. Close to a hydrant or another usable water supply.
    c. Downwind from the command post, SHIRT command, and staging areas.
    d. Upwind from the actual incident.
    e. Away from environmentally sensitive areas.

  2. The final portion of the decontamination area should be adjacent to the medical monitoring and rehabilitation area.

C. Decontamination procedures to be utilized shall be determined by the Decontamination Officer upon consultation with the Safety Officer. Methods that are available include:

  1. Dry decontamination involving careful removal and bagging of clothing and equipment, followed by disposal and/or other decontamination procedures at a later time.
  2. Washing and dilution to flush contaminants from clothing and equipment.
  3. Absorption of gross liquid contamination from surfaces before initiation of more thorough decontamination procedures.
  4. Brushing or other dry procedures to remove solid or particulate contamination from surfaces before initiation of more thorough decontamination procedures.
  5. Chemical degradation, which alters the chemical structure of the contamination, rendering it less harmful or more susceptible to other decontamination procedures.

D. Respiratory protection and protective clothing to be utilized by decontamination personnel shall be determined by the Decontamination Officer upon consultation with the Safety Officer.
E. All involved SHIRT personnel shall become familiar with the decontamination area and procedures prior to entering the Exclusion Zone.

  1. The Decontamination Officer shall inform the Safety Officer when the decontamination station is operational and SHIRT members are cleared for entry.
  2. SHIRT members shall report for medical monitoring and rehabilitation after completing decontamination.

F. Emergency decontamination may be performed on SHIRT or other affected personnel who may require immediate medical treatment and transport, or whose health may be endangered by the contaminants.

1. Contaminated persons will remove contaminated clothing and shower or wash using equipment available. Contaminated clothing, equipment, and personal articles shall be bagged for subsequent disposition.
2. Decontaminated persons may don uncoated Tyvek coveralls, or
3. If immediate medical transport is required, the decontaminated person may be wrapped in a clean sheet and transported.
4. The Decontamination Officer shall report to the Safety Officer if emergency decontamination procedures have been implemented as well as the disposition of the affected personnel.

G. Liquid or other by-products of decontamination shall be collected or controlled for subsequent disposal.

1. The SHIRT Supervisor, in consultation with the Safety Officer and the Decontamination Officer, and representatives of other agencies as appropriate, shall determine if such collection or control is required.
2. The SHIRT Supervisor, in consultation with the Safety Officer and Decontamination Officer, and representatives of other agencies as appropriate, shall determine the final disposition of materials generated or collected during the decontamination process.

H. Decontamination equipment available for use includes:

1. Plastic sheeting (4–5 mil).
2. Galvanized tubs.
3. Plastic pools (4'–5' diameter).
4. Plastic buckets.
5. Absorbent materials.
6. Applicators and sprayers.
7. Garden and fire hoses.
8. Hydrant adaptors.
9. Long- and short-handled scrub brushes.
10. Decontamination agents:

    a. Liquid detergent (Tide).
    b. Chlorine bleach.
    c. Lysol disinfectant.
    d. TSP.
    e. Acid/alkali-neutralizing solutions.

11. Plastic bags.
12. Overpack drums (various sizes).
13. Drum liners.
14. Transfer pumps for aqueous liquids.
15. Portable eye washes (5-gallon).

I. Diagrams for representative minimum decontamination stations may be found in Attachment 3 (at the end of this appendix).

IX. MEDICAL MONITORING AND REHABILITATION

A. The objective of medical monitoring is to rapidly assess the medical fitness of involved SHIRT members before and after site entry or the use of personal protective equipment, and to determine if any SHIRT members are displaying signs or symptoms of overexposure.

1. Medical monitoring shall be conducted prior to *and* after site entry or the use of personal protective equipment, and shall include:

    a. Heart rate; the pulse is taken for 30 seconds, then multiplied by 2.
    b. Blood pressure by auscultation; diastolic and systolic pressures must be recorded, and palpation is not acceptable.
    c. Skin temperature/appearance or
    d. Oral temperature, measured *before* drinking of any liquids.

2. Workers shall be considered medically unfit for duty if *any one* of the following is present:

    a. Pulse rate greater than 120 beats per minute.
    b. Irregular pulse rate.
    c. Blood pressure greater than 140/90.
    d. Evidence of extreme exertion, or heat or cold stress based upon skin temperature or appearance.
    e. Evidence of overexposure to the hazardous material of concern.

3. SHIRT members initially considered unfit for duty may be reexamined after a minimum rest interval of 10 minutes.

    a. If heat stress is suspected, the oral temperature should be determined. Personnel shall be considered medically unfit if the oral temperature is $\geq$ 100.6°F.
    b. Any SHIRT member displaying signs or symptoms of heat stress, cold

stress, overexertion, physical injury, or overexposure to a hazardous material shall receive immediate symptomatic medical treatment and be transported to a medical facility, as necessary.

c. Symptomatic medical treatment may include:

(1) Administration of fluids.
(2) Removal from the hot or cold environment, as appropriate.
(3) Administration of oxygen.
(4) Administration of hot or cold packs, as appropriate.
(5) Physical injury emergency field treatment, such as splinting, bandaging, or immobilization.

B. Transportation for additional medical treatment shall be provided by the local authority having jurisdiction. Such transportation, in the form of an ambulance or a rescue squad vehicle, *must* be at the scene for all Level III incidents that require the use of respiratory and other forms of personal protective equipment.

1. Responding ambulance or rescue personnel shall be informed of the potential signs and symptoms of overexposure and the recommended treatment, as appropriate.
2. SHIRT members shall *not* initiate any activity that involves the use of respiratory and personal protective equipment until:

a. Medical transportation is available.
b. SHIRT members know and understand the potential signs and symptoms of overexposure.

C. The medical monitoring area shall be positioned at the end of or adjacent to the end of the decontamination area.
D. The purpose of rehabilitation is to facilitate the rest and the recuperation of SHIRT members who have used personal protective equipment or otherwise exerted themselves at the scene of a SHIRT response.

1. SHIRT members who have completed medical monitoring shall remain in the rehabilitation area until rested.
2. Drinking of fluids is encouraged during rehabilitation.
3. Personal protective equipment should be completely removed during rehabilitation.
4. SHIRT members are responsible for overseeing their own rehabilitation. After rehabilitation, the member shall report to the Supervisor as available for assignment.

E.   The rehabilitation area shall be adjacent to the medical monitoring area.

F.   Equipment available for medical monitoring and first aid is as follows:

1. Medical monitoring:

a. Blood pressure cuff and stethoscope.
b. Oral thermometers with disposable shields.

2. First aid:

a. Oxygen delivery system with disposable masks.
b. Spare oxygen cylinders.
c. Bag valve mask.
d. Hot and cold packs.
e. Assorted wound dressings.
f. Assorted bandaging material and tape.
g. Anesthetic disinfectant solutions.
h. Various-size water gel dressings.
i. Portable eye washes.
j. Replacement eye wash solutions.

X.   INSTRUMENTATION

A.   The purpose of air monitoring instrumentation is to determine the presence and the relative concentration of airborne contaminants that may pose a hazard to human health or the environment.

1. Potential hazards may include toxic, flammable, and reactive materials, oxygen deficiency or excess, and radioactive materials.
2. Contaminated atmospheres may pose more than one hazard.

B.   A combination combustible gas %LEL/oxygen deficiency meter shall be used during:

1. Entry into any confined or semiconfined space.
2. Any response involving a spill or a release of a flammable material.
3. Any response involving a report of fire or an explosion.
4. Any response in a structure containing unknown or uncharacterized hazardous materials;

C.   A ppm combustible gas meter shall be used as indicated above in Section X.B.1 when:

1. The combustible gas indicator gives no LEL reading or a reading of less than 4% LEL, and

2. The oxygen concentration present in the tested atmosphere is sufficient for the meter to respond accurately.

D. Limitations and use considerations for combustible gas indicators are as follows:

1. Combustible gas readings are not accurate below an oxygen concentration specified by the manufacturer; unless otherwise noted by the manufacturer, there is usually insufficient oxygen for a combustible gas reading at oxygen concentrations of 10% or less.
2. Inhibitor filters shall be used, if available, to protect the sensor when the presence of leaded gasoline vapors is suspected.
3. A charcoal filter may be used, if available, to differentiate between heavy hydrocarbons (such as gasoline or petroleum hydrocarbons) and lighter hydrocarbon gases (such as methane).
4. A water trap shall be used, if available, to prevent inadvertent entry of liquids into the instrument.
5. A particulate filter shall be used, if available, to prevent inadvertent entry of dust and particulates into the instrument.
5. Combustible gas indicators shall not be used in corrosive environments. Wetted pH paper or detector tubes should be used to detect the presence of acid or alkaline vapors.

E. Ionization detectors are useful to detect low concentrations of organic contaminants. A flame ionization detector (FID) and a photoionization detector (PID) should be utilized when:

1. The combustible gas %LEL and/or the combustible gas ppm indicator gives no reading or a negligible reading.
2. The known or suspected organic contaminant is not expected to elicit a response from a combustible gas indicator (i.e., it is noncombustible).
3. Work-zone monitoring is required in order to maintain the current level of respiratory protection.

F. Limitations and use considerations when operating FIDs or PIDs are as follows:

1. PIDs use UV light to elicit a response. Limiting factors that affect the transmission of light and therefore the efficiency of the instrument include:

   a. Water vapor.
   b. Particulates or dust.
   c. Chemicals with ionization potentials greater than the eV of the UV lamp.

2. FIDs burn chemical contaminants in a hydrogen flame. Limiting factors that affect the flame include:

   a. An insufficient oxygen supply, which results in flameout or low flame height.
   b. The presence of flammable gases, which results in an oversized flame and subsequent flameout.

G. Colorimetric detector tubes may be useful in some instances.

   1. Detector tubes may be useful if a tube is available for a specific known or suspected chemical, and the actual approximate concentration can be determined. Such tubes are available, for example, for ammonia, halogenated hydrocarbons, alcohols, and petroleum hydrocarbons.
   2. Detector tubes may be useful to rule out the presence of a known or suspected contaminant, within the concentration limitations of the tube.
   3. A list of current detector tubes shall be maintained as part of the SHIRT inventory.
   4. The instructions for each tube shall be read and understood prior to its use. The detection range, temperature, humidity, and cross-specificity limitations should also be understood.
   5. Detector tubes respond to a variety of materials in addition to the specific chemical(s) for which they are calibrated. A positive result does not automatically indicate the presence of the chemical for which a tube was calibrated.

H. Limitations and use considerations for detector tubes are as follows:

   1. Tubes are sensitive to heat and light and should not be left in heat or direct sunlight for a prolonged period of time.
   2. Tubes shall not be reused unless the instructions clearly indicate that it is appropriate for them to be so used.
   3. Tubes shall be used only with the pump recommended by the manufacturer.
   4. Tubes shall immediately be removed from service when:

      a. They appear damaged, broken, or discolored.
      b. The expiration date, which is clearly printed on each box, is reached.
      c. They are subjected to extreme temperatures.

I. The Matheson Hazardous Gas Leak Detector is sensitive to concentrations of less than 1 ppm of specific inorganic gases that are used in the semiconductor industry. The leak detector shall be used when:

1. The response involves a suspected or an actual release of a specific gas or gases, including:

    a. Arsine (AsH_3)
    b. Phosphine (PH_3)
    c. Silane (SiH_4)
    d. Diborane (B_2H_4)
    e. Stibine (SbH_3)
    f. Germane (GeH_4)
    g. Hydrogen (H_2)

2. The response involves actual or potentially leaking compressed gas cylinders of unknown gas or gases.
3. The response involves a facility or a location where semiconductor process gases may be used, stored, or transported.

    Note that the leak detector responds to many organic gases, including acetone, benzene, toluene, ethanol, ethylene, and isopropyl alcohol.

J. Radiation detection instruments can detect and differentiate between beta and gamma radiation. Radiation detectors shall be used when:

1. The response is known to involve radiation.
2. The response involves unknown types of actual or suspected hazardous materials, including materials found in but not limited to the following:

    a. Abandoned laboratories or facilities containing mixed chemical storage or waste.
    b. Illegal laboratories or processing facilities.
    c. Suspicious packages.
    d. Locations or vehicles suspected of containing radiation.
    e. Municipal or private landfills.

3. The type of radiation emitted shall be determined, if possible, as well as the relative rate of emission measured in milliroentgens per hour (mR/hr).
4. Even when the meter gives no increased reading, low-level radiation may be present on the surfaces of objects. Always use gloves when checking for radiation, and discard them promptly after leaving the suspect area.

K. Chemical-specific monitoring instrumentation is available to measure ppm concentrations of toxic gases or vapors. These instruments include:

1. A mercury vapor analyzer.

    2. A carbon monoxide detector.

    3. A hydrogen sulfide detector.

L. pH paper shall be used to detect the presence of corrosive gases and vapors, and to determine the pH of liquids and solids.

    1. Water-wetted pH paper may be used in an unknown atmosphere to determine its relative pH.

    2. Water-wetted pH paper may be used to determine the relative pH of solid materials.
       Note that detector tubes also are available that can detect the presence of acid or alkaline vapors or gases. (See Section X.G. above.)

M. Detection kits may be used to detect the presence of chlorinated hydrocarbons in liquid or soil.

    1. These kits include: Chlor-N-Soil, Chlor-N-Oil, and Chlor-D-Tect.

    2. These kits may be used when:

        a. The material, or soil contaminated by the material, is suspected of containing chlorinated polybiphenyls, or other chlorinated hydrocarbons; or

        b. There is a need to demonstrate the lack of such contamination to the local authority having jurisdiction.

    3. Kit instructions must be read and understood prior to kit use. Latex gloves, and safety glasses or a face shield, must be worn; if gloves are contaminated with the suspect material, they shall be promptly changed before continuing.

N. Action levels may be defined as instrument response values that trigger specific actions.

    1. The following actions shall occur when an action level has been reached:

        a. The Safety Officer shall be notified, and

        b. All involved SHIRT members shall be notified, and

        c. The Safety Officer shall determine if a new action level is warranted, and/or

        d. The Safety Officer shall determine if personal protective equipment and safety precautions should be upgraded, and/or

        e. The Safety Officer shall determine if a portion of the affected area should be placed off-limits, or

        f. The Safety Officer shall determine if SHIRT members should leave the affected area.

2. The Safety Officer shall notify the SHIRT Supervisor of when an action level is reached, as well as the SHIRT response to the action level.

3. Action levels other than those defined below (in items 4–8) may be utilized as appropriate, and may be designated by the SHIRT Supervisor:

   a. Deviations from action levels defined below must be documented in writing by the SHIRT Supervisor; and

   b. The Safety Officer shall ensure that all involved SHIRT members know and understand these action levels.

4. Combustible gas %LEL/oxygen deficiency action levels are as follows:

   a. 10% LEL meter reading in a confined space.

   b. 20% LEL meter reading in excavations that may be occupied.

   c. 25% LEL meter reading for unknown gases and vapors in a work zone that is not a confined space; or

   d. 50% actual LEL of a known material in a work zone, based upon a response curve or response factor for the known contaminant; or

   e. 25% LEL meter reading for a known material in a work zone for which there is no response curve or response factor.

   f. 25% LEL meter reading for multiple known gases or vapors in a work zone.

   g. An increase or a decrease in oxygen concentration of 1% or more.

5. Radiation detector action levels are dealt with as follows:

   a. The action level is an increase in beta or gamma radiation of 1.0 mR/hr above normal background in the work zone. Background radiation levels may vary from 0.01 to 0.05 mR/hr; geographic location, proximity to radiation sources, and natural geologic formations can affect background radiation levels.

   b. SHIRT members shall attempt to define the area affected, and, if possible, shall indicate the source of the radiation and the distance from the source where radiation levels decrease to background or below the action level.

6. The Matheson Hazardous Gas Leak Detector action level is:

   a. A rapid alarm in any situation where other instrumentation fails to elicit a response.

7. Chemical-specific monitoring instrumentation action levels are dependent upon the OSHA or ACGIH TWA of the chemical of concern:

   a. Mercury vapor: 0.05 mg/m^3.

b. Hydrogen sulfide: 10 ppm.

c. Carbon monoxide: 35 ppm.

8. Ionization detector (FID, PID) action levels may be used as follows:

a. Action levels for ionization detectors must reflect the conditions encountered on-site as well the type of chemical hazards present, if known.

b. Monitoring for ppm-equivalent concentrations should be conducted only after other potentially dangerous conditions have been addressed and/or ruled out, such as flammable atmospheres, oxygen deficiency, ionizing radiation, and highly toxic inorganic vapors or gases.

c. In the absence of other monitoring instrument readings, or other indications of the presence of chemical hazards, the following PID or FID readings shall apply in dealing with unknowns:

(1) 0.2 to 5 units above background from an unknown chemical contaminant shall be the action level for donning respiratory protection (minimum Level C).

(2) Readings greater than 5 units above background from an unknown chemical contaminant shall be the action level for upgrading respiratory protection from Level C to Level B.

d. In dealing with a known chemical contaminant, ionization detectors should not be used to determine IDLH or exposure limit concentrations unless:

(1) It is known that only one contaminant is present, which the meter can detect; and

(2) The efficiency of the meter in detecting the contaminant is known; and

(3) There is a method available to verify the relative accuracy of the meter in detecting the contaminant that is present, such as a detector tube.

e. The absence of a meter response should be verified with other instruments before the area is declared to be relatively free of contamination.

9. In dealing with known materials, whenever possible, exposure limits and IDLH values should be determined by using:

a. Colorimetric detector tubes, and/or

b. Combustible gas indicators with response curves or factors for the known contaminant, and/or

c. Chemical-specific instrumentation, as appropriate.

d. Ionization detectors, as appropriate.

O.   Instrument checkout and use shall follow written procedures. Checkout procedures shall be documented in writing, using a checkoff format.

P.   Instruments shall be stored and maintained according to manufacturers' recommendations. The following procedures shall be performed on a monthly basis:

   1.   Rechargeable battery-operated instruments shall be run to exhaustion for at least 12 hours, and then recharged for at least 16 hours; and

   2.   After recharging, instrument calibration shall be checked by using a known concentration of calibrant gas or a check source.

   3.   Disposable battery-operated instruments shall be checked for battery life; and

   4.   Depleted batteries shall be replaced, as necessary; and

   5.   Instrument calibration shall be checked by using a known concentration of calibrant gas or a check source.

   6.   All instrument maintenance and calibration shall be documented in writing.

Q.   Additional considerations in using instrumentation are as follows:

   1.   Workers must use instruments that are appropriate for the situation.

   2.   The absence of a meter reading does not automatically mean that there is no contamination present.

   3.   Never assume that only one hazard is present.

   4.   Use one instrument to confirm another.

   5.   Interpret readings in terms of safety *and* health.

   6.   Do not use any instrument that has insufficient battery power.

XI.   TERMINATION

Termination may be defined as the process involved in ending the SHIRT emergency response to a hazardous materials incident.

A.   The SHIRT Supervisor shall determine, upon consultation with the IC, and the client if not the IC, when termination may begin.

   1.   The Supervisor shall inform the Safety Officer that termination may be initiated.

   2.   The Safety Officer shall ensure that all SHIRT members are notified that the emergency or remediation phase of the incident has ended, and that demobilization activities have been initiated.

   3.   The Decontamination Officer and the Medical Officer shall be responsible for demobilization of their respective areas.

4. The Equipment Coordinator shall ensure that all equipment is properly stowed. The Equipment Coordinator shall document all equipment that has been used during the incident, *and* any equipment that has been damaged or lost or requires replacement.

5. The Safety Officer and the Medical Officer shall finalize all documentation pertaining to entry, actual or suspected exposures, medical monitoring, or other information that is related to the health and safety of SHIRT members.

6. The Supervisor shall finalize all documentation relating to SHIRT response and operations. The SHIRT Supervisor shall collect and examine for completeness all other documentation, including:

    a. Personnel sign-in or manning sheets.
    b. Equipment use logs.
    c. Medical monitoring logs.
    d. Information collected from references.
    e. Instrument checkoff lists and response logs.

B. SHIRT members may be requested to return to ABCCI facilities to assist the Equipment Coordinator in replacing used equipment or disposables, cleaning of equipment and apparatus, filling used SCBA bottles, and any other activities required to make response equipment fully operational.

C. Debriefing shall occur when all pertinent documentation is collected and reviewed by the SHIRT membership. Errors and omissions shall be corrected during debriefing.

D. A post-incident critique shall be conducted with all members present. The critique shall include an analysis of the response from the point of view of the SHIRT Supervisor, Safety Officer, Sector Officers, Equipment Coordinator, and SHIRT members. Any changes to SOPs, or other suggestions, shall be documented.

E. Debriefing and critique activities shall be conducted as soon as possible after the end of an incident at the offices of Anything for a Buck Consultants and Contractors, Inc., or another suitable location.

F. The forms in Attachment 4 (below) shall be used to document activities throughout a SHIRT response.

XII. SHIRT RESPONDER CERTIFICATION

A. Personnel shall be certified to OSHA/NFPA response levels:

1. First Responder, Operations.
2. Hazardous Materials Technician.

B. Training for responder certification shall consist of:

1. Formal classroom and field evolutions.
2. Assigned reading and workbook completion.
3. Classroom discussion of workbook answers.

C. Response level competency shall be demonstrated by:

1. Achieving a passing grade (75%) on a written, closed-book examination; and
2. Successfully demonstrating hands-on competencies as defined in NFPA 472.

XIII. RECORDKEEPING

A. Documents and records pertaining to the following shall be maintained by the ABCCI Corporate Health and the Safety Manager, or his/her designee:

1. Annual physical examination and other medically related information.
2. Attendance at training sessions and field exercises.
3. Training and competency testing.
4. Respiratory fit-test results.
5. Potential exposures at hazardous materials incidents.
6. Incident response reports.
7. Response vehicle and equipment inventories.
8. SHIRT Standard Operating Procedures and Practices.
9. Site Emergency Response Plan.

XIV. DEFINITIONS

Unless otherwise noted, all definitions of terms are contained in the *MSDS Pocket Dictionary* (J. O. Accrocco, ed., Schenectady, NY: Genium Publishing Corporation, 1990). All SHIRT members have received a copy of the pocket dictionary; a copy is also available in the response trailer.

Other terms with specific SHIRT definitions are listed below:

SHIRT Supervisor: the individual responsible for coordinating the activities of the SHIRT.

Operational perimeter: the area within which SHIRT activities are conducted. The area includes the sector command area, the Restricted Zone, the Limited Access Zone, and the decontamination area, as well as areas where medical monitoring, rehabilitation, and dress-out are conducted. When the SHIRT is operational, the area around the response trailer shall be considered to be within the operational perimeter.

ATTACHMENTS

1. SHIRT APR Fit-Test Procedure
2. Instrument Procedure Summaries and Checklists
3. Representative Decontamination Stations
4. SHIRT Checklists and Forms

ATTACHMENT 1: SHIRT APR FIT-TEST PROCEDURE*

1. Three 500 ml jars are prepared, one jar containing 0.2 ml of isoamyl acetate (banana oil) in 250 ml distilled or odor-free water and the other two jars containing only 250 ml water. You, the testee, will select which jar contains isoamyl acetate, to ensure that you can actually detect isoamyl acetate. Proceed with the test *only* if you can detect the odor.

2. Select an APR, and ensure that it has organic vapor/particulate cartridges.

3. Don the respirator, and adjust it until it is comfortable yet snug, using a mirror to make sure the mask is on properly and that the straps are lying flat against your head.

4. Perform positive and negative pressure checks; if the mask leaks, readjust it and try again. If the mask is uncomfortable, select another and start over.

5. Enter the fit-test chamber containing isoamyl acetate (banana oil). Inside the chamber, do the following:

   - Breathe normally for about 10 seconds.
   - Breathe deeply for about 10 seconds.
   - Move your head from side to side for about 10 seconds.
   - Move your head up and down for about 10 seconds (do not bump the mask on your chest).
   - Recite the OSHA Rainbow Passage (given below).
   - Bend over and move your head around, and smile and frown, for 10 seconds.
   - Jog in place for 10 seconds.
   - Breathe normally for about 10 seconds.

6. If you detect no odor during the test, you pass it. Crack the mask slightly; you will detect a strong smell of banana oil.

7. If you pass the banana oil test, you enter the irritant smoke chamber and repeat Step 4.

8. If you detect no irritation, you pass the fit-test procedure for the make and the model of mask worn during the test. (Form 1 is the SHIRT fit-test form used to document the test results.)

Rainbow Passage**

When the sunlight strikes raindrops in the air, they act like a prism and form a rainbow. The rainbow is a division of white light into many beautiful colors. These take the shape of a long round arch, with its path high above, and its two ends apparently beyond the horizon.

---

*Modified USEPA method; for full-facepiece air-purifying respirators.

**If you have difficulty reading the passage, *slowly* and *loudly* recite the alphabet *and* then count from 1 to 20 instead.

There is, according to legend, a boiling pot of gold at one end. People look, but no one ever finds it. When a man looks for something beyond reach, his friends say he is looking for the pot of gold at the end of the rainbow.

Form 1. SHIRT fit-test form

NAME				DATE	
LOCATION OF FIT TEST					
NAME OF FIT-TESTER					

ISOAMYL ACETATE TEST RESULTS—FULL FACEPIECE RESPIRATOR ONLY					
MANUFACTURER	MODEL	SIZE	PASS	FAIL	REPEAT

IRRITANT SMOKE TEST RESULTS—FULL FACEPIECE RESPIRATORS ONLY					
MANUFACTURER	MODEL	SIZE	PASS	FAIL	REPEAT

TEST RESULTS ARE ACCURATE AS RECORDED ABOVE. FIT-TESTEE IS CLEAN SHAVEN (FOR MALES ONLY). SIGN BELOW ONLY IF CORRECT.	
FIT-TESTEE	DATE
FIT-TESTER	DATE
COMMENTS	

ATTACHMENT 2: INSTRUMENT PROCEDURE SUMMARIES AND CHECKLISTS

Instrument Operating Procedures:

      HNU Systems PI-101, GP-101, and ISPI-101 Series

      Photovac MicroTIP

      Foxboro Model 128 OVA (Survey Mode)

      MSA Model 360/361

      MSA Model 260

      Matheson 8057 Hazardous Gas Leak Detector

      Ludlum Model 14C Radiation Meter

Instrument Checklists:

      HNU Photoionization Detector (PID)

      Photovac MicroTIP

      Foxboro Model 128 Organic Vapor Analyzer (OVA)

      MSA Model 360/361

      MSA Model 260

      Matheson 8057 Hazardous Gas Leak Detector

      Ludlum Model 14C Radiation Meter

HNU Systems PI-101, GP-101, and ISPI-101 Series

1. Examine the cable plug to ensure that the 12 pins inside the socket are intact and not bent or missing.

2. With the instrument OFF, attach the probe assembly to the box by matching the alignment slot in the plug to the connector.

3. Turn the switch to BATT for a battery check; the needle display should go to the OK zone. If the red LED lights up, turn the instrument off immediately and recharge it. If there is insufficient battery power, the ISPI-101 will shut itself off.

4. Turn the switch to STANDBY; the unit is electronically on, but the UV lamp is still off. Use the ZERO knob to bring the needle to zero. Allow the instrument to warm up for a few minutes.

5. The instrument display should be stable. Turn it to the most sensitive range, 0–20; the lamp is now on. Slight needle movement should be noted; record the initial background reading (usually between 0.2 and 1.0 unit) and the span setting (usually around 9.8 for the 10.2 probe, 5.0 for the 11.7 probe). Allow the unit to warm up for 5 to 10 minutes.

6. Calibrate the unit with the manufacturer's span gas; adjust the SPAN knob if necessary, and record the setting (whole number in the window, decimal on the dial).

7. After calibration, check background readings on the 0–20 scale; there should be no significant change. If the reading is still less than 1, set the background reading to 1.0 with the ZERO adjust knob. Adjusting the background with the ZERO knob does not affect the calibration.

8. The HNU is now ready to use.

9. Remember that compounds with IPs greater than the UV lamp eV intensity will *not* be detected.

Photovac MicroTIP Operating Procedures

1. Press the button on the handle to turn the instrument on. The pump should be activated. The screen reads: WARMING UP NOW, PLEASE WAIT.

2. Press BATT (#1 button) to determine the battery voltage level. The normal operating voltage range is 6.0–8.5. A LoBat message indicates that the battery pack is low; shut off the instrument and replace or recharge the battery pack, or all stored data will be lost.

3. When the instrument is warmed up, the screen will read READY, and will indicate the time and the date.

4. When the meter indicates READY, press the light switch (#3) to determine the UV lamp intensity; backlighting will also be activated. The normal lamp intensity is between 1000 and 4000. Hold the light switch down to decrease the backlight intensity; then release it. Press the switch and hold it down again to turn the backlight off. Press CLEAR or EXIT to return to the READY mode.

5. Press the MAX switch; the maximum concentration previously encountered will be displayed for 15 seconds, and the time and the date when it was recorded will also be displayed. Press the MAX switch twice to clear the maximum reading; the message will read PRESS CLEAR TO RESET MAX. Press the CLEAR switch; the message will read MAX CLEARED.

6. Press the EVENT switch; use up/down arrows until DELETE EVENTS is displayed. Push the EVENT switch, and the display will ask DELETE WHAT? Use arrows to find ALL EVENTS; then push ENTER again. The EVENT datalogger is now clear to accept up to 255 recording events.

7. Fill a sample bag with zero (hydrocarbon-free) air and a second bag with a known concentration of calibration gas (isobutylene). Have the bags filled and ready before starting the calibration sequence. Do not use ambient air to zero the instrument; always use hydrocarbon-free air as specified by the manufacturer.

8. Press the CAL switch (#8); the display will read RESPONSE FACTOR 1.00? If using isobutylene, press ENTER as a yes answer. The screen will read CONNECT ZERO GAS AND PRESS ENTER. Do not press another key until the zero gas is attached to the meter inlet. Wait a few seconds, and then press ENTER.

9. The screen will briefly read CALIBRATING; then it will quickly switch to CONNECT SPAN GAS AND PRESS ENTER. Immediately attach the sample bag with isobutylene, and then press ENTER. The screen will indicate the concentration used when the last calibration was performed. If the current span gas concentration is the same, press ENTER; if not, enter the new concentration, and then press ENTER. When calibration is complete, the screen will indicate READY, and it will continue to give concentration readings as long as the sample bag is attached. The MicroTIP readings should be the same as the PPM concentration of calibration gas. If the readings are significantly different, repeat the entire calibration procedure.

10. Disconnect the sample bag. The meter is now calibrated and ready for use. Record the initial background readings.

11. It is possible to incorrectly calibrate the MicroTIP to ambient air instead of the span gas; this can happen when the ENTER key is pressed before the sample bag containing isobutylene is attached, or if the operator delays in attaching the bag to the meter inlet. If this occurs, the instrument will consistently give a reading of >2500; the entire calibration sequence must be repeated.

12. Previous EVENT data will remain stored until recorded over or deleted. To view previous data, press PLAY (#7), and then press * SETUP. The display message will ask PLAY FROM LAST EVENT? with a number. The number indicates the number of events that are available for review. To view all previous EVENT data, enter the number 1 on the keypad, and then press ENTER. The screen will read MAX? Use up/down arrows to select minimum (MIN?), average (AVG?), or maximum (MAX?) readings for each event, and then press ENTER. The time, date, and reading for each event will be displayed; and the sequence then will be repeated until the EXIT switch is pressed. To increase or decrease the speed of playback, press up or down arrows. To freeze the playback, press ENTER, and then press up or down arrows to go forward or backward until the desired event data are located.

13. To record data, press the EVENT key, use arrows until the screen reads ADVANCE EVENT, and then press ENTER to record event data. The bottom left of the display screen will indicate the event number (e.g., 003). To record another reading, press EVENT, and then press ENTER.

Foxboro Model 128 OVA (Survey Mode)

Start-up Procedure

1. Attach the readout assembly sample line and the readout jack to appropriate connectors on the instrument box (sidepack). Use a 1/2" wrench to *gently* tighten the sample connector.

2. Make sure the Teflon collar is inside the probe handle; insert the probe extension into the handle and hand-tighten it. (*Note:* The probe extension contains a particulate filter and should always be in place when the pump is on.)

3. Toggle the INSTRUMENT switch to the battery test position BATT (pull the toggle switch away from the sidepack; then move it up or down); hold it in position and check the readout. The needle should deflect into the OK range.

4. Toggle the INSTRUMENT switch to the ON position; allow the instrument to warm up *at least* 5 minutes.

5. Toggle the PUMP switch to ON.

6. Adjust the ALARM VOLUME knob to the lowest setting by turning it all the way counterclockwise.

7. The unit is now electronically on, but the flame has not been ignited yet. Allow it to warm up a few minutes.

8. Set the range switch to ×1; adjust the needle display to read 1.0. Turn the ALARM VOLUME knob all the way clockwise. Using the ADJUST knob, set the needle to the desired alarm level; turn the rheostat knob on the back of the readout assembly until the alarm comes on. The alarm is activated by the needle position, not the actual reading (e.g., if the alarm is set at 5 units, readings of 50 and 500 will elicit an alarm response).

9. Turn the ALARM VOLUME knob counterclockwise (off). Keep the range switch set at ×1; use the ADJUST knob to bring the needle to zero. Allow warm-up to continue for a few more minutes.

10. Open the H_2 TANK valve at least one full turn; the reading on the high pressure indicator should be at least 300 psi (the OVA burns hydrogen at about 150 psi/hr).

11. Open the H_2 SUPPLY valve at least one full turn; the reading on the low pressure indicator should be approximately 10 psi.

12. *Switch to the ×10 range to prevent damage to the needle display.*

13. *Wait a full 60 seconds after turning on the hydrogen supply.*

14. Press the red igniter button located above the readout connectors; hydrogen ignition will cause a faint pop and needle deflection. *Do not* keep the button depressed for more than 6 seconds. If ignition does not occur, wait 30 to 60 seconds and try.

15. Switch back to the ×1 setting; record the initial background reading and the span setting (usually around 3.0). Allow the instrument to warm up for 5 to

10 minutes. If the needle appears to be drifting, wait another 5 minutes, and then check for drift again.

16. If the background appears inordinately high, use hydrocarbon-free air to zero out the hydrogen fuel background.

17. Methane calibration gas must be placed in a sample bag and then drawn into the OVA under ambient pressure; *do not* feed calibration gas from a pressurized cylinder directly into the OVA. Adjust the range setting to accommodate the calibration gas concentration (e.g., if the methane concentration is 50 ppm, set the range to ×10-meter; now it reads in the 10–100 unit range).

18. Open the valve on the sample bag; then attach a hose or tubing to the OVA probe extension. Adjust GAS SELECT if necessary, and record the setting (whole number in window, decimal on dial). The optimum setting is 3.00. When calibration is complete, disconnect the tubing, and then shut the valve on the sample bag.

19. The GAS SELECT knob is the span potentiometer; it *does not* "select" the gas that will be detected.

20. After calibration, check the background reading on the ×1 range; there should be no significant change. If it is less than 1, set the background reading to at least 1.0 with the ADJUST knob. Adjusting the background reading does not affect the calibration.

21. Turn the ALARM VOLUME knob clockwise to the on position.

22. The OVA now is ready to use.

Shutdown Procedure

1. Check the fuel high pressure indicator, and record the pressure. The needle indicator will gradually fade toward zero over time, giving an inaccurate reading of the actual tank pressure. When the H_2 TANK valve is opened again, the actual pressure again will be correctly displayed on the indicator.

2. Close the H_2 TANK valve, and then the H_2 Supply valve.

3. Check the BATTery; replace or recharge it as necessary after shutdown.

4. When the H_2 SUPPLY low pressure indicator reading is zero, toggle the INSTRUMENT switch to OFF (the H_2 TANK high pressure indicator does not go to zero).

5. Wait a few seconds, then toggle the PUMP switch to OFF.

6. The OVA is shut off.

MSA Model 361/360 Procedures

1. Turn the Function knob to SCAN. A pulsating audible alarm should sound, and the alarm light should go on. Press the RESET button. The alarm light and the audible alarm should go off; if they do not, turn the Function knob to HORN OFF (an amber light should go on in the horn-off mode).

2. If a continuous alarm sounds, check the readout for BATT; if the BATT message is present, turn the instrument off and recharge the battery before use.

3. The flow indicator float should be visible and audible, indicating adequate airflow.

4. Press the SELECT button until the %LEL value is visible in the readout; adjust the LEL ZERO knob until the reading is 0% LEL. (Lift up on the knob, and then turn it to adjust it.)

5. Press the SELECT button to obtain %OXY (oxygen) value; adjust the OXY CALIBRATE knob until the reading is 20.9% oxygen.

6. Press the SELECT button to obtain the PPM TOX value; adjust the TOX ZERO knob until the reading is 0 ppm. (The MSA 361 TOX value is for 0–50 ppm H_2S; the MSA 360 TOX value is for 0–500 ppm CO.)

7. Turn the Function knob to SCAN for automatic scanning of all gases or to MANUAL for a continuous readout of only one gas. The readings should remain stable.

8. Check the calibration with MSA calibrant gases.

9. The instrument now is ready for use.

If you are unsure of alarm thresholds, check each sensor as follows:

10. Turn the Function knob to HORN OFF; press the SELECT button to view the %LEL reading. Slowly turn the %LEL ZERO knob until the alarm light comes on. Turn the knob back to 0%, and press the RESET button.

11. Repeat the procedure for OXY and PPM TOX values. Return to normal readings, and reset the unit after each procedure.

MSA Model 260 Procedures

1. Turn middle knob to ON position; a pulsating alarm should sound, and alarm lights should go on. Press the ALARM RESET button, and the audible alarm and the lights should go off; if not, turn the middle knob to the HORN OFF position (a green blinking light appears in the horn-off mode).
2. The flow indicator float should be visible and audible, indicating adequate airflow.
3. Press the CHECK button, and observe needle deflection in the %LEL display; the needle should be in the battery portion of the scale (80–100). If the battery reading is not in the OK range, change/recharge the battery before use.
4. Turn the ZERO LEL knob until the needle is at 0% LEL (lift up on the knob, and then turn it to adjust it).
5. Turn the CALIBRATE O_2 knob until the needle lies directly over the offset line, which indicates 20.9% oxygen (darkened line above N in the oxygen display window).
6. The readings should remain stable.
7. Check the calibration with MSA calibration gas.
8. The instrument now is ready for use.

If unsure of alarm thresholds, check each sensor as follows:

9. Slowly turn the %LEL ZERO knob, and watch the needle display until the audible alarm and the light are activated. Turn the knob back to 0%, and press the RESET button.
10. Repeat the procedure for oxygen, return to the normal setting, and reset the unit.

Matheson 8057 Hazardous Gas Leak
Detector Procedures

1. Turn the instrument on; the pump should start, and the alarm will sound. The battery light should be on. If the light is not on, turn off the instrument and recharge or replace the batteries.

2. Adjust the wheel setting clockwise until the alarm sounds; back off until the alarm just goes off.

3. Allow the unit to warm up 2 to 5 minutes; check the alarm setting again.

4. Test the unit by briefly exposing it to sample in the Check Vial; the alarm will sound. Allow the unit to run in clean air until the alarm shuts off (this may take several minutes or more).

5. If the meter is needed immediately, breathe toward the probe instead of using the Check Vial; the unit will alarm briefly and quickly return to the normal condition.

6. The unit now is ready for use.

Ludlum Model 14C Radiation Meter Procedures

1. Open the battery compartment lid (turn screw 1/4 turn to the left), and insert two D-size batteries. Close lid by turning screw 1/4 turn to the right.
2. Turn the instrument on by turning the range switch to ×1000. Push the BAT switch to check the battery. The needle should deflect into the BAT TEST portion of the scale.
3. Use a connector to attach the external detector to the box. Switch the response switch to F (fast). Turn the audio ON to activate the audible chirper response. Expose the detector to a check source on the ×1000 scale; then move to lower scales until a meter reading is obtained.
4. Press the RES button to reset the meter to zero.
5. The meter is ready for use.

HNU Photoionization Detector (PID) Checklist

———— Probe attached to box. UV lamp eV capacity is ————.

———— Battery in OK range; red LED not illuminated.

———— Zero unit on STANDBY.

———— Instrument readings stable.

———— Background reading at 0–20 range is———— units.

———— Calibrate with recommended span gas:

HNU should read ————with ———— eV probe.

Actual HNU reading ———— with SPAN at ————.

———— Background on 0–20 scale after calibration is ————.

———— Adjusted to 1.0 units if less than 1.0 after calibration.

Shutdown

———— Check battery status; recharge or replace it as needed.

———— Instrument off.

———— Instrument was used ———— hours.

Photovac MicroTIP Checklist

———— Instrument on; pump starts; display screen lit.

———— Battery voltage ————. LoBat message not displayed.

———— Screen reads READY; time and date displayed are correct.

———— UV lamp intensity is ————.

———— MAX cleared.

———— EVENTS cleared.

———— Sample bag with zero air and calibration gas ready.

Calibration gas concentration ———— ppm.

———— Meter reads ————. units while sampling calibration gas at end of calibration sequence.

———— Background reading after calibration ———— units.

———— EVENT key; arrow key to ADVANCE EVENT; then press ENTER. EVENT number indicated is————.

———— Instrument now ready for use.

Shutdown:

———— Battery voltage after use is ————; recharge or replace battery as needed.

———— Instrument off.

———— Instrument was used ———— hours.

Foxboro Model 128 Organic Vapor Analyzer (OVA) Checklist

Start-up Procedure

———— Readout assembly connected to sidepack; probe extension in place.

———— Battery in OK range.

———— Instrument turned on; allowed to warm up at least 5 minutes.

———— Pump switch on; alarm volume off.

———— Total warm-up time at least 5 minutes.

———— Alarm level set; then alarm volume turned back off.

———— Instrument zeroed on ×1 range.

———— Total warm-up time at least 5 minutes.

———— Hydrogen tank psi ————; hydrogen supply psi ————.

———— Ignite hydrogen on ×10 range.

———— Initial background on ×1 scale is ———— units.

———— Warm-up time with flame at least 5 minutes; readings stable.

———— Calibrate with ———— ppm methane.

          OVA reads ———— units with GAS SELECT at ————.

———— Background on ×1 scale is ———— units.

———— Adjusted to 1.0 if less than 1.0.

Shutdown Procedure:

———— H_2 TANK valve closed.

———— H_2 Supply valve closed.

———— Final H_2 TANK pressure is ———— psi.

———— Check battery status; recharge or replace battery as needed.

———— H_2 SUPPLY pressure at zero; toggle INSTRUMENT to OFF.

———— Toggle PUMP to OFF.

———— Instrument was used ———— hours.

MSA Model 360/361 Instrument and Calibration Checklist

—— Instrument on; pump starts.
—— Audible alarm sounds upon turn-on.
—— Alarm light goes on upon turn-on.
—— RESET button resets alarms; audible alarm and light off.
—— Horn-off mode; amber light on.
—— Check alarm levels on manual mode (reset after each):
  Alarm level for oxygen sensor is ——%.
  Alarm level for %LEL is ——%.
  Alarm level for toxic sensor (H_2S or CO) is —— ppm.
—— Set normal ambient levels for each sensor.
—— Instrument readings stable.
—— Calibrate instrument using MSA calibrant gases:
  —— % oxygen —— % LEL —— ppm CO or H_2S
—— Instrument response during calibration is:
  —— % oxygen —— % LEL —— ppm CO or H_2S
—— Instrument ready for use.

Shutdown:

—— Check battery status; recharge battery as needed.
—— Instrument off.
—— Instrument was used —— hours.

MSA Model 260 Instrument and Calibration Checklist

———— Instrument on; pump turns on.

———— Alarm sounds upon turn-on.

———— Alarm lights (2) go on upon turn-on.

———— RESET button resets alarm horn; lights off.

———— Horn-off mode; blinking green light.

———— Battery in OK range.

———— Check alarm levels on manual mode (reset after each):

Alarm level for oxygen sensor is ———— %.

Alarm level for %LEL is ———— %.

———— Set normal ambient levels for both sensors.

———— Instrument readings stable.

———— Calibrate instrument using MSA calibrant gas:

———— % oxygen ———— % LEL

———— Instrument response during calibration is:

———— % oxygen ———— % LEL

———— Instrument ready for use.

Shutdown:

———— Check battery status; recharge battery as needed.

———— Instrument off.

———— Instrument was used ———— hours.

Matheson 8057 Hazardous Gas Leak Detector Checklist

——— Instrument on; pump turns on.
——— Alarm sounds; alarm light goes on.
——— Red battery light is illuminated.
——— Adjust alarm to just off with wheel:
Wheel setting is ———— .
——— Allow unit to warm up 2–5 minutes.
——— Readjust alarm; wheel setting is ———— .
——— Test unit with Check Vial, or breathe into unit.
——— Unit returns to normal; now is ready for use.

Shutdown:

——— Check battery light; recharge or replace batteries as needed.
——— Instrument off.
——— Instrument was used ———— hours.

Ludlum Model 14C Radiation Meter

——— Batteries inserted; instrument on ×1000 scale.
——— Batteries in OK range.
——— Audio chirper on.
——— Switch to F for fast response
———— External detector connected, responds to check source on ———— range; approximate reading is ———— mR/hr.
——— Reset to zero; meter ready for use.

Shutdown:

——— Check battery status; replace batteries as needed.
——— Instrument off.
——— Instrument was used ———— hours.
——— Remove batteries; do not store instrument with disposable batteries in place.

ATTACHMENT 3: REPRESENTATIVE DECONTAMINATION STATIONS

The schematic diagrams show minimum decontamination layouts for Levels A and B protection and Level C protection. (See also Section VIII above.)

ATTACHMENT 4: SHIRT CHECKLISTS AND FORMS

Forms 2 through 9 are SHIRT emergency response worksheets.

Attachment 3. Representative decontamination station, Level A/Level B

MINIMUM DECONTAMINATION LAYOUT

LEVELS A & B PROTECTION

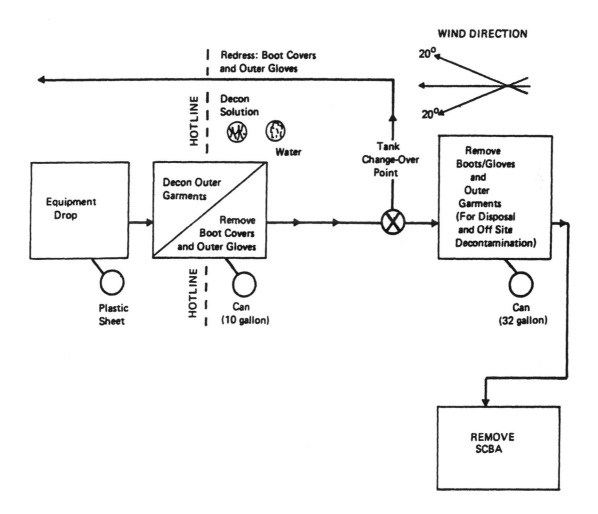

Attachment 3. Representative decontamination station, Level C

MINIMUM DECONTAMINATION LAYOUT

LEVEL C PROTECTION

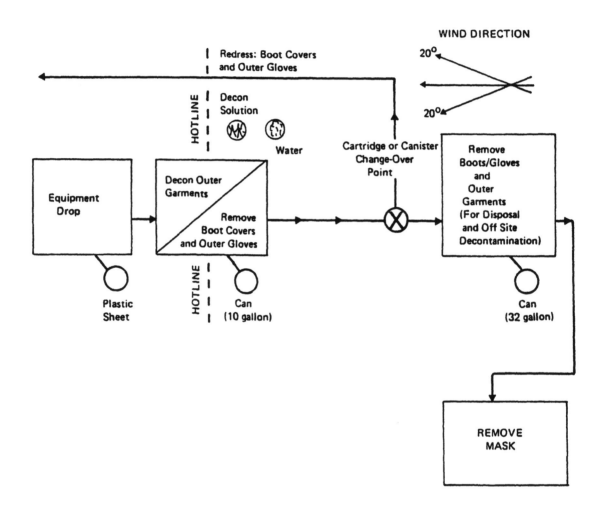

Form 2. Personnel assignment worksheet

DATE	LOCATION/FACILITY
TIME	

INCIDENT COMMANDER

SHIRT SUPERVISOR

SAFETY OFFICER (SO)	MEDICAL OFFICER	DECON OFFICER

EQUIPMENT COORDINATOR

ENTRY TEAM MEMBER	ENTRY TEAM MEMBER	ENTRY TEAM MEMBER

BACKUP TEAM MEMBER	BACKUP TEAM MEMBER	BACKUP TEAM MEMBER

DECON TEAM MEMBER	DECON TEAM MEMBER	DECON TEAM MEMBER

Form 3. Shirt Supervisor incident data sheet

CONTACT NAME				
FACILITY LOCATION				
INCIDENT COMMANDER				
PHONE NUMBER				
INCIDENT TYPE				
HAZARD INFORMATION				
NAME/TYPE OF MATERIAL				
ID #	GUIDE #	LIQUID	GAS	SOLID
FP	LEL	UEL	IDLH	TWA
SpG	VapD	Sol	Other	
REACTIVITY				
INCOMPATIBILITIES				

DRAWING OF INCIDENT

N
W — E
S

TIME	WEATHER

Form 4. Shirt Supervisor incident command worksheet

FACILITY		DATE
LOCATION		TIME
INCIDENT COMMANDER		
SHIRT PERSONNEL AT SCENE		
SUPPORT AGENCIES AT SCENE OR CALLED		
POLICE	FIRE	EMS
TRANS	HEALTH	PUB WORKS
CHEMTREC	MANUFACTURER	
OTHER		
FACILITY PERSONNEL AT SCENE OR CALLED		
CONTRACTOR PERSONNEL AT SCENE		
RESOURCES CONTACT		
ABCCI CONTACT		

Form 5. Safety Officer (SO) worksheet/checklist

FACILITY/LOCATION			DATE	
MEDICAL OFFICER		DECON OFFICER		
ENTRY				
BACKUP				
DECON				

HAZARD DESCRIPTION

ID #	GUIDE #	LIQUID	GAS	SOLID
FP	LEL	UEL	IDLH	TWA
SpG	VapD	Sol	Other	
REACTIVITY				
INCOMPATIBILITIES				

RESPIRATORY PROTECTION

ENTRY/BACKUP
DECON

PROTECTIVE CLOTHING

ENTRY/BACKUP
DECON
COMPATIBILITY CHECKED

SAFETY CHECKLIST

DECON READY	BACKUP READY
PRE-MEDICAL CHECK	SHIRT SUP NOTIFIED
ACTION PLAN REVIEWED	DECON REVIEWED
SAFETY PLAN REVIEWED	TIME OF ENTRY

Form 6. Medical monitoring checklist

MEDICAL OFFICER	
MEDICAL ASSISTANTS	
LOCATION	DATE
POSSIBLE SIGNS OF OVEREXPOSURE	

PRE AND POST ENTRY MEDICAL MONITORING RECORD

SHIRT MEMBER	TIME	PULSE	BP	RESP	SKIN*	STATUS
1.						
2.						
3.						
4.						
5.						
6.						

*Use skin temperature or oral temperature as appropriate

SHIRT READINESS CHECKLIST

ENTRY READY	BACKUP READY	DECON READY
SO NOTIFIED	ENTRY TIME	EXIT TIME

Form 7. Entry/backup team checklist

FACILITY/LOCATION			DATE	
HAZ SECTOR OFFICER		SAFETY OFFICER		
ENTRY				
AIR PRESSURE				
BACKUP				
AIR PRESSURE				
DECON				
AIR PRESSURE				
HAZARDS				
RESPIRATORY PROTECTION				
PROTECTIVE CLOTHING				
INSTRUMENTS				
ACTION PLAN/DIAGRAM				
COMMUNICATIONS				
SAFETY PLAN/ACTION LEVELS				
DECON REVIEWED		MEDICAL COMPLETED		
SO NOTIFIED		DECON NOTIFIED		
TIME ON AIR	ENTRY TIME		TIME OFF AIR	

Form 8. Decontamination officer checklist

FACILITY/LOCATION		DATE	
DECON OFFICER		SAFETY OFFICER	
DECON TEAM			
ANTICIPATED HAZARDS			
DECON METHOD			
DECON SET-UP (DIAGRAM)			
NUMBER PERSONNEL TO DECON			
DECON TEAM PROTECTIVE CLOTHING			
DECON TEAM RESPIRATORY PROTECTION			
ENTRY TEAM PROTECTIVE CLOTHING			
ENTRY TEAM RESPIRATORY PROTECTION			
EQUIPMENT TO DECON			
RETAIN DECON SOLUTION?		RETAIN PPE DISCARDS?	
SYMPTOMS OF OVEREXPOSURE			
MEDICAL/REHAB READY		DECON READY	
DECON REVIEWED		SAFETY PLAN REVIEWED	
SO NOTIFIED		MEDICAL NOTIFIED	
ENTRY TIME		EXIT TIME	
TIME DECON COMPLETED			

Form 9. Potential exposure/incident medical treatment record

LOCATION		DATE
SAFETY OFFICER (SO)		TIME
MEDICAL OFFICER		
SHIRT PERSONNEL		
TYPE/NAME OF HAZARD(S)		

VP	SpG	VapD	TWA	IDLH
INHALATION		SKIN	EYE	IRRITANT

SIGNS/SYMPTOMS OF EXPOSURE

TREATED SHIRT MEMBER	SIGNS/SYMPTOMS

TREATMENT GIVEN
FINAL OUTCOME

SO NOTIFIED	SHIRT SUP NOTIFIED

Appendix C

OSHA Hazardous Waste Operations and Emergency Response, Final Rule (29 CFR 1910.120)

§ 1910.120 HAZARDOUS WASTE OPERATIONS AND EMERGENCY RESPONSE

(a) Scope, application, and definitions-

  (1) Scope. This section covers the following operations, unless the employer can demonstrate that the operation does not involve employee exposure or the reasonable possibility for employee exposure to safety or health hazards:

  (i)   Clean-up operations required by a governmental body, whether Federal, state, local or other involving hazardous substances that are conducted at uncontrolled hazardous waste sites (including, but not limited to, the EPA's National Priority Site List (NPL), state priority site lists, sites recommended for the EPA NPL, and initial investigations of government identified sites which are conducted before the presence or absence of hazardous substances has been ascertained);

  (ii) Corrective actions involving clean-up operations at sites covered by the Resource Conservation and Recovery Act of 1976 (RCRA) as amended (42 U.S.C. 6901 et seq.);

  (iii) Voluntary clean-up operations at sites recognized by Federal, state, local or other governmental bodies as uncontrolled hazardous waste sites;

  (iv) Operations involving hazardous wastes that are conducted at treatment, storage, and disposal (TSD) facilities regulated by 40 CFR Parts 264 and 265 pursuant to RCRA; or by agencies under agreement with U.S.E.P.A. to implement RCRA regulations; and

(v) Emergency response operations for releases of, or substantial threats of releases of, hazardous substances without regard to the location of the hazard.

(2) Application.

(i) All requirements of part 1910 and part 1926 of title 29 of the Code of Federal Regulations apply pursuant to their terms to hazardous waste and emergency response operations whether covered by this section or not. If there is a conflict or overlap, the provision more protective of employee safety and health shall apply without regard to 29 CFR 1910.5(c)(1).

(ii) Hazardous substance clean-up operations within the scope of paragraphs (a)(1)(i) through (a)(1)(iii) of this section must comply with all paragraphs of this section except paragraphs (p) and (q).

(iii) Operations within the scope of paragraph (a)(1)(iv) of this section must comply only with the requirements of paragraph (p) of this section.

Notes and Exceptions: (A) All provisions of paragraph (p) of this section cover any treatment, storage or disposal (TSD) operation regulated by 40 CFR Parts 264 and 265 or by state law authorized under RCRA, and required to have a permit or interim status from EPA pursuant to 40 CFR 270.1 or from a state agency pursuant to RCRA.

(B) Employers who are not required to have a permit or interim status because they are conditionally exempt small quantity generators under 40 CFR 261.5 or are generators who qualify under 40 CFR 262.34 for exemptions from regulation under 40 CFR Parts 264, 265 and 270 ("excepted employers") are not covered by paragraphs (p)(1) through (p)(7) of this section.

Excepted employers who are required by the EPA or state agency to have their employees engage in emergency response or who direct their employees to engage in emergency response are covered by paragraph (p)(8) of this section, and cannot be exempted by (p)(8)(i) of this section. Excepted employers who are not required to have employees engage in emergency response, who direct their employees to evacuate in the case of such emergencies and who meet the requirements of paragraph (p)(8)(i) of this section are exempt from the balance of paragraph (p)(8) of this section.

(C) If an area is used primarily for treatment, storage or disposal, any emergency response operations in that area shall comply with paragraph (p)(8) of this section. In other areas not used primarily for treatment, storage, or disposal, any emergency response operations shall comply with paragraph (q) of this section. Compliance with the requirements of paragraph (q) of this section shall be deemed to be in compliance with the requirements of paragraph (p)(8) of this section.

(iv) Emergency response operations for releases of, or substantial threats of releases of, hazardous substances which are not covered by paragraphs (a)(1)(i) through (a)(1)(iv) of this section must only comply with the requirements of paragraph (q) of this section.

(3) Definitions-Buddy system means a system of organizing employees into work groups in such a manner that each employee of the work group is designated to be observed by at least one other employee in the work group. The purpose of the buddy system is to provide rapid assistance to employees in the event of an emergency.

Clean-up operation means an operation where hazardous substances are removed, contained, incinerated, neutralized, stabilized, cleared-up, or in any other manner processed or handled with the ultimate goal of making the site safer for people or the environment.

Decontamination means the removal of hazardous substances from employees and their equipment to the extent necessary to preclude the occurrence of foreseeable adverse health affects.

Emergency response or responding to emergencies means a response effort by employees from outside the immediate release area or by other designated responders (i.e., mutual-aid groups, local fire departments, etc.) to an occurrence which results, or is likely to result, in an uncontrolled release of a hazardous substance. Responses to incidental releases of hazardous substances where the substance can be absorbed, neutralized, or otherwise controlled at the time of release by employees in the immediate release area, or by maintenance personnel are not considered to be emergency responses within the scope of this standard.

Responses to releases of hazardous substances where there is no potential safety or health hazard (i.e., fire, explosion, or chemical exposure) are not considered to be emergency responses.

Facility means (A) any building, structure, installation, equipment, pipe or pipeline (including any pipe into a sewer or publicly owned treatment works), well, pit, pond, lagoon, impoundment, ditch, storage container, motor vehicle, rolling stock, or aircraft, or (B) any site or area where a hazardous substance has been deposited, stored, disposed of, or placed, or otherwise come to be located; but does not include any consumer product in consumer use or any water-borne vessel.

Hazardous materials response (HAZMAT) team means an organized group of employees, designated by the employer, who are expected to perform work to handle and control actual or potential leaks or spills of hazardous substances requiring possible close approach to the substance. The team members perform responses to releases or potential releases of hazardous substances for the purpose of control or stabilization of the incident. A HAZMAT team is not a fire brigade nor is a typical fire brigade a HAZMAT team.

A HAZMAT team, however, may be a separate component of a fire brigade or fire department.

Hazardous substance means any substance designated or listed under paragraphs (A) through (D) of this definition, exposure to which results or may result in adverse affects on the health or safety of employees:

(A) Any substance defined under section 101(14) of CERCLA;

(B) Any biological agent and other disease-causing agent which after release into the environment and upon exposure, ingestion, inhalation, or assimilation into any person, either directly from the environment or indirectly by ingestion through food chains, will or may reasonably be anticipated to cause death, disease, behavioral abnormalities, cancer, genetic mutation, physiological malfunctions (including malfunctions in reproduction) or physical deformations in such persons or their offspring;

(C) Any substance listed by the U.S. Department of Transportation as hazardous materials under 49 CFR 172.101 and appendices; and

(D) Hazardous waste as herein defined.

Hazardous waste means-

(A) A waste or combination of wastes as defined in 40 CFR 261.3, or

(B) Those substances defined as hazardous wastes in 49 CFR 171.8.

Hazardous waste operation means any operation conducted within the scope of this standard.

Hazardous waste site or Site means any facility or location within the scope of this standard at which hazardous waste operations take place.

Health hazard means a chemical, mixture of chemicals or a pathogen for which there is statistically significant evidence based on at least one study conducted in accordance with established scientific principles that acute or chronic health effects may occur in exposed employees. The term "health hazard" includes chemicals which are carcinogens, toxic or highly toxic agents, reproductive toxins, irritants, corrosives, sensitizers, heptaotoxins, nephrotoxins, neurotoxins, agents which act on the hematopoietic system, and agents which damage the lungs, skin, eyes, or mucous membranes. It also includes stress due to temperature extremes.

Further definition of the terms used above can be found in appendix A to 29 CFR 1910.1200.

IDLH or Immediately dangerous to life or health means an atmospheric concentration of any toxic, corrosive or asphyxiant substance that poses an immediate threat to life or would cause irreversible or delayed adverse health effects or would interfere with an individual's ability to escape from a dangerous atmosphere.

Oxygen deficiency means that concentration of oxygen by volume below which atmosphere supplying respiratory protection must be provided. It exists in atmospheres where the percentage of oxygen by volume is less than 19.5 percent oxygen.

Permissible exposure limit means the exposure, inhalation or dermal permissible exposure limit specified in 29 CFR Part 1910, subparts G and Z.

Published exposure level means the exposure limits published in "NIOSH Recommendations for Occupational Health Standards" dated 1986, which is incorporated by reference as specified in § 1910.6, or if none is specified, the exposure limits published in the standards specified by the American Conference of Governmental Industrial Hygienists in their publication "Threshold Limit Values and Biological Exposure Indices for 1987–88" dated 1987, which is incorporated by reference as specified in § 1910.6.

Post emergency response means that portion of an emergency response performed after the immediate threat of a release has been stabilized or eliminated and clean-up of the site has begun.

If post emergency response is performed by an employer's own employees who were part of the initial emergency response, it is considered to be part of the initial response and not post emergency response. However, if a group of an employer's own employees, separate from the group providing initial response, performs the clean-up operation, then the separate group of employees would be considered to be performing post-emergency response and subject to paragraph (q)(11) of this section.

Qualified person means a person with specific training, knowledge and experience in the area for which the person has the responsibility and the authority to control.

Site safety and health supervisor (or official) means the individual located on a hazardous waste site who is responsible to the employer and has the authority and knowledge necessary to implement the site safety and health plan and verify compliance with applicable safety and health requirements.

Small quantity generator means a generator of hazardous wastes who in any calendar month generates no more than 1,000 kilograms (2,205 pounds) of hazardous waste in that month.

Uncontrolled hazardous waste site means an area identified as an uncontrolled hazardous waste site by a governmental body, whether Federal, state, local or other where an accumulation of hazardous substances creates a threat to the health and safety of individuals or the environment or both. Some sites are found on public lands such as those created by former municipal, county or state landfills where illegal or poorly managed waste disposal has taken place. Other sites are found on private property, often belonging to generators or former generators of hazardous substance wastes. Examples of such sites include, but are not limited to, surface impoundments, landfills, dumps, and tank or drum farms. Normal operations at TSD sites are not covered by this definition.

(b) Safety and health program.

Note to (b): Safety and health programs developed and implemented to meet other

Federal, state, or local regulations are considered acceptable in meeting this requirement if they cover or are modified to cover the topics required in this paragraph. An additional or separate safety and health program is not required by this paragraph.

(1) General.

    (i)   Employers shall develop and implement a written safety and health program for their employees involved in hazardous waste operations. The program shall be designed to identify, evaluate, and control safety and health hazards, and provide for emergency response for hazardous waste operations.

    (ii)   The written safety and health program shall incorporate the following:

        (A) An organizational structure;

        (B) A comprehensive workplan;

        (C) A site-specific safety and health plan which need not repeat the employer's standard operating procedures required in paragraph (b)(1)(ii)(F) of this section;

        (D) The safety and health training program;

        (E) The medical surveillance program;

        (F) The employer's standard operating procedures for safety and health; and

        (G) Any necessary interface between general program and site specific activities.

    (iii)   Site excavation. Site excavations created during initial site preparation or during hazardous waste operations shall be shored or sloped as appropriate to prevent accidental collapse in accordance with subpart P of 29 CFR Part 1926.

    (iv)   Contractors and sub-contractors. An employer who retains contractor or sub-contractor services for work in hazardous waste operations shall inform those contractors, sub-contractors, or their representatives of the site emergency response procedures and any potential fire, explosion, health, safety or other hazards of the hazardous waste operation that have been identified by the employer, including those identified in the employer's information program.

    (v)   Program availability. The written safety and health program shall be made available to any contractor or subcontractor or their representative who will be involved with the hazardous waste operation; to employees; to employee designated representatives; to OSHA personnel, and to personnel of other Federal, state, or local agencies with regulatory authority over the site.

(2) Organizational structure part of the site program-

    (i)   The organizational structure part of the program shall establish the specific chain of command and specify the overall responsibilities of supervisors and employees. It shall include, at a minimum, the following elements:

(A) A general supervisor who has the responsibility and authority to direct all hazardous waste operations.

(B) A site safety and health supervisor who has the responsibility and authority to develop and implement the site safety and health plan and verify compliance.

(C) All other personnel needed for hazardous waste site operations and emergency response and their general functions and responsibilities.

(D) The lines of authority, responsibility, and communication.

(ii) The organizational structure shall be reviewed and updated as necessary to reflect the current status of waste site operations.

(3) Comprehensive workplan part of the site program. The comprehensive workplan part of the program shall address the tasks and objectives of the site operations and the logistics and resources required to reach those tasks and objectives.

(i) The comprehensive workplan shall address anticipated clean-up activities as well as normal operating procedures which need not repeat the employer's procedures available elsewhere.

(ii) The comprehensive workplan shall define work tasks and objectives and identify the methods for accomplishing those tasks and objectives.

(iii) The comprehensive workplan shall establish personnel requirements for implementing the plan.

(iv) The comprehensive workplan shall provide for the implementation of the training required in paragraph (e) of this section.

(v) The comprehensive workplan shall provide for the implementation of the required informational programs required in paragraph (i) of this section.

(vi) The comprehensive workplan shall provide for the implementation of the medical surveillance program described in paragraph (f) of this section.

(4) Site-specific safety and health plan part of the program-

(i) General. The site safety and health plan, which must be kept on site, shall address the safety and health hazards of each phase of site operation and include the requirements and procedures for employee protection.

(ii) Elements. The site safety and health plan, as a minimum, shall address the following:

(A) A safety and health risk or hazard analysis for each site task and operation found in the workplan.

(B) Employee training assignments to assure compliance with paragraph (e) of this section.

(C) Personal protective equipment to be used by employees for each of the site tasks and operations being conducted as required by the personal protective equipment program in paragraph (g)(5) of this section.

(D) Medical surveillance requirements in accordance with the program in paragraph (f) of this section.

(E) Frequency and types of air monitoring, personnel monitoring, and environmental sampling techniques and instrumentation to be used, including methods of maintenance and calibration of monitoring and sampling equipment to be used.

(F) Site control measures in accordance with the site control program required in paragraph (d) of this section.

(G) Decontamination procedures in accordance with paragraph (k) of this section.

(H) An emergency response plan meeting the requirements of paragraph (l) of this section for safe and effective responses to emergencies, including the necessary PPE and other equipment.

(I) Confined space entry procedures.

(J) A spill containment program meeting the requirements of paragraph (j) of this section.

(iii) Pre-entry briefing. The site specific safety and health plan shall provide for pre-entry briefings to be held prior to initiating any site activity, and at such other times as necessary to ensure that employees are apprised of the site safety and health plan and that this plan is being followed.

The information and data obtained from site characterization and analysis work required in paragraph (c) of this section shall be used to prepare and update the site safety and health plan.

(iv) Effectiveness of site safety and health plan. Inspections shall be conducted by the site safety and health supervisor or, in the absence of that individual, another individual who is knowledgeable in occupational safety and health, acting on behalf of the employer as necessary to determine the effectiveness of the site safety and health plan. Any deficiencies in the effectiveness of the site safety and health plan shall be corrected by the employer.

(c) Site characterization and analysis-

(1) General. Hazardous waste sites shall be evaluated in accordance with this paragraph to identify specific site hazards and to determine the appropriate safety and health control procedures needed to protect employees from the identified hazards.

(2) Preliminary evaluation. A preliminary evaluation of a site's characteristics shall be performed prior to site entry by a qualified person in order to aid in the selection of appropriate employee protection methods prior to site entry. Immediately after initial site entry, a more detailed evaluation of the site's specific characteristics shall be performed by a qualified person in order to further identify existing site hazards

and to further aid in the selection of the appropriate engineering controls and personal protective equipment for the tasks to be performed.

(3) Hazard identification. All suspected conditions that may pose inhalation or skin absorption hazards that are immediately dangerous to life or health (IDLH), or other conditions that may cause death or serious harm, shall be identified during the preliminary survey and evaluated during the detailed survey.

Examples of such hazards include, but are not limited to, confined space entry, potentially explosive or flammable situations, visible vapor clouds, or areas where biological indicators such as dead animals or vegetation are located.

(4) Required information. The following information to the extent available shall be obtained by the employer prior to allowing employees to enter a site:

    (i)   Location and approximate size of the site.

    (ii)   Description of the response activity and/or the job task to be performed.

    (iii)   Duration of the planned employee activity.

    (iv)   Site topography and accessibility by air and roads.

    (v)   Safety and health hazards expected at the site.

    (vi)   Pathways for hazardous substance dispersion.

    (vii)   Present status and capabilities of emergency response teams that would provide assistance to hazardous waste clean-up site employees at the time of an emergency.

    (viii)   Hazardous substances and health hazards involved or expected at the site, and their chemical and physical properties.

(5) Personal protective equipment. Personal protective equipment (PPE) shall be provided and used during initial site entry in accordance with the following requirements:

    (i)   Based upon the results of the preliminary site evaluation, an ensemble of PPE shall be selected and used during initial site entry which will provide protection to a level of exposure below permissible exposure limits and published exposure levels for known or suspected hazardous substances and health hazards, and which will provide protection against other known and suspected hazards identified during the preliminary site evaluation. If there is no permissible exposure limit or published exposure level, the employer may use other published studies and information as a guide to appropriate personal protective equipment.

    (ii)   If positive-pressure self-contained breathing apparatus is not used as part of the entry ensemble, and if respiratory protection is warranted by the potential hazards identified during the preliminary site evaluation, an escape self-contained breathing apparatus of at least five minute's duration shall be carried by employees during initial site entry.

    (iii)   If the preliminary site evaluation does not produce sufficient information to identify the hazards or suspected hazards of the site, an ensemble providing protection equivalent to Level B PPE shall be provided as minimum protection, and direct reading instruments shall be used as appropriate for identifying IDLH conditions. (See appendix B for a description of Level B hazards and the recommendations for Level B protective equipment.)

(iv) Once the hazards of the site have been identified, the appropriate PPE shall be selected and used in accordance with paragraph (g) of this section.

(6) Monitoring. The following monitoring shall be conducted during initial site entry when the site evaluation produces information that shows the potential for ionizing radiation or IDLH conditions, or when the site information is not sufficient reasonably to eliminate these possible conditions:

(i) Monitoring with direct reading instruments for hazardous levels of ionizing radiation.

(ii) Monitoring the air with appropriate direct reading test equipment (i.e., combustible gas meters, detector tubes) for IDLH and other conditions that may cause death or serious harm (combustible or explosive atmospheres, oxygen deficiency, toxic substances).

(iii) Visually observing for signs of actual or potential IDLH or other dangerous conditions.

(iv) An ongoing air monitoring program in accordance with paragraph (h) of this section shall be implemented after site characterization has determined the site is safe for the start-up of operations.

(7) Risk identification. Once the presence and concentrations of specific hazardous substances and health hazards have been established, the risks associated with these substances shall be identified. Employees who will be working on the site shall be informed of any risks that have been identified. In situations covered by the Hazard Communication Standard, 29 CFR 1910.1200, training required by that standard need not be duplicated.

Note to (c)(7).-Risks to consider include, but are not limited to:

(a) Exposures exceeding the permissible exposure limits and published exposure levels.

(b) IDLH concentrations.

(c) Potential skin absorption and irritation sources.

(d) Potential eye irritation sources.

(e) Explosion sensitivity and flammability ranges.

(f) Oxygen deficiency.

(8) Employee notification. Any information concerning the chemical, physical, and toxicologic properties of each substance known or expected to be present on site that is available to the employer and relevant to the duties an employee is expected to perform shall be made available to the affected employees prior to the commencement of their work activities. The employer may utilize information developed for the hazard communication standard for this purpose.

(d) Site control-

(1) General. Appropriate site control procedures shall be implemented to control employee exposure to hazardous substances before clean-up work begins.

(2) Site control program. A site control program for protecting employees which is part of the employer's site safety and health program required in paragraph (b) of this section shall be developed during the planning stages of a hazardous waste clean-up operation and modified as necessary as new information becomes available.

(3) Elements of the site control program. The site control program shall, as a minimum, include: A site map; site work zones; the use of a "buddy system"; site communications including alerting means for emergencies; the standard operating procedures or safe work practices; and, identification of the nearest medical assistance. Where these requirements are covered elsewhere, they need not be repeated.

(e) Training-

(1) General.

(i)   All employees working on site (such as but not limited to equipment operators, general laborers and others) exposed to hazardous substances, health hazards, or safety hazards and their supervisors and management responsible for the site shall receive training meeting the requirements of this paragraph before they are permitted to engage in hazardous waste operations that could expose them to hazardous substances, safety, or health hazards, and they shall receive review training as specified in this paragraph.

(ii)   Employees shall not be permitted to participate in or supervise field activities until they have been trained to a level required by their job function and responsibility.

(2) Elements to be covered. The training shall thoroughly cover the following:

(i)   Names of personnel and alternates responsible for site safety and health;

(ii)   Safety, health and other hazards present on the site;

(iii)   Use of personal protective equipment;

(iv)   Work practices by which the employee can minimize risks from hazards;

(v)   Safe use of engineering controls and equipment on the site;

(vi)   Medical surveillance requirements, including recognition of symptoms and signs which might indicate overexposure to hazards; and

(vii)   The contents of paragraphs (G) through (J) of the site safety and health plan set forth in paragraph (b)(4)(ii) of this section.

(3) Initial training.

(i)   General site workers (such as equipment operators, general laborers and supervisory personnel) engaged in hazardous substance removal or other activities which expose or potentially expose workers to hazardous substances and health hazards shall receive a minimum of 40 hours of instruction off the site, and a minimum of three days actual field experience under the direct supervision of a trained, experienced supervisor.

(ii) Workers on site only occasionally for a specific limited task (such as, but not limited to, ground water monitoring, land surveying, or geo-physical surveying) and who are unlikely to be exposed over permissible exposure limits and published exposure limits shall receive a minimum of 24 hours of instruction off the site, and the minimum of one day actual field experience under the direct supervision of a trained, experienced supervisor.

(iii) Workers regularly on site who work in areas which have been monitored and fully characterized indicating that exposures are under permissible exposure limits and published exposure limits where respirators are not necessary, and the characterization indicates that there are no health hazards or the possibility of an emergency developing, shall receive a minimum of 24 hours of instruction off the site and the minimum of one day actual field experience under the direct supervision of a trained, experienced supervisor.

(iv) Workers with 24 hours of training who are covered by paragraphs (e)(3)(ii) and (e)(3)(iii) of this section, and who become general site workers or who are required to wear respirators, shall have the additional 16 hours and two days of training necessary to total the training specified in paragraph (e)(3)(i).

(4) Management and supervisor training. On-site management and supervisors directly responsible for, or who supervise employees engaged in, hazardous waste operations shall receive 40 hours initial training, and three days of supervised field experience (the training may be reduced to 24 hours and one day if the only area of their responsibility is employees covered by paragraphs (e)(3)(ii) and (e)(3)(iii)) and at least eight additional hours of specialized training at the time of job assignment on such topics as, but not limited to, the employer's safety and health program and the associated employee training program, personal protective equipment program, spill containment program, and health hazard monitoring procedure and techniques.

(5) Qualifications for trainers. Trainers shall be qualified to instruct employees about the subject matter that is being presented in training. Such trainers shall have satisfactorily completed a training program for teaching the subjects they are expected to teach, or they shall have the academic credentials and instructional experience necessary for teaching the subjects.

Instructors shall demonstrate competent instructional skills and knowledge of the applicable subject matter.

(6) Training certification. Employees and supervisors that have received and successfully completed the training and field experience specified in paragraphs (e)(1) through (e)(4) of this section shall be certified by their instructor or the head instructor and trained supervisor as having successfully completed the necessary training. A written certificate shall be given to each person so certified. Any person who has not been so certified or who does not meet the requirements of paragraph

(e)(9) of this section shall be prohibited from engaging in hazardous waste operations.

(7) Emergency response. Employees who are engaged in responding to hazardous emergency situations at hazardous waste clean-up sites that may expose them to hazardous substances shall be trained in how to respond to such expected emergencies.

(8) Refresher training. Employees specified in paragraph (e)(1) of this section, and managers and supervisors specified in paragraph (e)(4) of this section, shall receive eight hours of refresher training annually on the items specified in paragraph (e)(2) and/or (e)(4) of this section, any critique of incidents that have occurred in the past year that can serve as training examples of related work, and other relevant topics.

(9) Equivalent training. Employers who can show by documentation or certification that an employee's work experience and/or training has resulted in training equivalent to that training required in paragraphs (e)(1) through (e)(4) of this section shall not be required to provide the initial training requirements of those paragraphs to such employees and shall provide a copy of the certification or documentation to the employee upon request.

However, certified employees or employees with equivalent training new to a site shall receive appropriate, site specific training before site entry and have appropriate supervised field experience at the new site. Equivalent training includes any academic training or the training that existing employees might have already received from actual hazardous waste site work experience.

(f) Medical surveillance-

(1) General. Employers engaged in operations specified in paragraphs (a)(1)(i) through (a)(1)(iv) of this section and not covered by (a)(2)(iii) exceptions and employers of employees specified in paragraph (q)(9) shall institute a medical surveillance program in accordance with this paragraph.

(2) Employees covered. The medical surveillance program shall be instituted by the employer for the following employees:

(i)   All employees who are or may be exposed to hazardous substances or health hazards at or above the permissible exposure limits or, if there is no permissible exposure limit, above the published exposure levels for these substances, without regard to the use of respirators, for 30 days or more a year;

(ii)   All employees who wear a respirator for 30 days or more a year or as required by § 1910.134;

(iii) All employees who are injured, become ill or develop signs or symptoms due to possible overexposure involving hazardous substances or health hazards from an emergency response or hazardous waste operation; and

(iv)   Members of HAZMAT teams.

(3) Frequency of medical examinations and consultations. Medical examinations and

consultations shall be made available by the employer to each employee covered under paragraph (f)(2) of this section on the following schedules:

(i)   For employees covered under paragraphs (f)(2)(i), (f)(2)(ii), and (f)(2)(iv):

    (A) Prior to assignment;

    (B) At least once every twelve months for each employee covered unless the attending physician believes a longer interval (not greater than biennially) is appropriate;

    (C) At termination of employment or reassignment to an area where the employee would not be covered if the employee has not had an examination within the last six months;

    (D) As soon as possible upon notification by an employee that the employee has developed signs or symptoms indicating possible overexposure to hazardous substances or health hazards, or that the employee has been injured or exposed above the permissible exposure limits or published exposure levels in an emergency situation;

    (E) At more frequent times, if the examining physician determines that an increased frequency of examination is medically necessary.

(ii)   For employees covered under paragraph (f)(2)(iii) and for all employees including those of employers covered by paragraph (a)(1)(v) who may have been injured, received a health impairment, developed signs or symptoms which may have resulted from exposure to hazardous substances resulting from an emergency incident, or exposed during an emergency incident to hazardous substances at concentrations above the permissible exposure limits or the published exposure levels without the necessary personal protective equipment being used:

    (A) As soon as possible following the emergency incident or development of signs or symptoms;

    (B) At additional times, if the examining physician determines that follow-up examinations or consultations are medically necessary.

(4) Content of medical examinations and consultations.

(i)   Medical examinations required by paragraph (f)(3) of this section shall include a medical and work history (or updated history if one is in the employee's file) with special emphasis on symptoms related to the handling of hazardous substances and health hazards, and to fitness for duty including the ability to wear any required PPE under conditions (i.e., temperature extremes) that may be expected at the work site.

(ii)   The content of medical examinations or consultations made available to employees pursuant to paragraph (f) shall be determined by the attending physician. The guidelines in the Occupational Safety and Health Guidance Manual for Hazardous Waste Site Activities (see appendix D, Reference #10) should be consulted.

(5) Examination by a physician and costs. All medical examinations and procedures shall be performed by or under the supervision of a licensed physician, preferably one knowledgeable in occupational medicine, and shall be provided without cost to the employee, without loss of pay, and at a reasonable time and place.

(6) Information provided to the physician. The employer shall provide one copy of this standard and its appendices to the attending physician, and in addition the following for each employee:

    (i)   A description of the employee's duties as they relate to the employee's exposures.

    (ii)   The employee's exposure levels or anticipated exposure levels.

    (iii)   A description of any personal protective equipment used or to be used.

    (iv)   Information from previous medical examinations of the employee which is not readily available to the examining physician.

    (v)   Information required by § 1910.134.

(7) Physician's written opinion.

    (i)   The employer shall obtain and furnish the employee with a copy of a written opinion from the attending physician containing the following:

        (A) The physician's opinion as to whether the employee has any detected medical conditions which would place the employee at increased risk of material impairment of the employee's health from work in hazardous waste operations or emergency response, or from respirator use.

        (B) The physician's recommended limitations upon the employee's assigned work.

        (C) The results of the medical examination and tests if requested by the employee.

        (D) A statement that the employee has been informed by the physician of the results of the medical examination and any medical conditions which require further examination or treatment.

    (ii)   The written opinion obtained by the employer shall not reveal specific findings or diagnoses unrelated to occupational exposures.

(8) Recordkeeping.

    (i)   An accurate record of the medical surveillance required by paragraph (f) of this section shall be retained.

        This record shall be retained for the period specified and meet the criteria of 29 CFR 1910.20.

    (ii)   The record required in paragraph (f)(8)(i) of this section shall include at least the following information:

        (A) The name and social security number of the employee;

        (B) Physician's written opinions, recommended limitations, and results of examinations and tests;

(C) Any employee medical complaints related to exposure to hazardous substances;

(D) A copy of the information provided to the examining physician by the employer, with the exception of the standard and its appendices.

(g) Engineering controls, work practices, and personal protective equipment for employee protection. Engineering controls, work practices, personal protective equipment, or a combination of these shall be implemented in accordance with this paragraph to protect employees from exposure to hazardous substances and safety and health hazards.

(1) Engineering controls, work practices and PPE for substances regulated in subparts G and Z. (i) Engineering controls and work practices shall be instituted to reduce and maintain employee exposure to or below the permissible exposure limits for substances regulated by 29 CFR Part 1910, to the extent required by subpart Z, except to the extent that such controls and practices are not feasible.

Note to (g)(1)(i): Engineering controls which may be feasible include the use of pressurized cabs or control booths on equipment, and/or the use of remotely operated material handling equipment.

Work practices which may be feasible are removing all non-essential employees from potential exposure during opening of drums, wetting down dusty operations and locating employees upwind of possible hazards.

(ii) Whenever engineering controls and work practices are not feasible or not required, any reasonable combination of engineering controls, work practices and PPE shall be used to reduce and maintain employee exposures to or below the permissible exposure limits or dose limits for substances regulated by 29 CFR Part 1910, subpart Z.

(iii) The employer shall not implement a schedule of employee rotation as a means of compliance with permissible exposure limits or dose limits except when there is no other feasible way of complying with the airborne or dermal dose limits for ionizing radiation.

(iv) The provisions of 29 CFR, subpart G, shall be followed.

(2) Engineering controls, work practices, and PPE for substances not regulated in subparts G and Z. An appropriate combination of engineering controls, work practices and personal protective equipment shall be used to reduce and maintain employee exposure to or below published exposure levels for hazardous substances and health hazards not regulated by 29 CFR Part 1910, subparts G and Z. The employer may use the published literature and MSDS as a guide in making the employer's determination as to what level of protection the employer believes is appropriate for hazardous substances and health hazards for which there is no permissible exposure limit or published exposure limit.

(3) Personal protective equipment selection.

(i) Personal protective equipment (PPE) shall be selected and used which will protect employees from the hazards and potential hazards they are likely to encounter as identified during the site characterization and analysis.

(ii) Personal protective equipment selection shall be based on an evaluation of the performance characteristics of the PPE relative to the requirements and limitations of the site, the task-specific conditions and duration, and the hazards and potential hazards identified at the site.

(iii) Positive pressure self-contained breathing apparatus, or positive pressure air-line respirators equipped with an escape air supply, shall be used when chemical exposure levels present will create a substantial possibility of immediate death, immediate serious illness or injury, or impair the ability to escape.

(iv) Totally-encapsulating chemical protective suits (protection equivalent to Level A protection as recommended in appendix B) shall be used in conditions where skin absorption of a hazardous substance may result in a substantial possibility of immediate death, immediate serious illness or injury, or impair the ability to escape.

(v) The level of protection provided by PPE selection shall be increased when additional information on site conditions indicates that increased protection is necessary to reduce employee exposures below permissible exposure limits and published exposure levels for hazardous substances and health hazards. (See appendix B for guidance on selecting PPE ensembles.)

Note to (g)(3): The level of employee protection provided may be decreased when additional information or site conditions show that decreased protection will not result in hazardous exposures to employees.

(vi) Personal protective equipment shall be selected and used to meet the requirements of 29 CFR Part 1910, subpart I, and additional requirements specified in this section.

(4) Totally-encapsulating chemical protective suits.

(i) Totally-encapsulating suits shall protect employees from the particular hazards which are identified during site characterization and analysis.

(ii) Totally-encapsulating suits shall be capable of maintaining positive air pressure. (See appendix A for a test method which may be used to evaluate this requirement.)

(iii) Totally-encapsulating suits shall be capable of preventing inward test gas leakage of more than 0.5 percent. (See appendix A for a test method which may be used to evaluate this requirement.)

(5) Personal protective equipment (PPE) program. A written personal protective equipment program, which is part of the employer's safety and health program required in paragraph (b) of this section or required in paragraph (p)(1) of this section and which is also a part of the site-specific safety and health plan, shall be established. The PPE program shall address the elements listed below. When elements, such as donning and doffing procedures, are provided by the manufacturer of a piece of equipment and are attached to the plan, they need not be rewritten into the plan as long as they adequately address the procedure or element.

(i)   PPE selection based upon site hazards,

(ii)   PPE use and limitations of the equipment,

(iii)   Work mission duration,

(iv)   PPE maintenance and storage,

(v)   PPE decontamination and disposal,

(vi)   PPE training and proper fitting,

(vii)   PPE donning and doffing procedures,

(viii)   PPE inspection procedures prior to, during, and after use,

(ix)   Evaluation of the effectiveness of the PPE program, and

(x)   Limitations during temperature extremes, heat stress, and other appropriate medical considerations.

(h) Monitoring-

(1) General.

(i)   Monitoring shall be performed in accordance with this paragraph where there may be a question of employee exposure to hazardous concentrations of hazardous substances in order to assure proper selection of engineering controls, work practices and personal protective equipment so that employees are not exposed to levels which exceed permissible exposure limits, or published exposure levels if there are no permissible exposure limits, for hazardous substances.

(ii)   Air monitoring shall be used to identify and quantify airborne levels of hazardous substances and safety and health hazards in order to determine the appropriate level of employee protection needed on site.

(2) Initial entry. Upon initial entry, representative air monitoring shall be conducted to identify any IDLH condition, exposure over permissible exposure limits or published exposure levels, exposure over a radioactive material's dose limits or other dangerous condition such as the presence of flammable atmospheres or oxygen-deficient environments.

(3) Periodic monitoring. Periodic monitoring shall be conducted when the possibility of an IDLH condition or flammable atmosphere has developed or when there is indication that exposures may have risen over permissible exposure limits or published exposure levels since prior monitoring. Situations where it shall be considered whether the possibility that exposures have risen are as follows:

(i)   When work begins on a different portion of the site.

(ii)   When contaminants other than those previously identified are being handled.

(iii)   When a different type of operation is initiated (e.g., drum opening as opposed to exploratory well drilling).

(iv) When employees are handling leaking drums or containers or working in areas with obvious liquid contamination (e.g., a spill or lagoon).

(4) Monitoring of high-risk employees. After the actual clean-up phase of any hazardous waste operation commences; for example, when soil, surface water or containers are moved or disturbed; the employer shall monitor those employees likely to have the highest exposures to hazardous substances and health hazards likely to be present above permissible exposure limits or published exposure levels by using personal sampling frequently enough to characterize employee exposures. If the employees likely to have the highest exposure are over permissible exposure limits or published exposure limits, then monitoring shall continue to determine all employees likely to be above those limits.

The employer may utilize a representative sampling approach by documenting that the employees and chemicals chosen for monitoring are based on the criteria stated above.

Note to (h): It is not required to monitor employees engaged in site characterization operations covered by paragraph (c) of this section.

(i) Informational programs. Employers shall develop and implement a program, which is part of the employer's safety and health program required in paragraph (b) of this section, to inform employees, contractors, and subcontractors (or their representative) actually engaged in hazardous waste operations of the nature, level and degree of exposure likely as a result of participation in such hazardous waste operations. Employees, contractors and subcontractors working outside of the operations part of a site are not covered by this standard.

(j) Handling drums and containers-

(1) General.

    (i) Hazardous substances and contaminated soils, liquids, and other residues shall be handled, transported, labeled, and disposed of in accordance with this paragraph.

    (ii) Drums and containers used during the clean-up shall meet the appropriate DOT, OSHA, and EPA regulations for the wastes that they contain.

    (iii) When practical, drums and containers shall be inspected and their integrity shall be assured prior to being moved. Drums or containers that cannot be inspected before being moved because of storage conditions (i.e., buried beneath the earth, stacked behind other drums, stacked several tiers high in a pile, etc.) shall be moved to an accessible location and inspected prior to further handling.

    (iv) Unlabelled drums and containers shall be considered to contain hazardous substances and handled accordingly until the contents are positively identified and labeled.

    (v) Site operations shall be organized to minimize the amount of drum or container movement.

(vi) Prior to movement of drums or containers, all employees exposed to the transfer operation shall be warned of the potential hazards associated with the contents of the drums or containers.

(vii) U.S. Department of Transportation specified salvage drums or containers and suitable quantities of proper absorbent shall be kept available and used in areas where spills, leaks, or ruptures may occur.

(viii) Where major spills may occur, a spill containment program, which is part of the employer's safety and health program required in paragraph (b) of this section, shall be implemented to contain and isolate the entire volume of the hazardous substance being transferred.

(ix) Drums and containers that cannot be moved without rupture, leakage, or spillage shall be emptied into a sound container using a device classified for the material being transferred.

(x) A ground-penetrating system or other type of detection system or device shall be used to estimate the location and depth of buried drums or containers.

(xi) Soil or covering material shall be removed with caution to prevent drum or container rupture.

(xii) Fire extinguishing equipment meeting the requirements of 29 CFR Part 1910, subpart L, shall be on hand and ready for use to control incipient fires.

(2) Opening drums and containers. The following procedures shall be followed in areas where drums or containers are being opened:

(i) Where an airline respirator system is used, connections to the source of air supply shall be protected from contamination and the entire system shall be protected from physical damage.

(ii) Employees not actually involved in opening drums or containers shall be kept a safe distance from the drums or containers being opened.

(iii) If employees must work near or adjacent to drums or containers being opened, a suitable shield that does not interfere with the work operation shall be placed between the employee and the drums or containers being opened to protect the employee in case of accidental explosion.

(iv) Controls for drum or container opening equipment, monitoring equipment, and fire suppression equipment shall be located behind the explosion-resistant barrier.

(v) When there is a reasonable possibility of flammable atmospheres being present, material handling equipment and hand tools shall be of the type to prevent sources of ignition.

(vi) Drums and containers shall be opened in such a manner that excess interior pressure will be safely relieved. If pressure can not be relieved from a remote location, appropriate shielding shall be placed between the employee and the drums or containers to reduce the risk of employee injury.

(vii) Employees shall not stand upon or work from drums or containers.

(3) Material handling equipment. Material handling equipment used to transfer drums and containers shall be selected, positioned and operated to minimize sources of ignition related to the equipment from igniting vapors released from ruptured drums or containers.

(4) Radioactive wastes. Drums and containers containing radioactive wastes shall not be handled until such time as their hazard to employees is properly assessed.

(5) Shock sensitive wastes. As a minimum, the following special precautions shall be taken when drums and containers containing or suspected of containing shock-sensitive wastes are handled:

(i)   All non-essential employees shall be evacuated from the area of transfer.

(ii)   Material handling equipment shall be provided with explosive containment devices or protective shields to protect equipment operators from exploding containers.

(iii)   An employee alarm system capable of being perceived above surrounding light and noise conditions shall be used to signal the commencement and completion of explosive waste handling activities.

(iv)   Continuous communications (i.e., portable radios, hand signals, telephones, as appropriate) shall be maintained between the employee-in-charge of the immediate handling area and both the site safety and health supervisor and the command post until such time as the handling operation is completed. Communication equipment or methods that could cause shock sensitive materials to explode shall not be used.

(v)   Drums and containers under pressure, as evidenced by bulging or swelling, shall not be moved until such time as the cause for excess pressure is determined and appropriate containment procedures have been implemented to protect employees from explosive relief of the drum.

(vi)   Drums and containers containing packaged laboratory wastes shall be considered to contain shock-sensitive or explosive materials until they have been characterized.

Caution: Shipping of shock sensitive wastes may be prohibited under U.S. Department of Transportation regulations. Employers and their shippers should refer to 49 CFR 173.21 and 173.50.

(6) Laboratory waste packs. In addition to the requirements of paragraph (j)(5) of this section, the following precautions shall be taken, as a minimum, in handling laboratory waste packs (lab packs):

(i)   Lab packs shall be opened only when necessary and then only by an individual knowledgeable in the inspection, classification, and segregation of the containers within the pack according to the hazards of the wastes.

(ii)   If crystalline material is noted on any container, the contents shall be handled as a shock-sensitive waste until the contents are identified.

(7) Sampling of drum and container contents. Sampling of containers and drums shall be done in accordance with a sampling procedure which is part of the site safety

and health plan developed for and available to employees and others at the specific worksite.

(8) Shipping and transport.

    (i)   Drums and containers shall be identified and classified prior to packaging for shipment.

    (ii)   Drum or container staging areas shall be kept to the minimum number necessary to identify and classify materials safely and prepare them for transport.

    (iii)   Staging areas shall be provided with adequate access and egress routes.

    (iv)   Bulking of hazardous wastes shall be permitted only after a thorough characterization of the materials has been completed.

(9) Tank and vault procedures.

    (i)   Tanks and vaults containing hazardous substances shall be handled in a manner similar to that for drums and containers, taking into consideration the size of the tank or vault.

    (ii)   Appropriate tank or vault entry procedures as described in the employer's safety and health plan shall be followed whenever employees must enter a tank or vault.

(k) Decontamination-

  (1) General. Procedures for all phases of decontamination shall be developed and implemented in accordance with this paragraph.

  (2) Decontamination procedures.

    (i)   A decontamination procedure shall be developed, communicated to employees and implemented before any employees or equipment may enter areas on site where potential for exposure to hazardous substances exists.

    (ii)   Standard operating procedures shall be developed to minimize employee contact with hazardous substances or with equipment that has contacted hazardous substances.

    (iii)   All employees leaving a contaminated area shall be appropriately decontaminated; all contaminated clothing and equipment leaving a contaminated area shall be appropriately disposed of or decontaminated.

    (iv)   Decontamination procedures shall be monitored by the site safety and health supervisor to determine their effectiveness.

          When such procedures are found to be ineffective, appropriate steps shall be Ftaken to correct any deficiencies.

  (3) Location. Decontamination shall be performed in geographical areas that will minimize the exposure of uncontaminated employees or equipment to contaminated employees or equipment.

  (4) Equipment and solvents. All equipment and solvents used for decontamination shall be decontaminated or disposed of properly.

(5) Personal protective clothing and equipment.

    (i)   Protective clothing and equipment shall be decontaminated, cleaned, laundered, maintained or replaced as needed to maintain their effectiveness.

    (ii)   Employees whose non-impermeable clothing becomes wetted with hazardous substances shall immediately remove that clothing and proceed to shower. The clothing shall be disposed of or decontaminated before it is removed from the work zone.

(6) Unauthorized employees. Unauthorized employees shall not remove protective clothing or equipment from change rooms.

(7) Commercial laundries or cleaning establishments. Commercial laundries or cleaning establishments that decontaminate protective clothing or equipment shall be informed of the potentially harmful effects of exposures to hazardous substances.

(8) Showers and change rooms. Where the decontamination procedure indicates a need for regular showers and change rooms outside of a contaminated area, they shall be provided and meet the requirements of 29 CFR 1910.141. If temperature conditions prevent the effective use of water, then other effective means for cleansing shall be provided and used.

(l) Emergency response by employees at uncontrolled hazardous waste sites-

(1) Emergency response plan. (i) An emergency response plan shall be developed and implemented by all employers within the scope of paragraphs (a)(1) (i)–(ii) of this section to handle anticipated emergencies prior to the commencement of hazardous waste operations. The plan shall be in writing and available for inspection and copying by employees, their representatives, OSHA personnel and other governmental agencies with relevant responsibilities.

    (ii)   Employers who will evacuate their employees from the danger area when an emergency occurs, and who do not permit any of their employees to assist in handling the emergency, are exempt from the requirements of this paragraph if they provide an emergency action plan complying with § 1910.38(a) of this part.

(2) Elements of an emergency response plan. The employer shall develop an emergency response plan for emergencies which shall address, as a minimum, the following:

    (i)   Pre-emergency planning.

    (ii)   Personnel roles, lines of authority, and communication.

    (iii)   Emergency recognition and prevention.

    (iv)   Safe distances and places of refuge.

    (v)   Site security and control.

    (vi)   Evacuation routes and procedures.

    (vii)   Decontamination procedures which are not covered by the site safety and health plan.

(viii) Emergency medical treatment and first aid.

(ix) Emergency alerting and response procedures.

(x) Critique of response and follow-up.

(xi) PPE and emergency equipment.

(3) Procedures for handling emergency incidents.

(i) In addition to the elements for the emergency response plan required in paragraph (1)(2) of this section, the following elements shall be included for emergency response plans:

(A) Site topography, layout, and prevailing weather conditions.

(B) Procedures for reporting incidents to local, state, and federal governmental agencies.

(ii) The emergency response plan shall be a separate section of the Site Safety and Health Plan.

(iii) The emergency response plan shall be compatible and integrated with the disaster, fire and/or emergency response plans of local, state, and federal agencies.

(iv) The emergency response plan shall be rehearsed regularly as part of the overall training program for site operations.

(v) The site emergency response plan shall be reviewed periodically and, as necessary, be amended to keep it current with new or changing site conditions or information.

(vi) An employee alarm system shall be installed in accordance with 29 CFR 1910.165 to notify employees of an emergency situation; to stop work activities if necessary; to lower background noise in order to speed communication; and to begin emergency procedures.

(vii) Based upon the information available at time of the emergency, the employer shall evaluate the incident and the site response capabilities and proceed with the appropriate steps to implement the site emergency response plan.

(m) Illumination. Areas accessible to employees shall be lighted to not less than the minimum illumination intensities listed in the following Table H-120.1 while any work is in progress:

Table H-120.1 Minimum Illumination Intensities in Foot-Candles

Foot-candles	Area or operations
5	General site areas.
3	Excavation and waste areas, accessways, active storage areas, loading platforms, refueling, and field maintenance areas.
5	Indoors: Warehouses, corridors, hallways, and exitways.
5	Tunnels, shafts, and general underground work areas. (Exception: Minimum of 10 foot-candles is required at tunnel and shaft heading during drilling mucking, and scaling. Mine Safety and Health Administration approved cap lights shall be acceptable for use in the tunnel heading.)
10	General shops (e.g., mechanical and electrical equipment rooms, active storerooms, barracks or living quarters, locker or dressing rooms, dining areas, and indoor toilets and workrooms.)
30	First aid stations, infirmaries, and offices.

(n) Sanitation at temporary workplaces-

  (1) Potable water.

    (i)   An adequate supply of potable water shall be provided on the site.

    (ii)   Portable containers used to dispense drinking water shall be capable of being tightly closed, and equipped with a tap. Water shall not be dipped from containers.

    (iii) Any container used to distribute drinking water shall be clearly marked as to the nature of its contents and not used for any other purpose.

    (iv) Where single service cups (to be used but once) are supplied, both a sanitary container for the unused cups and a receptacle for disposing of the used cups shall be provided.

  (2) Nonpotable water.

    (i)   Outlets for nonpotable water, such as water for firefighting purposes, shall be identified to indicate clearly that the water is unsafe and is not to be used for drinking, washing, or cooking purposes.

    (ii)   There shall be no cross-connection, open or potential, between a system furnishing potable water and a system furnishing nonpotable water.

  (3) Toilet facilities.

    (i)   Toilets shall be provided for employees according to the following Table H-120.2

Table H-120.2 Toilet Facilities

Number of employees	Minimum number of facilities
20 or fewer	One.
More than 20, fewer than 200	One toilet seat and one urinal per 40 employees.
More than 200	One toilet seat and one urinal per 50 employees.

    (ii)   Under temporary field conditions, provisions shall be made to assure that at least one toilet facility is available.

    (iii) Hazardous waste sites not provided with a sanitary sewer shall be provided with the following toilet facilities unless prohibited by local codes:

        (A) Chemical toilets;

        (B) Recirculating toilets;

        (C) Combustion toilets; or

        (D) Flush toilets.

    (iv) The requirements of this paragraph for sanitation facilities shall not apply to mobile crews having transportation readily available to nearby toilet facilities.

(v)   Doors entering toilet facilities shall be provided with entrance locks controlled from inside the facility.

(4) Food handling. All food service facilities and operations for employees shall meet the applicable laws, ordinances, and regulations of the jurisdictions in which they are located.

(5) Temporary sleeping quarters. When temporary sleeping quarters are provided, they shall be heated, ventilated, and lighted.

(6) Washing facilities. The employer shall provide adequate washing facilities for employees engaged in operations where hazardous substances may be harmful to employees. Such facilities shall be in near proximity to the worksite; in areas where exposures are below permissible exposure limits and published exposure levels and which are under the controls of the employer; and shall be so equipped as to enable employees to remove hazardous substances from themselves.

(7) Showers and change rooms. When hazardous waste clean-up or removal operations commence on a site and the duration of the work will require six months or greater time to complete, the employer shall provide showers and change rooms for all employees exposed to hazardous substances and health hazards involved in hazardous waste clean-up or removal operations.

(i)   Showers shall be provided and shall meet the requirements of 29 CFR 1910.141(d)(3).

(ii) Change rooms shall be provided and shall meet the requirements of 29 CFR 1910.141(e). Change rooms shall consist of two separate change areas separated by the shower area required in paragraph (n)(7)(i) of this section. One change area, with an exit leading off the worksite, shall provide employees with a clean area where they can remove, store, and put on street clothing. The second area, with an exit to the worksite, shall provide employees with an area where they can put on, remove and store work clothing and personal protective equipment.

(iii) Showers and change rooms shall be located in areas where exposures are below the permissible exposure limits and published exposure levels. If this cannot be accomplished, then a ventilation system shall be provided that will supply air that is below the permissible exposure limits and published exposure levels.

(iv) Employers shall assure that employees shower at the end of their work shift and when leaving the hazardous waste site.

(o) New technology programs.

(1) The employer shall develop and implement procedures for the introduction of effective new technologies and equipment developed for the improved protection of employees working with hazardous waste clean-up operations, and the same

shall be implemented as part of the site safety and health program to assure that employee protection is being maintained.

(2) New technologies, equipment or control measures available to the industry, such as the use of foams, absorbents, adsorbents, neutralizers, or other means to suppress the level of air contaminates while excavating the site or for spill control, shall be evaluated by employers or their representatives. Such an evaluation shall be done to determine the effectiveness of the new methods, materials, or equipment before implementing their use on a large scale for enhancing employee protection. Information and data from manufacturers or suppliers may be used as part of the employer's evaluation effort. Such evaluations shall be made available to OSHA upon request.

(p) Certain operations conducted under the Resource Conservation and Recovery Act of 1976 (RCRA). Employers conducting operations at treatment, storage and disposal (TSD) facilities specified in paragraph (a)(1)(iv) of this section shall provide and implement the programs specified in this paragraph. (See the "Notes and Exceptions" to paragraph (a)(2)(iii) of this section for employers not covered.)

(1) Safety and health program. The employer shall develop and implement a written safety and health program for employees involved in hazardous waste operations that shall be available for inspection by employees, their representatives and OSHA personnel. The program shall be designed to identify, evaluate and control safety and health hazards in their facilities for the purpose of employee protection, to provide for emergency response meeting the requirements of paragraph (p)(8) of this section and to address as appropriate site analysis, engineering controls, maximum exposure limits, hazardous waste handling procedures and uses of new technologies.

(2) Hazard communication program. The employer shall implement a hazard communication program meeting the requirements of 29 CFR 1910.1200 as part of the employer's safety and program.

Note to 1910.120-The exemption for hazardous waste provided in § 1910.1200 is applicable to this section.

(3) Medical surveillance program. The employer shall develop and implement a medical surveillance program meeting the requirements of paragraph (f) of this section.

(4) Decontamination program. The employer shall develop and implement a decontamination procedure meeting the requirements of paragraph (k) of this section.

(5) New technology program. The employer shall develop and implement procedures meeting the requirements of paragraph (o) of this section for introducing new and innovative equipment into the workplace.

(6) Material handling program. Where employees will be handling drums or containers, the employer shall develop and implement procedures meeting the requirements of paragraphs (j)(1)(ii) through (viii) and (xi) of this section, as well as (j)(3) and (j)(8) of this section prior to starting such work.

(7) Training program-

    (i)   New employees. The employer shall develop and implement a training program, which is part of the employer's safety and health program, for employees exposed to health hazards or hazardous substances at TSD operations to enable the employees to perform their assigned duties and functions in a safe and healthful manner so as not endanger themselves or other employees. The initial training shall be for 24 hours and refresher training shall be for eight hours annually. Employees who have received the initial training required by this paragraph shall be given a written certificate attesting that they have successfully completed the necessary training.

    (ii)   Current employees. Employers who can show by an employee's previous work experience and/or training that the employee has had training equivalent to the initial training required by this paragraph, shall be considered as meeting the initial training requirements of this paragraph as to that employee. Equivalent training includes the training that existing employees might have already received from actual site work experience. Current employees shall receive eight hours of refresher training annually.

    (iii)   Trainers. Trainers who teach initial training shall have satisfactorily completed a training course for teaching the subjects they are expected to teach or they shall have the academic credentials and instruction experience necessary to demonstrate a good command of the subject matter of the courses and competent instructional skills.

(8) Emergency response program-

    (i)   Emergency response plan. An emergency response plan shall be developed and implemented by all employers. Such plans need not duplicate any of the subjects fully addressed in the employer's contingency planning required by permits, such as those issued by the U.S. Environmental Protection Agency, provided that the contingency plan is made part of the emergency response plan. The emergency response plan shall be a written portion of the employer's safety and health program required in paragraph (p)(1) of this section. Employers who will evacuate their employees from the worksite location when an emergency occurs and who do not permit any of their employees to assist in handling the emergency are exempt from the requirements of paragraph (p)(8) if they provide an emergency action plan complying with § 1910.38(a) of this part.

    (ii)   Elements of an emergency response plan. The employer shall develop an emergency response plan for emergencies which shall address, as a minimum, the following areas to the extent that they are not addressed in any specific program required in this paragraph:

        (A) Pre-emergency planning and coordination with outside parties.

        (B) Personnel roles, lines of authority, and communication.

        (C) Emergency recognition and prevention.

(D) Safe distances and places of refuge.

(E) Site security and control.

(F) Evacuation routes and procedures.

(G) Decontamination procedures.

(H) Emergency medical treatment and first aid.

(I) Emergency alerting and response procedures.

(J) Critique of response and follow-up.

(K) PPE and emergency equipment.

(iii) Training.

(A) Training for emergency response employees shall be completed before they are called upon to perform in real emergencies. Such training shall include the elements of the emergency response plan, standard operating procedures the employer has established for the job, the personal protective equipment to be worn and procedures for handling emergency incidents.

Exception #1: An employer need not train all employees to the degree specified if the employer divides the work force in a manner such that a sufficient number of employees who have responsibility to control emergencies have the training specified, and all other employees, who may first respond to an emergency incident, have sufficient awareness training to recognize that an emergency response situation exists and that they are instructed in that case to summon the fully trained employees and not attempt control activities for which they are not trained.

Exception #2: An employer need not train all employees to the degree specified if arrangements have been made in advance for an outside fully-trained emergency response team to respond in a reasonable period, and all employees, who may come to the incident first, have sufficient awareness training to recognize that an emergency response situation exists and they have been instructed to call the designated outside fully-trained emergency response team for assistance.

(B) Employee members of TSD facility emergency response organizations shall be trained to a level of competence in the recognition of health and safety hazards to protect themselves and other employees. This would include training in the methods used to minimize the risk from safety and health hazards; in the safe use of control equipment; in the selection and use of appropriate personal protective equipment; in the safe operating procedures to be used at the incident scene; in the techniques of coordination with other employees to minimize risks; in the appropriate response to overexposure from health hazards or injury to themselves and other employees; and in the recognition of subsequent symptoms which may result from overexposures.

(C) The employer shall certify that each covered employee has attended and successfully completed the training required in paragraph (p)(8)(iii) of this section, or shall certify the employee's competency at least yearly. The method used to demonstrate competency for certification of training shall be recorded and maintained by the employer.

(iv) Procedures for handling emergency incidents.

(A) In addition to the elements for the emergency response plan required in paragraph (p)(8)(ii) of this section, the following elements shall be included for emergency response plans to the extent that they do not repeat any information already contained in the emergency response plan:

(1) Site topography, layout, and prevailing weather conditions.

(2) Procedures for reporting incidents to local, state, and federal governmental agencies.

(B) The emergency response plan shall be compatible and integrated with the disaster, fire and/or emergency response plans of local, state, and federal agencies.

(C) The emergency response plan shall be rehearsed regularly as part of the overall training program for site operations.

(D) The site emergency response plan shall be reviewed periodically and, as necessary, be amended to keep it current with new or changing site conditions or information.

(E) An employee alarm system shall be installed in accordance with 29 CFR 1910.165 to notify employees of an emergency situation; to stop work activities if necessary; to lower background noise in order to speed communication; and to begin emergency procedures.

(F) Based upon the information available at time of the emergency, the employer shall evaluate the incident and the site response capabilities and proceed with the appropriate steps to implement the site emergency response plan.

(q) Emergency response to hazardous substance releases. This paragraph covers employers whose employees are engaged in emergency response no matter where it occurs except that it does not cover employees engaged in operations specified in paragraphs (a)(1)(i) through (a)(1)(iv) of this section. Those emergency response organizations who have developed and implemented programs equivalent to this paragraph for handling releases of hazardous substances pursuant to section 303 of the Superfund Amendments and Reauthorization Act of 1986 (Emergency Planning and Community Right-to-Know Act of 1986, 42 U.S.C. 11003) shall be deemed to have met the requirements of this paragraph.

(1) Emergency response plan. An emergency response plan shall be developed and implemented to handle anticipated emergencies prior to the commencement of emergency response operations.

The plan shall be in writing and available for inspection and copying by employees, their representatives and OSHA personnel.

Employers who will evacuate their employees from the danger area when an emergency occurs, and who do not permit any of their employees to assist in handling the emergency, are exempt from the requirements of this paragraph if they provide an emergency action plan in accordance with § 1910.38(a) of this part.

(2) Elements of an emergency response plan. The employer shall develop an emergency response plan for emergencies which shall address, as a minimum, the following to the extent that they are not addressed elsewhere:

    (i)   Pre-emergency planning and coordination with outside parties.

    (ii)   Personnel roles, lines of authority, training, and communication.

    (iii)   Emergency recognition and prevention.

    (iv)   Safe distances and places of refuge.

    (v)   Site security and control.

    (vi)   Evacuation routes and procedures.

    (vii)   Decontamination.

    (viii) Emergency medical treatment and first aid.

    (ix)   Emergency alerting and response procedures.

    (x)   Critique of response and follow-up.

    (xi)   PPE and emergency equipment.

    (xii)   Emergency response organizations may use the local emergency response plan or the state emergency response plan or both, as part of their emergency response plan to avoid duplication.

            Those items of the emergency response plan that are being properly addressed by the SARA Title III plans may be substituted into their emergency plan or otherwise kept together for the employer and employee's use.

(3) Procedures for handling emergency response.

    (i)   The senior emergency response official responding to an emergency shall become the individual in charge of a site-specific Incident Command System (ICS). All emergency responders and their communications shall be coordinated and controlled through the individual in charge of the ICS assisted by the senior official present for each employer.

            Note to (q)(3)(i).-The "senior official" at an emergency response is the most senior official on the site who has the responsibility for controlling the operations at the site. Initially it is the senior officer on the first-due piece of responding emergency apparatus to arrive on the incident scene. As more senior officers arrive (i.e., battalion chief, fire chief, state law enforcement official, site coordinator, etc.), the position is passed up the line of authority which has been previously established.

(ii) The individual in charge of the ICS shall identify, to the extent possible, all hazardous substances or conditions present and shall address as appropriate site analysis, use of engineering controls, maximum exposure limits, hazardous substance handling procedures, and use of any new technologies.

(iii) Based on the hazardous substances and/or conditions present, the individual in charge of the ICS shall implement appropriate emergency operations, and assure that the personal protective equipment worn is appropriate for the hazards to be encountered. However, personal protective equipment shall meet, at a minimum, the criteria contained in 29 CFR 1910.156(e) when worn while performing fire fighting operations beyond the incipient stage for any incident.

(iv) Employees engaged in emergency response and exposed to hazardous substances presenting an inhalation hazard or potential inhalation hazard shall wear positive pressure self-contained breathing apparatus while engaged in emergency response, until such time that the individual in charge of the ICS determines through the use of air monitoring that a decreased level of respiratory protection will not result in hazardous exposures to employees.

(v) The individual in charge of the ICS shall limit the number of emergency response personnel at the emergency site, in those areas of potential or actual exposure to incident or site hazards, to those who are actively performing emergency operations. However, operations in hazardous areas shall be performed using the buddy system in groups of two or more.

(vi) Back-up personnel shall stand by with equipment ready to provide assistance or rescue. Advance first aid support personnel, as a minimum, shall also stand by with medical equipment and transportation capability.

(vii) The individual in charge of the ICS shall designate a safety official, who is knowledgable in the operations being implemented at the emergency response site, with specific responsibility to identify and evaluate hazards and to provide direction with respect to the safety of operations for the emergency at hand.

(viii) When activities are judged by the safety official to be an IDLH condition and/or to involve an imminent danger condition, the safety official shall have the authority to alter, suspend, or terminate those activities. The safety official shall immediately inform the individual in charge of the ICS of any actions needed to be taken to correct these hazards at the emergency scene.

(ix) After emergency operations have terminated, the individual in charge of the ICS shall implement appropriate decontamination procedures.

(x) When deemed necessary for meeting the tasks at hand, approved self-contained compressed air breathing apparatus may be used with approved cylinders from other approved self-contained compressed air breathing apparatus provided that such cylinders are of the same capacity and pressure rating. All compressed air cylinders used with self-contained breathing apparatus shall meet U.S. Department of Transportation and National Institute for Occupational Safety and Health criteria.

(4) Skilled support personnel. Personnel, not necessarily an employer's own employees, who are skilled in the operation of certain equipment, such as mechanized earth moving or digging equipment or crane and hoisting equipment, and who are needed temporarily to perform immediate emergency support work that cannot reasonably be performed in a timely fashion by an employer's own employees, and who will be or may be exposed to the hazards at an emergency response scene, are not required to meet the training required in this paragraph for the employer's regular employees. However, these personnel shall be given an initial briefing at the site prior to their participation in any emergency response. The initial briefing shall include instruction in the wearing of appropriate personal protective equipment, what chemical hazards are involved, and what duties are to be performed.

All other appropriate safety and health precautions provided to the employer's own employees shall be used to assure the safety and health of these personnel.

(5) Specialist employees. Employees who, in the course of their regular job duties, work with and are trained in the hazards of specific hazardous substances, and who will be called upon to provide technical advice or assistance at a hazardous substance release incident to the individual in charge, shall receive training or demonstrate competency in the area of their specialization annually.

(6) Training. Training shall be based on the duties and function to be performed by each responder of an emergency response organization.

The skill and knowledge levels required for all new responders, those hired after the effective date of this standard, shall be conveyed to them through training before they are permitted to take part in actual emergency operations on an incident.

Employees who participate, or are expected to participate, in emergency response, shall be given training in accordance with the following paragraphs:

(i)   First responder awareness level. First responders at the awareness level are individuals who are likely to witness or discover a hazardous substance release and who have been trained to initiate an emergency response sequence by notifying the proper authorities of the release. They would take no further action beyond notifying the authorities of the release. First responders at the awareness level shall have sufficient training or have had sufficient experience to objectively demonstrate competency in the following areas:

(A) An understanding of what hazardous substances are, and the risks associated with them in an incident.

(B) An understanding of the potential outcomes associated with an emergency created when hazardous substances are present.

(C) The ability to recognize the presence of hazardous substances in an emergency.

(D) The ability to identify the hazardous substances, if possible.

(E) An understanding of the role of the first responder awareness individual in the employer's emergency response plan including site security and control and the U.S. Department of Transportation's Emergency Response Guidebook.

(F) The ability to realize the need for additional resources, and to make appropriate notifications to the communication center.

(ii) First responder operations level. First responders at the operations level are individuals who respond to releases or potential releases of hazardous substances as part of the initial response to the site for the purpose of protecting nearby persons, property, or the environment from the effects of the release. They are trained to respond in a defensive fashion without actually trying to stop the release. Their function is to contain the release from a safe distance, keep it from spreading, and prevent exposures. First responders at the operational level shall have received at least eight hours of training or have had sufficient experience to objectively demonstrate competency in the following areas in addition to those listed for the awareness level and the employer shall so certify:

(A) Knowledge of the basic hazard and risk assessment techniques.

(B) Know how to select and use proper personal protective equipment provided to the first responder operational level.

(C) An understanding of basic hazardous materials terms.

(D) Know how to perform basic control, containment and/or confinement operations within the capabilities of the resources and personal protective equipment available with their unit.

(E) Know how to implement basic decontamination procedures.

(F) An understanding of the relevant standard operating procedures and termination procedures.

(iii) Hazardous materials technician. Hazardous materials technicians are individuals who respond to releases or potential releases for the purpose of stopping the release. They assume a more aggressive role than a first responder at the operations level in that they will approach the point of release in order to plug, patch or otherwise stop the release of a hazardous substance. Hazardous materials technicians shall have received at least 24 hours of training equal to the first responder operations level and in addition have competency in the following areas and the employer shall so certify:

(A) Know how to implement the employer's emergency response plan.

(B) Know the classification, identification and verification of known and unknown materials by using field survey instruments and equipment.

(C) Be able to function within an assigned role in the Incident Command System.

(D) Know how to select and use proper specialized chemical personal protective equipment provided to the hazardous materials technician.

(E) Understand hazard and risk assessment techniques.

(F) Be able to perform advance control, containment, and/or confinement operations within the capabilities of the resources and personal protective equipment available with the unit.

(G) Understand and implement decontamination procedures.

(H) Understand termination procedures.

(I) Understand basic chemical and toxicological terminology and behavior.

(iv) Hazardous materials specialist. Hazardous materials specialists are individuals who respond with and provide support to hazardous materials technicians. Their duties parallel those of the hazardous materials technicians; however, those duties require a more directed or specific knowledge of the various substances they may be called upon to contain. The hazardous materials specialist would also act as the site liaison with Federal, state, local and other government authorities in regards to site activities.

Hazardous materials specialists shall have received at least 24 hours of training equal to the technician level and in addition have competency in the following areas and the employer shall so certify:

(A) Know how to implement the local emergency response plan.

(B) Understand classification, identification and verification of known and unknown materials by using advanced survey instruments and equipment.

(C) Know of the state emergency response plan.

(D) Be able to select and use proper specialized chemical personal protective equipment provided to the hazardous materials specialist.

(E) Understand in-depth hazard and risk techniques.

(F) Be able to perform specialized control, containment, and/or confinement operations within the capabilities of the resources and personal protective equipment available.

(G) Be able to determine and implement decontamination procedures.

(H) Have the ability to develop a site safety and control plan.

(I) Understand chemical, radiological and toxicological terminology and behavior.

(v) On scene incident commander. Incident commanders, who will assume control of the incident scene beyond the first responder awareness level, shall receive at least 24 hours of training equal to the first responder operations level and in addition have competency in the following areas and the employer shall so certify:

(A) Know and be able to implement the employer's incident command system.

(B) Know how to implement the employer's emergency response plan.

(C) Know and understand the hazards and risks associated with employees working in chemical protective clothing.

(D) Know how to implement the local emergency response plan.

(E) Know of the state emergency response plan and of the Federal Regional Response Team.

(F) Know and understand the importance of decontamination procedures.

(7) Trainers. Trainers who teach any of the above training subjects shall have satisfactorily completed a training course for teaching the subjects they are expected to teach, such as the courses offered by the U.S. National Fire Academy, or they shall have the training and/or academic credentials and instructional experience necessary to demonstrate competent instructional skills and a good command of the subject matter of the courses they are to teach.

(8) Refresher training.

(i) Those employees who are trained in accordance with paragraph (q)(6) of this section shall receive annual refresher training of sufficient content and duration to maintain their competencies, or shall demonstrate competency in those areas at least yearly.

(ii) A statement shall be made of the training or competency, and if a statement of competency is made, the employer shall keep a record of the methodology used to demonstrate competency.

(9) Medical surveillance and consultation.

(i) Members of an organized and designated HAZMAT team and hazardous materials specialists shall receive a baseline physical examination and be provided with medical surveillance as required in paragraph (f) of this section.

(ii) Any emergency response employees who exhibits signs or symptoms which may have resulted from exposure to hazardous substances during the course of an emergency incident, either immediately or subsequently, shall be provided with medical consultation as required in paragraph (f)(3)(ii) of this section.

(10) Chemical protective clothing. Chemical protective clothing and equipment to be used by organized and designated HAZMAT team members, or to be used by hazardous materials specialists, shall meet the requirements of paragraphs (g)(3) through (5) of this section.

(11) Post-emergency response operations. Upon completion of the emergency response, if it is determined that it is necessary to remove hazardous substances, health hazards, and materials contaminated with them (such as contaminated soil or other elements of the natural environment) from the site of the incident, the employer conducting the clean-up shall comply with one of the following:

(i) Meet all of the requirements of paragraphs (b) through (o) of this section; or

(ii) Where the clean-up is done on plant property using plant or workplace employees, such employees shall have completed the training requirements of the following: 29 CFR 1910.38(a); 1910.134; 1910.1200, and other appropriate safety and health training made necessary by the tasks that they are expected to perform such as personal protective equipment and decontamination procedures. All equipment to be used in the performance of the clean-up work shall be in serviceable condition and shall have been inspected prior to use.

APPENDICES TO § 1910.120—HAZARDOUS WASTE OPERATIONS AND EMERGENCY RESPONSE

Note: The following appendices serve as non-mandatory guidelines to assist employees and employers in complying with the appropriate requirements of this section. However, paragraph 1910.120(g) makes mandatory in certain circumstances the use of Level A and Level B PPE protection.

Appendix A to § 1910.120—Personal Protective Equipment Test Methods

This appendix sets forth the non-mandatory examples of tests which may be used to evaluate compliance with § 1910.120 (g)(4) (ii) and (iii). Other tests and other challenge agents may be used to evaluate compliance.

A. Totally-encapsulating chemical protective suit pressure test

1.0-Scope

11.1 This practice measures the ability of a gas tight totally-encapsulating chemical protective suit material, seams, and closures to maintain a fixed positive pressure. The results of this practice allow the gas tight integrity of a totally-encapsulating chemical protective suit to be evaluated.

1.2 Resistance of the suit materials to permeation, penetration, and degradation by specific hazardous substances is not determined by this test method.

2.0-Definition of Terms

2.1 Totally-encapsulated chemical protective suit (TECP suit) means a full body garment which is constructed of protective clothing materials; covers the wearer's torso, head, arms, legs and respirator; may cover the wearer's hands and feet with tightly attached gloves and boots; completely encloses the wearer and respirator by itself or in combination with the wearer's gloves and boots.

2.2 Protective clothing material means any material or combination of materials used in an item of clothing for the purpose of isolating parts of the body from direct contact with a potentially hazardous liquid or gaseous chemicals.

2.3 Gas tight means, for the purpose of this test method, the limited flow of a gas under pressure from the inside of a TECP suit to atmosphere at a prescribed pressure and time interval.

3.0-Summary of Test Method

3.1 The TECP suit is visually inspected and modified for the test. The test apparatus is attached to the suit to permit inflation to the pre-test suit expansion pressure for removal of suit wrinkles and creases. The pressure is lowered to the test pressure and monitored for three minutes. If the pressure drop is excessive, the TECP suit fails the test and is removed from service. The test is repeated after leak location and repair.

4.0-Required Supplies

4.1 Source of compressed air.

4.2 Test apparatus for suit testing, including a pressure measurement device with a sensitivity of at least 1/4 inch water gauge.

4.3 Vent valve closure plugs or sealing tape.

4.4 Soapy water solution and soft brush.

4.5 Stop watch or appropriate timing device.

5.0-Safety Precautions

5.1 Care shall be taken to provide the correct pressure safety devices required for the source of compressed air used.

6.0-Test Procedure

6.1 Prior to each test, the tester shall perform a visual inspection of the suit. Check the suit for seam integrity by visually examining the seams and gently pulling on the seams. Ensure that all air supply lines, fittings, visor, zippers, and valves are secure and show no signs of deterioration.

6.1.1 Seal off the vent valves along with any other normal inlet or exhaust points (such as umbilical air line fittings or face piece opening) with tape or other appropriate means (caps, plugs, fixture, etc.). Care should be exercised in the sealing process not to damage any of the suit components.

6.1.2 Close all closure assemblies.

6.1.3 Prepare the suit for inflation by providing an improvised connection point on the suit for connecting an airline. Attach the pressure test apparatus to the suit to permit suit inflation from a compressed air source equipped with a pressure indicating regulator. The leak tightness of the pressure test apparatus should be tested before and after each test by closing off the end of the tubing attached to the suit and assuring a pressure of three inches water gauge for three minutes can be maintained.

If a component is removed for the test, that component shall be replaced and a second test conducted with another component removed to permit a complete test of the ensemble.

6.1.4 The pre-test expansion pressure (A) and the suit test pressure (B) shall be supplied by the suit manufacturer, but in no case shall they be less than: (A) = three inches water gauge; and (B) = two inches water gauge. The ending suit pressure (C) shall be no less than 80 percent of the test pressure (B); i.e., the pressure drop shall not exceed 20 percent of the test pressure (B).

6.1.5 Inflate the suit until the pressure inside is equal to pressure (A), the pre-test expansion suit pressure. Allow at least one minute to fill out the wrinkles in the suit. Release sufficient air to reduce the suit pressure to pressure (B), the suit test pressure. Begin timing. At the end of three minutes, record the suit pressure as pressure (C), the ending suit pressure.

The difference between the suit test pressure and the ending suit test pressure (B − C) shall be defined as the suit pressure drop.

6.1.6 If the suit pressure drop is more than 20 percent of the suit test pressure (B) during the three-minute test period, the suit fails the test and shall be removed from service.

7.0-Retest Procedure

7.1 If the suit fails the test, check for leaks by inflating the suit to pressure (A) and brushing or wiping the entire suit (including seams, closures, lens gaskets, glove-to-sleeve joints, etc.) with a mild soap and water solution. Observe the suit for the formation of soap bubbles, which is an indication of a leak. Repair all identified leaks.

7.2 Retest the TECP suit as outlined in Test procedure 6.0.

8.0-Report

8.1 Each TECP suit tested by this practice shall have the following information recorded:

8.1.1 Unique identification number, identifying brand name, date of purchase, material of construction, and unique fit features, e.g., special breathing apparatus.

8.1.2 The actual values for test pressures (A), (B), and (C) shall be recorded along with the specific observation times.

If the ending pressure (C) is less than 80 percent of the test pressure (B), the suit shall be identified as failing the test.

When possible, the specific leak location shall be identified in the test records. Retest pressure data shall be recorded as an additional test.

8.1.3 The source of the test apparatus used shall be identified and the sensitivity of the pressure gauge shall be recorded.

8.1.4 Records shall be kept for each pressure test even if repairs are being made at the test location.

CAUTION

Visually inspect all parts of the suit to be sure they are positioned correctly and secured tightly before putting the suit back into service. Special care should be taken to examine each exhaust valve to make sure it is not blocked.

Care should also be exercised to assure that the inside and outside of the suit is completely dry before it is put into storage.

B. Totally-encapsulating chemical protective suit qualitative leak test

1.0-Scope

1.1 This practice semi-qualitatively tests gas tight totally-encapsulating chemical protective suit integrity by detecting inward leakage of ammonia vapor. Since no modifications are made to the suit to carry out this test, the results from this practice provide a realistic test for the integrity of the entire suit.

1.2 Resistance of the suit materials to permeation, penetration, and degradation is not determined by this test method. ASTM test methods are available to test suit materials for these characteristics and the tests are usually conducted by the manufacturers of the suits.

2.0-Definition of Terms

2.1 Totally-encapsulated chemical protective suit (TECP suit) means a full body garment which is constructed of protective clothing materials; covers the wearer's torso, head, arms, legs and respirator; may cover the wearer's hands and feet with tightly attached gloves and boots; completely encloses the wearer and respirator by itself or in combination with the wearer's gloves, and boots.

2.2 Protective clothing material means any material or combination of materials used in an item of clothing for the purpose of isolating parts of the body from direct contact with a potentially hazardous liquid or gaseous chemicals.

2.3 Gas tight means, for the purpose of this test method, the limited flow of a gas under pressure from the inside of a TECP suit to atmosphere at a prescribed pressure and time interval.

2.4 Intrusion coefficient means a number expressing the level of protection provided by a gas tight totally-encapsulating chemical protective suit. The intrusion coefficient is calculated by dividing the test room challenge agent concentration by the concentration of challenge agent found inside the suit. The accuracy of the intrusion coefficient is dependent on the challenge agent monitoring methods. The larger the intrusion coefficient, the greater the protection provided by the TECP suit.

3.0-Summary of Recommended Practice

3.1 The volume of concentrated aqueous ammonia solution (ammonia hydroxide NH_4OH) required to generate the test atmosphere is determined using the directions outlined in 6.1. The suit is donned by a person wearing the appropriate respiratory equipment (either a positive pressure self-contained breathing apparatus or a positive pressure supplied air respirator) and worn inside the enclosed test room. The concentrated aqueous ammonia solution is taken by the suited individual into the test room and poured into an open plastic pan. A two-minute evaporation period is observed before the test room concentration is measured, using a high range ammonia length of stain detector tube.

When the ammonia vapor reaches a concentration of between 1000 and 1200 ppm, the suited individual starts a standardized exercise protocol to stress and flex the suit.

After this protocol is completed, the test room concentration is measured again. The suited individual exits the test room and his stand-by person measures the ammonia concentration inside the suit using a low range ammonia length of stain detector tube or other more sensitive ammonia detector.

A stand-by person is required to observe the test individual during the test procedure; aid the person in donning and doffing the TECP suit; and monitor the suit interior. The intrusion coefficient of the suit can be calculated by dividing the average test area concentration by the interior suit concentration.

A colorimetric ammonia indicator strip of bromophenol blue or equivalent is placed on the inside of the suit face piece lens so that the suited individual is able to detect a color change and know if the suit has a significant leak. If a color change is observed, the individual shall leave the test room immediately.

4.0-Required Supplies

4.1 A supply of concentrated aqueous ammonium hydroxide (58% by weight).

4.2 A supply of bromophenol/blue indicating paper or equivalent, sensitive to 5–10 ppm ammonia or greater over a two-minute period of exposure. [pH 3.0 (yellow) to pH 4.6 (blue)]

4.3 A supply of high range (0.5–10 volume percent) and low range (5–700 ppm) detector tubes for ammonia and the corresponding sampling pump. More sensitive ammonia detectors can be substituted for the low range detector tubes to improve the sensitivity of this practice.

4.4 A shallow plastic pan (PVC) at least 12":14":1" and a half pint plastic container (PVC) with tightly closing lid.

4.5 A graduated cylinder or other volumetric measuring device of at least 50 milliliters in volume with an accuracy of at least ±1 milliliter.

5.0-Safety Precautions

5.1 Concentrated aqueous ammonium hydroxide, NH_4OH, is a corrosive volatile liquid requiring eye, skin, and respiratory protection. The person conducting the test shall review the MSDS for aqueous ammonia.

5.2 Since the established permissible exposure limit for ammonia is 35 ppm as a 15 minute STEL, only persons wearing a positive pressure self-contained breathing apparatus or a positive pressure supplied air respirator shall be in the chamber. Normally only the person wearing the totally-encapsulating suit will be inside the chamber. A stand-by person shall have a positive pressure self-contained breathing apparatus, or a positive pressure supplied air respirator available to enter the test area should the suited individual need assistance.

5.3 A method to monitor the suited individual must be used during this test. Visual contact is the simplest but other methods using communication devices are acceptable.

5.4 The test room shall be large enough to allow the exercise protocol to be carried out and then to be ventilated to allow for easy exhaust of the ammonia test atmosphere after the test(s) are completed.

5.5 Individuals shall be medically screened for the use of respiratory protection and checked for allergies to ammonia before participating in this test procedure.

6.0-Test Procedure

6.1.1 Measure the test area to the nearest foot and calculate its volume in cubic feet. Multiply the test area volume by 0.2 milliliter of concentrated aqueous ammonia solution per cubic foot of test area volume to determine the approximate volume of concentrated aqueous ammonia required to generate 1000 ppm in the test area.

6.1.2 Measure this volume from the supply of concentrated aqueous ammonia and place it into a closed plastic container.

6.1.3 Place the container, several high range ammonia detector tubes, and the pump in the clean test pan and locate it near the test area entry door so that the suited individual has easy access to these supplies.

6.2.1 In a non-contaminated atmosphere, open a pre-sealed ammonia indicator strip and fasten one end of the strip to the inside of the suit face shield lens where it can be seen by the wearer.

Moisten the indicator strip with distilled water. Care shall be taken not to contaminate the detector part of the indicator paper by touching it. A small piece of masking tape or equivalent should be used to attach the indicator strip to the interior of the suit face shield.

6.2.2 If problems are encountered with this method of attachment, the indicator strip can be attached to the outside of the respirator face piece lens being used during the test.

6.3 Don the respiratory protective device normally used with the suit, and then don the TECP suit to be tested. Check to be sure all openings which are intended to be sealed (zippers, gloves, etc.) are completely sealed. DO NOT, however, plug off any venting valves.

6.4 Step into the enclosed test room such as a closet, bathroom, or test booth, equipped with an exhaust fan. No air should be exhausted from the chamber during the test because this will dilute the ammonia challenge concentrations.

6.5 Open the container with the pre-measured volume of concentrated aqueous ammonia within the enclosed test room, and pour the liquid into the empty plastic test pan. Wait two minutes to allow for adequate volatilization of the concentrated aqueous ammonia. A small mixing fan can be used near the evaporation pan to increase the evaporation rate of the ammonia solution.

6.6 After two minutes a determination of the ammonia concentration within the chamber should be made using the high range colorimetric detector tube. A concentration of 1000 ppm ammonia or greater shall be generated before the exercises are started.

6.7 To test the integrity of the suit the following four minute exercise protocol should be followed:

6.7.1 Raising the arms above the head with at least 15 raising motions completed in one minute.

6.7.2 Walking in place for one minute with at least 15 raising motions of each leg in a one-minute period.

6.7.3 Touching the toes with a least 10 complete motions of the arms from above the head to touching of the toes in a one-minute period.

6.7.4 Knee bends with at least 10 complete standing and squatting motions in a one-minute period.

6.8 If at any time during the test the colorimetric indicating paper should change colors, the test should be stopped and section 6.10 and 6.12 initiated (See #4.2).

6.9 After completion of the test exercise, the test area concentration should be measured again using the high range colorimetric detector tube.

6.10 Exit the test area.

6.11 The opening created by the suit zipper or other appropriate suit penetration should be used to determine the ammonia concentration in the suit with the low range length of stain detector tube or other ammonia monitor. The internal TECP suit air should be sampled far enough from the enclosed test area to prevent a false ammonia reading.

6.12 After completion of the measurement of the suit interior ammonia concentration the test is concluded and the suit is doffed and the respirator removed.

6.13 The ventilating fan for the test room should be turned on and allowed to run for enough time to remove the ammonia gas. The fan shall be vented to the outside of the building.

6.14 Any detectable ammonia in the suit interior (five ppm ammonia (NH_3) or more for the length of stain detector tube) indicates that the suit has failed the test. When other ammonia detectors are used a lower level of detection is possible, and it should be specified as the pass/fail criteria.

6.15 By following this test method, an intrusion coefficient of approximately 200 or more can be measured with the suit in a completely operational condition. If the intrusion coefficient is 200 or more, then the suit is suitable for emergency response and field use.

7.0-Retest Procedures

7.1 If the suit fails this test, check for leaks by following the pressure test in test A above.

7.2 Retest the TECP suit as outlined in the test procedure 6.0.

8.0-Report

8.1 Each gas tight totally-encapsulating chemical protective suit tested by this practice shall have the following information recorded.

8.1.1 Unique identification number, identifying brand name, date of purchase, material of construction, and unique suit features; e.g., special breathing apparatus.

8.1.2 General description of test room used for test.

8.1.3 Brand name and purchase date of ammonia detector strips and color change data.

8.1.4 Brand name, sampling range, and expiration date of the length of stain ammonia detector tubes. The brand name and model of the sampling pump should also be recorded. If another type of ammonia detector is used, it should be identified along with its minimum detection limit for ammonia.

8.1.5 Actual test results shall list the two test area concentrations, their average, the interior suit concentration, and the calculated intrusion coefficient. Retest data shall be recorded as an additional test.

8.2 The evaluation of the data shall be specified as "suit passed" or "suit failed," and the date of the test. Any detectable ammonia (five ppm or greater for the length of stain detector tube) in the suit interior indicates the suit has failed this test. When other ammonia detectors are used, a lower level of detection is possible and it should be specified as the pass fail criteria.

CAUTION

Visually inspect all parts of the suit to be sure they are positioned correctly and secured tightly before putting the suit back into service. Special care should be taken to examine each exhaust valve to make sure it is not blocked.

Care should also be exercised to assure that the inside and outside of the suit is completely dry before it is put into storage.

Appendix B to § 1910.120—General Description and Discussion of the Levels of Protection and Protective Gear

This appendix sets forth information about personal protective equipment (PPE) protection levels which may be used to assist employers in complying with the PPE requirements of this section.

As required by the standard, PPE must be selected which will protect employees from the specific hazards which they are likely to encounter during their work on-site.

Selection of the appropriate PPE is a complex process which should take into consideration a variety of factors. Key factors involved in this process are identification of the hazards, or suspected hazards; their routes of potential hazard to employees (inhalation, skin absorption, ingestion, and eye or skin contact); and the performance of the PPE materials (and seams) in providing a barrier to these hazards. The amount of protection provided by PPE is material-hazard specific. That is, protective equipment materials will protect well against some hazardous substances and poorly, or not at all, against others. In many instances, protective equipment materials cannot be found which will provide continuous protection from the particular hazardous substance.

In these cases the breakthrough time of the protective material should exceed the work durations.

Other factors in this selection process to be considered are matching the PPE to the employee's work requirements and task-specific conditions. The durability of PPE materials, such as tear strength and seam strength, should be considered in relation to the employee's tasks. The effects of PPE in relation to heat stress and task duration are a factor in selecting and using PPE. In some cases layers of PPE may be necessary to provide sufficient protection, or to protect expensive PPE inner garments, suits or equipment.

The more that is known about the hazards at the site, the easier the job of PPE selection becomes. As more information about the hazards and conditions at the site becomes available, the site supervisor can make decisions to up-grade or down-grade the level of PPE protection to match the tasks at hand.

The following are guidelines which an employer can use to begin the selection of the appropriate PPE. As noted above, the site information may suggest the use of combinations of PPE selected from the different protection levels (i.e., A, B, C, or D) as being more suitable to the hazards of the work. It should be cautioned that the listing below does not fully address the performance of the specific PPE material in relation to the specific hazards at the job site, and that PPE selection, evaluation and re-selection is an ongoing process until sufficient information about the hazards and PPE performance is obtained.

Part A. Personal protective equipment is divided into four categories based on the degree of protection afforded. (See Part B of this appendix for further explanation of Levels A, B, C, and D hazards.)

I. Level A-To be selected when the greatest level of skin, respiratory, and eye protection is required.

The following constitute Level A equipment; it may be used as appropriate.

1. Positive pressure, full face-piece self-contained breathing apparatus (SCBA), or positive pressure supplied air respirator with escape SCBA, approved by the National Institute for Occupational Safety and Health (NIOSH).

2. Totally-encapsulating chemical-protective suit.

3. Coveralls.[1]

4. Long underwear.[1]

5. Gloves, outer, chemical-resistant.

6. Gloves, inner, chemical-resistant.

7. Boots, chemical-resistant, steel toe and shank.

8. Hard hat (under suit).[1]

9. Disposable protective suit, gloves and boots (depending on suit construction, may be worn over totally-encapsulating suit).

---

[1]Optional, as applicable.

II. Level B-The highest level of respiratory protection is necessary but a lesser level of skin protection is needed.

The following constitute Level B equipment; it may be used as appropriate.

1. Positive pressure, full-facepiece self-contained breathing apparatus (SCBA), or positive pressure supplied air respirator with escape SCBA (NIOSH approved).

2. Hooded chemical-resistant clothing (overalls and long-sleeved jacket; coveralls; one or two-piece chemical-splash suit; disposable chemical-resistant overalls).

3. Coveralls.[1]

4. Gloves, outer, chemical-resistant.

5. Gloves, inner, chemical-resistant.

6. Boots, outer, chemical-resistant steel toe and shank.

7. Boot-covers, outer, chemical-resistant (disposable).[1]

8. Hard hat.[1]

9. [Reserved]

10. Face shield.[1]

III. Level C-The concentration(s) and type(s) of airborne substance(s) are known and the criteria for using air purifying respirators are met.

The following constitute Level C equipment; it may be used as appropriate.

1. Full-face or half-mask, air purifying respirators (NIOSH approved).

2. Hooded chemical-resistant clothing (overalls; two-piece chemical-splash suit; disposable chemical-resistant overalls).

3. Coveralls.[1]

4. Gloves, outer, chemical-resistant.

5. Gloves, inner, chemical-resistant.

6. Boots (outer), chemical-resistant steel toe and shank.[1]

7. Boot-covers, outer, chemical-resistant (disposable)[1].

8. Hard hat.[1]

9. Escape mask.[1]

10. Face shield.[1]

IV. Level D-A work uniform affording minimal protection, used for nuisance contamination only.

The following constitute Level D equipment; it may be used as appropriate:

1. Coveralls.

2. Gloves.[1]

3. Boots/shoes, chemical-resistant steel toe and shank.

4. Boots, outer, chemical-resistant (disposable).[1]

5. Safety glasses or chemical splash goggles*.

6. Hard hat.[1]

7. Escape mask.[1]

8. Face shield.[1]

Part B. The types of hazards for which levels A, B, C, and D protection are appropriate are described below:

I. Level A-Level A protection should be used when:

1. The hazardous substance has been identified and requires the highest level of protection for skin, eyes, and the respiratory system based on either the measured (or potential for) high concentration of atmospheric vapors, gases, or particulates; or the site operations and work functions involve a high potential for splash, immersion, or exposure to unexpected vapors, gases, or particulates of materials that are harmful to skin or capable of being absorbed through the skin;

2. Substances with a high degree of hazard to the skin are known or suspected to be present, and skin contact is possible; or

3. Operations are being conducted in confined, poorly ventilated areas, and the absence of conditions requiring Level A has not yet been determined.

II. Level B-Level B protection should be used when:

1. The type and atmospheric concentration of substances have been identified and require a high level of respiratory protection, but less skin protection;

2. The atmosphere contains less than 19.5 percent oxygen; or

3. The presence of incompletely identified vapors or gases is indicated by a direct-reading organic vapor detection instrument, but vapors and gases are not suspected of containing high levels of chemicals harmful to skin or capable of being absorbed through the skin.

Note: This involves atmospheres with IDLH concentrations of specific substances that present severe inhalation hazards and that do not represent a severe skin hazard; or that do not meet the criteria for use of air-purifying respirators.

III. Level C-Level C protection should be used when:

1. The atmospheric contaminants, liquid splashes, or other direct contact will not adversely affect or be absorbed through any exposed skin;

2. The types of air contaminants have been identified, concentrations measured, and an air-purifying respirator is available that can remove the contaminants; and

3. All criteria for the use of air-purifying respirators are met.

IV. Level D-Level D protection should be used when:

1. The atmosphere contains no known hazard; and

2. Work functions preclude splashes, immersion, or the potential for unexpected inhalation of or contact with hazardous levels of any chemicals.

Note: As stated before, combinations of personal protective equipment other than those described for Levels A, B, C, and D protection may be more appropriate and may be used to provide the proper level of protection.

As an aid in selecting suitable chemical protective clothing, it should be noted that the National Fire Protection Association (NFPA) has developed standards on chemical protective clothing.

The standards that have been adopted include:

NFPA 1991-Standard on Vapor-Protective Suits for Hazardous Chemical Emergencies (EPA Level A Protective Clothing).

NFPA 1992-Standard on Liquid Splash-Protective Suits for Hazardous Chemical Emergencies (EPA Level B Protective Clothing).

NFPA 1993-Standard on Liquid Splash-Protective Suits for Non-emergency, Non-flammable Hazardous Chemical Situations (EPA Level B Protective Clothing).

These standards apply documentation and performance requirements to the manufacture of chemical protective suits. Chemical protective suits meeting these requirements are labelled as compliant with the appropriate standard. It is recommended that chemical protective suits that meet these standards be used.

Appendix C to § 1910.120—Compliance Guidelines

1. Occupational Safety and Health Program. Each hazardous waste site clean-up effort will require an occupational safety and health program headed by the site coordinator or the employer's representative. The purpose of the program will be the protection of employees at the site and will be an extension of the employer's overall safety and health program. The program will need to be developed before work begins on the site and implemented as work proceeds as stated in paragraph (b). The program is to facilitate coordination and communication of safety and health issues among personnel responsible for the various activities which will take place at the site. It will provide the overall means for planning and implementing the needed safety and health training and job orientation of employees who will be working at the site. The program will provide the means for identifying and controlling worksite hazards and the means for monitoring program effectiveness. The program will need to cover the responsibilities and authority of the site coordinator or the employer's manager on the site for the safety and health of employees at the site, and the relationships with contractors or support services as to what each employer's safety and health responsibilities are for their employees on the site. Each contractor on the site needs to have its own safety and health program so structured that it will smoothly interface with the program of the site coordinator or principal contractor.

Also those employers involved with treating, storing or disposal of hazardous waste as covered in paragraph (p) must have implemented a safety and health program for their employees. This program is to include the hazard communication program required in paragraph (p)(1) and the training required in paragraphs (p)(7) and (p)(8) as parts of the employers comprehensive overall safety and health program. This program is to be in writing.

Each site or workplace safety and health program will need to include the following: (1) Policy statements of the line of authority and accountability for implementing the program, the objectives of the program and the role of the site safety and health supervisor or manager and staff; (2) means or methods for the development of procedures for identifying and controlling workplace hazards at the site; (3) means or methods for the development and communication to employees of the various plans, work rules, standard operating procedures and practices that pertain to individual employees and supervisors; (4) means for the training of supervisors and employees to develop the needed skills and knowledge to perform their work in a safe and healthful manner; (5) means to anticipate and prepare for emergency situations; and (6) means for obtaining information feedback to aid in evaluating the program and for improving the effectiveness of the program.

The management and employees should be trying continually to improve the effectiveness of the program, thereby enhancing the protection being afforded those working on the site.

Accidents on the site or workplace should be investigated to provide information on how such occurrences can be avoided in the future. When injuries or illnesses occur on the site or workplace, they will need to be investigated to determine what needs to be done to prevent this incident from occurring again.

Such information will need to be used as feedback on the effectiveness of the program and the information turned into positive steps to prevent any reoccurrence. Receipt of employee suggestions or complaints relating to safety and health issues involved with site or workplace activities is also a feedback mechanism that can be used effectively to improve the program and may serve in part as an evaluative tool(s).

For the development and implementation of the program to be the most effective, professional safety and health personnel should be used. Certified Safety Professionals, Board Certified Industrial Hygienists or Registered Professional Safety Engineers are good examples of professional stature for safety and health managers who will administer the employer's program.

2. Training. The training programs for employees subject to the requirements of paragraph (e) of this standard should address: the safety and health hazards employees should expect to find on hazardous waste clean-up sites; what control measures or techniques are effective for those hazards; what monitoring procedures are effective in characterizing exposure levels; what makes an effective employer's safety and health program; what a site safety and health plan should include; hands-on training with personal protective equipment and clothing they may be expected to use; the contents of the OSHA standard relevant to the employee's duties and function; and, employee's responsibilities under OSHA and other regulations. Supervisors will need training in their responsibilities under the safety and health program and its subject areas such as the spill containment program, the personal protective equipment pro-

gram, the medical surveillance program, the emergency response plan and other areas.

The training programs for employees subject to the requirements of paragraph (p) of this standard should address: the employer's safety and health program elements impacting employees; the hazard communication program; the medical surveillance program; the hazards and the controls for such hazards that employees need to know for their job duties and functions. All require annual refresher training.

The training programs for employees covered by the requirements of paragraph (q) of this standard should address those competencies required for the various levels of response such as: the hazards associated with hazardous substances; hazard identification and awareness; notification of appropriate persons; the need for and use of personal protective equipment including respirators; the decontamination procedures to be used; preplanning activities for hazardous substance incidents including the emergency response plan; company standard operating procedures for hazardous substance emergency responses; the use of the incident command system and other subjects. Hands-on training should be stressed whenever possible. Critiques done after an incident which include an evaluation of what worked and what did not and how could the incident be better handled the next time may be counted as training time.

For hazardous materials specialists (usually members of hazardous materials teams), the training should address the care, use and/or testing of chemical protective clothing including totally encapsulating suits, the medical surveillance program, the standard operating procedures for the hazardous materials team including the use of plugging and patching equipment and other subject areas.

Officers and leaders who may be expected to be in charge at an incident should be fully knowledgeable of their company's incident command system. They should know where and how to obtain additional assistance and be familiar with the local district's emergency response plan and the state emergency response plan.

Specialist employees such as technical experts, medical experts or environmental experts that work with hazardous materials in their regular jobs, who may be sent to the incident scene by the shipper, manufacturer or governmental agency to advise and assist the person in charge of the incident should have training on an annual basis. Their training should include the care and use of personal protective equipment including respirators; knowledge of the incident command system and how they are to relate to it; and those areas needed to keep them current in their respective field as it relates to safety and health involving specific hazardous substances.

Those skilled support personnel, such as employees who work for public works departments or equipment operators who operate bulldozers, sand trucks, backhoes, etc., who may be called to the incident scene to provide emergency support assistance, should have at least a safety and health briefing before entering the area of potential or actual exposure. These skilled support personnel, who have not been a part of the emergency response plan and do not meet the training requirements, should be made aware of the hazards they face and should be provided all necessary protective clothing and equipment required for their tasks.

There are two National Fire Protection Association standards, NFPA 472-"Standard for Professional Competence of Responders to Hazardous Material Incidents" and NFPA 471-"Recommended Practice for Responding to Hazardous Material Incidents", which are excellent resource documents to aid fire departments and other emergency response organizations in developing their training program materials. NFPA 472 provides guidance on the skills and knowledge needed for first responder awareness level, first responder operations level, hazmat technicians, and hazmat specialist.

It also offers guidance for the officer corp who will be in charge of hazardous substance incidents.

3. Decontamination. Decontamination procedures should be tailored to the specific hazards of the site, and may vary in complexity and number of steps, depending on the level of hazard and the employee's exposure to the hazard. Decontamination procedures and PPE decontamination methods will vary depending upon the specific substance, since one procedure or method may not work for all substances. Evaluation of decontamination methods and procedures should be performed, as necessary, to assure that employees are not exposed to hazards by reusing PPE. References in appendix D may be used for guidance in establishing an effective decontamination program. In addition, the U.S. Coast Guard's Manual, "Policy Guidance for Response to Hazardous Chemical Releases," U.S. Department of Transportation, Washington, DC (COMDTINST M16465.30) is a good reference for establishing an effective decontamination program.

4. Emergency response plans. States, along with designated districts within the states, will be developing or have developed local emergency response plans. These state and district plans should be utilized in the emergency response plans called for in the standard. Each employer should assure that its emergency response plan is compatible with the local plan. The major reference being used to aid in developing the state and local district plans is the Hazardous Materials Emergency Planning Guide, NRT-1. The current Emergency Response Guidebook from the U.S. Department of Transportation, CMA's CHEMTREC and the Fire Service Emergency Management Handbook may also be used as resources.

Employers involved with treatment, storage, and disposal facilities for hazardous waste, which have the required contingency plan called for by their permit, would not need to duplicate the same planning elements. Those items of the emergency response plan that are properly addressed in the contingency plan may be substituted into the emergency response plan required in 1910.120 or otherwise kept together for employer and employee use.

5. Personal protective equipment programs. The purpose of personal protective clothing and equipment (PPE) is to shield or isolate individuals from the chemical, physical, and biologic hazards that may be encountered at a hazardous substance site.

As discussed in appendix B, no single combination of protective equipment and clothing is capable of protecting against all hazards. Thus PPE should be used in conjunction with other protective methods and its effectiveness evaluated periodically.

The use of PPE can itself create significant worker hazards, such as heat stress, physical and psychological stress, and impaired vision, mobility, and communication. For any given situation, equipment and clothing should be selected that provide an adequate level of protection. However, over-protection, as well as under-protection, can be hazardous and should be avoided where possible.

Two basic objectives of any PPE program should be to protect the wearer from safety and health hazards, and to prevent injury to the wearer from incorrect use and/or malfunction of the PPE.

To accomplish these goals, a comprehensive PPE program should include hazard identification, medical monitoring, environmental surveillance, selection, use, maintenance, and decontamination of PPE and its associated training.

The written PPE program should include policy statements, procedures, and guidelines. Copies should be made available to all employees, and a reference copy should be made available at the worksite.

Technical data on equipment, maintenance manuals, relevant regulations, and other essential information should also be collected and maintained.

6. Incident command system (ICS). Paragraph 1910.120(q)(3)(ii) requires the implementation of an ICS. The ICS is an organized approach to effectively control and manage operations at an emergency incident. The individual in charge of the ICS is the senior official responding to the incident. The ICS is not much different than the "command post" approach used for many years by the fire service. During large complex fires involving several companies and many pieces of apparatus, a command post would be established. This enabled one individual to be in charge of managing the incident, rather than having several officers from different companies making separate, and sometimes conflicting, decisions. The individual in charge of the command post would delegate responsibility for performing various tasks to subordinate officers. Additionally, all communications were routed through the command post to reduce the number of radio transmissions and eliminate confusion. However, strategy, tactics, and all decisions were made by one individual.

The ICS is a very similar system, except it is implemented for emergency response to all incidents, both large and small, that involve hazardous substances.

For a small incident, the individual in charge of the ICS may perform many tasks of the ICS. There may not be any, or little, delegation of tasks to subordinates. For example, in response to a small incident, the individual in charge of the ICS, in addition to normal command activities, may become the safety officer and may designate only one employee (with proper equipment) as a back-up to provide assistance if needed. OSHA does recommend, however, that at least two employees be designated as back-up personnel since the assistance needed may include rescue.

To illustrate the operation of the ICS, the following scenario might develop during a small incident, such as an overturned tank truck with a small leak of flammable liquid.

The first responding senior officer would implement and take command of the ICS.

That person would size-up the incident and determine if additional personnel and apparatus were necessary; would determine what actions to take to control the leak; and, determine the proper level of personal protective equipment.

If additional assistance is not needed, the individual in charge of the ICS would implement actions to stop and control the leak using the fewest number of personnel that can effectively accomplish the tasks. The individual in charge of the ICS then would designate himself as the safety officer and two other employees as a back-up in case rescue may become necessary. In this scenario, decontamination procedures would not be necessary.

A large complex incident may require many employees and difficult, time-consuming efforts to control. In these situations, the individual in charge of the ICS will want to delegate different tasks to subordinates in order to maintain a span of control that will keep the number of subordinates, that are reporting, to a manageable level.

Delegation of task at large incidents may be by location, where the incident scene is divided into sectors, and subordinate officers coordinate activities within the sector that they have been assigned.

Delegation of tasks can also be by function. Some of the functions that the individual in charge of the ICS may want to delegate at a large incident are: medical services; evacuation; water supply; resources (equipment, apparatus); media relations; safety; and, site control (integrate activities with police for crowd and traffic control). Also for a large incident, the individual in charge of the ICS will designate several employees as back-up personnel; and a number of safety officers to monitor conditions and recommend safety precautions.

Therefore, no matter what size or complexity an incident may be, by implementing an ICS there will be one individual in charge who makes the decisions and gives directions; and, all actions, and communications are coordinated through one central point of command. Such a system should reduce confusion, improve safety, organize and coordinate actions, and should facilitate effective management of the incident.

7. Site Safety and Control Plans. The safety and security of response personnel and others in the area of an emergency response incident site should be of primary concern to the incident commander.

The use of a site safety and control plan could greatly assist those in charge of assuring the safety and health of employees on the site.

A comprehensive site safety and control plan should include the following: summary analysis of hazards on the site and a risk analysis of those hazards; site map or sketch; site work zones (clean zone, transition or decontamination zone, work or hot zone); use of the buddy system; site communications; command post or command center; standard operating procedures and safe work practices; medical assistance and triage area; hazard monitoring plan (air contaminate monitoring, etc.); decontamination procedures and area; and other relevant areas. This plan should be a part of the employer's emergency response plan or an extension of it to the specific site.

8. Medical surveillance programs. Workers handling hazardous substances may be

exposed to toxic chemicals, safety hazards, biologic hazards, and radiation. Therefore, a medical surveillance program is essential to assess and monitor workers' health and fitness for employment in hazardous waste operations and during the course of work; to provide emergency and other treatment as needed; and to keep accurate records for future reference.

The Occupational Safety and Health Guidance Manual for Hazardous Waste Site Activities developed by the National Institute for Occupational Safety and Health (NIOSH), the Occupational Safety and Health Administration (OSHA), the U.S. Coast Guard (USCG), and the Environmental Protection Agency (EPA); October 1985 provides an excellent example of the types of medical testing that should be done as part of a medical surveillance program.

9. New Technology and Spill Containment Programs. Where hazardous substances may be released by spilling from a container that will expose employees to the hazards of the materials, the employer will need to implement a program to contain and control the spilled material. Diking and ditching, as well as use of absorbents like diatomaceous earth, are traditional techniques which have proven to be effective over the years. However, in recent years new products have come into the marketplace, the uses of which complement and increase the effectiveness of these traditional methods. These new products also provide emergency responders and others with additional tools or agents to use to reduce the hazards of spilled materials.

These agents can be rapidly applied over a large area and can be uniformly applied or otherwise can be used to build a small dam, thus improving the workers' ability to control spilled material. These application techniques enhance the intimate contact between the agent and the spilled material allowing for the quickest effect by the agent or quickest control of the spilled material. Agents are available to solidify liquid spilled materials, to suppress vapor generation from spilled materials, and to do both. Some special agents, when applied as recommended by the manufacturer, will react in a controlled manner with the spilled material to neutralize acids or caustics, or greatly reduce the level of hazard of the spilled material.

There are several modern methods and devices for use by emergency response personnel or others involved with spill control efforts to safely apply spill control agents to control spilled material hazards. These include portable pressurized applicators similar to hand-held portable fire extinguishing devices, and nozzle and hose systems similar to portable fire fighting foam systems which allow the operator to apply the agent without having to come into contact with the spilled material. The operator is able to apply the agent to the spilled material from a remote position.

The solidification of liquids provides for rapid containment and isolation of hazardous substance spills. By directing the agent at run-off points or at the edges of the spill, the reactant solid will automatically create a barrier to slow or stop the spread of the material. Clean-up of hazardous substances is greatly improved when solidifying agents, acid or caustic neutralizers, or activated carbon adsorbents are used. Properly applied, these agents can totally solidify liquid hazardous substances or neutralize or absorb them, which results in materials which are less hazardous and easier to handle, transport, and dispose of. The concept of spill treatment, to create less hazardous substances, will improve the safety and level of protection of employees working at spill clean-up operations or emergency response operations to spills of hazardous substances.

The use of vapor suppression agents for volatile hazardous substances, such as flammable liquids and those substances which present an inhalation hazard, is important for protecting workers. The rapid and uniform distribution of the agent over the surface of the spilled material can provide quick vapor knockdown. There are temporary and long-term foam-type agents which are effective on vapors and dusts, and activated carbon adsorption agents which are effective for vapor control and soaking-up of the liquid. The proper use of hose lines or hand-held portable pressurized applicators provides good mobility and permits the worker to deliver the agent from a safe distance without having to step into the untreated spilled material. Some of these systems can be recharged in the field to provide coverage of larger spill areas than the design limits of a single charged applicator unit. Some of the more effective agents can solidify the liquid flammable hazardous substances and at the same time elevate the flashpoint above 140°F so the resulting substance may be handled as a nonhazardous waste material if it meets the U.S. Environmental Protection Agency's 40 CFR Part 261 requirements (See particularly § 261.21).

All workers performing hazardous substance spill control work are expected to wear the proper protective clothing and equipment for the materials present and to follow the employer's established standard operating procedures for spill control. All involved workers need to be trained in the established operating procedures; in the use and care of spill control equipment; and in the associated hazards and control of such hazards of spill containment work.

These new tools and agents are the things that employers will want to evaluate as part of their new technology program. The treatment of spills of hazardous substances or wastes at an emergency incident as part of the immediate spill containment and control efforts is sometimes acceptable to EPA and a permit exception is described in 40 CFR 264.1(g)(8) and 265.1(c)(11).

Appendix D to § 1910.120—References. The Following References May Be Consulted for Further Information on the Subject of This Standard:

1. OSHA Instruction DFO CPL 2.70-January 29, 1986, Special Emphasis Program: Hazardous Waste Sites.

2. OSHA Instruction DFO CPL 2-2.37A-January 29, 1986, Technical Assistance and Guidelines for Superfund and Other Hazardous Waste Site Activities.

3. OSHA Instruction DTS CPL 2.74-January 29, 1986, Hazardous Waste Activity Form, OSHA 175.

4. Hazardous Waste Inspections Reference Manual, U.S. Department of Labor, Occupational Safety and Health Administration, 1986.

5. Memorandum of Understanding Among the National Institute for Occupational Safety and Health, the Occupational Safety and Health Administration, the United States Coast Guard, and the United States Environmental Protection Agency, Guidance for Worker Protection During Hazardous Waste Site Investigations and Clean-up and Hazardous Substance Emergencies. December 18, 1980.

6. National Priorities List, 1st Edition, October 1984; U.S. Environmental Protection Agency, Revised periodically.

7. The Decontamination of Response Personnel, Field Standard Operating Procedures (F.S.O.P.) 7; U.S. Environmental Protection Agency, Office of Emergency and Remedial Response, Hazardous Response Support Division, December 1984.

8. Preparation of a Site Safety Plan, Field Standard Operating Procedures (F.S.O.P.) 9; U.S. Environmental Protection Agency, Office of Emergency and Remedial Response, Hazardous Response Support Division, April 1985.

9. Standard Operating Safety Guidelines; U.S. Environmental Protection Agency, Office of Emergency and Remedial Response, Hazardous Response Support Division, Environmental Response Team; November 1984.

10. Occupational Safety and Health Guidance Manual for Hazardous Waste Site Activities, National Institute for Occupational Safety and Health (NIOSH), Occupational Safety and Health Administration (OSHA), U.S. Coast Guard (USCG), and Environmental Protection Agency (EPA); October 1985.

11. Protecting Health and Safety at Hazardous Waste Sites: An Overview, U.S. Environmental Protection Agency, EPA/625/9-85/006; September 1985.

12. Hazardous Waste Sites and Hazardous Substance Emergencies, NIOSH Worker Bulletin, U.S. Department of Health and Human Services, Public Health Service, Centers for Disease Control, National Institute for Occupational Safety and Health; December 1982.

13. Personal Protective Equipment for Hazardous Materials Incidents: A Selection Guide; U.S. Department of Health and Human Services, Public Health Service, Centers for Disease Control, National Institute for Occupational Safety and Health; October 1984.

14. Fire Service Emergency Management Handbook, International Association of Fire Chiefs Foundation, 101 East Holly Avenue, Unit 10B, Sterling, VA 22170, January 1985.

15. Emergency Response Guidebook, U.S. Department of Transportation, Washington, DC, 1987.

16. Report to the Congress on Hazardous Materials Training, Planning and Preparedness, Federal Emergency Management Agency, Washington, DC, July 1986.

17. Workbook for Fire Command, Alan V. Brunacini and J. David Beageron, National Fire Protection Association, Batterymarch Park, Quincy, MA 02269, 1985.

18. Fire Command, Alan V. Brunacini, National Fire Protection Association, Battery-march Park, Quincy, MA 02269, 1985.

19. Incident Command System, Fire Protection Publications, Oklahoma State University, Stillwater, OK 74078, 1983.

20. Site Emergency Response Planning, Chemical Manufacturers Association, Washington, DC 20037, 1986.

21. Hazardous Materials Emergency Planning Guide, NRT-1, Environmental Protection Agency, Washington, DC, March 1987.

22. Community Teamwork: Working Together to Promote Hazardous Materials Transportation Safety. U.S. Department of Transportation, Washington, DC, May 1983.

23. Disaster Planning Guide for Business and Industry, Federal Emergency Management Agency, Publication No. FEMA 141, August 1987.

(The Office of Management and Budget has approved the information collection requirements in this section under control number 1218-0139)

Appendix E to §1910.120—Training Curriculum Guidelines.

The following non-mandatory general criteria may be used for assistance in developing site-specific training curriculum used to meet the training requirements of 29 CFR 1910.120(e); 29 CFR 1910.120(p)(7), (p)(8)(iii); and 29 CFR 1910.120(q)(6), (q)(7), and (q)(8). These are generic guidelines and they are not presented as a complete training curriculum for any specific employer. Site-specific training programs must be developed on the basis of a needs assessment of the hazardous waste site, RCRA/TSDF, or emergency response operation in accordance with 29 CFR 1910.120.

It is noted that the legal requirements are set forth in the regulatory text of §1910.120. The guidance set forth here presents a highly effective program that in the areas covered would meet or exceed the regulatory requirements. In addition, other approaches could meet the regulatory requirements.

Suggested General Criteria:

Definitions:

"Competent" means possessing the skills, knowledge, experience, and judgment to perform assigned tasks or activities satisfactorily as determined by the employer.

"Demonstration" means the showing by actual use of equipment or procedures.

"Hands-on training" means training in a simulated work environment that permits each student to have experience performing tasks, making decisions, or using equipment appropriate to the job assignment for which the training is being conducted.

"Initial training" means training required prior to beginning work.

"Lecture" means an interactive discourse with a class led by an instructor.

"Proficient" means meeting a stated level of achievement.

"Site-specific" means individual training directed to the operations of a specific job site.

"Training hours" means the number of hours devoted to lecture, learning activities, small group work sessions, demonstration, evaluations, or hands-on experience.

Suggested Core Criteria:

1. Training facility. The training facility should have available sufficient resources, equipment, and site locations to perform didactic and hands-on training when appropriate. Training facilities should have sufficient organization, support staff, and services to conduct training in each of the courses offered.

2. Training Director. Each training program should be under the direction of a training director who is responsible for the program. The Training Director should have a minimum of two years of employee education experience.

3. Instructors. Instructors should be deemed competent on the basis of previous documented experience in their area of instruction, successful completion of a "train-the-trainer" program specific to the topics they will teach, and an evaluation of instructional competence by the Training Director.

Instructors should be required to maintain professional competency by participating in continuing education or professional development programs or by completing successfully an annual refresher course and having an annual review by the Training Director.

The annual review by the Training Director should include observation of an instructor's delivery, a review of those observations with the trainer, and an analysis of any instructor or class evaluations completed by the students during the previous year.

4. Course materials. The Training Director should approve all course materials to be used by the training provider. Course materials should be reviewed and updated at least annually.

Materials and equipment should be in good working order and maintained properly.

All written and audio-visual materials in training curricula should be peer reviewed by technically competent outside reviewers or by a standing advisory committee.

Reviewers should possess expertise in the following disciplines where applicable: occupational health, industrial hygiene and safety, chemical/environmental engineering, employee education, or emergency response. One or more of the peer reviewers should be an employee experienced in the work activities to which the training is directed.

5. Students. The program for accepting students should include:

a. Assurance that the student is or will be involved in work where chemical exposures are likely and that the student possesses the skills necessary to perform the work.

b. A policy on the necessary medical clearance.

6. Ratios. Student-instructor ratios should not exceed 30 students per instructor. Hands-on activity requiring the use of personal protective equipment should have the following student instructor ratios. For Level C or Level D personal protective equipment the ratio should be 10 students per instructor. For Level A or Level B personal protective equipment the ratio should be 5 students per instructor.

7. Proficiency assessment. Proficiency should be evaluated and documented by the use of a written assessment and a skill demonstration selected and developed by the Training Director and training staff. The assessment and demonstration should evaluate the knowledge and individual skills developed in the course of training. The level of minimum achievement necessary for proficiency shall be specified in writing by the Training Director.

If a written test is used, there should be a minimum of 50 questions. If a written test is used in combination with a skills demonstration, a minimum of 25 questions should be used. If a skills demonstration is used, the tasks chosen and the means to rate successful completion should be fully documented by the Training Director.

The content of the written test or of the skill demonstration shall be relevant to the objectives of the course. The written test and skill demonstration should be updated as necessary to reflect changes in the curriculum and any update should be approved by the Training Director.

The proficiency assessment methods, regardless of the approach or combination of approaches used, should be justified, documented and approved by the Training Director.

The proficiency of those taking the additional courses for supervisors should be evaluated and documented by using proficiency assessment methods acceptable to the Training Director. These proficiency assessment methods must reflect the additional responsibilities borne by supervisory personnel in hazardous waste operations or emergency response.

8. Course certificate. Written documentation should be provided to each student who satisfactorily completes the training course.

The documentation should include:

a. Student's name.

b. Course title.

c. Course date.

d. Statement that the student has successfully completed the course.

e. Name and address of the training provider.

f. An individual identification number for the certificate.

g. List of the levels of personal protective equipment used by the student to complete the course.

This documentation may include a certificate and an appropriate wallet-sized laminated card with a photograph of the student and the above information. When such course certificate cards are used, the individual identification number for the training certificate should be shown on the card.

9. Recordkeeping. Training providers should maintain records listing the dates courses were presented, the names of the individual course attenders, the names of those students successfully completing each course, and the number of training certificates issued to each successful student. These records should be maintained for a minimum of five years after the date an individual participated in a training program offered by the training provider. These records should be available and provided upon the student's request or as mandated by law.

10. Program quality control. The Training Director should conduct or direct an annual written audit of the training program.

Program modifications to address deficiencies, if any, should be documented, approved, and implemented by the training provider.

The audit and the program modification documents should be maintained at the training facility.

Suggested Program Quality Control Criteria Factors listed here are suggested criteria for determining the quality and appropriateness of employee health and safety training for hazardous waste operations and emergency response.

A. Training plan.

Adequacy and appropriateness of the training program's curriculum development, instructor training, distribution of course materials, and direct student training should be considered, including:

1. The duration of training, course content, and course schedules/agendas;

2. The different training requirements of the various target populations, as specified in the appropriate generic training curriculum;

3. The process for the development of curriculum, which includes appropriate technical input, outside review, evaluation, program pretesting.

4. The adequate and appropriate inclusion of hands-on, demonstration, and instruction methods;

5. Adequate monitoring of student safety, progress, and performance during the training.

B. Program management, Training Director, staff, and consultants.

Adequacy and appropriateness of staff performance and delivering an effective training program should be considered, including:

1. Demonstration of the training director's leadership in assuring quality of health and safety training.

2. Demonstration of the competency of the staff to meet the demands of delivering high quality hazardous waste employee health and safety training.

3. Organization charts establishing clear lines of authority.

4. Clearly defined staff duties including the relationship of the training staff to the overall program.

5. Evidence that the training organizational structure suits the needs of the training program.

6. Appropriateness and adequacy of the training methods used by the instructors.

7. Sufficiency of the time committed by the training director and staff to the training program.

8. Adequacy of the ratio of training staff to students.

9. Availability and commitment of the training program of adequate human and equipment resources in the areas of

a. Health effects,

b. Safety,

c. Personal protective equipment (PPE),

d. Operational procedures,

e. Employee protection practices/procedures.

10. Appropriateness of management controls.

11. Adequacy of the organization and appropriate resources assigned to assure appropriate training.

12. In the case of multiple-site training programs, adequacy of satellite centers management.

C. Training facilities and resources.

Adequacy and appropriateness of the facilities and resources for supporting the training program should be considered, including:

1. Space and equipment to conduct the training.

2. Facilities for representative hands-on training.

3. In the case of multiple-site programs, equipment and facilities at the satellite centers.

4. Adequacy and appropriateness of the quality control and evaluations program to account for instructor performance.

5. Adequacy and appropriateness of the quality control and evaluation program to ensure appropriate course evaluation, feedback, updating, and corrective action.

6. Adequacy and appropriateness of disciplines and expertise being used within the quality control and evaluation program.

7. Adequacy and appropriateness of the role of student evaluations to provide feedback for training program improvement.

D. Quality control and evaluation.

Adequacy and appropriateness of quality control and evaluation plans for training programs should be considered, including:

1. A balanced advisory committee and/or competent outside reviewers to give overall policy guidance;

2. Clear and adequate definition of the composition and active programmatic role of the advisory committee or outside reviewers.

3. Adequacy of the minutes or reports of the advisory committee or outside reviewers' meetings or written communication.

4. Adequacy and appropriateness of the quality control and evaluations program to account for instructor performance.

5. Adequacy and appropriateness of the quality control and evaluation program to ensure appropriate course evaluation, feedback, updating, and corrective action.

6. Adequacy and appropriateness of disciplines and expertise being used within the quality control and evaluation program.

7. Adequacy and appropriateness of the role of student evaluations to provide feedback for training program improvement.

E. Students.

Adequacy and appropriateness of the program for accepting students should be considered, including:

1. Assurance that the students already possess the necessary skills for their job, including necessary documentation.

2. Appropriateness of methods the program uses to ensure that recruits are capable of satisfactorily completing training.

3. Review and compliance with any medical clearance policy.

F. Institutional environment and administrative support.

The adequacy and appropriateness of the institutional environment and administrative support system for the training program should be considered, including:

1. Adequacy of the institutional commitment to the employee training program.

2. Adequacy and appropriateness of the administrative structure and administrative support.

G. Summary of evaluation questions.

Key questions for evaluating the quality and appropriateness of an overall training program should include the following:

1. Are the program objectives clearly stated?

2. Is the program accomplishing its objectives?

3. Are appropriate facilities and staff available?

4. Is there an appropriate mix of classroom, demonstration, and hands-on training?

5. Is the program providing quality employee health and safety training that fully meets the intent of regulatory requirements?

6. What are the program's main strengths?

7. What are the program's main weaknesses?

8. What is recommended to improve the program?

9. Are instructors instructing according to their training outlines?

10. Is the evaluation tool current and appropriate for the program content?

11. Is the course material current and relevant to the target group?

Suggested Training Curriculum Guidelines:

The following training curriculum guidelines are for those operations specifically identified in 29 CFR 1910.120 as requiring training. Issues such as qualifications of instructors, training certification, and similar criteria appropriate to all categories of operations addressed in 1910.120 have been covered in the preceding section and are not re-addressed in each of the generic guidelines. Basic core requirements for training programs that are addressed include:

1. General Hazardous Waste Operations.

2. RCRA Operations-Treatment, Storage, and Disposal Facilities.

3. Emergency Response.

A. General Hazardous Waste Operations and Site-Specific Training.

1. Off-site training. Training course content for hazardous waste operations, required by 29 CFR 1910.120(e), should include the following topics or procedures:

a. Regulatory knowledge.

(1) A review of 29 CFR 1910.120 and the core elements of an occupational safety and health program.

(2) The content of a medical surveillance program as outlined in 29 CFR 1910.120(f).

(3) The content of an effective site safety and health plan consistent with the requirements of 29 CFR 1910.120(b)(4)(ii).

(4) Emergency response plan and procedures as outlined in 29 CFR 1910.38 and 29 CFR 1910.120(l).

(5) Adequate illumination.

(6) Sanitation recommendation and equipment.

(7) Review and explanation of OSHA's hazard-communication standard (29 CFR 1910.1200) and lock-out-tag-out standard (29 CFR 1910.147).

(8) Review of other applicable standards including but not limited to those in the construction standards (29 CFR Part 1926).

(9) Rights and responsibilities of employers and employees under applicable OSHA and EPA laws.

b. Technical knowledge.

(1) Type of potential exposures to chemical, biological, and radiological hazards; types of human responses to these hazards and recognition of those responses; principles of toxicology and information about acute and chronic hazards; health and safety considerations of new technology.

(2) Fundamentals of chemical hazards including but not limited to vapor pressure, boiling points, flash points, pH, other physical and chemical properties.

(3) Fire and explosion hazards of chemicals.

(4) General safety hazards such as but not limited to electrical hazards, powered equipment hazards, motor vehicle hazards, walking-working surface hazards, excavation hazards, and hazards associated with working in hot and cold temperature extremes.

(5) Review and knowledge of confined space entry procedures in 29 CFR 1910.146.

(6) Work practices to minimize employee risk from site hazards.

(7) Safe use of engineering controls, equipment, and any new relevant safety technology or safety procedures.

(8) Review and demonstration of competency with air sampling and monitoring equipment that may be used in a site monitoring program.

(9) Container sampling procedures and safeguarding; general drum and container handling procedures including special requirement for laboratory waste packs, shock-sensitive wastes, and radioactive wastes.

(10) The elements of a spill control program.

(11) Proper use and limitations of material handling equipment.

(12) Procedures for safe and healthful preparation of containers for shipping and transport.

(13) Methods of communication including those used while wearing respiratory protection.

c. Technical skills.

(1) Selection, use maintenance, and limitations of personal protective equipment including the components and procedures for carrying out a respirator program to comply with 29 CFR 1910.134.

(2) Instruction in decontamination programs including personnel, equipment, and hardware; hands-on training including level A, B, and C ensembles and appropriate decontamination lines; field activities including the donning and doffing of protective equipment to a level commensurate with the employee's anticipated job function and responsibility and to the degree required by potential hazards.

(3) Sources for additional hazard information; exercises using relevant manuals and hazard coding systems.

d. Additional suggested items.

(1) A laminated, dated card or certificate with photo, denoting limitations and level of protection for which the employee is trained should be issued to those students successfully completing a course.

(2) Attendance should be required at all training modules, with successful completion of exercises and a final written or oral examination with at least 50 questions.

(3) A minimum of one-third of the program should be devoted to hands-on exercises.

(4) A curriculum should be established for the 8-hour refresher training required by 29 CFR 1910.120(e)(8), with delivery of such courses directed toward those areas of previous training that need improvement or reemphasis.

(5) A curriculum should be established for the required 8-hour training for supervisors. Demonstrated competency in the skills and knowledge provided in a 40-hour course should be a prerequisite for supervisor training.

2. Refresher training.

The 8-hour annual refresher training required in 29 CFR 1910.120(e)(8) should be conducted by qualified training providers. Refresher training should include at a minimum the following topics and procedures:

(a) Review of and retraining on relevant topics covered in the 40-hour program, as appropriate, using reports by the students on their work experiences.

(b) Update on developments with respect to material covered in the 40-hour course.

(c) Review of changes to pertinent provisions of EPA or OSHA standards or laws.

(d) Introduction of additional subject areas as appropriate.

(e) Hands-on review of new or altered PPE or decontamination equipment or procedures. Review of new developments in personal protective equipment.

(f) Review of newly developed air and contaminant monitoring equipment.

3. On-site training.

a. The employer should provide employees engaged in hazardous waste site activities with information and training prior to initial assignment into their work area, as follows:

(1) The requirements of the hazard communication program including the location and availability of the written program, required lists of hazardous chemicals, and material safety data sheets.

(2) Activities and locations in their work area where hazardous substance may be present.

(3) Methods and observations that may be used to detect the present or release of a hazardous chemical in the work area (such as monitoring conducted by the employer, continuous monitoring devices, visual appearances, or other evidence (sight, sound or smell) of hazardous chemicals being released, and applicable alarms from monitoring devices that record chemical releases).

(4) The physical and health hazards of substances known or potentially present in the work area.

(5) The measures employees can take to help protect themselves from work-site hazards, including specific procedures the employer has implemented.

(6) An explanation of the labeling system and material safety data sheets and how employees can obtain and use appropriate hazard information.

(7) The elements of the confined space program including special PPE, permits, monitoring requirements, communication procedures, emergency response, and applicable lock-out procedures.

b. The employer should provide hazardous waste employees information and training and should provide a review and access to the site safety and plan as follows:

(1) Names of personnel and alternate responsible for site safety and health.

(2) Safety and health hazards present on the site.

(3) Selection, use, maintenance, and limitations of personal protective equipment specific to the site.

(4) Work practices by which the employee can minimize risks from hazards.

(5) Safe use of engineering controls and equipment available on site.

(6) Safe decontamination procedures established to minimize employee contact with hazardous substances, including:

(A) Employee decontamination,

(B) Clothing decontamination, and

(C) Equipment decontamination.

(7) Elements of the site emergency response plan, including:

(A) Pre-emergency planning.

(B) Personnel roles and lines of authority and communication.

(C) Emergency recognition and prevention.

(D) Safe distances and places of refuge.

(E) Site security and control.

(F) Evacuation routes and procedures.

(G) Decontamination procedures not covered by the site safety and health plan.

(H) Emergency medical treatment and first aid.

(I) Emergency equipment and procedures for handling emergency incidents.

c. The employer should provide hazardous waste employees information and training on personal protective equipment used at the site, such as the following:

(1) PPE to be used based upon known or anticipated site hazards.

(2) PPE limitations of materials and construction; limitations during temperature extremes, heat stress, and other appropriate medical considerations; use and limitations of respirator equipment as well as documentation procedures as outlined in 29 CFR 1910.134.

(3) PPE inspection procedures prior to, during, and after use.

(4) PPE donning and doffing procedures.

(5) PPE decontamination and disposal procedures.

(6) PPE maintenance and storage.

(7) Task duration as related to PPE limitations.

d. The employer should instruct the employee about the site medical surveillance program relative to the particular site, including

(1) Specific medical surveillance programs that have been adapted for the site.

(2) Specific signs and symptoms related to exposure to hazardous materials on the site.

(3) The frequency and extent of periodic medical examinations that will be used on the site.

(4) Maintenance and availability of records.

(5) Personnel to be contacted and procedures to be followed when signs and symptoms of exposures are recognized.

e. The employees will review and discuss the site safety plan as part of the training program. The location of the site safety plan and all written programs should be discussed with employees including a discussion of the mechanisms for access, review, and references described.

B. RCRA Operations Training for Treatment, Storage and Disposal Facilities.

1. As a minimum, the training course required in 29 CFR 1910.120 (p) should include the following topics:

(a) Review of the applicable paragraphs of 29 CFR 1910.120 and the elements of the employer's occupational safety and health plan.

(b) Review of relevant hazards such as, but not limited to, chemical, biological, and radiological exposures; fire and explosion hazards; thermal extremes; and physical hazards.

(c) General safety hazards including those associated with electrical hazards, powered equipment hazards, lock-out-tag-out procedures, motor vehicle hazards and walking-working surface hazards.

(d) Confined-space hazards and procedures.

(e) Work practices to minimize employee risk from workplace hazards.

(f) Emergency response plan and procedures including first aid meeting the requirements of paragraph (p)(8).

(g) A review of procedures to minimize exposure to hazardous waste and various type of waste streams, including the materials handling program and spill containment program.

(h) A review of hazard communication programs meeting the requirements of 29 CFR 1910.1200.

(i) A review of medical surveillance programs meeting the requirements of 29 CFR

1910.120(p)(3) including the recognition of signs and symptoms of overexposure to hazardous substances including known synergistic interactions.

(j) A review of decontamination programs and procedures meeting the requirements of 29 CFR 1910.120(p)(4).

(k) A review of an employer's requirements to implement a training program and its elements.

(l) A review of the criteria and programs for proper selection and use of personal protective equipment, including respirators.

(m) A review of the applicable appendices to 29 CFR 1910.120.

(n) Principles of toxicology and biological monitoring as they pertain to occupational health.

(o) Rights and responsibilities of employees and employers under applicable OSHA and EPA laws.

(p) Hands-on exercises and demonstrations of competency with equipment to illustrate the basic equipment principles that may be used during the performance of work duties, including the donning and doffing of PPE.

(q) Sources of reference, efficient use of relevant manuals, and knowledge of hazard coding systems to include information contained in hazardous waste manifests.

(r) At least 8 hours of hands-on training.

(s) Training in the job skills required for an employee's job function and responsibility before they are permitted to participate in or supervise field activities.

2. The individual employer should provide hazardous waste employees with information and training prior to an employee's initial assignment into a work area. The training and information should cover the following topics:

(a) The Emergency response plan and procedures including first aid.

(b) A review of the employer's hazardous waste handling procedures including the materials handling program and elements of the spill containment program, location of spill response kits or equipment, and the names of those trained to respond to releases.

(c) The hazardous communication program meeting the requirements of 29 CFR 1910.1200.

(d) A review of the employer's medical surveillance program including the recognition of signs and symptoms of exposure to relevant hazardous substances including known synergistic interactions.

(e) A review of the employer's decontamination program and procedures.

(f) An review of the employer's training program and the parties responsible for that program.

(g) A review of the employer's personal protective equipment program including the proper selection and use of PPE based upon specific site hazards.

(h) All relevant site-specific procedures addressing potential safety and health hazards. This may include, as appropriate, biological and radiological exposures, fire and explosion hazards, thermal hazards, and physical hazards such as electrical hazards, powered equipment hazards, lock-out-tag-out hazards, motor vehicle hazards, and walking-working surface hazards.

(i) Safe use engineering controls and equipment on site.

(j) Names of personnel and alternates responsible for safety and health.

C. Emergency Response Training.

Federal OSHA standards in 29 CFR 1910.120(q) are directed toward private sector emergency responders. Therefore, the guidelines provided in this portion of the appendix are directed toward that employee population. However, they also impact indirectly through State OSHA or USEPA regulations some public sector emergency responders. Therefore, the guidelines provided in this portion of the appendix may be applied to both employee populations.

States with OSHA state plans must cover their employees with regulations at least as effective as the Federal OSHA standards.

Public employees in states without approved state OSHA programs covering hazardous waste operations and emergency response are covered by the U.S. EPA under 40 CFR 311, a regulation virtually identical to §1910.120.

Since this is a non-mandatory appendix and therefore not an enforceable standard, OSHA recommends that those employers, employees or volunteers in public sector emergency response organizations outside Federal OSHA jurisdiction consider the following criteria in developing their own training programs.

A unified approach to training at the community level between emergency response organizations covered by Federal OSHA and those not covered directly by Federal OSHA can help ensure an effective community response to the release or potential release of hazardous substances in the community.

a. General considerations.

Emergency response organizations are required to consider the topics listed in §1910.120(q)(6). Emergency response organizations may use some or all of the following topics to supplement those mandatory topics when developing their response training programs.

Many of the topics would require an interaction between the response provider and the individuals responsible for the site where the response would be expected.

(1) Hazard recognition, including:

(A) Nature of hazardous substances present,

(B) Practical applications of hazard recognition, including presentations on biology, chemistry, and physics.

(2) Principles of toxicology, biological monitoring, and risk assessment.

(3) Safe work practices and general site safety.

(4) Engineering controls and hazardous waste operations.

(5) Site safety plans and standard operating procedures.

(6) Decontamination procedures and practices.

(7) Emergency procedures, first aid, and self-rescue.

(8) Safe use of field equipment.

(9) Storage, handling, use and transportation of hazardous substances.

(10) Use, care, and limitations of personal protective equipment.

(11) Safe sampling techniques.

(12) Rights and responsibilities of employees under OSHA and other related laws concerning right-to-know, safety and health, compensations and liability.

(13) Medical monitoring requirements.

(14) Community relations.

b. Suggested criteria for specific courses.

(1) First responder awareness level.

(A) Review of and demonstration of competency in performing the applicable skills of 29 CFR 1910.120(q).

(B) Hands-on experience with the U.S. Department of Transportation's Emergency Response Guidebook (ERG) and familiarization with OSHA standard 29 CFR 1910.1201.

(C) Review of the principles and practices for analyzing an incident to determine both the hazardous substances present and the basic hazard and response information for each hazardous substance present.

(D) Review of procedures for implementing actions consistent with the local emergency response plan, the organization's standard operating procedures, and the current edition of DOT's ERG including emergency notification procedures and follow-up communications.

(E) Review of the expected hazards including fire and explosions hazards, confined space hazards, electrical hazards, powered equipment hazards, motor vehicle hazards, and walking-working surface hazards.

(F) Awareness and knowledge of the competencies for the First Responder at the Awareness Level covered in the National Fire Protection Association's Standard No. 472, Professional Competence of Responders to Hazardous Materials Incidents.

(2) First responder operations level.

(A) Review of and demonstration of competency in performing the applicable skills of 29 CFR 1910.120(q)

(B) Hands-on experience with the U.S. Department of Transportation's Emergency Response Guidebook (ERG), manufacturer material safety data sheets, CHEMTREC/CANUTEC, shipper or manufacturer contacts, and other relevant sources of information addressing hazardous substance releases. Familiarization with OSHA standard 29 CFR 1910.1201.

(C) Review of the principles and practices for analyzing an incident to determine the hazardous substances present, the likely behavior of the hazardous substance and its container, the types of hazardous substance transportation containers and vehicles, the types and selection of the appropriate defensive strategy for containing the release.

(D) Review of procedures for implementing continuing response actions consistent with the local emergency response plan, the organization's standard operating procedures, and the current edition of DOT's ERG including extended emergency notification procedures and follow-up communications.

(E) Review of the principles and practice for proper selection and use of personal protective equipment.

(F) Review of the principles and practice of personnel and equipment decontamination.

(G) Review of the expected hazards including fire and explosions hazards, confined space hazards, electrical hazards, powered equipment hazards, motor vehicle hazards, and walking-working surface hazards.

(H) Awareness and knowledge of the competencies for the First Responder at the Operations Level covered in the National Fire Protection Association's Standard No. 472, Professional Competence of Responders to Hazardous Materials Incidents.

(3) Hazardous materials technician.

(A) Review of and demonstration of competency in performing the applicable skills of 29 CFR 1910.120(q).

(B) Hands-on experience with written and electronic information relative to response decision making including but not limited to the U.S. Department of Transportation's Emergency Response Guidebook (ERG), manufacturer material safety data sheets, CHEMTREC/CANUTEC, shipper or manufacturer contacts, computer data bases and response models, and other relevant sources of information addressing hazardous substance releases. Familiarization with OSHA standard 29 CFR 1910.1201.

(C) Review of the principles and practices for analyzing an incident to determine the hazardous substances present, their physical and chemical properties, the likely behavior of the hazardous substance and its container, the types of hazardous substance transportation containers and vehicles involved in the release, the appropriate strategy for approaching release sites and containing the release.

(D) Review of procedures for implementing continuing response actions consistent with the local emergency response plan, the organization's standard operating procedures, and the current edition of DOT's ERG including extended emergency notification procedures and follow-up communications.

(E) Review of the principles and practice for proper selection and use of personal protective equipment.

(F) Review of the principles and practices of establishing exposure zones, proper decontamination and medical surveillance stations and procedures.

(G) Review of the expected hazards including fire and explosions hazards, confined space hazards, electrical hazards, powered equipment hazards, motor vehicle hazards, and walking-working surface hazards.

(H) Awareness and knowledge of the competencies for the Hazardous Materials Technician covered in the National Fire Protection Association's Standard No. 472, Professional Competence of Responders to Hazardous Materials Incidents.

(4) Hazardous materials specialist.

(A) Review of and demonstration of competency in performing the applicable skills of 29 CFR 1910.120(q).

(B) Hands-on experience with retrieval and use of written and electronic information relative to response decision making including but not limited to the U.S. Department of Transportation's Emergency Response Guidebook (ERG), manufacturer material safety data sheets, CHEMTREC/CANUTEC, shipper or manufacturer contacts, computer data bases and response models, and other relevant sources of information addressing hazardous substance releases.

Familiarization with OSHA standard 29 CFR 1910.1201.

(C) Review of the principles and practices for analyzing an incident to determine the hazardous substances present, their physical and chemical properties, and the likely behavior of the hazardous substance and its container, vessel, or vehicle.

(D) Review of the principles and practices for identification of the types of hazardous substance transportation containers, vessels and vehicles involved in the release; selecting and using the various types of equipment available for plugging or patching transportation containers, vessels or vehicles; organizing and directing the use of multiple teams of hazardous material technicians and selecting the appropriate strategy for approaching release sites and containing or stopping the release.

(E) Review of procedures for implementing continuing response actions consistent with the local emergency response plan, the organization's standard operating procedures, including knowledge of the available public and private response resources, establishment of an incident command post, direction of hazardous material technician teams, and extended emergency notification procedures and follow-up communications.

(F) Review of the principles and practice for proper selection and use of personal protective equipment.

(G) Review of the principles and practices of establishing exposure zones and proper decontamination, monitoring and medical surveillance stations and procedures.

(H) Review of the expected hazards including fire and explosions hazards, confined space hazards, electrical hazards, powered equipment hazards, motor vehicle hazards, and walking-working surface hazards.

(I) Awareness and knowledge of the competencies for the Off site Specialist Employee covered in the National Fire Protection Association's Standard No. 472, Professional Competence of Responders to Hazardous Materials Incidents.

(5) Incident commander.

The incident commander is the individual who, at any one time, is responsible for and in control of the response effort.

This individual is the person responsible for the direction and coordination of the response effort. An incident commander's position should be occupied by the most senior, appropriately trained individual present at the response site. Yet, as necessary and appropriate by the level of response provided, the position may be occupied by many individuals during a particular response as the need for greater authority, responsibility, or training increases. It is possible for the first responder at the awareness level to assume the duties of incident commander until a more senior and appropriately trained individual arrives at the response site.

Therefore, any emergency responder expected to perform as an incident commander should be trained to fulfill the obligations of the position at the level of response they will be providing including the following:

(A) Ability to analyze a hazardous substance incident to determine the magnitude of the response problem.

(B) Ability to plan and implement an appropriate response plan within the capabilities of available personnel and equipment.

(C) Ability to implement a response to favorably change the outcome of the incident in a manner consistent with the local emergency response plan and the organization's standard operating procedures. .

(D) Ability to evaluate the progress of the emergency response to ensure that the response objectives are being met safely, effectively, and efficiently.

(E) Ability to adjust the response plan to the conditions of the response and to notify higher levels of response when required by the changes to the response plan.

Appendix D

OSHA Permit-Required Confined Space for General Industry, Final Rule (29 CFR 1910.146)

§ 1910.146 PERMIT-REQUIRED CONFINED SPACES

(a) Scope and application. This section contains requirements for practices and procedures to protect employees in general industry from the hazards of entry into permit-required confined spaces. This section does not apply to agriculture, to construction, or to shipyard employment (Parts 1928, 1926, and 1915 of this chapter, respectively).

(b) Definitions.

Acceptable entry conditions means the conditions that must exist in a permit space to allow entry and to ensure that employees involved with a permit-required confined space entry can safely enter into and work within the space.

Attendant means an individual stationed outside one or more permit spaces who monitors the authorized entrants and who performs all attendant's duties assigned in the employer's permit space program.

Authorized entrant means an employee who is authorized by the employer to enter a permit space.

Blanking or blinding means the absolute closure of a pipe, line, or duct by the fastening of a solid plate (such as a spectacle blind or a skillet blind) that completely covers the bore and that is capable of withstanding the maximum pressure of the pipe, line, or duct with no leakage beyond the plate.

Confined space means a space that:

(1) Is large enough and so configured that an employee can bodily enter and perform assigned work; and

(2) Has limited or restricted means for entry or exit (for example, tanks, vessels, silos, storage bins, hoppers, vaults, and pits are spaces that may have limited means of entry.); and

(3) Is not designed for continuous employee occupancy.

Double block and bleed means the closure of a line, duct, or pipe by closing and locking or tagging two in-line valves and by opening and locking or tagging a drain or vent valve in the line between the two closed valves.

Emergency means any occurrence (including any failure of hazard control or monitoring equipment) or event internal or external to the permit space that could endanger entrants.

Engulfment means the surrounding and effective capture of a person by a liquid or finely divided (flowable) solid substance that can be aspirated to cause death by filling or plugging the respiratory system or that can exert enough force on the body to cause death by strangulation, constriction, or crushing.

Entry means the action by which a person passes through an opening into a permit-required confined space. Entry includes ensuing work activities in that space and is considered to have occurred as soon as any part of the entrant's body breaks the plane of an opening into the space.

Entry permit (permit) means the written or printed document that is provided by the employer to allow and control entry into a permit space and that contains the information specified in paragraph (f) of this section.

Entry supervisor means the person (such as the employer, foreman, or crew chief) responsible for determining if acceptable entry conditions are present at a permit space where entry is planned, for authorizing entry and overseeing entry operations, and for terminating entry as required by this section.

Note: An entry supervisor also may serve as an attendant or as an authorized entrant, as long as that person is trained and equipped as required by this section for each role he or she fills. Also, the duties of entry supervisor may be passed from one individual to another during the course of an entry operation.

Hazardous atmosphere means an atmosphere that may expose employees to the risk of death, incapacitation, impairment of ability to self-rescue (that is, escape unaided from a permit space), injury, or acute illness from one or more of the following causes:

(1) Flammable gas, vapor, or mist in excess of 10 percent of its lower flammable limit (LFL);

(2) Airborne combustible dust at a concentration that meets or exceeds its LFL;

Note: This concentration may be approximated as a condition in which the dust obscures vision at a distance of 5 feet (1.52 m) or less.

(3) Atmospheric oxygen concentration below 19.5 percent or above 23.5 percent;

(4) Atmospheric concentration of any substance for which a dose or a permissible exposure limit is published in Subpart G, Occupational Health and Environmental Control, or in Subpart Z, Toxic and Hazardous Substances, of this part and which

could result in employee exposure in excess of its dose or permissible exposure limit;

Note: An atmospheric concentration of any substance that is not capable of causing death, incapacitation, impairment of ability to self-rescue, injury, or acute illness due to its health effects is not covered by this provision.

(5) Any other atmospheric condition that is immediately dangerous to life or health.

Note: For air contaminants for which OSHA has not determined a dose or permissible exposure limit, other sources of information, such as Material Safety Data Sheets that comply with the Hazard Communication Standard, §1910.1200 of this part, published information, and internal documents can provide guidance in establishing acceptable atmospheric conditions.

Hot work permit means the employer's written authorization to perform operations (for example, riveting, welding, cutting, burning, and heating) capable of providing a source of ignition.

Immediately dangerous to life or health (IDLH) means any condition that poses an immediate or delayed threat to life or that would cause irreversible adverse health effects or that would interfere with an individual's ability to escape unaided from a permit space.

Note: Some materials-hydrogen fluoride gas and cadmium vapor, for example-may produce immediate transient effects that, even if severe, may pass without medical attention, but are followed by sudden, possibly fatal collapse 12–72 hours after exposure.

The victim "feels normal" from recovery from transient effects until collapse. Such materials in hazardous quantities are considered to be "immediately" dangerous to life or health.

Inerting means the displacement of the atmosphere in a permit space by a noncombustible gas (such as nitrogen) to such an extent that the resulting atmosphere is noncombustible.

Note: This procedure produces an IDLH oxygen-deficient atmosphere.

Isolation means the process by which a permit space is removed from service and completely protected against the release of energy and material into the space by such means as: blanking or blinding; misaligning or removing sections of lines, pipes, or ducts; a double block and bleed system; lockout or tagout of all sources of energy; or blocking or disconnecting all mechanical linkages.

Line breaking means the intentional opening of a pipe, line, or duct that is or has been carrying flammable, corrosive, or toxic material, an inert gas, or any fluid at a volume, pressure, or temperature capable of causing injury.

Non-permit confined space means a confined space that does not contain or, with respect to atmospheric hazards, have the potential to contain any hazard capable of causing death or serious physical harm.

Oxygen deficient atmosphere means an atmosphere containing less than 19.5 percent oxygen by volume.

Oxygen enriched atmosphere means an atmosphere containing more than 23.5 percent oxygen by volume.

Permit-required confined space (permit space) means a confined space that has one or more of the following characteristics:

(1) Contains or has a potential to contain a hazardous atmosphere;

(2) Contains a material that has the potential for engulfing an entrant;

(3) Has an internal configuration such that an entrant could be trapped or asphyxiated by inwardly converging walls or by a floor which slopes downward and tapers to a smaller cross-section; or

(4) Contains any other recognized serious safety or health hazard.

Permit-required confined space program (permit space program) means the employer's overall program for controlling, and, where appropriate, for protecting employees from, permit space hazards and for regulating employee entry into permit spaces.

Permit system means the employer's written procedure for preparing and issuing permits for entry and for returning the permit space to service following termination of entry.

Prohibited condition means any condition in a permit space that is not allowed by the permit during the period when entry is authorized.

Rescue service means the personnel designated to rescue employees from permit spaces.

Retrieval system means the equipment (including a retrieval line, chest or full-body harness, wristlets, if appropriate, and a lifting device or anchor) used for non-entry rescue of persons from permit spaces.

Testing means the process by which the hazards that may confront entrants of a permit space are identified and evaluated. Testing includes specifying the tests that are to be performed in the permit space.

Note: Testing enables employers both to devise and implement adequate control measures for the protection of authorized entrants and to determine if acceptable entry conditions are present immediately prior to, and during, entry.

(c) General requirements.

(1) The employer shall evaluate the workplace to determine if any spaces are permit-required confined spaces.

Note: Proper application of the decision flow chart in Appendix A to §1910.146 would facilitate compliance with this requirement.

(2) If the workplace contains permit spaces, the employer shall inform exposed employees, by posting danger signs or by any other equally effective means, of the existence and location of and the danger posed by the permit spaces.

Note: A sign reading "DANGER-PERMIT-REQUIRED CONFINED SPACE, DO NOT ENTER" or using other similar language would satisfy the requirement for a sign.

(3) If the employer decides that its employees will not enter permit spaces, the employer shall take effective measures to prevent its employees from entering the permit spaces and shall comply with paragraphs (c)(1), (c)(2), (c)(6), and (c)(8) of this section.

(4) If the employer decides that its employees will enter permit spaces, the employer shall develop and implement a written permit space program that complies with this section. The written program shall be available for inspection by employees and their authorized representatives.

(5) An employer may use the alternate procedures specified in paragraph (c)(5)(ii) of this section for entering a permit space under the conditions set forth in paragraph (c)(5)(i) of this section.

   (i)   An employer whose employees enter a permit space need not comply with paragraphs (d) through (f) and (h) through (k) of this section, provided that:

      (A) The employer can demonstrate that the only hazard posed by the permit space is an actual or potential hazardous atmosphere;

      (B) The employer can demonstrate that continuous forced air ventilation alone is sufficient to maintain that permit space safe for entry;

      (C) The employer develops monitoring and inspection data that supports the demonstrations required by paragraphs (c)(5)(i)(A) and (c)(5)(i)(B) of this section;

      (D) If an initial entry of the permit space is necessary to obtain the data required by paragraph (c)(5)(i)(C) of this section, the entry is performed in compliance with paragraphs (d) through (k) of this section;

      (E) The determinations and supporting data required by paragraphs (c)(5)(i)(A), (c)(5)(i)(B), and (c)(5)(i)(C) of this section are documented by the employer and are made available to each employee who enters the permit space under the terms of paragraph (c)(5) of this section; and

      (F) Entry into the permit space under the terms of paragraph (c)(5)(i) of this section is performed in accordance with the requirements of paragraph (c)(5)(ii) of this section.

            Note: See paragraph (c)(7) of this section for reclassification of a permit space after all hazards within the space have been eliminated.

   (ii)   The following requirements apply to entry into permit spaces that meet the conditions set forth in paragraph (c)(5)(i) of this section.

      (A) Any conditions making it unsafe to remove an entrance cover shall be eliminated before the cover is removed.

      (B) When entrance covers are removed, the opening shall be promptly guarded by a railing, temporary cover, or other temporary barrier that will prevent an accidental fall through the opening and that will protect each employee working in the space from foreign objects entering the space.

(C) Before an employee enters the space, the internal atmosphere shall be tested, with a calibrated direct-reading instrument, for the following conditions in the order given:

(1) Oxygen content,

(2) Flammable gases and vapors, and

(3) Potential toxic air contaminants.

(D) There may be no hazardous atmosphere within the space whenever any employee is inside the space.

(E) Continuous forced air ventilation shall be used, as follows:

(1) An employee may not enter the space until the forced air ventilation has eliminated any hazardous atmosphere;

(2) The forced air ventilation shall be so directed as to ventilate the immediate areas where an employee is or will be present within the space and shall continue until all employees have left the space;

(3) The air supply for the forced air ventilation shall be from a clean source and may not increase the hazards in the space.

(F) The atmosphere within the space shall be periodically tested as necessary to ensure that the continuous forced air ventilation is preventing the accumulation of a hazardous atmosphere.

(G) If a hazardous atmosphere is detected during entry:

(1) Each employee shall leave the space immediately;

(2) The space shall be evaluated to determine how the hazardous atmosphere developed; and

(3) Measures shall be implemented to protect employees from the hazardous atmosphere before any subsequent entry takes place.

(H) The employer shall verify that the space is safe for entry and that the pre-entry measures required by paragraph (c)(5)(ii) of this section have been taken, through a written certification that contains the date, the location of the space, and the signature of the person providing the certification. The certification shall be made before entry and shall be made available to each employee entering the space.

(6) When there are changes in the use or configuration of a non-permit confined space that might increase the hazards to entrants, the employer shall reevaluate that space and, if necessary, reclassify it as a permit-required confined space.

(7) A space classified by the employer as a permit-required confined space may be reclassified as a non-permit confined space under the following procedures:

(i) If the permit space poses no actual or potential atmospheric hazards and if all hazards within the space are eliminated without entry into the space, the permit space may be reclassified as a non-permit confined space for as long as the non-atmospheric hazards remain eliminated.

    (ii)   If it is necessary to enter the permit space to eliminate hazards, such entry shall be performed under paragraphs (d) through (k) of this section. If testing and inspection during that entry demonstrate that the hazards within the permit space have been eliminated, the permit space may be reclassified as a non-permit confined space for as long as the hazards remain eliminated.

           Note: Control of atmospheric hazards through forced air ventilation does not constitute elimination of the hazards. Paragraph (c)(5) covers permit space entry where the employer can demonstrate that forced air ventilation alone will control all hazards in the space.

    (iii)   The employer shall document the basis for determining that all hazards in a permit space have been eliminated, through a certification that contains the date, the location of the space, and the signature of the person making the determination.

           The certification shall be made available to each employee entering the space.

    (iv)   If hazards arise within a permit space that has been declassified to a non-permit space under paragraph (c)(7) of this section, each employee in the space shall exit the space.

           The employer shall then reevaluate the space and determine whether it must be reclassified as a permit space, in accordance with other applicable provisions of this section.

(8) When an employer (host employer) arranges to have employees of another employer (contractor) perform work that involves permit space entry, the host employer shall:

    (i)   Inform the contractor that the workplace contains permit spaces and that permit space entry is allowed only through compliance with a permit space program meeting the requirements of this section;

    (ii)   Apprise the contractor of the elements, including the hazards identified and the host employer's experience with the space, that make the space in question a permit space;

    (iii)   Apprise the contractor of any precautions or procedures that the host employer has implemented for the protection of employees in or near permit spaces where contractor personnel will be working;

    (iv)   Coordinate entry operations with the contractor, when both host employer personnel and contractor personnel will be working in or near permit spaces, as required by paragraph (d)(11) of this section; and

    (v)   Debrief the contractor at the conclusion of the entry operations regarding the permit space program followed and regarding any hazards confronted or created in permit spaces during entry operations.

(9) In addition to complying with the permit space requirements that apply to all employers, each contractor who is retained to perform permit space entry operations shall:

(i)   Obtain any available information regarding permit space hazards and entry operations from the host employer;

(ii)   Coordinate entry operations with the host employer, when both host employer personnel and contractor personnel will be working in or near permit spaces, as required by paragraph (d)(11) of this section; and

(iii) Inform the host employer of the permit space program that the contractor will follow and of any hazards confronted or created in permit spaces, either through a debriefing or during the entry operation.

(d) Permit-required confined space program (permit space program).

Under the permit space program required by paragraph (c)(4) of this section, the employer shall:

(1) Implement the measures necessary to prevent unauthorized entry;

(2) Identify and evaluate the hazards of permit spaces before employees enter them;

(3) Develop and implement the means, procedures, and practices necessary for safe permit space entry operations, including, but not limited to, the following:

(i)   Specifying acceptable entry conditions;

(ii)   Isolating the permit space;

(iii) Purging, inerting, flushing, or ventilating the permit space as necessary to eliminate or control atmospheric hazards;

(iv)   Providing pedestrian, vehicle, or other barriers as necessary to protect entrants from external hazards; and

(v)   Verifying that conditions in the permit space are acceptable for entry throughout the duration of an authorized entry.

(4) Provide the following equipment (specified in paragraphs (d)(4)(i) through (d)(4)(ix) of this section) at no cost to employees, maintain that equipment properly, and ensure that employees use that equipment properly:

(i)   Testing and monitoring equipment needed to comply with paragraph (d)(5) of this section;

(ii)   Ventilating equipment needed to obtain acceptable entry conditions;

(iii) Communications equipment necessary for compliance with paragraphs (h)(3) and (i)(5) of this section;

(iv)   Personal protective equipment insofar as feasible engineering and work practice controls do not adequately protect employees;

(v)   Lighting equipment needed to enable employees to see well enough to work safely and to exit the space quickly in an emergency;

(vi)   Barriers and shields as required by paragraph (d)(3)(iv) of this section;

(vii) Equipment, such as ladders, needed for safe ingress and egress by authorized entrants;

(viii) Rescue and emergency equipment needed to comply with paragraph (d)(9) of this section, except to the extent that the equipment is provided by rescue services; and

(ix) Any other equipment necessary for safe entry into and rescue from permit spaces.

(5) Evaluate permit space conditions as follows when entry operations are conducted:

(i) Test conditions in the permit space to determine if acceptable entry conditions exist before entry is authorized to begin, except that, if isolation of the space is infeasible because the space is large or is part of a continuous system (such as a sewer), pre-entry testing shall be performed to the extent feasible before entry is authorized and, if entry is authorized, entry conditions shall be continuously monitored in the areas where authorized entrants are working;

(ii) Test or monitor the permit space as necessary to determine if acceptable entry conditions are being maintained during the course of entry operations; and

(iii) When testing for atmospheric hazards, test first for oxygen, then for combustible gases and vapors, and then for toxic gases and vapors.

Note: Atmospheric testing conducted in accordance with Appendix B to § 1910.146 would be considered as satisfying the requirements of this paragraph. For permit space operations in sewers, atmospheric testing conducted in accordance with Appendix B, as supplemented by Appendix E to § 1910.146, would be considered as satisfying the requirements of this paragraph.

(6) Provide at least one attendant outside the permit space into which entry is authorized for the duration of entry operations; Note: Attendants may be assigned to monitor more than one permit space provided the duties described in paragraph (i) of this section can be effectively performed for each permit space that is monitored. Likewise, attendants may be stationed at any location outside the permit space to be monitored as long as the duties described in paragraph (i) of this section can be effectively performed for each permit space that is monitored.

(7) If multiple spaces are to be monitored by a single attendant, include in the permit program the means and procedures to enable the attendant to respond to an emergency affecting one or more of the permit spaces being monitored without distraction from the attendant's responsibilities under paragraph (i) of this section;

(8) Designate the persons who are to have active roles (as, for example, authorized entrants, attendants, entry supervisors, or persons who test or monitor the atmosphere in a permit space) in entry operations, identify the duties of each such employee, and provide each such employee with the training required by paragraph (g) of this section;

(9) Develop and implement procedures for summoning rescue and emergency services, for rescuing entrants from permit spaces, for providing necessary emergency

services to rescued employees, and for preventing unauthorized personnel from attempting a rescue;

(10) Develop and implement a system for the preparation, issuance, use, and cancellation of entry permits as required by this section;

(11) Develop and implement procedures to coordinate entry operations when employees of more than one employer are working simultaneously as authorized entrants in a permit space, so that employees of one employer do not endanger the employees of any other employer;

(12) Develop and implement procedures (such as closing off a permit space and canceling the permit) necessary for concluding the entry after entry operations have been completed;

(13) Review entry operations when the employer has reason to believe that the measures taken under the permit space program may not protect employees and revise the program to correct deficiencies found to exist before subsequent entries are authorized; and

Note: Examples of circumstances requiring the review of the permit space program are: any unauthorized entry of a permit space, the detection of a permit space hazard not covered by the permit, the detection of a condition prohibited by the permit, the occurrence of an injury or near-miss during entry, a change in the use or configuration of a permit space, and employee complaints about the effectiveness of the program.

(14) Review the permit space program, using the canceled permits retained under paragraph (e)(6) of this section within 1 year after each entry and revise the program as necessary, to ensure that employees participating in entry operations are protected from permit space hazards.

Note: Employers may perform a single annual review covering all entries performed during a 12-month period. If no entry is performed during a 12-month period, no review is necessary.

Appendix C to § 1910.146 presents examples of permit space programs that are considered to comply with the requirements of paragraph (d) of this section.

(e) Permit system.

(1) Before entry is authorized, the employer shall document the completion of measures required by paragraph (d)(3) of this section by preparing an entry permit.

Note: Appendix D to § 1910.146 presents examples of permits whose elements are considered to comply with the requirements of this section.

(2) Before entry begins, the entry supervisor identified on the permit shall sign the entry permit to authorize entry.

(3) The completed permit shall be made available at the time of entry to all authorized entrants, by posting it at the entry portal or by any other equally effective means, so that the entrants can confirm that pre-entry preparations have been completed.

(4) The duration of the permit may not exceed the time required to complete the assigned task or job identified on the permit in accordance with paragraph (f)(2) of this section.

(5) The entry supervisor shall terminate entry and cancel the entry permit when:

(i)   The entry operations covered by the entry permit have been completed; or

(ii)   A condition that is not allowed under the entry permit arises in or near the permit space.

(6) The employer shall retain each canceled entry permit for at least 1 year to facilitate the review of the permit-required confined space program required by paragraph (d)(14) of this section. Any problems encountered during an entry operation shall be noted on the pertinent permit so that appropriate revisions to the permit space program can be made.

(f) Entry permit. The entry permit that documents compliance with this section and authorizes entry to a permit space shall identify:

(1) The permit space to be entered;

(2) The purpose of the entry;

(3) The date and the authorized duration of the entry permit;

(4) The authorized entrants within the permit space, by name or by such other means (for example, through the use of rosters or tracking systems) as will enable the attendant to determine quickly and accurately, for the duration of the permit, which authorized entrants are inside the permit space;

Note: This requirement may be met by inserting a reference on the entry permit as to the means used, such as a roster or tracking system, to keep track of the authorized entrants within the permit space.

(5) The personnel, by name, currently serving as attendants;

(6) The individual, by name, currently serving as entry supervisor, with a space for the signature or initials of the entry supervisor who originally authorized entry;

(7) The hazards of the permit space to be entered;

(8) The measures used to isolate the permit space and to eliminate or control permit space hazards before entry;

Note: Those measures can include the lockout or tagging of equipment and procedures for purging, inerting, ventilating, and flushing permit spaces.

(9) The acceptable entry conditions;

(10) The results of initial and periodic tests performed under paragraph (d)(5) of this section, accompanied by the names or initials of the testers and by an indication of when the tests were performed;

(11) The rescue and emergency services that can be summoned and the means (such as the equipment to use and the numbers to call) for summoning those services;

(12) The communication procedures used by authorized entrants and attendants to maintain contact during the entry;

(13) Equipment, such as personal protective equipment, testing equipment, communications equipment, alarm systems, and rescue equipment, to be provided for compliance with this section;

(14) Any other information whose inclusion is necessary, given the circumstances of the particular confined space, in order to ensure employee safety; and

(15) Any additional permits, such as for hot work, that have been issued to authorize work in the permit space.

(g) Training.

(1) The employer shall provide training so that all employees whose work is regulated by this section acquire the understanding, knowledge, and skills necessary for the safe performance of the duties assigned under this section.

(2) Training shall be provided to each affected employee:

(i)   Before the employee is first assigned duties under this section;

(ii)   Before there is a change in assigned duties;

(iii) Whenever there is a change in permit space operations that presents a hazard about which an employee has not previously been trained;

(iv)   Whenever the employer has reason to believe either that there are deviations from the permit space entry procedures required by paragraph (d)(3) of this section or that there are inadequacies in the employee's knowledge or use of these procedures.

(3) The training shall establish employee proficiency in the duties required by this section and shall introduce new or revised procedures, as necessary, for compliance with this section.

(4) The employer shall certify that the training required by paragraphs (g)(1) through (g)(3) of this section has been accomplished. The certification shall contain each employee's name, the signatures or initials of the trainers, and the dates of training. The certification shall be available for inspection by employees and their authorized representatives.

(h) Duties of authorized entrants. The employer shall ensure that all authorized entrants:

(1) Know the hazards that may be faced during entry, including information on the mode, signs or symptoms, and consequences of the exposure;

(2) Properly use equipment as required by paragraph (d)(4) of this section;

(3) Communicate with the attendant as necessary to enable the attendant to monitor entrant status and to enable the attendant to alert entrants of the need to evacuate the space as required by paragraph (i)(6) of this section;

(4) Alert the attendant whenever:

(i)   The entrant recognizes any warning sign or symptom of exposure to a dangerous situation, or

(ii)   The entrant detects a prohibited condition; and

(5) Exit from the permit space as quickly as possible whenever:

(i)   An order to evacuate is given by the attendant or the entry supervisor,

(ii)   The entrant recognizes any warning sign or symptom of exposure to a dangerous situation,

(iii)   The entrant detects a prohibited condition, or

(iv)   An evacuation alarm is activated.

(i) Duties of attendants. The employer shall ensure that each attendant:

(1) Knows the hazards that may be faced during entry, including information on the mode, signs or symptoms, and consequences of the exposure;

(2) Is aware of possible behavioral effects of hazard exposure in authorized entrants;

(3) Continuously maintains an accurate count of authorized entrants in the permit space and ensures that the means used to identify authorized entrants under paragraph (f)(4) of this section accurately identifies who is in the permit space;

(4) Remains outside the permit space during entry operations until relieved by another attendant;

Note: When the employer's permit entry program allows attendant entry for rescue, attendants may enter a permit space to attempt a rescue if they have been trained and equipped for rescue operations as required by paragraph (k)(1) of this section and if they have been relieved as required by paragraph (i)(4) of this section.

(5) Communicates with authorized entrants as necessary to monitor entrant status and to alert entrants of the need to evacuate the space under paragraph (i)(6) of this section;

(6) Monitors activities inside and outside the space to determine if it is safe for entrants to remain in the space and orders the authorized entrants to evacuate the permit space immediately under any of the following conditions;

(i)   If the attendant detects a prohibited condition;

(ii)   If the attendant detects the behavioral effects of hazard exposure in an authorized entrant;

(iii)   If the attendant detects a situation outside the space that could endanger the authorized entrants; or

(iv)   If the attendant cannot effectively and safely perform all the duties required under paragraph (i) of this section;

(7) Summon rescue and other emergency services as soon as the attendant determines that authorized entrants may need assistance to escape from permit space hazards;

(8) Takes the following actions when unauthorized persons approach or enter a permit space while entry is underway:

(i)   Warn the unauthorized persons that they must stay away from the permit space;

(ii)   Advise the unauthorized persons that they must exit immediately if they have entered the permit space; and

(iii)   Inform the authorized entrants and the entry supervisor if unauthorized persons have entered the permit space;

(9) Performs non-entry rescues as specified by the employer's rescue procedure; and

(10) Performs no duties that might interfere with the attendant's primary duty to monitor and protect the authorized entrants.

(j)   Duties of entry supervisors. The employer shall ensure that each entry supervisor:

(1) Knows the hazards that may be faced during entry, including information on the mode, signs or symptoms, and consequences of the exposure;

(2) Verifies, by checking that the appropriate entries have been made on the permit, that all tests specified by the permit have been conducted and that all procedures and equipment specified by the permit are in place before endorsing the permit and allowing entry to begin;

(3) Terminates the entry and cancels the permit as required by paragraph (e)(5) of this section;

(4) Verifies that rescue services are available and that the means for summoning them are operable;

(5) Removes unauthorized individuals who enter or who attempt to enter the permit space during entry operations; and

(6) Determines, whenever responsibility for a permit space entry operation is transferred and at intervals dictated by the hazards and operations performed within the space, that entry operations remain consistent with terms of the entry permit and that acceptable entry conditions are maintained.

(k)   Rescue and emergency services.

(1) The following requirements apply to employers who have employees enter permit spaces to perform rescue services.

(i)   The employer shall ensure that each member of the rescue service is provided with, and is trained to use properly, the personal protective equipment and rescue equipment necessary for making rescues from permit spaces.

(ii)   Each member of the rescue service shall be trained to perform the assigned rescue duties. Each member of the rescue service shall also receive the training required of authorized entrants under paragraph (g) of this section.

(iii)   Each member of the rescue service shall practice making permit space rescues at least once every 12 months, by means of simulated rescue operations in which they remove dummies, manikins, or actual persons from the actual permit spaces or from representative permit spaces. Representative permit spaces shall, with respect to opening size, configuration, and accessibility, simulate the types of permit spaces from which rescue is to be performed.

   (iv)   Each member of the rescue service shall be trained in basic first-aid and in cardiopulmonary resuscitation (CPR).

         At least one member of the rescue service holding current certification in first aid and in CPR shall be available.

(2) When an employer (host employer) arranges to have persons other than the host employer's employees perform permit space rescue, the host employer shall:

   (i)   Inform the rescue service of the hazards they may confront when called on to perform rescue at the host employer's facility, and

   (ii)   Provide the rescue service with access to all permit spaces from which rescue may be necessary so that the rescue service can develop appropriate rescue plans and practice rescue operations.

(3) To facilitate non-entry rescue, retrieval systems or methods shall be used whenever an authorized entrant enters a permit space, unless the retrieval equipment would increase the overall risk of entry or would not contribute to the rescue of the entrant.

Retrieval systems shall meet the following requirements.

   (i)   Each authorized entrant shall use a chest or full body harness, with a retrieval line attached at the center of the entrant's back near shoulder level, or above the entrant's head.

         Wristlets may be used in lieu of the chest or full body harness if the employer can demonstrate that the use of a chest or full body harness is infeasible or creates a greater hazard and that the use of wristlets is the safest and most effective alternative.

   (ii)   The other end of the retrieval line shall be attached to a mechanical device or fixed point outside the permit space in such a manner that rescue can begin as soon as the rescuer becomes aware that rescue is necessary. A mechanical device shall be available to retrieve personnel from vertical type permit spaces more than 5 feet (1.52 m) deep.

(4) If an injured entrant is exposed to a substance for which a Material Safety Data Sheet (MSDS) or other similar written information is required to be kept at the worksite, that MSDS or written information shall be made available to the medical facility treating the exposed entrant.

[58 FR 4549, Jan. 14, 1993; 58 FR 34885, June 29, 1993, as amended at 59 FR 26114, May 19, 1994]

APPENDICES TO § 1910.146—PERMIT-REQUIRED CONFINED SPACES

Note: Appendices A through E serve to provide information and nonmandatory guidelines to assist employers and employees in complying with the appropriate requirements of this section.

APPENDIX A TO § 1910.146—PERMIT-REQUIRED CONFINED SPACE DECISION FLOW CHART

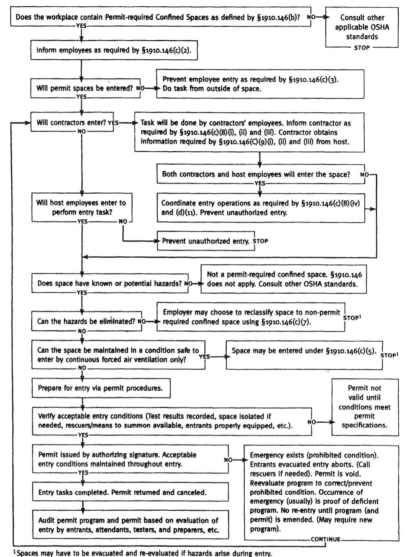

¹Spaces may have to be evacuated and re-evaluated if hazards arise during entry.

APPENDIX B TO § 1910.146—PROCEDURES FOR ATMOSPHERIC TESTING

Atmospheric testing is required for two distinct purposes: evaluation of the hazards of the permit space and verification that acceptable entry conditions for entry into that space exist.

(1) Evaluation testing. The atmosphere of a confined space should be analyzed using equipment of sufficient sensitivity and specificity to identify and evaluate any hazardous atmospheres that may exist or arise, so that appropriate permit entry procedures can be developed and acceptable entry conditions stipulated for that space. Evaluation and interpretation of these data, and development of the entry procedure, should be done by, or reviewed by, a technically qualified professional (e.g., OSHA consultation service, or certified industrial hygienist, registered safety engineer, certified safety professional, certified marine chemist, etc.) based on evaluation of all serious hazards.

(2) Verification testing. The atmosphere of a permit space which may contain a hazardous atmosphere should be tested for residues of all contaminants identified by evaluation testing using permit specified equipment to determine that residual concentrations at the time of testing and entry are within the range of acceptable entry conditions. Results of testing (i.e., actual concentration, etc.) should be recorded on the permit in the space provided adjacent to the stipulated acceptable entry condition.

(3) Duration of testing. Measurement of values for each atmospheric parameter should be made for at least the minimum response time of the test instrument specified by the manufacturer.

(4) Testing stratified atmospheres. When monitoring for entries involving a descent into atmospheres that may be stratified, the atmospheric envelope should be tested a distance of approximately 4 feet (1.22 m) in the direction of travel and to each side.

If a sampling probe is used, the entrant's rate of progress should be slowed to accommodate the sampling speed and detector response.

(5) Order of testing. A test for oxygen is performed first because most combustible gas meters are oxygen dependent and will not provide reliable readings in an oxygen deficient atmosphere.

Combustible gasses are tested for next because the threat of fire or explosion is both more immediate and more life threatening, in most cases, than exposure to toxic gasses and vapors. If tests for toxic gasses and vapors are necessary, they are performed last.

APPENDIX C TO § 1910.146—EXAMPLES OF PERMIT-REQUIRED CONFINED SPACE PROGRAMS

Example 1.

Workplace: Sewer entry.

Potential hazards. The employees could be exposed to the following:

Engulfment.

Presence of toxic gases. Equal to or more than 10 ppm hydrogen sulfide measured as an 8-hour time-weighted average. If the presence of other toxic contaminants is suspected, specific monitoring programs will be developed.

Presence of explosive/flammable gases. Equal to or greater than 10% of the lower flammable limit (LFL).

Oxygen Deficiency. A concentration of oxygen in the atmosphere equal to or less than 19.5% by volume.

A. Entry Without Permit/Attendant

Certification. Confined spaces may be entered without the need for a written permit or attendant provided that the space can be maintained in a safe condition for entry by mechanical ventilation alone, as provided in §1910.146(c)(5). All spaces shall be considered permit-required confined spaces until the pre-entry procedures demonstrate otherwise. Any employee required or permitted to pre-check or enter an enclosed/confined space shall have successfully completed, as a minimum, the training as required by the following sections of these procedures. A written copy of operating and rescue procedures as required by these procedures shall be at the work site for the duration of the job. The Confined Space Pre-Entry Check List must be completed by the LEAD WORKER before entry into a confined space. This list verifies completion of items listed below. This check list shall be kept at the job site for duration of the job. If circumstances dictate an interruption in the work, the permit space must be re-evaluated and a new check list must be completed.

Control at atmospheric and engulfment hazards:

Pumps and Lines. All pumps and lines which may reasonably cause contaminants to flow into the space shall be disconnected, blinded and locked out, or effectively isolated by other means to prevent development of dangerous air contamination or engulfment. Not all laterals to sewers or storm drains require blocking. However, where experience or knowledge of industrial use indicates there is a reasonable potential for contamination of air or engulfment into an occupied sewer, then all affected laterals shall be blocked. If blocking and/or isolation requires entry into the space the provisions for entry into a permit-required confined space must be implemented.

Surveillance. The surrounding area shall be surveyed to avoid hazards such as drifting vapors from the tanks, piping, or sewers.

Testing. The atmosphere within the space will be tested to determine whether dangerous air contamination and/or oxygen deficiency exists. Detector tubes, alarm only gas monitors and explosion meters are examples of monitoring equipment that may be used to test permit space atmospheres. Testing shall be performed by the LEAD WORKER who has successfully completed the Gas Detector training for the monitor he will use. The minimum parameters to be monitored are oxygen deficiency, LFL, and hydrogen sulfide concentration. A written record of the pre-entry test results shall be made and kept at the work site for the duration of the job. The supervisor will certify in writing, based upon the results of the pre-entry testing, that all hazards have been eliminated. Affected employees shall be able to review the testing results. The most hazardous conditions shall govern when work is being performed in two adjoining, connecting spaces.

Entry Procedures. If there are no non-atmospheric hazards present and if the pre-entry tests show there is no dangerous air contamination and/or oxygen deficiency within the space and there is no reason to believe that any is likely to develop, entry into and work within may proceed. Continuous testing of the atmosphere in the immediate vicinity of the workers within the space shall be accomplished. The workers will immediately leave the permit space when any of the gas monitor alarm set points are reached as defined. Workers will not return to the area until a SUPERVISOR who has completed the gas detector training has used a direct reading gas detector to evaluate the situation and has determined that it is safe to enter.

Rescue. Arrangements for rescue services are not required where there is no attendant. See the rescue portion of section B., below, for instructions regarding rescue planning where an entry permit is required.

B. Entry Permit Required

Permits. Confined Space Entry Permit. All spaces shall be considered permit-required confined spaces until the pre-entry procedures demonstrate otherwise. Any employee required or permitted to pre-check or enter a permit-required confined space shall have successfully completed, as a minimum, the training as required by the following sections of these procedures. A written copy of operating and rescue procedures as required by these procedures shall be at the work site for the duration of the job. The Confined Space Entry Permit must be completed before approval can be given to enter a permit-required confined space. This permit verifies completion of items listed below. This permit shall be kept at the job site for the duration of the job. If circumstances cause an interruption in the work or a change in the alarm conditions for which entry was approved, a new Confined Space Entry Permit must be completed.

Control of atmospheric and engulfment hazards:

Surveillance. The surrounding area shall be surveyed to avoid hazards such as drifting vapors from tanks, piping or sewers.

Testing. The confined space atmosphere shall be tested to determine whether dangerous air contamination and/or oxygen deficiency exists. A direct reading gas monitor shall be used. Testing shall be performed by the SUPERVISOR who has successfully completed the gas detector training for the monitor he will use. The minimum parameters to be monitored are oxygen deficiency, LFL and hydrogen sulfide concentration. A written record of the preentry test results shall be made and kept at the work site for the duration of the job. Affected employees shall be able to review the testing results. The most hazardous conditions shall govern when work is being performed in two adjoining, connected spaces.

Space Ventilation. Mechanical ventilation systems, where applicable, shall be set at 100% outside air. Where possible, open additional manholes to increase air circulation. Use portable blowers to augment natural circulation if needed. After a suitable ventilating period, repeat the testing. Entry may not begin until testing has demonstrated that the hazardous atmosphere has been eliminated.

Entry Procedures. The following procedure shall be observed under any of the following conditions: 1.) Testing demonstrates the existence of dangerous or deficient conditions and additional ventilation cannot reduce concentrations to safe levels; 2.) The atmosphere tests as safe but unsafe conditions can reasonably be expected to develop; 3.) It is not feasible to provide for ready exit from spaces equipped with automatic fire suppression systems and it is not practical or safe to deactivate such systems; or 4.) An emergency exists and it is not feasible to wait for preentry procedures to take effect.

All personnel must be trained. A self contained breathing apparatus shall be worn by any person entering the space. At least one worker shall stand by the outside of the space ready to give assistance in case of emergency. The standby worker shall have a self contained breathing apparatus available for immediate use. There shall be at least one additional worker within sight or call of the standby worker. Continuous powered communications shall be maintained between the worker within the confined space and standby personnel.

If at any time there is any questionable action or non-movement by the worker inside, a verbal check will be made. If there is no response, the worker will be moved immediately. Exception: If the worker is disabled due to falling or impact, he/she shall not be removed from the confined space unless there is immediate danger to his/her life. Local fire department rescue personnel shall be notified immediately. The standby worker may only enter the confined space in case of an emergency (wearing the self contained breathing apparatus) and only after being relieved by another worker. Safety belt or harness with attached lifeline shall be used by all workers entering the space with the free end of the line secured outside the entry opening. The standby worker shall attempt to remove a disabled worker via his lifeline before entering the space.

When practical, these spaces shall be entered through side openings—those within 3 1/2 feet (1.07 m) of the bottom. When entry must be through a top opening, the safety belt shall be of the harness type that suspends a person upright and a hoisting device or similar apparatus shall be available for lifting workers out of the space.

In any situation where their use may endanger the worker, use of a hoisting device or safety belt and attached lifeline may be discontinued.

When dangerous air contamination is attributable to flammable and/or explosive substances, lighting and electrical equipment shall be Class 1, Division 1 rated per National Electrical Code and no ignition sources shall be introduced into the area.

Continuous gas monitoring shall be performed during all confined space operations. If alarm conditions change adversely, entry personnel shall exit the confined space and a new confined space permit shall be issued.

Rescue. Call the fire department services for rescue. Where immediate hazards to injured personnel are present, workers at the site shall implement emergency procedures to fit the situation.

Example 2.

Workplace: Meat and poultry rendering plants.

Cookers and dryers are either batch or continuous in their operation.

Multiple batch cookers are operated in parallel. When one unit of a multiple set is shut down for repairs, means are available to isolate that unit from the others which remain in operation.

Cookers and dryers are horizontal, cylindrical vessels equipped with a center, rotating shaft and agitator paddles or discs.

If the inner shell is jacketed, it is usually heated with steam at pressures up to 150 psig (1034.25 kPa). The rotating shaft assembly of the continuous cooker or dryer is also steam heated.

Potential Hazards. The recognized hazards associated with cookers and dryers are the risk that employees could be:

1. Struck or caught by rotating agitator;
2. Engulfed in raw material or hot, recycled fat;
3. Burned by steam from leaks into the cooker/dryer steam jacket or the condenser duct system if steam valves are not properly closed and locked out;
4. Burned by contact with hot metal surfaces, such as the agitator shaft assembly, or inner shell of the cooker/dryer;
5. Suffer heat stress caused by warm atmosphere inside cooker/dryer;
6. Slipping and falling on grease in the cooker/dryer;
7. Electrically shocked by faulty equipment taken into the cooker/dryer;
8. Burned or overcome by fire or products of combustion; of
9. Overcome by fumes generated by welding or cutting done on grease covered surfaces.

Permits. The supervisor in this case is always present at the cooker/dryer or other permit entry confined space when entry is made. The supervisor must follow the pre-entry isolation procedures described in the entry permit in preparing for entry, and ensure that the protective clothing, ventilating equipment and any other equipment required by the permit are at the entry site.

Control of hazards: Mechanical. Lock out main power switch to agitator motor at main power panel. Affix tag to the lock to inform others that a permit entry confined space entry is in progress.

Engulfment. Close all valves in the raw material blow line.

Secure each valve in its closed position using chain and lock.

Attach a tag to the valve and chain, warning that a permit entry confined space entry is in progress. The same procedure shall be used for securing the fat recycle valve.

Burns and heat stress. Close steam supply valves to jacket and secure with chains and tags. Insert solid blank at flange in cooker vent line to condenser manifold duct system. Vent cooker/dryer by opening access door at discharge end and top center door to allow natural ventilation throughout the entry. If faster cooling is needed, use a portable ventilation fan to increase ventilation. Cooling water may be circulated through the jacket to reduce both outer and inner surface temperatures of cooker/dryers faster. Check air and inner surface temperatures in cooker/dryer to assure they are within acceptable limits before entering, or use proper protective clothing.

Fire and fume hazards. Careful site preparation, such as cleaning the area within 4 inches (10.16 cm) of all welding or torch cutting operations, and proper ventilation are the preferred controls. All welding and cutting operations shall be done in accordance with the requirements of 29 CFR Part 1910, Subpart Q, OSHA's welding standard. Proper ventilation may be achieved by local exhaust ventilation, or the use of portable ventilation fans, or a combination of the two practices.

Electrical shock. Electrical equipment used in cooker/dryers shall be in serviceable condition.

Slips and falls. Remove residual grease before entering cooker/dryer.

Attendant. The supervisor shall be the attendant for employees entering cooker/dryers.

Permit. The permit shall specify how isolation shall be done and any other preparations needed before making entry. This is especially important in parallel arrangements of cooker/dryers so that the entire operation need not be shut down to allow safe entry into one unit.

Rescue. When necessary, the attendant shall call the fire department as previously arranged.

Example 3.

Workplace: Workplaces where tank cars, trucks, and trailers, dry bulk tanks and trailers, railroad tank cars, and similar portable tanks are fabricated or serviced.

A. During fabrication. These tanks and dry-bulk carriers are entered repeatedly throughout the fabrication process. These products are not configured identically, but the manufacturing processes by which they are made are very similar.

Sources of hazards. In addition to the mechanical hazards arising from the risks that an entrant would be injured due to contact with components of the tank or the tools being used, there is also the risk that a worker could be injured by breathing fumes from welding materials or mists or vapors from materials used to coat the tank interior. In addition, many of these vapors and mists are flammable, so the failure to properly ventilate a tank could lead to a fire or explosion.

Control of hazards:

Welding. Local exhaust ventilation shall be used to remove welding fumes once the tank or carrier is completed to the point that workers may enter and exit only through a manhole. (Follow the requirements of 29 CFR 1910, Subpart Q, OSHA's welding standard, at all times.) Welding gas tanks may never be brought into a tank or carrier that is a permit entry confined space.

Application of interior coatings/linings. Atmospheric hazards shall be controlled by forced air ventilation sufficient to keep the atmospheric concentration of flammable materials below 10% of the lower flammable limit (LFL) (or lower explosive limit

(LEL), whichever term is used locally). The appropriate respirators are provided and shall be used in addition to providing forced ventilation if the forced ventilation does not maintain acceptable respiratory conditions.

Permits. Because of the repetitive nature of the entries in these operations, an "Area Entry Permit" will be issued for a 1 month period to cover those production areas where tanks are fabricated to the point that entry and exit are made using manholes.

Authorization. Only the area supervisor may authorize an employee to enter a tank within the permit area. The area supervisor must determine that conditions in the tank trailer, dry bulk trailer or truck, etc. meet permit requirements before authorizing entry.

Attendant. The area supervisor shall designate an employee to maintain communication by employer specified means with employees working in tanks to ensure their safety. The attendant may not enter any permit entry confined space to rescue an entrant or for any other reason, unless authorized by the rescue procedure and, and even then, only after calling the rescue team and being relieved as attendant by another worker.

Communications and observation. Communications between attendant and entrant(s) shall be maintained throughout entry. Methods of communication that may be specified by the permit include voice, voice powered radio, tapping or rapping codes on tank walls, signalling tugs on a rope, and the attendant's observation that work activities such as chipping, grinding, welding, spraying, etc., which require deliberate operator control continue normally.

These activities often generate so much noise that the necessary hearing protection makes communication by voice difficult.

Rescue procedures. Acceptable rescue procedures include entry by a team of employee-rescuers, use of public emergency services, and procedures for breaching the tank. The area permit specifies which procedures are available, but the area supervisor makes the final decision based on circumstances. (Certain injuries may make it necessary to breach the tank to remove a person rather than risk additional injury by removal through an existing manhole.) However, the supervisor must ensure that no breaching procedure used for rescue would violate terms of the entry permit.

For instance, if the tank must be breached by cutting with a torch, the tank surfaces to be cut must be free of volatile or combustible coatings within 4 inches (10.16 cm) of the cutting line and the atmosphere within the tank must be below the LFL.

Retrieval line and harnesses. The retrieval lines and harnesses generally required under this standard are usually impractical for use in tanks because the internal configuration of the tanks and their interior baffles and other structures would prevent rescuers from hauling out injured entrants. However, unless the rescue procedure calls for breaching the tank for rescue, the rescue team shall be trained in the use of retrieval lines and harnesses for removing injured employees through manholes.

B. Repair or service of "used" tanks and bulk trailers.

Sources of hazards. In addition to facing the potential hazards encountered in fabrication or manufacturing, tanks or trailers which have been in service may contain residues of dangerous materials, whether left over from the transportation of hazardous cargoes or generated by chemical or bacterial action on residues of non-hazardous cargoes.

Control of atmospheric hazards. A "used" tank shall be brought into areas where tank entry is authorized only after the tank has been emptied, cleansed (without employee entry) of any residues, and purged of any potential atmospheric hazards.

Welding. In addition to tank cleaning for control of atmospheric hazards, coating and surface materials shall be removed 4 inches (10.16 cm) or more from any surface area where welding or other torch work will be done and care taken that the atmosphere within the tank remains well below the LFL. (Follow the requirements of 29 CFR 1910, Subpart Q, OSHA's welding standard, at all times.)

Permits. An entry permit valid for up to 1 year shall be issued prior to authorization of entry into used tank trailers, dry bulk trailers or trucks. In addition to the pre-entry cleaning requirement, this permit shall require the employee safeguards specified for new tank fabrication or construction permit areas.

Authorization. Only the area supervisor may authorize an employee to enter a tank trailer, dry bulk trailer or truck within the permit area. The area supervisor must determine that the entry permit requirements have been met before authorizing entry.

APPENDIX D TO § 1910.146—
SAMPLE PERMITS

Appendix D - 1A   Sewer Entry Permit

Confined Space Pre-Entry Check List

See Safety Procedure.
A confined space either is entered through an opening other than a
door (such as manhole or side port) or requires the use of a ladder
or rungs to reach the working level and test results are
satisfactory. This check list must be filled out whenever the job
site meets this criteria.

		Yes	No
1.	Did your survey of the surrounding area show it to be free of hazards such as drifting vapors from tanks, piping or sewers?	( )	( )
2.	Does your knowledge of industrial or other discharges indicate this area is likely to remain free of dangerous air contaminants while occupied?	( )	( )
3.	Are you certified in operation of the gas monitor to be used?	( )	( )
4.	Has a gas monitor functional test (Bump Test) been performed this shift on the gas monitor to be used?	( )	( )
5.	Did you test the atmosphere of the confined space prior to entry?	( )	( )
6.	Did the atmosphere check as acceptable (no alarms given)?	( )	( )
7.	Will the atmosphere be continuously monitored while the space is occupied?	( )	( )

Contact County Centrex for personnel rescue by local fire
department in the event of an emergency. If on-site at the
Regional Treatment Plant, contact the Plant Control Center (PCC).

Notice: If any of the above questions are answered "no" do not
enter. Contact your immediate supervisor.

Job
Location_____
LEAD MAN
signature_____Date_____

Appendix D - 1B

Confined Space Entry Permit (Pre-Entry/Entry Check List)

Date and Time Issued: _____

Job site: _____

Equipment to be worked on: _____

Pre-Entry (See Safety Procedure)

1. Atmospheric Checks: Time _____

Oxygen	_____	%
Explosive	_____	% L.F.L.
Toxic	_____	PPM

2. Source isolation (No Entry):

	N/A	Yes	No
Pumps or lines blinded, disconnected, or blocked	( )	( )	( )

3. Ventilation Modification:

	N/A	Yes	No
Mechanical	( )	( )	( )
Natural Ventilation only	( )	( )	( )

4. Atmospheric check after isolation and Ventilation:

Oxygen	_____	%	> 19.5 %
Explosive	_____	% L.F.L.	< 10 %
Toxic	_____	PPM	< 10 PPM H$_2$S
Time	_____		

If conditions are in compliance with the above requirements and there is no reason to believe conditions may change adversely, then proceed to the Permit Space Pre-Entry Check List. Complete and post with this permit. If conditions are not in compliance with the above requirements or there is reason to believe that conditions may change adversely, proceed to the Entry Check-List portion of this permit.

Permit and Check List Prepared By: (Supervisor) _____

Approved By: (Unit Supervisor) _____

Reviewed By (Confined Space Operations Personnel):(printed name & signature) _____

Date and Time Expires: _____

Job Supervisor _____

Work to be performed: _____

Entry (See Safety Procedure)

1. Entry, standby, and back up persons:

	Yes	No
Successfully completed required training?		

	N/A	Yes	No
Is it current?	( )	( )	( )

2. Equipment:

	N/A	Yes	No
Direct reading gas monitor - tested	( )	( )	( )
Safety harnesses and lifelines for entry and standby persons	( )	( )	( )
Hoisting equipment	( )	( )	( )
Powered communications	( )	( )	( )
SCBA's for entry and standby persons	( )	( )	( )
Protective Clothing	( )	( )	( )
All electric equipment listed Class I, Division I, Group D and Non-sparking tools	( )	( )	( )

3. Rescue Procedure: _____

We have reviewed the work authorized by this permit and the information contained here-in. Written instructions and safety procedures have been received and are understood. Entry cannot be approved if any squares are marked in the "No" column. This permit is not valid unless all appropriate items are completed.

This permit to be kept at job site. Return job site copy to Safety Office following job completion.

Copies: White Original (Safety Office) Yellow (Unit Supervisor) Hard(Job site)

Appendix D - 2

ENTRY PERMIT

CONFINED SPACE _____ HAZARDOUS AREA
PERMIT VALID FOR 8 HOURS ONLY. ALL COPIES OF PERMIT WILL REMAIN AT JOB SITE UNTIL JOB IS COMPLETED
SITE LOCATION and DESCRIPTION _____
PURPOSE OF ENTRY _____
SUPERVISOR(S) in charge of crews _____ Type of Crew Phone # _____

* **BOLD DENOTES MINIMUM REQUIREMENTS TO BE COMPLETED AND REVIEWED PRIOR TO ENTRY***
REQUIREMENTS COMPLETED   DATE TIME   REQUIREMENTS COMPLETED   DATE TIME
Lock Out/De-energize/Try-out _____   **Full Body Harness w/"D" ring** _____
Line(s) Broken-Capped-Blanked _____   **Emergency Escape Retrieval Equip** _____
Purge-Flush and Vent _____   Lifelines _____
Ventilation _____   Fire Extinguishers _____
Secure Area (Post and Flag) _____   Lighting (Explosive Proof) _____
Breathing Apparatus _____   Protective Clothing _____
Resuscitator - Inhalator _____   Respirator(s) (Air Purifying) _____
Standby Safety Personnel _____   Burning and Welding Permit _____
Note: Items that do not apply enter N/A in the blank.
** RECORD CONTINUOUS MONITORING RESULTS EVERY 2 HOURS

CONTINUOUS MONITORING**   Permissible
TEST(S) TO BE TAKEN   Entry Level
PERCENT OF OXYGEN   19.5% to 23.5%
LOWER FLAMMABLE LIMIT   Under 10%   _____
CARBON MONOXIDE   +35 PPM   _____
Aromatic Hydrocarbon   + 1 PPM * 5PPM   _____
Hydrogen Cyanide   (Skin) * 4PPM   _____
Hydrogen Sulfide   +10 PPM *15PPM   _____
Sulfur Dioxide   + 2 PPM * 5PPM   _____
Ammonia   *35PPM   _____
* Short-term exposure limit:Employee can work in the area up to 15 minutes.
+ 8 hr. Time Weighted Avg.:Employee can work in area 8 hrs (longer with appropriate respiratory protection)
REMARKS: _____
GAS TESTER NAME & CHECK # INSTRUMENT(S) USED MODEL &/OR TYPE SERIAL &/OR UNIT # _____

SAFETY STANDBY PERSON IS REQUIRED FOR ALL CONFINED SPACE WORK
SAFETY STANDBY PERSON(S) CHECK # NAME OF SAFETY STANDBY PERSON(S)   CHECK # _____

SUPERVISOR AUTHORIZING ENTRY _____   AMBULANCE 2800   FIRE 2900
ALL ABOVE CONDITIONS SATISFIED _____   Safety   4901   Gas Coordinator 4529/5387
DEPARTMENT _____ Phone _____   Original to Department Pink Copy to Safety

APPENDIX E TO § 1910.146—SEWER SYSTEM ENTRY

Sewer entry differs in three vital respects from other permit entries; first, there rarely exists any way to completely isolate the space (a section of a continuous system) to be entered; second, because isolation is not complete, the atmosphere may suddenly and unpredictably become lethally hazardous (toxic, flammable or explosive) from causes beyond the control of the entrant or employer, and third, experienced sewer workers are especially knowledgeable in entry and work in their permit spaces because of their frequent entries. Unlike other employments where permit space entry is a rare and exceptional event, sewer workers' usual work environment is a permit space.

(1) Adherence to procedure. The employer should designate as entrants only employees who are thoroughly trained in the employer's sewer entry procedures and who demonstrate that they follow these entry procedures exactly as prescribed when performing sewer entries.

(2) Atmospheric monitoring. Entrants should be trained in the use of, and be equipped with, atmospheric monitoring equipment which sounds an audible alarm, in addition to its visual readout, whenever one of the following conditions is encountered: Oxygen concentration less than 19.5 percent; flammable gas or vapor at 10 percent or more of the lower flammable limit (LFL); or hydrogen sulfide or carbon monoxide at or above 10 ppm or 35 ppm, respectively, measured as an 8-hour time-weighted average.

Atmospheric monitoring equipment needs to be calibrated according to the manufacturer's instructions. The oxygen sensor/broad range sensor is best suited for initial use in situations where the actual or potential contaminants have not been identified, because broad range sensors, unlike substance-specific sensors, enable employers to obtain an overall reading of the hydrocarbons (flammables) present in the space. However, such sensors only indicate that a hazardous threshold of a class of chemicals has been exceeded. They do not measure the levels of contamination of specific substances. Therefore, substance-specific devices, which measure the actual levels of specific substances, are best suited for use where actual and potential contaminants have been identified. The measurements obtained with substance-specific devices are of vital importance to the employer when decisions are made concerning the measures necessary to protect entrants (such as ventilation or personal protective equipment) and the setting and attainment of appropriate entry conditions.

However, the sewer environment may suddenly and unpredictably change, and the substance-specific devices may not detect the potentially lethal atmospheric hazards which may enter the sewer environment.

Although OSHA considers the information and guidance provided above to be appropriate and useful in most sewer entry situations, the Agency emphasizes that each employer must consider the unique circumstances, including the predictability of the atmosphere, of the sewer permit spaces in the employer's workplace in preparing for entry. Only the employer can decide, based upon his or her knowledge of, and experience with permit spaces in sewer systems, what the best type of testing instrument may be for any specific entry operation.

The selected testing instrument should be carried and used by the entrant in sewer line work to monitor the atmosphere in the entrant's environment, and in advance of the entrant's direction of movement, to warn the entrant of any deterioration in atmospheric conditions. Where several entrants are working together in the same immediate location, one instrument, used by the lead entrant, is acceptable.

(3) Surge flow and flooding. Sewer crews should develop and maintain liaison, to the extent possible, with the local weather bureau and fire and emergency services in their area so that sewer work may be delayed or interrupted and entrants withdrawn whenever sewer lines might be suddenly flooded by rain or fire suppression activities, or whenever flammable or other hazardous materials are released into sewers during emergencies by industrial or transportation accidents.

(4) Special Equipment. Entry into large bore sewers may require the use of special equipment. Such equipment might include such items as atmosphere monitoring devices with automatic audible alarms, escape self-contained breathing apparatus (ESCBA) with at least 10 minute air supply (or other NIOSH approved self-rescuer), and waterproof flashlights, and may also include boats and rafts, radios and rope stand-offs for pulling around bends and corners as needed.

[58 FR 4549, Jan. 14, 1993; 58 FR 34885, June 29, 1993, as amended at 59 FR 26114, May 19, 1994]

Appendix E

OSHA Excavations, Final Rule (29 CFR, Subpart P)

SUBPART P EXCAVATIONS

Authority: Sec. 107, Contract Worker Hours and Safety Standards Act (Construction Safety Act) (40 U.S.C. 333); Secs. 4, 6, 8, Occupational Safety and Health Act of 1970 (29 U.S.C. 653, 655, 657); Secretary of Labor's Order No. 12-71 (36 FR 8754), 8-76 (41 FR 25059), or 9-83 (48 FR 35736), as applicable, and 29 CFR part 1911. Source: 54 FR 45959, Oct. 31, 1989, unless otherwise noted.

§ 1926.650 SCOPE, APPLICATION, AND DEFINITIONS APPLICABLE TO THIS SUBPART.

(a) Scope and application. This subpart applies to all open excavations made in the earth's surface. Excavations are defined to include trenches.

(b) Definitions applicable to this subpart.

Accepted engineering practices means those requirements which are compatible with standards of practice required by a registered professional engineer.

Aluminum Hydraulic Shoring means a pre-engineered shoring system comprised of aluminum hydraulic cylinders (crossbraces) used in conjunction with vertical rails (uprights) or horizontal rails (walers). Such system is designed, specifically to support the sidewalls of an excavation and prevent cave-ins.

Bell-bottom pier hole means a type of shaft or footing excavation, the bottom of which is made larger than the cross section above to form a belled shape.

Benching (Benching system) means a method of protecting employees from cave-ins by excavating the sides of an excavation to form one or a series of horizontal levels or steps, usually with vertical or near-vertical surfaces between levels.

Cave-in means the separation of a mass of soil or rock material from the side of an excavation, or the loss of soil from under a trench shield or support system, and its sudden movement into the excavation, either by falling or sliding, in sufficient quantity so that it could entrap, bury, or otherwise injure and immobilize a person.

Competent person means one who is capable of identifying existing and predictable hazards in the surroundings, or working conditions which are unsanitary, hazardous, or dangerous to employees, and who has authorization to take prompt corrective measures to eliminate them.

Cross braces mean the horizontal members of a shoring system installed perpendicular to the sides of the excavation, the ends of which bear against either uprights or wales.

Excavation means any man-made cut, cavity, trench, or depression in an earth surface, formed by earth removal.

Faces or sides means the vertical or inclined earth surfaces formed as a result of excavation work.

Failure means the breakage, displacement, or permanent deformation of a structural member or connection so as to reduce its structural integrity and its supportive capabilities.

Hazardous atmosphere means an atmosphere which by reason of being explosive, flammable, poisonous, corrosive, oxidizing, irritating, oxygen deficient, toxic, or otherwise harmful, may cause death, illness, or injury.

Kickout means the accidental release or failure of a cross brace.

Protective system means a method of protecting employees from cave-ins, from material that could fall or roll from an excavation face or into an excavation, or from the collapse of adjacent structures. Protective systems include support systems, sloping and benching systems, shield systems, and other systems that provide the necessary protection.

Ramp means an inclined walking or working surface that is used to gain access to one point from another, and is constructed from earth or from structural materials such as steel or wood.

Registered Professional Engineer means a person who is registered as a professional engineer in the state where the work is to be performed. However, a professional engineer, registered in any state is deemed to be a "registered professional engineer" within the meaning of this standard when approving designs for "manufactured protective systems" or "tabulated data" to be used in interstate commerce.

Sheeting means the members of a shoring system that retain the earth in position and in turn are supported by other members of the shoring system.

Shield (Shield system) means a structure that is able to withstand the forces imposed on it by a cave-in and thereby protect employees within the structure. Shields can be permanent structures or can be designed to be portable and moved along as work progresses.

Additionally, shields can be either premanufactured or job-built in accordance with § 1926.652 (c)(3) or (c)(4). Shields used in trenches are usually referred to as "trench boxes" or "trench shields."

Shoring (Shoring system) means a structure such as a metal hydraulic, mechanical or timber shoring system that supports the sides of an excavation and which is designed to prevent cave-ins.

Sides. See "Faces."

Sloping (Sloping system) means a method of protecting employees from cave-ins by excavating to form sides of an excavation that are inclined away from the excavation so as to prevent cave-ins. The angle of incline required to prevent a cave-in varies with differences in such factors as the soil type, environmental conditions of exposure, and application of surcharge loads.

Stable rock means natural solid mineral material that can be excavated with vertical sides and will remain intact while exposed. Unstable rock is considered to be stable when the rock material on the side or sides of the excavation is secured against caving-in or movement by rock bolts or by another protective system that has been designed by a registered professional engineer.

Structural ramp means a ramp built of steel or wood, usually used for vehicle access. Ramps made of soil or rock are not considered structural ramps.

Support system means a structure such as underpinning, bracing, or shoring, which provides support to an adjacent structure, underground installation, or the sides of an excavation.

Tabulated data means tables and charts approved by a registered professional engineer and used to design and construct a protective system.

Trench (Trench excavation) means a narrow excavation (in relation to its length) made below the surface of the ground. In general, the depth is greater than the width, but the width of a trench (measured at the bottom) is not greater than 15 feet (4.6 m).

If forms or other structures are installed or constructed in an excavation so as to reduce the dimension measured from the forms or structure to the side of the excavation to 15 feet (4.6 m) or less (measured at the bottom of the excavation), the excavation is also considered to be a trench.

Trench box. See "Shield."

Trench shield. See "Shield."

Uprights means the vertical members of a trench shoring system placed in contact with the earth and usually positioned so that individual members do not contact each other. Uprights placed so that individual members are closely spaced, in contact with or interconnected to each other, are often called "sheeting." Wales means horizontal members of a shoring system placed parallel to the excavation face whose sides bear against the vertical members of the shoring system or earth.

§ 1926.651 SPECIFIC EXCAVATION REQUIREMENTS.

(a) Surface encumbrances. All surface encumbrances that are located so as to create a hazard to employees shall be removed or supported, as necessary, to safeguard employees.

(b) Underground installations.

(1) The estimated location of utility installations, such as sewer, telephone, fuel, electric, water lines, or any other underground installations that reasonably may be expected to be encountered during excavation work, shall be determined prior to opening an excavation.

(2) Utility companies or owners shall be contacted within established or customary local response times, advised of the proposed work, and asked to establish the location of the utility underground installations prior to the start of actual excavation.

When utility companies or owners cannot respond to a request to locate underground utility installations within 24 hours (unless a longer period is required by state or local law), or cannot establish the exact location of these installations, the employer may proceed, provided the employer does so with caution, and provided detection equipment or other acceptable means to locate utility installations are used.

(3) When excavation operations approach the estimated location of underground installations, the exact location of the installations shall be determined by safe and acceptable means.

(4) While the excavation is open, underground installations shall be protected, supported or removed as necessary to safeguard employees.

(c) Access and egress-

(1) Structural ramps.

(i) Structural ramps that are used solely by employees as a means of access or egress from excavations shall be designed by a competent person. Structural ramps used for access or egress of equipment shall be designed by a competent person qualified in structural design, and shall be constructed in accordance with the design.

(ii) Ramps and runways constructed of two or more structural members shall have the structural members connected together to prevent displacement.

(iii) Structural members used for ramps and runways shall be of uniform thickness.

(iv) Cleats or other appropriate means used to connect runway structural members shall be attached to the bottom of the runway or shall be attached in a manner to prevent tripping.

(v) Structural ramps used in lieu of steps shall be provided with cleats or other surface treatments on the top surface to prevent slipping.

(2) Means of egress from trench excavations. A stairway, ladder, ramp or other safe means of egress shall be located in trench excavations that are 4 feet (1.22 m) or more in depth so as to require no more than 25 feet (7.62 m) of lateral travel for employees.

(d) Exposure to vehicular traffic. Employees exposed to public vehicular traffic shall be provided with, and shall wear, warning vests or other suitable garments marked with or made of reflectorized or high-visibility material.

(e) Exposure to falling loads. No employee shall be permitted underneath loads handled by lifting or digging equipment. Employees shall be required to stand away from any vehicle being loaded or unloaded to avoid being struck by any spillage or falling materials. Operators may remain in the cabs of vehicles being loaded or unloaded when the vehicles are equipped, in accordance with § 1926.601(b)(6), to provide adequate protection for the operator during loading and unloading operations.

(f) Warning system for mobile equipment. When mobile equipment is operated adjacent to an excavation, or when such equipment is required to approach the edge of an excavation, and the operator does not have a clear and direct view of the edge of the excavation, a warning system shall be utilized such as barricades, hand or mechanical signals, or stop logs. If possible, the grade should be away from the excavation.

(g) Hazardous atmospheres-

(1) Testing and controls. In addition to the requirements set forth in subparts D and E of this part (29 CFR 1926.50–1926.107) to prevent exposure to harmful levels of atmospheric contaminants and to assure acceptable atmospheric conditions, the following requirements shall apply:

(i) Where oxygen deficiency (atmospheres containing less than 19.5 percent oxygen) or a hazardous atmosphere exists or could reasonably be expected to exist, such as in excavations in landfill areas or excavations in areas where hazardous substances are stored nearby, the atmospheres in the excavation shall be tested before employees enter excavations greater than 4 feet (1.22 m) in depth.

(ii) Adequate precautions shall be taken to prevent employee exposure to atmospheres containing less than 19.5 percent oxygen and other hazardous atmospheres. These precautions include providing proper respiratory protection or ventilation in accordance with subparts D and E of this part respectively.

(iii) Adequate precaution shall be taken such as providing ventilation, to prevent employee exposure to an atmosphere containing a concentration of a flammable gas in excess of 20 percent of the lower flammable limit of the gas.

(iv) When controls are used that are intended to reduce the level of atmospheric contaminants to acceptable levels, testing shall be conducted as often as necessary to ensure that the atmosphere remains safe.

(2) Emergency rescue equipment.

(i) Emergency rescue equipment, such as breathing apparatus, a safety harness and line, or a basket stretcher, shall be readily available where hazardous atmospheric conditions exist or may reasonably be expected to develop during work in an excavation. This equipment shall be attended when in use.

(ii)   Employees entering bell-bottom pier holes, or other similar deep and confined footing excavations, shall wear a harness with a life-line securely attached to it. The lifeline shall be separate from any line used to handle materials, and shall be individually attended at all times while the employee wearing the lifeline is in the excavation.

(h)   Protection from hazards associated with water accumulation.

(1)   Employees shall not work in excavations in which there is accumulated water, or in excavations in which water is accumulating, unless adequate precautions have been taken to protect employees against the hazards posed by water accumulation. The precautions necessary to protect employees adequately vary with each situation, but could include special support or shield systems to protect from cave-ins, water removal to control the level of accumulating water, or use of a safety harness and lifeline.

(2)   If water is controlled or prevented from accumulating by the use of water removal equipment, the water removal equipment and operations shall be monitored by a competent person to ensure proper operation.

(3)   If excavation work interrupts the natural drainage of surface water (such as streams), diversion ditches, dikes, or other suitable means shall be used to prevent surface water from entering the excavation and to provide adequate drainage of the area adjacent to the excavation. Excavations subject to runoff from heavy rains will require an inspection by a competent person and compliance with paragraphs (h)(1) and (h)(2) of this section.

(i)   Stability of adjacent structures.

(1)   Where the stability of adjoining buildings, walls, or other structures is endangered by excavation operations, support systems such as shoring, bracing, or underpinning shall be provided to ensure the stability of such structures for the protection of employees.

(2)   Excavation below the level of the base or footing of any foundation or retaining wall that could be reasonably expected to pose a hazard to employees shall not be permitted except when:

(i)   A support system, such as underpinning, is provided to ensure the safety of employees and the stability of the structure; or

(ii)   The excavation is in stable rock; or

(iii)   A registered professional engineer has approved the determination that the structure is sufficiently removed from the excavation so as to be unaffected by the excavation activity; or

(iv)   A registered professional engineer has approved the determination that such excavation work will not pose a hazard to employees.

(3)   Sidewalks, pavements, and appurtenant structure shall not be undermined unless a support system or another method of protection is provided to protect employees from the possible collapse of such structures.

(j) Protection of employees from loose rock or soil.

(1) Adequate protection shall be provided to protect employees from loose rock or soil that could pose a hazard by falling or rolling from an excavation face. Such protection shall consist of scaling to remove loose material; installation of protective barricades at intervals as necessary on the face to stop and contain falling material; or other means that provide equivalent protection.

(2) Employees shall be protected from excavated or other materials or equipment that could pose a hazard by falling or rolling into excavations. Protection shall be provided by placing and keeping such materials or equipment at least 2 feet (.61 m) from the edge of excavations, or by the use of retaining devices that are sufficient to prevent materials or equipment from falling or rolling into excavations, or by a combination of both if necessary.

(k) Inspections.

(1) Daily inspections of excavations, the adjacent areas, and protective systems shall be made by a competent person for evidence of a situation that could result in possible cave-ins, indications of failure of protective systems, hazardous atmospheres, or other hazardous conditions. An inspection shall be conducted by the competent person prior to the start of work and as needed throughout the shift. Inspections shall also be made after every rainstorm or other hazard increasing occurrence.

These inspections are only required when employee exposure can be reasonably anticipated.

(2) Where the competent person finds evidence of a situation that could result in a possible cave-in, indications of failure of protective systems, hazardous atmospheres, or other hazardous conditions, exposed employees shall be removed from the hazardous area until the necessary precautions have been taken to ensure their safety.

(l) Fall protection.

(1) Walkways shall be provided where employees or equipment are required or permitted to cross over excavations. Guardrails which comply with § 1926.502(b) shall be provided where walkways are 6 feet (1.8 m) or more above lower levels.

(2) Adequate barrier physical protection shall be provided at all remotely located excavations. All wells, pits, shafts, etc., shall be barricaded or covered. Upon completion of exploration and similar operations, temporary wells, pits, shafts, etc., shall be backfilled.

§ 1926.652 REQUIREMENTS FOR PROTECTIVE SYSTEMS.

(a) Protection of employees in excavations.

(1) Each employee in an excavation shall be protected from cave-ins by an adequate protective system designed in accordance with paragraph (b) or (c) of this section except when:

    (i)   Excavations are made entirely in stable rock; or

    (ii)   Excavations are less than 5 feet (1.52 m) in depth and examination of the ground by a competent person provides no indication of a potential cave-in.

(2) Protective systems shall have the capacity to resist without failure all loads that are intended or could reasonably be expected to be applied or transmitted to the system.

(b) Design of sloping and benching systems. The slopes and configurations of sloping and benching systems shall be selected and constructed by the employer or his designee and shall be in accordance with the requirements of paragraph (b)(1); or, in the alternative, paragraph (b)(2); or, in the alternative, paragraph (b)(3), or, in the alternative, paragraph (b)(4), as follows:

(1) Option (1)-Allowable configurations and slopes:

    (i)   Excavations shall be sloped at an angle not steeper than one and one-half horizontal to one vertical (34 degrees measured from the horizontal), unless the employer uses one of the other options listed below.

    (ii)   Slopes specified in paragraph (b)(1)(i) of this section, shall be excavated to form configurations that are in accordance with the slopes shown for Type C soil in Appendix B to this subpart.

(2) Option (2)-Determination of slopes and configurations using Appendices A and B. Maximum allowable slopes, and allowable configurations for sloping and benching systems, shall be determined in accordance with the conditions and requirements set forth in appendices A and B to this subpart.

(3) Option (3)-Designs using other tabulated data.

    (i)   Designs of sloping or benching systems shall be selected from and be in accordance with tabulated data, such as tables and charts.

    (ii)   The tabulated data shall be in written form and shall include all of the following:

        (A) Identification of the parameters that affect the selection of a sloping or benching system drawn from such data;

        (B) Identification of the limits of use of the data, to include the magnitude and configuration of slopes determined to be safe;

        (C) Explanatory information as may be necessary to aid the user in making a correct selection of a protective system from the data.

    (iii)   At least one copy of the tabulated data which identifies the registered professional engineer who approved the data, shall be maintained at the jobsite during construction of the protective system. After that time the data may be stored off the jobsite, but a copy of the data shall be made available to the Secretary upon request.

(4) Option (4)-Design by a registered professional engineer.

    (i)   Sloping and benching systems not utilizing Option (1) or Option (2) or Option (3) under paragraph (b) of this section shall be approved by a registered professional engineer.

(ii) Designs shall be in written form and shall include at least the following:

(A) The magnitude of the slopes that were determined to be safe for the particular project;

(B) The configurations that were determined to be safe for the particular project; and

(C) The identity of the registered professional engineer approving the design.

(iii) At least one copy of the design shall be maintained at the jobsite while the slope is being constructed. After that time the design need not be at the jobsite, but a copy shall be made available to the Secretary upon request.

(c) Design of support systems, shield systems, and other protective systems. Designs of support systems, shield systems, and other protective systems shall be selected and constructed by the employer or his designee and shall be in accordance with the requirements of paragraph (c)(1); or, in the alternative, paragraph (c)(2); or, in the alternative, paragraph (c)(3); or, in the alternative, paragraph (c)(4) as follows:

(1) Option (1)-Designs using appendices A, C and D. Designs for timber shoring in trenches shall be determined in accordance with the conditions and requirements set forth in appendices A and C to this subpart. Designs for aluminum hydraulic shoring shall be in accordance with paragraph (c)(2) of this section, but if manufacturer's tabulated data cannot be utilized, designs shall be in accordance with appendix D.

(2) Option (2)-Designs Using Manufacturer's Tabulated Data.

(i) Design of support systems, shield systems, or other protective systems that are drawn from manufacturer's tabulated data shall be in accordance with all specifications, recommendations, and limitations issued or made by the manufacturer.

(ii) Deviation from the specifications, recommendations, and limitations issued or made by the manufacturer shall only be allowed after the manufacturer issues specific written approval.

(iii) Manufacturer's specifications, recommendations, and limitations, and manufacturer's approval to deviate from the specifications, recommendations, and limitations shall be in written form at the jobsite during construction of the protective system. After that time this data may be stored off the jobsite, but a copy shall be made available to the Secretary upon request.

(3) Option (3)-Designs using other tabulated data.

(i) Designs of support systems, shield systems, or other protective systems shall be selected from and be in accordance with tabulated data, such as tables and charts.

(ii) The tabulated data shall be in written form and include all of the following:

   (A) Identification of the parameters that affect the selection of a protective system drawn from such data;

   (B) Identification of the limits of use of the data;

   (C) Explanatory information as may be necessary to aid the user in making a correct selection of a protective system from the data.

(iii) At least one copy of the tabulated data, which identifies the registered professional engineer who approved the data, shall be maintained at the jobsite during construction of the protective system. After that time the data may be stored off the jobsite, but a copy of the data shall be made available to the Secretary upon request.

(4) Option (4)-Design by a registered professional engineer.

   (i) Support systems, shield systems, and other protective systems not utilizing Option 1, Option 2 or Option 3, above, shall be approved by a registered professional engineer.

   (ii) Designs shall be in written form and shall include the following:

      (A) A plan indicating the sizes, types, and configurations of the materials to be used in the protective system; and

      (B) The identity of the registered professional engineer approving the design.

   (iii) At least one copy of the design shall be maintained at the jobsite during construction of the protective system.

      After that time, the design may be stored off the jobsite, but a copy of the design shall be made available to the Secretary upon request.

(d) Materials and equipment.

   (1) Materials and equipment used for protective systems shall be free from damage or defects that might impair their proper function.

   (2) Manufactured materials and equipment used for protective systems shall be used and maintained in a manner that is consistent with the recommendations of the manufacturer, and in a manner that will prevent employee exposure to hazards.

   (3) When material or equipment that is used for protective systems is damaged, a competent person shall examine the material or equipment and evaluate its suitability for continued use.

      If the competent person cannot assure the material or equipment is able to support the intended loads or is otherwise suitable for safe use, then such material or equipment shall be removed from service, and shall be evaluated and approved by a registered professional engineer before being returned to service.

(e) Installation and removal of support-

   (1) General.

      (i) Members of support systems shall be securely connected together to prevent sliding, falling, kickouts, or other predictable failure.

(ii)   Support systems shall be installed and removed in a manner that protects employees from cave-ins, structural collapses, or from being struck by members of the support system.

(iii)   Individual members of support systems shall not be subjected to loads exceeding those which those members were designed to withstand.

(iv)   Before temporary removal of individual members begins, additional precautions shall be taken to ensure the safety of employees, such as installing other structural members to carry the loads imposed on the support system.

(v)   Removal shall begin at, and progress from, the bottom of the excavation. Members shall be released slowly so as to note any indication of possible failure of the remaining members of the structure or possible cave-in of the sides of the excavation.

(vi)   Backfilling shall progress together with the removal of support systems from excavations.

(2) Additional requirements for support systems for trench excavations.

(i)   Excavation of material to a level no greater than 2 feet (.61 m) below the bottom of the members of a support system shall be permitted, but only if the system is designed to resist the forces calculated for the full depth of the trench, and there are no indications while the trench is open of a possible loss of soil from behind or below the bottom of the support system.

(ii)   Installation of a support system shall be closely coordinated with the excavation of trenches.

(f) Sloping and benching systems. Employees shall not be permitted to work on the faces of sloped or benched excavations at levels above other employees except when employees at the lower levels are adequately protected from the hazard of falling, rolling, or sliding material or equipment.

(g) Shield systems-

(1) General.

(i)   Shield systems shall not be subjected to loads exceeding those which the system was designed to withstand.

(ii)   Shields shall be installed in a manner to restrict lateral or other hazardous movement of the shield in the event of the application of sudden lateral loads.

(iii)   Employees shall be protected from the hazard of cave-ins when entering or exiting the areas protected by shields.

(iv)   Employees shall not be allowed in shields when shields are being installed, removed, or moved vertically.

(2) Additional requirement for shield systems used in trench excavations. Excavations of earth material to a level not greater than 2 feet (.61 m) below the bottom of a shield shall be permitted, but only if the shield is designed to resist the forces calculated for the full depth of the trench, and there are no indications while the trench is open of a possible loss of soil from behind or below the bottom of the shield.

APPENDIX A TO SUBPART P—SOIL CLASSIFICATION

(a) Scope and application-

  (1) Scope. This appendix describes a method of classifying soil and rock deposits based on site and environmental conditions, and on the structure and composition of the earth deposits. The appendix contains definitions, sets forth requirements, and describes acceptable visual and manual tests for use in classifying soils.

  (2) Application. This appendix applies when a sloping or benching system is designed in accordance with the requirements set forth in § 1926.652(b)(2) as a method of protection for employees from cave-ins. This appendix also applies when timber shoring for excavations is designed as a method of protection from cave-ins in accordance with appendix C to subpart P of part 1926, and when aluminum hydraulic shoring is designed in accordance with appendix D. This Appendix also applies if other protective systems are designed and selected for use from data prepared in accordance with the requirements set forth in § 1926.652(c), and the use of the data is predicated on the use of the soil classification system set forth in this appendix.

(b) Definitions. The definitions and examples given below are based on, in whole or in part, the following: American Society for Testing Materials (ASTM) Standards D653-85 and D2488; The Unified Soils Classification System, The U.S. Department of Agriculture (USDA) Textural Classification Scheme; and The National Bureau of Standards Report BSS-121.

Cemented soil means a soil in which the particles are held together by a chemical agent, such as calcium carbonate, such that a hand-size sample cannot be crushed into powder or individual soil particles by finger pressure.

Cohesive soil means clay (fine grained soil), or soil with a high clay content, which has cohesive strength. Cohesive soil does not crumble, can be excavated with vertical sideslopes, and is plastic when moist. Cohesive soil is hard to break up when dry, and exhibits significant cohesion when submerged.

Cohesive soils include clayey silt, sandy clay, silty clay, clay and organic clay.

Dry soil means soil that does not exhibit visible signs of moisture content.

Fissured means a soil material that has a tendency to break along definite planes of fracture with little resistance, or a material that exhibits open cracks, such as tension cracks, in an exposed surface.

Granular soil means gravel, sand, or silt, (coarse grained soil) with little or no clay content. Granular soil has no cohesive strength. Some moist granular soils exhibit apparent cohesion.

Granular soil cannot be molded when moist and crumbles easily when dry.

Layered system means two or more distinctly different soil or rock types arranged in layers. Micaceous seams or weakened planes in rock or shale are considered layered.

Moist soil means a condition in which a soil looks and feels damp. Moist cohesive soil can easily be shaped into a ball and rolled into small diameter threads before crumbling. Moist granular soil that contains some cohesive material will exhibit signs of cohesion between particles.

Plastic means a property of a soil which allows the soil to be deformed or molded without cracking, or appreciable volume change.

Saturated soil means a soil in which the voids are filled with water. Saturation does not require flow. Saturation, or near saturation, is necessary for the proper use of instruments such as a pocket penetrometer or sheer vane.

Soil classification system means, for the purpose of this subpart, a method of categorizing soil and rock deposits in a hierarchy of Stable Rock, Type A, Type B, and Type C, in decreasing order of stability. The categories are determined based on an analysis of the properties and performance characteristics of the deposits and the environmental conditions of exposure.

Stable rock means natural solid mineral matter that can be excavated with vertical sides and remain intact while exposed.

Submerged soil means soil which is underwater or is free seeping.

Type A means cohesive soils with an unconfined compressive strength of 1.5 ton per square foot (tsf) (144 kPa) or greater.

Examples of cohesive soils are: clay, silty clay, sandy clay, clay loam and, in some cases, silty clay loam and sandy clay loam. Cemented soils such as caliche and hardpan are also considered Type A. However, no soil is Type A if:

(i)   The soil is fissured; or

(ii)   The soil is subject to vibration from heavy traffic, pile driving, or similar effects; or

(iii)   The soil has been previously disturbed; or

(iv)   The soil is part of a sloped, layered system where the layers dip into the excavation on a slope of four horizontal to one vertical (4H:1V) or greater; or

(v)   The material is subject to other factors that would require it to be classified as a less stable material.

Type B means:

(i)   Cohesive soil with an unconfined compressive strength greater than 0.5 tsf (48 kPa) but less than 1.5 tsf (144 kPa); or

(ii)   Granular cohesionless soils including: angular gravel (similar to crushed rock), silt, silt loam, sandy loam and, in some cases, silty clay loam and sandy clay loam.

(iii)   Previously disturbed soils except those which would otherwise be classed as Type C soil.

(iv)   Soil that meets the unconfined compressive strength or cementation requirements for Type A, but is fissured or subject to vibration; or

(v)   Dry rock that is not stable; or

(vi)   Material that is part of a sloped, layered system where the layers dip into the excavation on a slope less steep than four horizontal to one vertical (4H:1V), but only if the material would otherwise be classified as Type B.

Type C means:

(i)   Cohesive soil with an unconfined compressive strength of 0.5 tsf (48 kPa) or less; or

(ii)   Granular soils including gravel, sand, and loamy sand; or

(iii)   Submerged soil or soil from which water is freely seeping; or

(iv)   Submerged rock that is not stable, or

(v)   Material in a sloped, layered system where the layers dip into the excavation or a slope of four horizontal to one vertical (4H:1V) or steeper.

Unconfined compressive strength means the load per unit area at which a soil will fail in compression. It can be determined by laboratory testing, or estimated in the field using a pocket penetrometer, by thumb penetration tests, and other methods.

Wet soil means soil that contains significantly more moisture than moist soil, but in such a range of values that cohesive material will slump or begin to flow when vibrated. Granular material that would exhibit cohesive properties when moist will lose those cohesive properties when wet.

(c)   Requirements-

(1) Classification of soil and rock deposits.

Each soil and rock deposit shall be classified by a competent person as Stable Rock, Type A, Type B, or Type C in accordance with the definitions set forth in paragraph (b) of this appendix.

(2) Basis of classification. The classification of the deposits shall be made based on the results of at least one visual and at least one manual analysis. Such analyses shall be conducted by a competent person using tests described in paragraph (d) below, or in other recognized methods of soil classification and testing such as those adopted by the America Society for Testing and Materials, or the U.S. Department of Agriculture textural classification system.

(3) Visual and manual analyses. The visual and manual analyses, such as those noted as being acceptable in paragraph (d) of this appendix, shall be designed and con-ducted to provide sufficient quantitative and qualitative information as may be necessary to identify properly the properties, factors, and conditions affecting the classification of the deposits.

(4) Layered systems. In a layered system, the system shall be classified in accordance with its weakest layer. However, each layer may be classified individually where a more stable layer lies under a less stable layer.

(5) Reclassification. If, after classifying a deposit, the properties, factors, or conditions affecting its classification change in any way, the changes shall be evaluated by a competent person. The deposit shall be reclassified as necessary to reflect the changed circumstances.

(d) Acceptable visual and manual tests.-

(1) Visual tests. Visual analysis is conducted to determine qualitative information regarding the excavation site in general, the soil adjacent to the excavation, the soil forming the sides of the open excavation, and the soil taken as samples from excavated material.

  (i) Observe samples of soil that are excavated and soil in the sides of the excavation. Estimate the range of particle sizes and the relative amounts of the particle sizes. Soil that is primarily composed of fine-grained material is cohesive material.

  Soil composed primarily of coarse-grained sand or gravel is granular material.

  (ii) Observe soil as it is excavated. Soil that remains in clumps when excavated is cohesive. Soil that breaks up easily and does not stay in clumps is granular.

  (iii) Observe the side of the opened excavation and the surface area adjacent to the excavation. Crack-like openings such as tension cracks could indicate fissured material. If chunks of soil spall off a vertical side, the soil could be fissured.

  Small spalls are evidence of moving ground and are indications of potentially hazardous situations.

  (iv) Observe the area adjacent to the excavation and the excavation itself for evidence of existing utility and other underground structures, and to identify previously disturbed soil.

  (v) Observe the opened side of the excavation to identify layered systems. Examine layered systems to identify if the layers slope toward the excavation. Estimate the degree of slope of the layers.

  (vi) Observe the area adjacent to the excavation and the sides of the opened excavation for evidence of surface water, water seeping from the sides of the excavation, or the location of the level of the water table.

  (vii) Observe the area adjacent to the excavation and the area within the excavation for sources of vibration that may affect the stability of the excavation face.

(2) Manual tests. Manual analysis of soil samples is conducted to determine quantitative as well as qualitative properties of soil and to provide more information in order to classify soil properly.

  (i) Plasticity. Mold a moist or wet sample of soil into a ball and attempt to roll it into threads as thin as 1/8-inch in diameter. Cohesive material can be successfully rolled into threads without crumbling. For example, if at least a two inch (50 mm) length of 1/8-inch thread can be held on one end without tearing, the soil is cohesive.

  (ii) Dry strength. If the soil is dry and crumbles on its own or with moderate pressure into individual grains or fine powder, it is granular (any combination of gravel, sand, or silt). If the soil is dry and falls into clumps which break up into smaller clumps, but the smaller clumps can only be broken up with difficulty, it may be clay in any combination with gravel, sand or silt. If the dry soil breaks into clumps which do not break up into small clumps and which can only be broken with difficulty, and there is no visual indication the soil is fissured, the soil may be considered unfissured.

(iii)   Thumb penetration. The thumb penetration test can be used to estimate the unconfined compressive strength of cohesive soils. (This test is based on the thumb penetration test described in American Society for Testing and Materials (ASTM) Standard designation D2488-"Standard Recommended Practice for Description of Soils (Visual-Manual Procedure).") Type A soils with an unconfined compressive strength of 1.5 tsf can be readily indented by the thumb; however, they can be penetrated by the thumb only with very great effort. Type C soils with an unconfined compressive strength of 0.5 tsf can be easily penetrated several inches by the thumb, and can be molded by light finger pressure. This test should be conducted on an undisturbed soil sample, such as a large clump of spoil, as soon as practicable after excavation to keep to a minimum the effects of exposure to drying influences.

If the excavation is later exposed to wetting influences (rain, flooding), the classification of the soil must be changed accordingly.

(iv)   Other strength tests. Estimates of unconfined compressive strength of soils can also be obtained by use of a pocket penetrometer or by using a hand-operated shearvane.

(v)   Drying test. The basic purpose of the drying test is to differentiate between cohesive material with fissures, unfissured cohesive material, and granular material. The procedure for the drying test involves drying a sample of soil that is approximately one inch thick (2.54 cm) and six inches (15.24 cm) in diameter until it is thoroughly dry:

(A)   If the sample develops cracks as it dries, significant fissures are indicated.

(B)   Samples that dry without cracking are to be broken by hand. If considerable force is necessary to break a sample, the soil has significant cohesive material content. The soil can be classified as an unfissured cohesive material and the unconfined compressive strength should be determined.

(C)   If a sample breaks easily by hand, it is either a fissured cohesive material or a granular material. To distinguish between the two, pulverize the dried clumps of the sample by hand or by stepping on them. If the clumps do not pulverize easily, the material is cohesive with fissures. If they pulverize easily into very small fragments, the material is granular.

APPENDIX B TO SUBPART P—SLOPING AND BENCHING

(a) Scope and application. This appendix contains specifications for sloping and benching when used as methods of protecting employees working in excavations from cave-ins. The requirements of this appendix apply when the design of sloping and benching protective systems is to be performed in accordance with the requirements set forth in § 1926.652(b)(2).

(b) Definitions.

Actual slope means the slope to which an excavation face is excavated.

Distress means that the soil is in a condition where a cave-in is imminent or is likely to occur. Distress is evidenced by such phenomena as the development of fissures in the face of or adjacent to an open excavation; the subsidence of the edge of an excavation; the slumping of material from the face or the bulging or heaving of material from the bottom of an excavation; the spalling of material from the face of an excavation; and ravelling, i.e., small amounts of material such as pebbles or little clumps of material suddenly separating from the face of an excavation and trickling or rolling down into the excavation.

Maximum allowable slope means the steepest incline of an excavation face that is acceptable for the most favorable site conditions as protection against cave-ins, and is expressed as the ratio of horizontal distance to vertical rise (H:V).

Short term exposure means a period of time less than or equal to 24 hours that an excavation is open.

(c) Requirements-

(1) Soil classification. Soil and rock deposits shall be classified in accordance with appendix A to subpart P of part 1926.

(2) Maximum allowable slope. The maximum allowable slope for a soil or rock deposit shall be determined from Table B-1 of this appendix.

(3) Actual slope.

    (i)   The actual slope shall not be steeper than the maximum allowable slope.

    (ii)   The actual slope shall be less steep than the maximum allowable slope, when there are signs of distress. If that situation occurs, the slope shall be cut back to an actual slope which is at least 1/2 horizontal to one vertical (1/2H:1V) less steep than the maximum allowable slope.

    (iii)   When surcharge loads from stored material or equipment, operating equipment, or traffic are present, a competent person shall determine the degree to which the actual slope must be reduced below the maximum allowable slope, and shall assure that such reduction is achieved. Surcharge loads from adjacent structures shall be evaluated in accordance with § 1926.651(i).

(4) Configurations. Configurations of sloping and benching systems shall be in accordance with Figure B–1.

Table B-1. Maximum allowable slopes

Soil or rock type	Maximum allowable slopes (H:V)[1] for excavations less than 20 feet deep3	
Stable rock	Vertical	(90°)
Type A[2]	3/4:1	(53°)
Type B	1:1	(45°)
Type C	1 1/2:1	(34°)

Notes:

[1]Numbers shown in parentheses next to maximum allowable slopes are angles expressed in degrees from the horizontal. Angles have been rounded off.

[2]A short-term maximum allowable slope of 1/2H:1V (63°) is allowed in excavations in Type A soil that are 12 feet (3.67 m) or less in depth. Short-term maximum allowable slopes for excavations greater than 12 feet (3.67 m) in depth shall be 3/4H:1V (53°).

[3]Sloping or benching for excavations greater than 20 feet deep shall be designed by a registered professional engineer.

Figure B-1 Slope Configurations

(All slopes stated below are in the horizontal to vertical ratio)

B-1.1 Excavations made in Type A soil.

1. All simple slope excavation 20 feet or less in depth shall have a maximum allowable slope of 3/4:1.

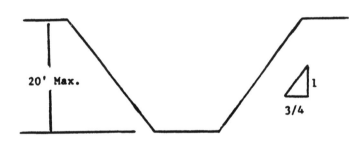

Simple Slope-General

Exception: Simple slope excavations which are open 24 hours or less (short term) and which are 12 feet or less in depth shall have a maximum allowable slope of 1/2:1.

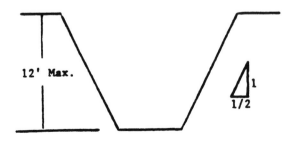

Simple Slope-Short Term

2. All benched excavations 20 feet or less in depth shall have a maximum allowable slope of to 1 and maximum bench dimensions as follows:

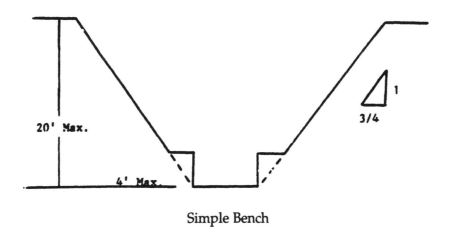

Simple Bench

Multiple Bench

3. All excavations 8 feet or less in depth which have unsupported vertically sided lower portions shall have a maximum vertical side of 3 1/2 feet.

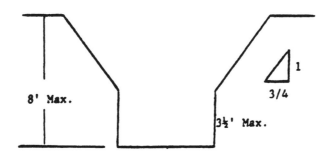

Unsupported Vertically Sided Lower
Portion-Maximum 8 Feet in Depth

All excavations more than 8 feet but not more than 12 feet in depth which have unsupported vertically sided lower portions shall have a maximum allowable slope of 1:1 and a maximum vertical side of 3 1/2 feet.

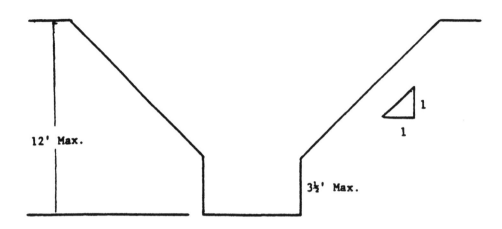

Unsupported Vertically Sided Lower Portion-Maximum
12 Feet in Depth

All excavations 20 feet or less in depth which have vertically sided lower portions that are supported or shielded shall have a maximum allowable slope of 3/4:1. The support or shield system must extend at least 18 inches above the top of the vertical side.

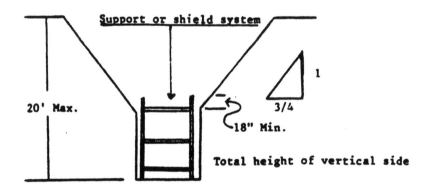

Supported or Shielded Vertically Sided
Lower Portion

4. All other simple slope, compound slope, and vertically sided lower portion excavations shall be in accordance with the other options permitted under § 1926.652(b).

B-1.2 Excavations Made in Type B Soil

1. All simple slope excavations 20 feet or less in depth shall have a maximum allowable slope of 1:1.

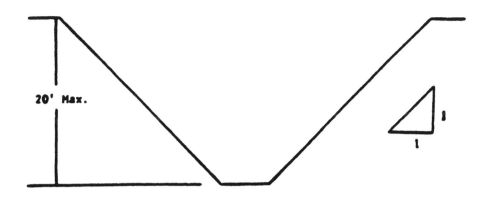

Simple Slope

2. All benched excavations 20 feet or less in depth shall have a maximum allowable slope of 1:1 and maximum bench dimensions as follows:

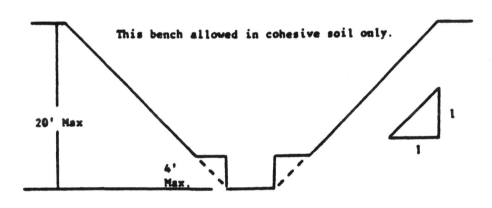

Single Bench

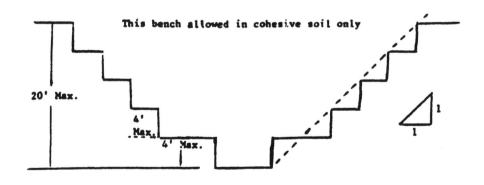

Multiple Bench

3. All excavations 20 feet or less in depth which have vertically sided lower portions shall be shielded or supported to a height at least 18 inches above the top of the vertical side. All such excavations shall have a maximum allowable slope of 1:1.

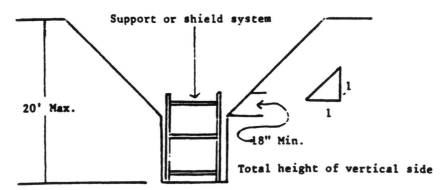

Vertically Sided Lower Portion

4. All other sloped excavations shall be in accordance with the other options permitted in § 1926.652(b).

B-1.3 Excavations Made in Type C Soil

1. All simple slope excavations 20 feet or less in depth shall have a maximum allowable slope of 1 1/2:1.

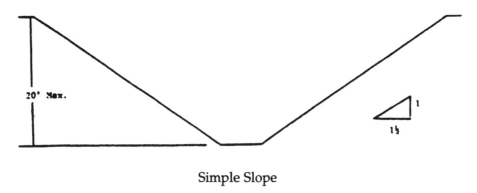

Simple Slope

2. All excavations 20 feet or less in depth which have vertically sided lower portions shall be shielded or supported to a height at least 18 inches above the top of the vertical side. All such excavations shall have a maximum allowable slope of 1 1/2:1.

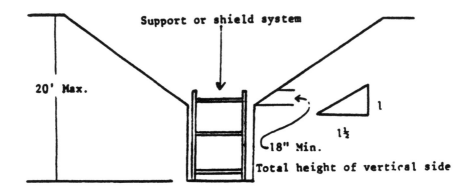

Vertical Sided Lower Portion

3. All other sloped excavations shall be in accordance with the other options permitted in § 1926.652(b).

B-1.4 Excavations Made in Layered Soils

1. All excavations 20 feet or less in depth made in layered soils shall have a maximum allowable slope for each layer as set forth below.

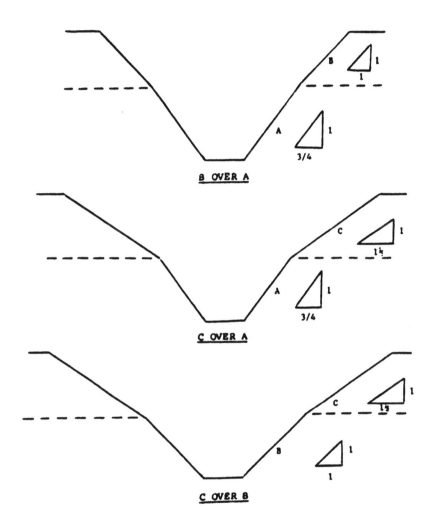

2. All other sloped excavations shall be in accordance with the other options permitted in § 1926.652(b).

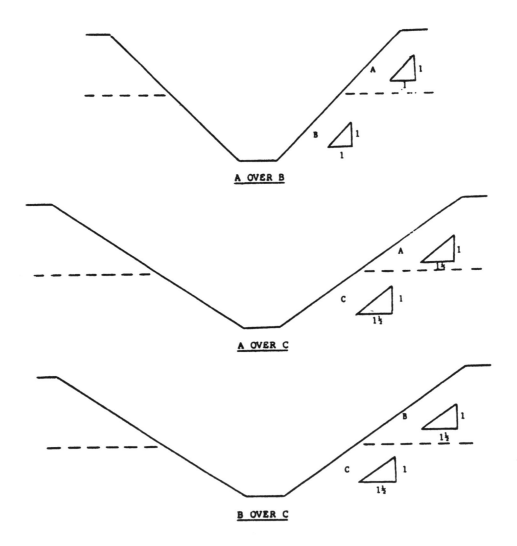

APPENDIX C TO SUBPART P—TIMBER SHORING FOR TRENCHES

(a) Scope. This appendix contains information that can be used when timber shoring is provided as a method of protection from cave-ins in trenches that do not exceed 20 feet (6.1 m) in depth.

This appendix must be used when the design of timber shoring protective systems is to be performed in accordance with § 1926.652(c)(1).

Other timber shoring configurations; other systems of support such as hydraulic and pneumatic systems; and other protective systems such as sloping, benching, shielding, and freezing systems must be designed in accordance with the requirements set forth in § 1926.652(b) and § 1926.652(c).

(b) Soil Classification. In order to use the data presented in this appendix, the soil type or types in which the excavation is made must first be determined using the soil classification method set forth in appendix A of subpart P of this part.

(c) Presentation of Information. Information is presented in several forms as follows:

(1) Information is presented in tabular form in Tables C-1.1, C-1.2, and C-1.3, and

Tables C-2.1, C-2.2 and C-2.3 following paragraph (g) of the appendix. Each table presents the minimum sizes of timber members to use in a shoring system, and each table contains data only for the particular soil type in which the excavation or portion of the excavation is made. The data are arranged to allow the user the flexibility to select from among several acceptable configurations of members based on varying the horizontal spacing of the crossbraces. Stable rock is exempt from shoring requirements and therefore, no data are presented for this condition.

(2) Information concerning the basis of the tabular data and the limitations of the data is presented in paragraph (d) of this appendix, and on the tables themselves.

(3) Information explaining the use of the tabular data is presented in paragraph (e) of this appendix.

(4) Information illustrating the use of the tabular data is presented in paragraph (f) of this appendix.

(5) Miscellaneous notations regarding Tables C-1.1 through C-1.3 and Tables C-2.1 through C-2.3 are presented in paragraph (g) of this Appendix.

(d) Basis and limitations of the data.-

(1) Dimensions of timber members.

(i) The sizes of the timber members listed in Tables C-1.1 through C-1.3 are taken from the National Bureau of Standards (NBS) report, "Recommended Technical Provisions for Construction Practice in Shoring and Sloping of Trenches and Excavations." In addition, where NBS did not recommend specific sizes of members, member sizes are based on an analysis of the sizes required for use by existing codes and on empirical practice.

(ii) The required dimensions of the members listed in Tables C-1.1 through C-1.3 refer to actual dimensions and not nominal dimensions of the timber. Employers wanting to use nominal size shoring are directed to Tables C-2.1 through C-2.3, or have this choice under § 1926.652(c)(3), and are referred to The Corps of Engineers, The Bureau of Reclamation or data from other acceptable sources.

(2) Limitation of application.

(i) It is not intended that the timber shoring specification apply to every situation that may be experienced in the field. These data were developed to apply to the situations that are most commonly experienced in current trenching practice. Shoring systems for use in situations that are not covered by the data in this appendix must be designed as specified in § 1926.652(c).

(ii) When any of the following conditions are present, the members specified in the tables are not considered adequate.

Either an alternate timber shoring system must be designed or another type of protective system designed in accordance with § 1926.652.

(A) When loads imposed by structures or by stored material adjacent to the trench weigh in excess of the load imposed by a two-foot soil surcharge. The term "adjacent" as used here means the area within a horizontal distance from the edge of the trench equal to the depth of the trench.

(B) When vertical loads imposed on cross braces exceed a 240-pound gravity load distributed on a one-foot section of the center of the crossbrace.

(C) When surcharge loads are present from equipment weighing in excess of 20,000 pounds.

(D) When only the lower portion of a trench is shored and the remaining portion of the trench is sloped or benched unless: The sloped portion is sloped at an angle less steep than three horizontal to one vertical; or the members are selected from the tables for use at a depth which is determined from the top of the overall trench, and not from the toe of the sloped portion.

(e) Use of Tables. The members of the shoring system that are to be selected using this information are the cross braces, the uprights, and the wales, where wales are required. Minimum sizes of members are specified for use in different types of soil. There are six tables of information, two for each soil type. The soil type must first be determined in accordance with the soil classification system described in appendix A to subpart P of part 1926. Using the appropriate table, the selection of the size and spacing of the members is then made. The selection is based on the depth and width of the trench where the members are to be installed and, in most instances, the selection is also based on the horizontal spacing of the crossbraces. Instances where a choice of horizontal spacing of crossbracing is available, the horizontal spacing of the crossbraces must be chosen by the user before the size of any member can be determined. When the soil type, the width and depth of the trench, and the horizontal spacing of the crossbraces are known, the size and vertical spacing of the crossbraces, the size and vertical spacing of the wales, and the size and horizontal spacing of the uprights can be read from the appropriate table.

(f) Examples to Illustrate the Use of Tables C-1.1 through C-1.3.

(1) Example 1.

A trench dug in Type A soil is 13 feet deep and five feet wide.

From Table C-1.1, four acceptable arrangements of timber can be used.

Arrangement #1

Space 4×4 crossbraces at six feet horizontally and four feet vertically.

Wales are not required.

Space 3×8 uprights at six feet horizontally. This arrangement is commonly called "skip shoring."

Arrangement #2

Space 4×6 crossbraces at eight feet horizontally and four feet vertically.

Space 8×8 wales at four feet vertically.

Space 2×6 uprights at four feet horizontally.

Arrangement #3

Space 6×6 crossbraces at 10 feet horizontally and four feet vertically.

Space 8×10 wales at four feet vertically.

Space 2×6 uprights at five feet horizontally.

Arrangement #4

Space 6×6 crossbraces at 12 feet horizontally and four feet vertically.

Space 10×10 wales at four feet vertically.

Spaces 3×8 uprights at six feet horizontally.

(2) Example 2.

A trench dug in Type B soil in 13 feet deep and five feet wide. From Table C-1.2 three acceptable arrangements of members are listed.

Arrangement #1

Space 6×6 crossbraces at six feet horizontally and five feet vertically.

Space 8×8 wales at five feet vertically.

Space 2×6 uprights at two feet horizontally.

Arrangement #2

Space 6×8 crossbraces at eight feet horizontally and five feet vertically.

Space 10×10 wales at five feet vertically.

Space 2×6 uprights at two feet horizontally.

Arrangement #3

Space 8×8 crossbraces at 10 feet horizontally and five feet vertically.

Space 10×12 wales at five feet vertically.

Space 2×6 uprights at two feet vertically.

(3) Example 3.

A trench dug in Type C soil is 13 feet deep and five feet wide.

From Table C-1.3 two acceptable arrangements of members can be used.

Arrangement #1

Space 8×8 crossbraces at six feet horizontally and five feet vertically.

Space 10×12 wales at five feet vertically.

Position 2×6 uprights as closely together as possible.

If water must be retained, use special tongue and groove uprights to form tight sheeting.

Arrangement #2

Space 8×10 crossbraces at eight feet horizontally and five feet vertically.

Space 12×12 wales at five feet vertically.

Position 2×6 uprights in a close sheeting configuration unless water pressure must be resisted. Tight sheeting must be used where water must be retained.

(4) Example 4.

A trench dug in Type C soil is 20 feet deep and 11 feet wide.

The size and spacing of members for the section of trench that is over 15 feet in depth is determined using Table C-1.3. Only one arrangement of members is provided.

Space 8×10 crossbraces at six feet horizontally and five feet vertically.

Space 12×12 wales at five feet vertically.

Use 3×6 tight sheeting.

Use of Tables C-2.1 through C-2.3 would follow the same procedures.

(g) Notes for all Tables.

1. Member sizes at spacings other than indicated are to be determined as specified in § 1926.652(c), "Design of Protective Systems."

2. When conditions are saturated or submerged, use Tight Sheeting.

Tight Sheeting refers to the use of specially-edged timber planks (e.g., tongue and groove) at least three inches thick, steel sheet piling, or similar construction that when driven or placed in position provide a tight wall to resist the lateral pressure of water and to prevent the loss of backfill material. Close Sheeting refers to the placement of planks side-by-side allowing as little space as possible between them.

3. All spacing indicated is measured center to center.

4. Wales to be installed with greater dimension horizontal.

5. If the vertical distance from the center of the lowest crossbrace to the bottom of the trench exceeds two and one-half feet, uprights shall be firmly embedded or a mudsill shall be used. Where uprights are embedded, the vertical distance from the center of the lowest crossbrace to the bottom of the trench shall not exceed 36 inches. When mudsills are used, the vertical distance shall not exceed 42 inches. Mudsills are wales that are installed at the toe of the trench side.

6. Trench jacks may be used in lieu of or in combination with timber crossbraces.

7. Placement of crossbraces. When the vertical spacing of crossbraces is four feet, place the top crossbrace no more than two feet below the top of the trench. When the vertical spacing of crossbraces is five feet, place the top crossbrace no more than 2.5 feet below the top of the trench.

TABLE C-1.1

TIMBER TRENCH SHORING -- MINIMUM TIMBER REQUIREMENTS *

SOIL TYPE A $P_a = 25 \times H + 72$ psf (2 ft Surcharge)

SIZE (ACTUAL) AND SPACING OF MEMBERS **

DEPTH OF TRENCH (FEET)	HORIZ. SPACING (FEET)	CROSS BRACES — WIDTH OF TRENCH (FEET)						WALES		UPRIGHTS — MAXIMUM ALLOWABLE HORIZONTAL SPACING (FEET)				
		UP TO 4	UP TO 6	UP TO 9	UP TO 12	UP TO 15	VERT. SPACING (FEET)	SIZE (IN)	VERT. SPACING (FEET)	CLOSE	4	5	6	8
5 TO 10	UP TO 6	4X4	4X4	4X6	6X6	6X6	4	Not Req'd	---				2X6	
	UP TO 8	4X4	4X4	4X6	6X6	6X6	4	Not Req'd	---					2X8
	UP TO 10	4X6	4X6	4X6	6X6	6X6	4	8X8	4			2X6		
	UP TO 12	4X6	4X6	6X6	6X6	6X6	4	8X8	4				2X6	
10 TO 15	UP TO 6	4X4	4X4	4X6	6X6	6X6	4	Not Req'd	---				3X8	
	UP TO 8	4X6	4X6	6X6	6X6	6X6	4	8X8	4		2X6			
	UP TO 10	6X6	6X5	6X6	6X6	6X8	4	8X10	4			2X6		
	UP TO 12	6X6	6X6	6X6	6X8	6X8	4	10X10	4				3X8	
15 TO 20	UP TO 6	6X6	6X6	6X6	6X8	6X8	4	6X8	4	3X6				
	UP TO 8	6X6	6X6	6X6	6X8	6X8	4	8X8	4	3X6				
	UP TO 10	8X8	8X8	8X8	8X8	8X10	4	8X10	4	3X6				
	UP TO 12	8X8	8X8	8X8	8X8	8X10	4	10X10	4	3X6				
OVER 20	SEE NOTE 1													

* Mixed oak or equivalent with a bending strength not less than 850 psi.
** Manufactured members of equivalent strength may by substituted for wood.

TABLE C-1.2

TIMBER TRENCH SHORING — MINIMUM TIMBER REQUIREMENTS *

SOIL TYPE B $P_a = 45 \times H + 72$ psf (2 ft. Surcharge)

SIZE (ACTUAL) AND SPACING OF MEMBERS **

DEPTH OF TRENCH (FEET)	HORIZ. SPACING (FEET)	CROSS BRACES — WIDTH OF TRENCH (FEET)					VERT. SPACING (FEET)	WALES SIZE (IN)	WALES VERT. SPACING (FEET)	UPRIGHTS — MAXIMUM ALLOWABLE HORIZONTAL SPACING (FEET)		
		UP TO 4	UP TO 6	UP TO 9	UP TO 12	UP TO 15				CLOSE	2	3
5 TO 10	UP TO 6	4X6	4X6	6X6	6X6	6X6	5	6X8	5			2X6
	UP TO 8	6X6	6X6	6X6	6X6	6X8	5	8X10	5			2X6
	UP TO 10	6X6	6X6	6X6	6X8	6X8	5	10X10	5			2X6
	See Note 1											
10 TO 15	UP TO 6	6X6	6X6	6X6	6X8	6X8	5	8X8	5		2X6	
	UP TO 8	6X8	6X8	6X8	6X8	6X8	5	10X10	5		2X6	
	UP TO 10	8X8	8X8	8X8	8X8	8X10	5	10X12	5		2X6	
	See Note 1											
15 TO 20	UP TO 6	6X8	6X8	6X8	8X8	8X8	5	8X10	5	3X6		
	UP TO 8	8X8	8X8	8X8	8X8	8X10	5	10X12	5	3X6		
	UP TO 10	8X10	8X10	8X10	8X10	10X10	5	12X12	5	3X6		
	See Note 1											
OVER 20	SEE NOTE 1											

* Mixed oak or equivalent with a bending strength not less than 850 psi.
** Manufactured members of equivalent strength may by substituted for wood.

TABLE C-1.3

TIMBER TRENCH SHORING -- MINIMUM TIMBER REQUIREMENTS *

SOIL TYPE C P_a = 80 X H + 72 psf (2 ft. Surcharge)

SIZE (ACTUAL) AND SPACING OF MEMBERS**

DEPTH OF TRENCH (FEET)	HORIZ. SPACING (FEET)	CROSS BRACES — WIDTH OF TRENCH (FEET) UP TO 4	UP TO 6	UP TO 9	UP TO 12	UP TO 15	VERT. SPACING (FEET)	WALES SIZE (IN)	VERT. SPACING (FEET)	UPRIGHTS — MAXIMUM ALLOWABLE HORIZONTAL SPACING (FEET) (See Note 2) CLOSE
5 TO 10	UP TO 6	6X8	6X8	6X8	8X8	8X8	5	8X10	5	2X6
	UP TO 8	8X8	8X8	8X8	8X8	8X10	5	10X12	5	2X6
	UP TO 10	8X10	8X10	8X10	8X10	10X10	5	12X12	5	2X6
	See Note 1									
10 TO 15	UP TO 6	8X8	8X8	8X8	8X8	8X10	5	10X12	5	2X6
	UP TO 8	8X10	8X10	8X10	8X10	10X10	5	12X12	5	2X6
	See Note 1									
	See Note 1									
15 TO 20	UP TO 6	8X10	8X10	8X10	8X10	10X10	5	12X12	5	2X6
	See Note 1									
	See Note 1									
	See Note 1									
OVER 20	SEE NOTE 1									

* Mixed Oak or equivalent with a bending strength not less than 850 psi.
** Manufactured members of equivalent strength may be substituted for wood.

TABLE C-2.1

TIMBER TRENCH SHORING -- MINIMUM TIMBER REQUIREMENTS *
SOIL TYPE A $P_a = 25 \times H + 72$ psf (2 ft. Surcharge)

SIZE (S4S) AND SPACING OF MEMBERS **

DEPTH OF TRENCH (FEET)	HORIZ. SPACING (FEET)	CROSS BRACES WIDTH OF TRENCH (FEET) UP TO 4	UP TO 6	UP TO 9	UP TO 12	UP TO 15	VERT. SPACING (FEET)	WALES SIZE (IN)	VERT. SPACING (FEET)	UPRIGHTS MAXIMUM ALLOWABLE HORIZONTAL SPACING (FEET) CLOSE	4	5	6	8
5 TO 10	UP TO 6	4X4	4X4	4X4	4X4	4X6	4	Not Req'd	Not Req'd				4X6	
	UP TO 8	4X4	4X4	4X4	4X6	4X6	4	Not Req'd	Not Req'd					4X8
	UP TO 10	4X6	4X6	4X6	6X6	6X6	4	8X8	4			4X6		
	UP TO 12	4X6	4X6	4X6	6X6	6X6	4	8X8	4				4X6	
10 TO 15	UP TO 6	4X4	4X4	4X4	4X6	6X6	4	Not Req'd	Not Req'd				4X10	
	UP TO 8	4X6	4X6	4X6	6X6	6X6	4	6X8	4		4X6			
	UP TO 10	6X6	6X6	6X6	6X6	6X6	4	8X8	4			4X8		
	UP TO 12	6X6	6X6	6X6	6X6	6X6	4	8X10	4		4X6		4X10	
15 TO 20	UP TO 6	6X6	6X6	6X6	6X6	6X6	4	6X8	4	3X6				
	UP TO 8	6X6	6X6	6X6	6X6	6X6	4	8X8	4	3X6	4X12			
	UP TO 10	6X6	6X6	6X6	6X6	6X8	4	8X10	4	3X6				
	UP TO 12	6X6	6X6	6X8	6X8	6X8	4	8X12	4	3X6	4X12			
OVER 20	SEE NOTE 1													

* Douglas fir or equivalent with a bending strength not less than 1500 psi.
** Manufactured members of equivalent strength may be substituted for wood.

TABLE C-2.2

TIMBER TRENCH SHORING -- MINIMUM TIMBER REQUIREMENTS *

SOIL TYPE B P_a = 45 X H + 72 psf (2 ft. Surcharge)

SIZE (S4S) AND SPACING OF MEMBERS **

DEPTH OF TRENCH (FEET)	HORIZ. SPACING (FEET)	CROSS BRACES — WIDTH OF TRENCH (FEET)					VERT. SPACING (FEET)	WALES SIZE (IN)	VERT. SPACING (FEET)	UPRIGHTS — MAXIMUM ALLOWABLE HORIZONTAL SPACING (FEET)				
		UP TO 4	UP TO 6	UP TO 9	UP TO 12	UP TO 15				CLOSE	2	3	4	6
5 TO 10	UP TO 6	4X6	4X6	4X6	6X6	6X6	5	6X8	5					4X12
	UP TO 8	4X6	4X6	6X6	6X6	6X6	5	8X8	5		3X8	3X12 / 4X8	4X8	
	UP TO 10	4X6	4X6	6X6	6X6	6X8	5	8X10	5			4X8		
	See Note 1													
10 TO 15	UP TO 6	6X6	6X6	6X6	6X8	6X8	5	8X8	5	3X6	4X10			
	UP TO 8	6X8	6X8	6X8	8X8	8X8	5	10X10	5	3X6	4X10			
	UP TO 10	6X8	6X8	8X8	8X8	8X8	5	10X12	5	3X6	4X10			
	See Note 1													
15 TO 20	UP TO 6	6X8	6X8	6X8	6X8	8X8	5	8X10	5	4X6				
	UP TO 8	6X8	6X8	6X8	8X8	8X8	5	10X12	5	4X6				
	UP TO 10	8X8	8X8	8X8	8X8	8X8	5	12X12	5	4X6				
	See Note 1													
OVER 20	SEE NOTE 1													

* Douglas fir or equivalent with a bending strength not less than 1500 psi.
** Manufactured members of equivalent strength may be substituted for wood.

TABLE C-2.3

TIMBER TRENCH SHORING -- MINIMUM TIMBER REQUIREMENTS *

SOIL TYPE C P_a = 80 X H + 72 psf (2 ft. Surcharge)

SIZE (S4S) AND SPACING OF MEMBERS **

DEPTH OF TRENCH (FEET)	HORIZ. SPACING (FEET)	CROSS BRACES WIDTH OF TRENCH (FEET) UP TO 4	UP TO 6	UP TO 9	UP TO 12	UP TO 15	VERT. SPACING (FEET)	WALES SIZE (IN)	VERT. SPACING (FEET)	UPRIGHTS MAXIMUM ALLOWABLE HORIZONTAL SPACING (FEET) CLOSE
5 TO 10	UP TO 6	6X6	6X6	6X6	6X6	8X8	5	8X8	5	3X6
	UP TO 8	6X6	6X6	6X6	8X8	8X8	5	10X10	5	3X6
	UP TO 10	6X6	6X6	8X8	8X8	8X8	5	10X12	5	3X6
	See Note 1									
10 TO 15	UP TO 6	6X8	6X8	6X8	8X8	8X8	5	10X10	5	4X6
	UP TO 8	8X8	8X8	8X8	8X8	8X8	5	12X12	5	4X6
	See Note 1									
	See Note 1									
15 TO 20	UP TO 6	8X8	8X8	8X8	8X10	8X10	5	10X12	5	4X6
	See Note 1									
	See Note 1									
	See Note 1									
OVER 20	SEE NOTE 1									

* Douglas fir or equivalent with a bending strength not less than 1500 psi.

** Manufactured members of equivalent strength may be substituted for wood.

APPENDIX D TO SUBPART P—ALUMINUM HYDRAULIC SHORING FOR TRENCHES

(a) Scope. This appendix contains information that can be used when aluminum hydraulic shoring is provided as a method of protection against cave-ins in trenches that do not exceed 20 feet (6.1m) in depth. This appendix must be used when design of the aluminum hydraulic protective system cannot be performed in accordance with § 1926.652(c)(2).

(b) Soil Classification. In order to use data presented in this appendix, the soil type or types in which the excavation is made must first be determined using the soil classification method set forth in appendix A of subpart P of part 1926.

(c) Presentation of Information. Information is presented in several forms as follows:

(1) Information is presented in tabular form in Tables D-1.1, D-1.2, D-1.3 and D-1.4. Each table presents the maximum vertical and horizontal spacings that may be used with various aluminum member sizes and various hydraulic cylinder sizes. Each table contains data only for the particular soil type in which the excavation or portion of the excavation is made. Tables D-1.1 and D-1.2 are for vertical shores in Types A and B soil. Tables D-1.3 and D1.4 are for horizontal waler systems in Types B and C soil.

(2) Information concerning the basis of the tabular data and the limitations of the data is presented in paragraph (d) of this appendix.

(3) Information explaining the use of the tabular data is presented in paragraph (e) of this appendix.

(4) Information illustrating the use of the tabular data is presented in paragraph (f) of this appendix.

(5) Miscellaneous notations (footnotes) regarding Table D-1.1 through D-1.4 are presented in paragraph (g) of this appendix.

(6) Figures, illustrating typical installations of hydraulic shoring, are included just prior to the Tables. The illustrations page is entitled "Aluminum Hydraulic Shoring; Typical Installations."

(d) Basis and limitations of the data.

(1) Vertical shore rails and horizontal wales are those that meet the Section Modulus requirements in the D-1 Tables. Aluminum material is 6061-T6 or material of equivalent strength and properties.

(2) Hydraulic cylinders specifications.

   (i) 2-inch cylinders shall be a minimum 2-inch inside diameter with a minimum safe working capacity of no less than 18,000 pounds axial compressive load at maximum extension. Maximum extension is to include full range of cylinder extensions as recommended by product manufacturer.

   (ii) 3-inch cylinders shall be a minimum 3-inch inside diameter with a safe working capacity of not less than 30,000 pounds axial compressive load at extensions as recommended by product manufacturer.

(3) Limitation of application.

(i)   It is not intended that the aluminum hydraulic specification apply to every situation that may be experienced in the field.

These data were developed to apply to the situations that are most commonly experienced in current trenching practice. Shoring systems for use in situations that are not covered by the data in this appendix must be otherwise designed as specified in § 1926.652(c).

(ii)   When any of the following conditions are present, the members specified in the Tables are not considered adequate.

In this case, an alternative aluminum hydraulic shoring system or other type of protective system must be designed in accordance with § 1926.652.

(A)   When vertical loads imposed on cross braces exceed a 100 pound gravity load distributed on a one foot section of the center of the hydraulic cylinder.

(B)   When surcharge loads are present from equipment weighing in excess of 20,000 pounds.

(C)   When only the lower portion or a trench is shored and the remaining portion of the trench is sloped or benched unless: The sloped portion is sloped at an angle less steep than three horizontal to one vertical; or the members are selected from the tables for use at a depth which is determined from the top of the overall trench, and not from the toe of the sloped portion.

(e)   Use of Tables D-1.1, D-1.2, D-1.3 and D-1.4. The members of the shoring system that are to be selected using this information are the hydraulic cylinders, and either the vertical shores or the horizontal wales. When a waler system is used, the vertical timber sheeting to be used is also selected from these tables.

The Tables D-1.1 and D-1.2 for vertical shores are used in Type A and B soils that do not require sheeting. Type B soils that may require sheeting, and Type C soils that always require sheeting are found in the horizontal wale Tables D-1.3 and D-1.4. The soil type must first be determined in accordance with the soil classification system described in appendix A to subpart P of part 1926. Using the appropriate table, the selection of the size and spacing of the members is made. The selection is based on the depth and width of the trench where the members are to be installed. In these tables the vertical spacing is held constant at four feet on center. The tables show the maximum horizontal spacing of cylinders allowed for each size of wale in the waler system tables, and in the vertical shore tables, the hydraulic cylinder horizontal spacing is the same as the vertical shore spacing.

(f)   Example to Illustrate the Use of the Tables:

(1) Example 1:

A trench dug in Type A soil is 6 feet deep and 3 feet wide.

From Table D-1.1: Find vertical shores and 2 inch diameter cylinders spaced 8 feet on center (o.c.) horizontally and 4 feet on center (o.c.) vertically. (See Figures 1 & 3 for typical installations.)

(2) Example 2:

A trench is dug in Type B soil that does not require sheeting, 13 feet deep and 5 feet wide. From Table D-1.2: Find vertical shores and 2 inch diameter cylinders spaced 6.5 feet o.c. horizontally and 4 feet o.c. vertically. (See Figures 1 & 3 for typical installations.)

(3) A trench is dug in Type B soil that does not require sheeting, but does experience some minor raveling of the trench face.

The trench is 16 feet deep and 9 feet wide. From Table D-1.2: Find vertical shores and 2 inch diameter cylinder (with special oversleeves as designated by footnote #2) spaced 5.5 feet o.c. horizontally and 4 feet o.c. vertically, plywood (per footnote (g)(7) to the D-1 Table) should be used behind the shores. (See Figures 2 & 3 for typical installations.)

(4) Example 4: A trench is dug in previously disturbed Type B soil, with characteristics of a Type C soil, and will require sheeting. The trench is 18 feet deep and 12 feet wide. 8 foot horizontal spacing between cylinders is desired for working space. From Table D-1.3: Find horizontal wale with a section modulus of 14.0 spaced at 4 feet o.c. vertically and 3 inch diameter cylinder spaced at 9 feet maximum o.c. horizontally. 3×12 timber sheeting is required at close spacing vertically. (See Figure 4 for typical installation.)

(5) Example 5: A trench is dug in Type C soil, 9 feet deep and 4 feet wide. Horizontal cylinder spacing in excess of 6 feet is desired for working space. From Table D-1.4: Find horizontal wale with a section modulus of 7.0 and 2 inch diameter cylinders spaced at 6.5 feet o.c. horizontally. Or, find horizontal wale with a 14.0 section modulus and 3 inch diameter cylinder spaced at 10 feet o.c. horizontally. Both wales are spaced 4 feet o.c. vertically. 3×12 timber sheeting is required at close spacing vertically. (See Figure 4 for typical installation.)

(g)Footnotes, and general notes, for Tables D-1.1, D-1.2, D-1.3, and D-1.4.

(1) For applications other than those listed in the tables, refer to § 1926.652(c)(2) for use of manufacturer's tabulated data. For trench depths in excess of 20 feet, refer to § 1926.652(c)(2) and § 1926.652(c)(3).

(2) 2 inch diameter cylinders, at this width, shall have structural steel tube (3.5 × 3.5 × 0.1875) oversleeves, or structural oversleeves of manufacturer's specification, extending the full, collapsed length.

(3) Hydraulic cylinders capacities.

   (i)   2 inch cylinders shall be a minimum 2-inch inside diameter with a safe working capacity of not less than 18,000 pounds axial compressive load at maximum extension. Maximum extension is to include full range of cylinder extensions as recommended by product manufacturer.

   (ii)   3-inch cylinders shall be a minimum 3-inch inside diameter with a safe work capacity of not less than 30,000 pounds axial compressive load at maximum extension. Maximum extension is to include full range of cylinder extensions as recommended by product manufacturer.

(4) All spacing indicated is measured center to center.

(5) Vertical shoring rails shall have a minimum section modulus of 0.40 inch.

(6) When vertical shores are used, there must be a minimum of three shores spaced equally, horizontally, in a group.

(7) Plywood shall be 1.125 in. thick softwood or 0.75 in. thick, 14 ply, arctic white birch (Finland form). Please note that plywood is not intended as a structural member, but only for prevention of local raveling (sloughing of the trench face) between shores.

(8) See appendix C for timber specifications.

(9) Wales are calculated for simple span conditions.

(10) See appendix D, item (d), for basis and limitations of the data.

ALUMINUM HYDRAULIC SHORING
TYPICAL INSTALLATIONS

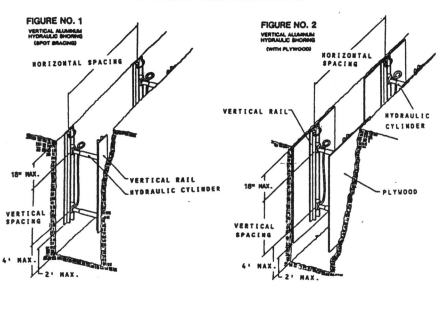

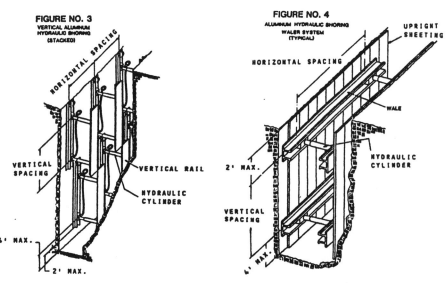

TABLE D - 1.1
ALUMINUM HYDRAULIC SHORING
VERTICAL SHORES
FOR SOIL TYPE A

DEPTH OF TRENCH (FEET)	HYDRAULIC CYLINDERS		WIDTH OF TRENCH (FEET)		
	MAXIMUM HORIZONTAL SPACING (FEET)	MAXIMUM VERTICAL SPACING (FEET)	UP TO 8	OVER 8 UP TO 12	OVER 12 UP TO 15
OVER 5 UP TO 10	8	4	2 INCH DIAMETER	2 INCH DIAMETER NOTE (2)	3 INCH DIAMETER
OVER 10 UP TO 15	8				
OVER 15 UP TO 20	7				
OVER 20	NOTE (1)				

Footnotes to tables, and general notes on hydraulic shoring, are found in Appendix D, Item (g)
Note (1): See Appendix D, Item (g) (1)
Note (2): See Appendix D, Item (g) (2)

TABLE D - 1.2
ALUMINUM HYDRAULIC SHORING
VERTICAL SHORES
FOR SOIL TYPE B

DEPTH OF TRENCH (FEET)	HYDRAULIC CYLINDERS				
	MAXIMUM HORIZONTAL SPACING (FEET)	MAXIMUM VERTICAL SPACING (FEET)	WIDTH OF TRENCH (FEET)		
			UP TO 8	OVER 8 UP TO 12	OVER 12 UP TO 15
OVER 5 UP TO 10	8	4	2 INCH DIAMETER	2 INCH DIAMETER NOTE (2)	3 INCH DIAMETER
OVER 10 UP TO 15	6.5				
OVER 15 UP TO 20	5.5				
OVER 20	NOTE (1)				

Footnotes to tables, and general notes on hydraulic shoring, are found in Appendix D, Item (g)

Note (1): See Appendix D, Item (g) (1)
Note (2): See Appendix D, Item (g) (2)

TABLE D - 1.3
ALUMINUM HYDRAULIC SHORING
WALER SYSTEMS
FOR SOIL TYPE B

DEPTH OF TRENCH (FEET)	WALES VERTICAL SPACING (FEET)	SECTION MODULUS* (IN³)	HYDRAULIC CYLINDERS — WIDTH OF TRENCH (FEET) UP TO 8 HORIZ. SPACING	CYLINDER DIAMETER	OVER 8 UP TO 12 HORIZ. SPACING	CYLINDER DIAMETER	OVER 12 UP TO 15 HORIZ. SPACING	CYLINDER DIAMETER	TIMBER UPRIGHTS MAX. HORIZ. SPACING (ON CENTER) SOLID SHEET	2 FT.	3 FT.
OVER 5 UP TO 10	4	3.5	8.0	2 IN	8.0	2 IN NOTE(2)	8.0	3 IN			3x12
		7.0	9.0	2 IN	9.0	2 IN NOTE(2)	9.0	3 IN		—	
		14.0	12.0	3 IN	12.0	3 IN	12.0	3 IN			
OVER 10 UP TO 15	4	3.5	6.0	2 IN	6.0	2 IN NOTE(2)	6.0	3 IN		3x12	
		7.0	8.0	3 IN	8.0	3 IN	8.0	3 IN			—
		14.0	10.0	3 IN	10.0	3 IN	10.0	3 IN			
OVER 15 UP TO 20	4	3.5	5.5	2 IN	5.5	2 IN NOTE(2)	5.5	3 IN	3x12		
		7.0	6.0	3 IN	6.0	3 IN	6.0	3 IN			—
		14.0	9.0	3 IN	9.0	3 IN	9.0	3 IN			
OVER 20	NOTE (1)										

Footnotes to tables, and general notes on hydraulic shoring, are found in Appendix D, Item (g)

Notes (1): See Appendix D, item (g) (1)

Notes (2): See Appendix D, Item (g) (2)

* Consult product manufacturer and/or qualified engineer for Section Modulus of available wales.

TABLE D - 1.4
ALUMINUM HYDRAULIC SHORING
WALER SYSTEMS
FOR SOIL TYPE C

DEPTH OF TRENCH (FEET)	WALES		HYDRAULIC CYLINDERS						TIMBER UPRIGHTS		
	VERTICAL SPACING (FEET)	SECTION MODULUS * (IN³)	WIDTH OF TRENCH (FEET)						MAX. HORIZ SPACING (ON CENTER)		
			UP TO 8		OVER 8 UP TO 12		OVER 12 UP TO 15		SOLID SHEET	2 FT.	3 FT.
			HORIZ. SPACING	CYLINDER DIAMETER	HORIZ. SPACING	CYLINDER DIAMETER	HORIZ. SPACING	CYLINDER DIAMETER			
OVER 5 UP TO 10	4	3.5	6.0	2 IN	6.0	2 IN NOTE(2)	6.0	3 IN	3x12	—	—
		7.0	6.5	2 IN	6.5	2 IN NOTE(2)	6.5	3 IN			
		14.0	10.0	3 IN	10.0	3 IN	10.0	3 IN			
OVER 10 UP TO 15	4	3.5	4.0	2 IN	4.0	2 IN NOTE(2)	4.0	3 IN	3x12	—	—
		7.0	5.5	3 IN	5.5	3 IN	5.5	3 IN			
		14.0	8.0	3 IN	8.0	3 IN	8.0	3 IN			
OVER 15 UP TO 20	4	3.5	3.5	2 IN	3.5	2 IN NOTE(2)	3.5	3 IN	3x12	—	—
		7.0	5.0	3 IN	5.0	3 IN	5.0	3 IN			
		14.0	6.0	3 IN	6.0	3 IN	6.0	3 IN			
OVER 20		NOTE (1)									

Footnotes to tables, and general notes on hydraulic shoring, are found in Appendix D, Item (g)

Notes (1): See Appendix D, item (g) (1)

Notes (2): See Appendix D, Item (g) (2)

* Consult product manufacturer and/or qualified engineer for Section Modulus of available wales.

APPENDIX E TO SUBPART P—ALTERNATIVES TO TIMBER SHORING

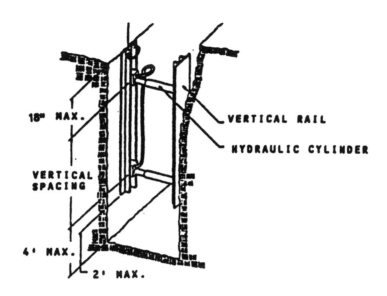

Figure 1. Aluminum Hydraulic Shoring

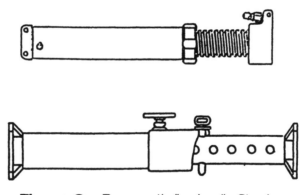

Figure 2. Pneumatic/hydraulic Shoring

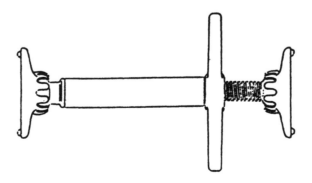

Figure 3. Trench Jacks (Screw Jacks)

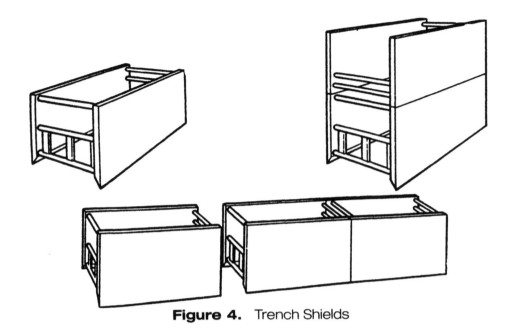

Figure 4. Trench Shields

APPENDIX F TO SUBPART P—SELECTION OF PROTECTIVE SYSTEMS

The following figures are a graphic summary of the requirements contained in subpart P for excavations 20 feet or less in depth.

Protective systems for use in excavations more than 20 feet in depth must be designed by a registered professional engineer in accordance with § 1926.652 (b) and (c).

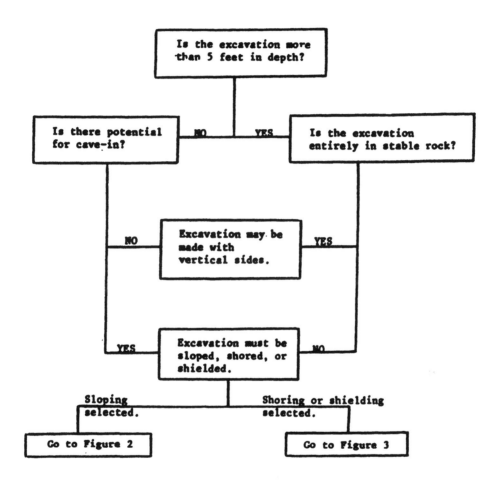

Figure 1. Preliminary Decisions

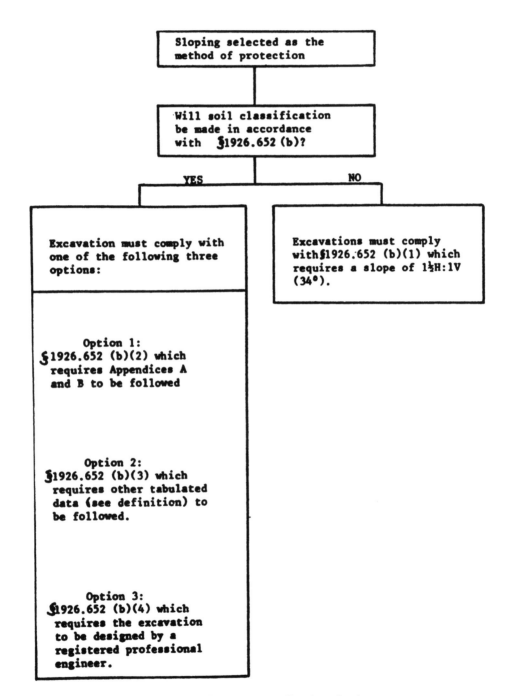

Figure 2. Sloping Options

```
┌─────────────────────────────────────┐
│  Shoring or shielding selected      │
│  as the method of protection.       │
└─────────────────────────────────────┘
            │
┌─────────────────────────────────────┐
│  Soil classification is required    │
│  when shoring or shielding is       │
│  used.  The excavation must comply  │
│  with one of the following four     │
│  options:                           │
├─────────────────────────────────────┤
│                                     │
│         Option 1                    │
│  §1926.652 (c)(1) which requires    │
│   Appendices A and C to be followed │
│   (e.g.  timber shoring).           │
│                                     │
│                                     │
│                                     │
│         Option 2                    │
│  §1926.652 (c)(2) which requires    │
│   manufacturers data to be followed │
│   (e.g.  hydraulic shoring,trench   │
│   jacks, air shores, shields).      │
│                                     │
│                                     │
│                                     │
│         Option 3                    │
│  §1926.652 (c)(3) which requires    │
│   tabulated data (see definition)   │
│   to be followed (e.g.  any system  │
│   as per the tabulated data).       │
│                                     │
│                                     │
│                                     │
│         Option 4                    │
│  §1926.652 (c)(4) which requires    │
│   the excavation to be designed     │
│   by a registered professional      │
│   engineer (e.g. any designed       │
│   system).                          │
│                                     │
└─────────────────────────────────────┘
```

Figure 3. Shoring and Shielding Options

SELECT BIBLIOGRAPHY

Accrocco, J. O. (Ed.). 1990. *MSDS Pocket Dictionary.* Schenectady, NY: Genium Publishing Corporation.

Accrocco, J. O. and Cinquanti, M. (Eds.). 1991. *Right-to-Know Pocket Guide for Laboratory Employees.* Schenectady, NY: Genium Publishing Corporation.

ACGIH (American Conference of Governmental Industrial Hygienists). 1993. *Documentation of the Threshold Limit Values and Biological Exposure Indices,* 6th edition. Cincinnati, OH: ACGIH.

ACGIH. 1996. *1996–1997 Threshold Limit Values for Chemical Substances and Physical Agents and Biological Exposure Indices.* Cincinnati, OH: ACGIH.

Amdur, M. O., Doull, J., and Klaassen, C. D. (Eds.) 1991. *Casarett and Doull's Toxicology—The Basic Science of Poisons,* 4th edition. New York: Pergamon Press.

Amoore, J. E. and Hautala, E. 1983. Odor as an aid to chemical safety: odor thresholds compared with threshold limit values and volatilities for 214 industrial chemicals in air and water dilution. *Journal of Applied Toxicology* 3: 279–290.

Andrews, L. P. (Ed.). 1990. *Worker Protection during Hazardous Waste Remediation.* New York: Van Nostrand Reinhold.

ANSI (American National Standards Institute). 1967. *ANSI Z41.1-1967, Men's Safety-Toe Footwear.* New York: ANSI.

ANSI. 1968. *ANSI Z87.1-1968, Eye and Face Protection.*

ANSI. 1969a. *ANSI Z89.1-1969, Safety Requirements for Industrial Head Protection.*

ANSI. 1969b. *ANSI Z88.2-1969, Standard Practice for Respiratory Protection.*

ANSI. 1974. *ANSI S3.19-1974, Measurement of Real-Ear Protection of Hearing Protectors and Physical Attenuation of Earmuffs.*

ANSI. 1986. *ANSI Z89.1-1986, Protective Headwear for Industrial Workers—Requirements.*

ANSI. 1989. *ANSI Z87.1-1989, Occupational and Educational Eye and Face Protection.*

ANSI. 1991. *ANSI Z41-1991, American National Standard for Personal Protection—Protective Footwear.*

ANSI. 1992. *ANSI Z88.2-1992, American National Standard for Respiratory Protection.*

ANSI. 1993. *ANSI Z400.1-1993, Hazardous Industrial Chemicals, Material Safety Data Sheets, Preparation.*

Arena, J. M. and Drew, R. H. 1986. *Poisoning,* 5th edition. Springfield, IL: Charles Thomas Publishers.

Arthur D. Little, Inc. 1968. *Research on Chemical Odors, Part I—Odor Thresholds for 53 Commercial Chemicals.* Washington, DC: Manufacturing Chemists Association.

Association of American Railroads. 1990. *Emergency Action Guides.* Washington, D.C.: Association of American Railroads.

ASTM (American Society for Testing and Materials). 1970. *ASTM D 1292-65(70), Standard Method of Test for Odor in Water.* Philadelphia: ASTM.

ASTM. 1977. *ASTM D-1003, Test Method for Haze and Luminous Transmittance of Transparent Plastics.*

ASTM. 1979a. *ASTM D-751, Methods of Testing Coated Fabrics.*

ASTM. 1979b. *ASTM D-1605, Standard Recommended Practices for Sampling Atmospheres for the Analysis of Gases and Vapors.*

ASTM. 1983. *A Guide to Safe Handling of Hazardous Materials Incidents.* Special Technical Publication 825.

ASTM. 1985. *ASTM F-739, Test Method for Resistance of Protective Clothing Materials to Permeation by Liquids and Gases.*

ASTM. 1987a. *ASTM D-4598-87, Standard Practice for Sampling Workplace Atmospheres to Collect Gases or Vapors with Liquid Sorbent Diffusional Samplers.*

ASTM. 1987b. *ASTM F-903, Test Method for Resistance of Protective Clothing Materials to Penetration by Liquids.*

ASTM. 1987c. *ASTM F-1052, Practice for Pressure Testing of Gas-Tight Totally-Encapsulated Chemical Protective Suits.*

ASTM. 1989. *ASTM F-1001, Standard Guide for Chemicals to Evaluate Protective Clothing Materials.*

Billings, C. E. and Jonas, L. C. 1981. Odor thresholds in air compared to threshold limit values. *American Industrial Hygiene Association Journal* 42: 479–480.

Budavari, S. (Ed.). 1996. *The Merck Index,* 12 Edition. New Jersey: Merck & Co., Inc.

Compressed Gas Association (CGA). 1988. *Compressed Air for Human Respiration.* Publication G-7-1988. Arlington, VA: CGA.

Dresbach, R. H. *Handbook of Toxic and Hazardous Chemicals and Carcinogens,* 2nd edition. Park Ridge, NJ: Noyes Publications.

Emergency Resource Inc. 1990. *Surviving the Hazardous Materials Incident—Student Workbook Series 1.* Fort Collins, CO: Emergency Resource Inc.

Fire, F. L. 1986. *A Common Sense Approach to Hazardous Materials.* New York: Fire Engineering.

Forsberg, K. and Mansdorf, S. Z. 1997. *Quick Selection Guide to Chemical Protective Clothing,* 3rd edition. New York: Van Nostrand Reinhold.

Goldfrank, L. R., Flomenbaum, N. E., Lewin, N. A., et al. 1990. *Goldfrank's Toxicologic Emergencies,* 4th edition. Norwalk, CT: Appleton and Lange.

Hall, A. H. and Maslansky, C. J. 1994. *Medical Response to Chemical Emergencies. Module 5—Toxicology.* Washington, DC: Chemical Manufacturers Association.

Hering, S. V. (Ed.). 1989. *Air Sampling Instruments for Evaluation of Atmospheric Contaminants,* 7th edition. Cincinnati, OH: ACGIH.

Kisner, S. M. and Fosbroke, D. E. 1994. Injury hazards in the construction industry. *Journal Occupational Medicine* 36: 137–143.

Levine, S. P. and Martin, W. F. (Eds.). 1985. *Protecting Personnel at Hazardous Waste Sites.* Stoneham, MA: Butterworth Publishers.

Lewis, R. J. Sr. (Ed.). 1993. *Hawley's Condensed Chemical Dictionary,* 12th Edition. New York: Van Nostrand Reinhold.

Lewis, R. J. Sr. (Ed.). 1996. *Hazardous Chemicals Desk Reference,* 4th Edition. New York: Van Nostrand Reinhold.

Lisella, F. S. (Ed.). 1994. *The VNR Dictionary of Environmental Health and Safety.* New York: Van Nostrand Reinhold.

Martin, W. F., Lippitt, J. M., and Prothero, T. G. 1987. *Hazardous Waste Handbook for Health and Safety.* Stoneham, MA: Butterworth Publishers.

Maslansky, C. J. 1987. Decontamination. In: *The EMS Response to a Hazardous Material Incident.* New York: Institute for Trauma and Emergency Care, New York Medical College.

Maslansky, C. J. and Maslansky, S. P. 1989. Air monitoring instrumentation. *Water Well Journal* 43: 79–83.

Maslansky, C. J. and Maslansky, S. P. 1993. *Air Monitoring Instrumentation—A Manual for Emergency, Investigatory, and Remedial Responders.* New York: Van Nostrand Reinhold.

Maslansky, S. P. 1981. Proceed with caution—waste disposal site ahead. *Pollution Engineering* 13: 25–26.

Maslansky, S. P. 1982a. Characterization of abandoned hazardous waste land burial sites through field investigations. In: *Hazardous Waste Management for the 80's*. Ann Arbor, MI: Ann Arbor Science Publishers, pp. 173–181.

Maslansky, S. P. 1982b. Hazardous waste site investigations: safety training, how much is enough? In: *Proceedings of the Third National Conference on Management of Uncontrolled Hazardous Waste Sites*, Washington, DC, pp. 319–320.

Maslansky, S. P. 1983. Well drilling and hazardous waste sites. *Water Well Journal* 37: 46–50.

Maslansky, S. P. 1984. Drilling at hazardous waste sites. *Drill Bits*, Fall Issue: 1–14.

Maslansky, S. P. 1987. Landfill drilling—think safety first. *Ground Water Monitoring Review* 8: 40–41.

Maslansky, S. P. 1993. Site safety plans—who needs them? *Water Well Journal* 47: 59–62.

Maslansky, S. P. 1994. Instituting an effective safety plan. In: *Workshop Proceedings of the NGWA Eighth Outdoor Action Conference on Aquifer Remediation, Ground Water Monitoring and Geophysical Methods*. Dublin, OH: National Ground Water Association pp. 117–125.

Maslansky, S. P. and Maslansky, C. J. 1988. Are drillers and respirators compatible? *Water Well Journal* 42: 49–53.

Maslansky, S. P. and Maslansky, C. J. 1991. Safety considerations in ground water monitoring investigations. In: *Practical Handbook on Ground Water Monitoring*, pp. 589–623. Chelsea, MI: Lewis Publishers, Inc.

Maslansky, S. P. and Maslansky, C. J. 1994. Gearing up for hazardous site conditions. *Ground Water Age* 28: 14–16.

Maslansky, S. P. and Maslansky, C. J. 1995. Permit-required confined spaces and the ground water industry. In: *Workshop Proceedings of the NGWA Ninth Outdoor Action Conference and Exposition*, Dublin, OH: NGWA, pp. 167–179.

Maslansky, S. P. and Maslansky, C. J. 1996. Drilling in flammable environments. In: *Workshop Proceedings of the NGWA Tenth Outdoor Action Conference and Exposition*. Dublin, OH: NGWA, pp. 47–56.

Maslansky, S. P., Dillon, J. D., and Williams, T. A. 1985. Decontaminate! A perspective for first responders. *Fire Engineering* 138: 16–20.

Matsumura, F. 1985. *Toxicology of Insecticides*, 2nd edition. New York: Plenum Press.

National Safety Council. 1991. *Supervisor's Safety Manual*, 7th Edition. Chicago: NSC.

NCRP (National Council on Radiation Protection and Measurements). 1975. *Review of the Current State of Radiation Protection*. NCRP Report No. 43. Bethesda, MD: NCRP.

NCRP. 1987. *Recommendations on Limits for Exposure to Ionizing Radiation*. NCRP Report No. 91. Bethesda, MD: NCRP.

Ness, S. A. 1991. *Air Monitoring for Toxic Exposures*. New York: Van Nostrand Reinhold.

NFPA (National Fire Protection Association). 1990a. *NFPA 1991—Standard on Vapor-Protective Suits for Hazardous Chemical Emergencies*. Quincy, MA: NFPA.

NFPA. 1990b. *NFPA 1992—Standard on Liquid Splash-Protective Suits for Hazardous Chemical Emergencies*. Quincy, MA: NFPA.

NFPA. 1990c. *NFPA 1993—Standard on Support Function Protective Garments for Hazardous Chemical Operations*. Quincy, MA: NFPA.

NFPA. 1992. *Hazardous Materials Response Handbook*, 2nd edition. Quincy, MA: NFPA.

NFPA. 1994. *Fire Protection Handbook*, 11th edition. Quincy, MA: NFPA.

NIOSH (National Institute for Safety and Occupational Health). 1973. *The Industrial Environment—Its Evaluation and Control*. Washington, DC: Government Printing Office.

NIOSH. 1984. *NIOSH Manual of Analytical Methods*, 3rd Edition. Publication No. 84–100.

NIOSH. 1987. *NIOSH Guide to Industrial Respiratory Protection*. Publication No. 87–116.

NIOSH. 1987. *Registry of Toxic Effects of Chemical Substances (RTECS)*. NIOSH Publication No. 87–114.

NIOSH. 1989. *Occupational Safety and Health Guidelines for Chemical Hazards*. NIOSH Publication No. 89–104.

NIOSH. 1992. *NIOSH Recommendations for Occupational Safety and Health: Compendium of Policy Documents and Statements*. Publication No. 92–100.

NIOSH. 1993. *Certified Equipment List as of September 30, 1993*. GPO Stock No. 017-033-00469-2.

NIOSH. 1994. *NIOSH Manual of Analytical Methods*, 4th Edition. Publication No. 94–113.

NIOSH. 1994. *NIOSH Pocket Guide to Chemical Hazards*. Publication No. 94–116.

NIOSH/OSHA/USCG/EPA. 1985. *Occupational Safety and Health Guidance Manual for Hazardous Waste Site Activities*. NIOSH Publication 85–115. Washington, DC: Government Printing Office.

Noll, G. G., Hildebrand, M. S., and Yvorra, J. G. 1988. *Hazardous Materials—Managing the Incident*. Stillwater, OK: Fire Protection Publications.

Olsen, K. 1990. *Poisoning and Drug Overdose*. San Mateo, CA: Appleton and Lange.

OSHA (Occupational Safety and Health Administration). 1984. *OSHA Industrial Hygiene Technical Manual*, Washington, DC: Government Printing Office.

OSHA. 1986. *Respiratory Protection*. Publication No. 3079.

OSHA. 1989a. 29 CFR 1910.120, Hazardous Waste Operations and Emergency Response; Final Rule. *Federal Register*, Vol. 54, No. 42, pp. 9294–9336.

OSHA. 1989b. 29 CFR 1910. Subpart P, Occupational Safety and Health Standards—Excavations; Final Rule. *Federal Register*, Vol. 54, No. 209, pp. 45959–45991.

OSHA. 1990. *OSHA Analytical Methods Manual*.

OSHA. 1993. 29 CFR 1910.146, Permit-Required Confined Spaces for General Industry; Final Rule. *Federal Register*, Vol. 58, No. 9, pp. 4462–4568.

OSHA. 1994a. 29 CFR 1910. Subpart I, Personal Protective Equipment for General Industry; Final Rule. *Federal Register*, Vol. 59, No. 66, pp. 16334–16364.

OSHA. 1994b. 29 CFR 1910.120 and 1926.65. Hazardous Waste Operations and Emergency Response; Final Rule. *Federal Register*, Vol. 59, No. 161, pp. 43268–43280.

Ottoboni, M. A. 1991. *The Dose Makes the Poison*, 2nd edition. New York: Van Nostrand Reinhold.

Plog, B. A. (Ed). 1988. *Fundamentals of Industrial Hygiene*, 3rd edition. Chicago: National Safety Council.

Ruth, J. H. 1986. Odor thresholds and irritation levels of several chemical substances: a review. *American Industrial Hygiene Association Journal* 47: A142–A151.

Sax, N. I. and Lewis, R. J. Sr. (Eds.). 1995. *Dangerous Properties of Industrial Materials*, 9th Edition. New York: Van Nostrand Reinhold.

Schwope, A. D., Costas, P. P., Jacson, J. O., et al. 1987. *Guidelines for Selection of Chemical Protective Clothing*, 3rd edition. Cincinnati, OH: ACGIH.

Shapiro, J. 1990. *Radiation Protection—A Guide for Scientists and Physicians*, 2nd edition. Cambridge, MA: Harvard University Press.

Sullivan, J. B. and Krieger, G. R. 1992. *Hazardous Materials Toxicology*. Baltimore, MD: Williams & Wilkins.

Tolke, G. (Ed.). 1993. *Hazardous Materials Response Handbook*. Qunicy, MA: National Fire Protection Association.

U.S. Coast Guard. 1992. *Chemical Hazard Response Information System (CHRIS)*. GPO Stock No. 050-012-00329-7.

USDOT (U.S. Department of Transportation). *DOT Specification Cylinders Shipping Containers*, revised August 1986. Available through Training Unit, DMH-51, Office of Hazardous Materials Transportation, Washington, DC.

USDOT. 1990. 49 CFR Part 107, et al. Performance-Oriented Packaging Standards; Changes to Classification, Hazardous Communication, Packaging and Handling Requirements Based on UN Standards and Agency Initiative; Final Rule. *Federal Register* Vol. 55, No. 246, pp. 52402–52729.

USDOT. 1996. North American Emergency Response Guidebook. Stock No. 050-000-00561-5.

U.S. Environmental Protection Agency (USEPA). 1984. *Compendium of Methods for the Determination of Toxic Organic Compounds in Ambient Air.* Publication EPA-600/4-84-041. Washington, DC. Government Printing Office.

USEPA. 1985a. *Standard Operating Guidelines for Site Entry.* Publication 9285.2-01A.

USEPA. 1985b. *Standard Operating Guidelines for Establishing Work Zones.* Publication 9285.2-04A.

USEPA. 1987a. *Course Manual, Response Safety Decision-Making Workshop,* Course No. 165.8.

USEPA. 1987b. *A Compendium of Superfund Field Operations Methods.* Publication EPA/540/p-87/001.

USEPA. 1988. *Standard Operating Safety Guides.*

USEPA. 1992a. *Standard Operating Safety Guides.* Publication 9285.1-03.

USEPA. 1992b. *Limited-Use Chemical Protective Clothing for EPA Superfund Activities.* Publication EPA/600/R-92/014.

Verschueren, K. 1996. *Handbook of Environmental Data on Organic Chemicals,* 3rd. Edition. New York: Van Nostrand Reinhold.

York, K. J. and Grey, G. L. 1989. *Hazardous Material/Waste Handling for the Emergency Responder.* New York: Fire Engineering.

INDEX